Additive Manufacturing of Advanced Composites

Additive Manufacturing of Advanced Composites

Editor

Yuan Chen

Basel • Beijing • Wuhan • Barcelona • Belgrade • Novi Sad • Cluj • Manchester

Editor
Yuan Chen
School of System Design and Intelligent Manufacturing (SDIM)
Southern University of Science and Technology
Shenzhen
China

Editorial Office
MDPI AG
Grosspeteranlage 5
4052 Basel, Switzerland

This is a reprint of articles from the Special Issue published online in the open access journal *Journal of Composites Science* (ISSN 2504-477X) (available at: https://www.mdpi.com/journal/jcs/special_issues/additive_manufacturing_advanced_composites).

For citation purposes, cite each article independently as indicated on the article page online and as indicated below:

Lastname, A.A.; Lastname, B.B. Article Title. *Journal Name* **Year**, *Volume Number*, Page Range.

ISBN 978-3-7258-2185-3 (Hbk)
ISBN 978-3-7258-2186-0 (PDF)
doi.org/10.3390/books978-3-7258-2186-0

© 2024 by the authors. Articles in this book are Open Access and distributed under the Creative Commons Attribution (CC BY) license. The book as a whole is distributed by MDPI under the terms and conditions of the Creative Commons Attribution-NonCommercial-NoDerivs (CC BY-NC-ND) license.

Contents

About the Editor vii

Preface ix

Yuan Chen
Editorial for the Special Issue on Additive Manufacturing of Advanced Composites
Reprinted from: *J. Compos. Sci.* **2024**, *8*, 344, doi:10.3390/jcs8090344 1

Theodor Florian Zach and Mircea Cristian Dudescu
The Three-Dimensional Printing of Composites: A Review of the Finite Element/Finite Volume Modelling of the Process
Reprinted from: *J. Compos. Sci.* **2024**, *8*, 146, doi:10.3390/jcs8040146 5

Gabriele Marabello, Mohamed Chairi and Guido Di Bella
Optimising Additive Manufacturing to Produce PLA Sandwich Structures by Varying Cell Type and Infill: Effect on Flexural Properties
Reprinted from: *J. Compos. Sci.* **2024**, *8*, 360, doi:10.3390/jcs8090360 38

Ahmed Ali Farhan Ogaili, Ali Basem, Mohammed Salman Kadhim, Zainab T. Al-Sharify, Alaa Abdulhady Jaber, Emad Kadum Njim, et al.
The Effect of Chopped Carbon Fibers on the Mechanical Properties and Fracture Toughness of 3D-Printed PLA Parts: An Experimental and Simulation Study
Reprinted from: *J. Compos. Sci.* **2024**, *8*, 273, doi:10.3390/jcs8070273 59

Sven Meißner, Jiri Kafka, Hannah Isermann, Susanna Labisch, Antonia Kesel, Oliver Eberhardt, et al.
Development and Evaluation of a Novel Method for Reinforcing Additively Manufactured Polymer Structures with Continuous Fiber Composites
Reprinted from: *J. Compos. Sci.* **2024**, *8*, 272, doi:10.3390/jcs8070272 79

Benny Susanto, Vishnu Vijay Kumar, Leonard Sean, Murni Handayani, Farid Triawan, Yosephin Dewiani Rahmayanti, et al.
Investigating Microstructural and Mechanical Behavior of DLP-Printed Nickel Microparticle Composites
Reprinted from: *J. Compos. Sci.* **2024**, *8*, 247, doi:10.3390/jcs8070247 106

Mohamed Daly, Mostapha Tarfaoui, Mountasar Bouali and Amine Bendarma
Effects of Infill Density and Pattern on the Tensile Mechanical Behavior of 3D-Printed Glycolyzed Polyethylene Terephthalate Reinforced with Carbon-Fiber Composites by the FDM Process
Reprinted from: *J. Compos. Sci.* **2024**, *8*, 115, doi:10.3390/jcs8040115 124

Mariana P. Salgueiro, Fábio A. M. Pereira, Carlos L. Faria, Eduardo B. Pereira, João A. P. P. Almeida, Teresa D. Campos, et al.
Numerical and Experimental Characterisation of Polylactic Acid (PLA) Processed by Additive Manufacturing (AM): Bending and Tensile Tests
Reprinted from: *J. Compos. Sci.* **2024**, *8*, 55, doi:10.3390/jcs8020055 139

Christopher Billings, Ridwan Siddique, Benjamin Sherwood, Joshua Hall and Yingtao Liu
Additive Manufacturing and Characterization of Sustainable Wood Fiber-Reinforced Green Composites
Reprinted from: *J. Compos. Sci.* **2023**, *7*, 489, doi:10.3390/jcs7120489 159

Anthonin Demarbaix, Imi Ochana, Julien Levrie, Isaque Coutinho, Sebastião Simões Cunha, Jr. and Marc Moonens
Additively Manufactured Multifunctional Composite Parts with the Help of Coextrusion Continuous Carbon Fiber: Study of Feasibility to Print Self-Sensing without Doped Raw Material
Reprinted from: *J. Compos. Sci.* **2023**, *7*, 355, doi:10.3390/jcs7090355 173

Behzad Sadeghi, Pasquale Cavaliere and Behzad Sadeghian
Enhancing Strength and Toughness of Aluminum Laminated Composites through Hybrid Reinforcement Using Dispersion Engineering
Reprinted from: *J. Compos. Sci.* **2023**, *7*, 332, doi:10.3390/jcs7080332 185

S. Sharafi, M. H. Santare, J. Gerdes and S. G. Advani
Extrusion-Based Additively Manufactured PAEK and PAEK/CF Polymer Composites Performance: Role of Process Parameters on Strength, Toughness and Deflection at Failure
Reprinted from: *J. Compos. Sci.* **2023**, *7*, 157, doi:10.3390/jcs7040157 199

Laurent Spitaels, Hugo Dantinne, Julien Bossu, Edouard Rivière-Lorphèvre and François Ducobu
A Systematic Approach to Determine the Cutting Parameters of AM Green Zirconia in Finish Milling
Reprinted from: *J. Compos. Sci.* **2023**, *7*, 112, doi:10.3390/jcs7030112 215

Elena Astafurova, Galina Maier, Evgenii Melnikov, Sergey Astafurov, Marina Panchenko, Kseniya Reunova, et al.
Temperature-Dependent Deformation Behavior of "γ-austenite/δ-ferrite" Composite Obtained through Electron Beam Additive Manufacturing with Austenitic Stainless-Steel Wire
Reprinted from: *J. Compos. Sci.* **2023**, *7*, 45, doi:10.3390/jcs7020045 237

Ana Paulo, Jorge Santos, João da Rocha, Rui Lima and João Ribeiro
Mechanical Properties of PLA Specimens Obtained by Additive Manufacturing Process Reinforced with Flax Fibers
Reprinted from: *J. Compos. Sci.* **2023**, *7*, 27, doi:10.3390/jcs7010027 248

Olusanmi Adeniran, Norman Osa-uwagboe, Weilong Cong and Monsuru Ramoni
Fabrication Temperature-Related Porosity Effects on the Mechanical Properties of Additively Manufactured CFRP Composites
Reprinted from: *J. Compos. Sci.* **2023**, *7*, 12, doi:10.3390/jcs7010012 265

Corson L. Cramer, Bola Yoon, Michael J. Lance, Ercan Cakmak, Quinn A. Campbell and David J. Mitchell
Additive Manufacturing of C/C-SiC Ceramic Matrix Composites by Automated Fiber Placement of Continuous Fiber Tow in Polymer with Pyrolysis and Reactive Silicon Melt Infiltration
Reprinted from: *J. Compos. Sci.* **2022**, *6*, 359, doi:10.3390/jcs6120359 279

Kajogbola R. Ajao, Segun E. Ibitoye, Adedire D. Adesiji and Esther T. Akinlabi
Design and Construction of a Low-Cost-High-Accessibility 3D Printing Machine for Producing Plastic Components
Reprinted from: *J. Compos. Sci.* **2022**, *6*, 265, doi:10.3390/jcs6090265 293

About the Editor

Yuan Chen

Dr. Yuan Chen serves as Assistant Professor at Southern University of Science and Technology. He obtained his PhD degree from the University of Sydney in 2019. From 2019 to 2022, he was an ARC postdoctoral research associate at the University of Sydney. His research interests are mostly concentrated on the additive manufacturing of composite materials, the design and computational analysis of composite structures, topological and structural optimization for metamaterials and advanced composites, etc. As the first or correspondent author, he has published 37 papers in renowned international journals such as Comput Meth Appl Mech Eng, Compos Sci Technol, etc., and his research outcomes have been well recognized and appraised by world-renowned experts, academic institutions, international organizations and so forth—He has been elected as Editor-in-Chief for Advanced Manufacturing: Polymer & Composites Science (Taylor & Francis), Early Career Editorial board member for famous journals *Composites Communications and Chinese Journal of Mechanical Engineering*, and Editorial Board Member and Topical advisory panel member for other SCI journals. Currently, he has been supported by the National Natural Science Foundation of China, National Key Research & Development Program of China, Guangdong University Key-Area Special Program, etc., and was a national project reviewer for National Science Centre (NCN), Poland, etc. He has won several international awards and is also a professional member of several renowned academic societies and organizations, such as a senior member of the Chinese Society of Theoretical and Applied Mechanics, etc.

Preface

Advanced composites, e.g., continuous, or discontinuous fibre reinforced composites, nanocomposites, etc., are attracting increasing attention in industrial applications due to excellent performance, i.e., high mechanical properties in terms of stiffness- and strength-to-weight ratios, when compared to their counterparts. As such, the development of advanced composites can fulfil many special but important engineering missions, such as the safety improvement, weight reduction, energy-absorption enhancement, and so forth. Meanwhile, additive manufacturing or 3D printing has undergone massive development, opening new horizons for manufacturing small-scale and complex composite structural parts that cannot be appropriately made using conventional techniques. In recent years, big advances have been witnessed in additive manufacture of advanced composites with novel design, fabrication, and analysis methods, indicating a huge potential and a promising future for 3D printed advanced composites.

The studies published here give the reader insight into the state-of-the-art of current advances and new developments in fabrication technique and process, modelling and characterization, structure-property performance, design and improvement, etc. with respect to AM of advanced composites. Overall, increasing attentions are drawn and significant improvement is witnessed on the AM of advanced composites, the excellence of advanced composites by AM is still undergoing a long way, on account that their mechanical properties still cannot compare to those by conventional fabrications. With these in mind, more and more cutting-edge techniques and in-depth investigations are desired in the future.

Yuan Chen
Editor

Editorial

Editorial for the Special Issue on Additive Manufacturing of Advanced Composites

Yuan Chen [1,2]

[1] Shenzhen Key Laboratory of Intelligent Manufacturing for Continuous Carbon Fiber Reinforced Composites, Southern University of Science and Technology, Shenzhen 518055, China; chenyuan@sustech.edu.cn

[2] School of System Design and Intelligent Manufacturing, Southern University of Science and Technology, Shenzhen 518055, China

Advanced composites are attracting increasing attention in industrial applications due to their excellent performance, i.e., high mechanical properties in terms of stiffness- and strength-to-weight ratios when compared to their counterparts. Meanwhile, the rapid enhancement of additive manufacturing (AM) techniques has been leading to satisfactory printing of advanced composites, accelerating many highly specific but important engineering productions for advanced materials and structures with complex geometries, unique functions, excellent performance, etc. [1–5]. The importance of AM of advanced composites is evident, and in this regard, investigations into fabrication techniques and processes, modeling and characterization, structure–property performance, design and optimization of AM or 3D printing are urgent and crucial.

Regarding the AM fabrication technology for advanced composites, usually, it creates objects directly by stacking layers of material on each other until the required product is obtained [6–8]. As such, the extrusion-based method has currently been becoming the most used approach in the fabrication of advanced composites [9,10]. Ajao et al. [6] employed extrusion-based AM to design and construct a low-cost and high-accessibility 3D printing machine to manufacture plastic objects, where the distance between the nozzle tip and the bed is recommended as 0.1 mm during the AM process. In addition to this, Sharafi et al. [9] applied fused filament fabrication (FFF), a key approach of extrusion-based AM, to produce intricate PAEK and PAEK composite parts and to tailor their mechanical properties such as stiffness, strength, and deflection at failure. A multiscale modeling framework was used to identify the layer design and process parameters that may significantly affect the mechanical properties of polyetherketoneketone (PAEK) and PAEK composites. The results show that the mechanical properties of AM-produced parts are comparable to those of injection-molded parts [9]. Cramer et al. [11] developed an additive manufacturing process for fabricating ceramic matrix composites based on the C/C-SiC system. In this study, automated fiber placement of the continuous carbon fibers in a polyether ether ketone matrix was performed to consolidate the carbon fibers into a printed preform, and pyrolysis was performed to convert the polymer matrix to porous carbon. Then, Si was introduced by reactive melt infiltration to convert a portion of the carbon matrix to silicon carbide. Astafurova et al. [12] used the electron-beam AM method to obtain dual-phase specimens, their temperature dependence of tensile deformation behavior, and their mechanical properties (yield strength, ultimate tensile strength, and an elongation-to-failure). Susanto et al. [13] applied the digital light processing (DLP) technique to fabricate nickel microparticle-reinforced composites. Their mechanical properties were evaluated based on tensile strength, surface roughness, and hardness, and the findings demonstrate the potential of DLP-fabricated Ni-reinforced composites for applications demanding enhanced mechanical performance while maintaining favorable printability, paving the way for further exploration in this domain.

For modeling and characterization of structure–property performance, Zach et al. [14] reviewed the AM process across different spectrums of finite element analyses (FEA) and

Citation: Chen, Y. Editorial for the Special Issue on Additive Manufacturing of Advanced Composites. *J. Compos. Sci.* **2024**, *8*, 344. https://doi.org/10.3390/jcs8090344

Received: 19 July 2024
Accepted: 2 September 2024
Published: 3 September 2024

Copyright: © 2024 by the author. Licensee MDPI, Basel, Switzerland. This article is an open access article distributed under the terms and conditions of the Creative Commons Attribution (CC BY) license (https://creativecommons.org/licenses/by/4.0/).

summarized the building definition (support definition) with the optimization of deposition trajectories and the multi-physics of melting/solidification using computational fluid dynamics. The process modeling continues with the displacement/temperature distribution. Salgueiro et al. [15] investigated the effects of infill percentage and filament orientation on the mechanical properties of 3D-printed polylactic acid (PLA) structures, where both the finite element method and experimental tests were utilized in characterization. The results demonstrated that the process parameters have a significant impact on mechanical performance, particularly when the infill percentage is less than 100%. Meanwhile, Daly et al. [16] studied the impacts of infill patterns and densities on the mechanical characteristics of polyethylene terephthalate specimens reinforced with carbon fibers, fabricated by extrusion-based AM. The results show that the design with a 75% honeycomb and 100% infill density has the highest Young's modulus and tensile strength. In addition, the honeycomb was the ideal infill pattern, with 75% and 100% densities, providing significant strength and stiffness. Additionally, Ogaili et al. [17] evaluated the tensile and fracture behaviors of 3D-printed PLA composites reinforced with chopped carbon fibers through experimental characterization and finite element analysis (FEA). An inverse correlation between tensile strength and fracture toughness was observed, attributed to mechanisms such as crack deflection, fiber bridging, and fiber pull-out facilitated by multi-directional fiber orientations. Paulo et al. [18] developed 3D-printed PLA composites reinforced with flax fibers to evaluate the improvement in tensile and flexural strengths. An experimental design was utilized to study the effects of extruder temperature, number of strands, infill percentage of the specimens, and whether surface chemical treatment with flax fiber had an impact on the mechanical properties. The results show that the surface chemical treatment with the NaOH in the fiber does not have any influence on the mechanical properties of the composites; in contrast, the infill density demonstrated a huge influence on the improvement in mechanical strength. The maximum values of tensile and bending stress were 50 MPa and 73 MPa, respectively. In addition to the flax fiber-reinforced composites, Billings et al. [19] integrated wood fibers as a versatile renewable resource of cellulose, within bio-based PLA polymer, for the development and AM of sustainable and recyclable green composites using FFF technology. The 3D-printed composites were comprehensively characterized to understand the critical materials' properties, including density, porosity, microstructures, tensile modulus, and ultimate strength. Adeniran et al. [20] employed micro-CT scan analysis to quantitatively compare the fabrication temperature's effect on the mechanical properties of AM-fabricated carbon fiber-reinforced plastic (CFRP) composites. Additionally, SEM evaluation was used to determine the temperature effects on interlayer and intralayer porosity generation, and it was found that the porosity volume was related to the mechanical properties. Therefore, it was determined that temperatures influence porosity volumes.

To obtain improved mechanical performance in the AM-fabricated advanced composites, several studies were conducted. For example, for the fabrication of advanced composites, hybrid machines can solve this problem by combining the advantages of both additive and subtractive processes. However, little information is currently available to determine the milling parameters; hence, Spitaels et al. [21] proposed a systematic approach to experimentally determine the cutting parameters of green AM zirconia parts. Sadeghi et al. [22] developed a hybrid approach to solve the challenge of balancing strength and ductility in aluminum matrix composites. Owing to their studies, potential applications include lightweight and high-strength components for use in the aerospace and automotive industries. Structural materials for use in advanced mechanical systems that require both high strength and toughness are also expected. Demarbaix et al. [23] aimed to explore the feasibility of printing parts in continuous carbon fiber and using this fiber as an indicator, thanks to the electrical properties of the carbon fiber. The results show that the resistivity evolves linearly during the elastic period. The gauge factor increases when the number of passes in the manufacturing plane is low; however, repeatability is impacted. Meißner et al. [24] proposed a novel two-stage AM method in which the process steps of AM and continuous fiber integration are decoupled from each other. By using this method of fabrication, a significant improvement in the mechanical properties of the 3D-printed specimen was achieved compared to unreinforced polymer structures. The Young's modulus and tensile strength were increased by factors of 9.1 and 2.7, respectively.

Although increased attention is being paid to the AM of advanced composites, and the performance of AM-fabricated composites has seen significant improvement, the advanced composites made by AM still have a long way to go before they can be considered excellent, because their mechanical properties still cannot compare to those of composited created with conventional fabrication methods [25–27]. The collection of studies in this Issue may help advance technology and bring industry closer to promoting the wider application of AM for advanced composites.

Conflicts of Interest: The authors declare no conflict of interest.

References

1. Chen, Y.; Ye, L.; Kinloch, A.; Zhang, Y.X. 3D printed carbon-fibre reinforced composite lattice structures with good thermal-dimensional stability. *Compos. Sci. Technol.* **2022**, *227*, 109599. [CrossRef]
2. Li, G.X.; Chen, Y.; Wei, G.K. Continuous fiber reinforced meta-composites with tailorable Poisson's ratio and effective elastic modulus: Design and experiment. *Compos. Struct.* **2024**, *329*, 117768. [CrossRef]
3. Li, N.Y.; Link, G.; Wang, T.; Ramopoulos, V.; Neumaier, D.; Hofele, T.; Walter, M.; Jelonnek, J. Path-designed 3D printing for topological optimized continuous carbon fibre reinforced composite structures. *Compos. Part B Eng.* **2020**, *182*, 107612. [CrossRef]
4. Chen, Y.; Ye, L. Path-dependent progressive failure analysis for 3D-printed continuous carbon fibre reinforced composites. *Chin. J. Mech. Eng.* **2024**, *37*, 72. [CrossRef]
5. Chen, Y.; Klingler, A.; Fu, K.K.; Ye, L. 3D printing and modelling of continuous fibre reinforced composite grids with enhanced shear modulus. *Eng. Struct.* **2023**, *286*, 116165. [CrossRef]
6. Ajao, K.R.; Ibitoye, S.E.; Adesiji, A.D.; Akinlabi, E.T. Design and Construction of a Low-Cost-High-Accessibility 3D Printing Machine for Producing Plastic Components. *J. Compos. Sci.* **2022**, *6*, 265. [CrossRef]
7. Cai, H.; Chen, Y. A review on print head for fused filament fabrication of continuous carbon fiber reinforced composites. *Micromachines* **2024**, *15*, 432. [CrossRef]
8. Zhang, H.; Huang, T.; Jiang, Q.; He, L.; Bismarck, A.; Hu, Q. Recent progress of 3D printed continuous fiber reinforced polymer composites based on fused deposition modeling: A review. *J. Mater. Sci.* **2021**, *56*, 12999–13022. [CrossRef]
9. Sharafi, S.; Santare, M.H.; Gerdes, J.; Advani, S.G. Extrusion-Based Additively Manufactured PAEK and PAEK/CF Polymer Composites Performance: Role of Process Parameters on Strength, Toughness and Deflection at Failure. *J. Compos. Sci.* **2023**, *7*, 157. [CrossRef]
10. Chen, Y.; Ye, L. Topological design for 3D-printing of carbon fibre reinforced composite structural parts. *Compos. Sci. Technol.* **2021**, *204*, 108644. [CrossRef]
11. Cramer, C.L.; Yoon, B.; Lance, M.J.; Cakmak, E.; Campbell, Q.A.; Mitchell, D.J. Additive Manufacturing of C/C-SiC Ceramic Matrix Composites by Automated Fiber Placement of Continuous Fiber Tow in Polymer with Pyrolysis and Reactive Silicon Melt Infiltration. *J. Compos. Sci.* **2022**, *6*, 359. [CrossRef]
12. Astafurova, E.; Maier, G.; Melnikov, E.; Astafurov, S.; Panchenko, M.; Reunova, K.; Luchin, A.; Kolubaev, E. Temperature-Dependent Deformation Behavior of "γ austenite/δ-ferrite" Composite Obtained through Electron Beam Additive Manufacturing with Austenitic Stainless-Steel Wire. *J. Compos. Sci.* **2023**, *7*, 45. [CrossRef]
13. Susanto, B.; Kumar, V.V.; Sean, L.; Handayani, M.; Triawan, F.; Rahmayanti, Y.D.; Ardianto, H.; Muflikhun, M.A. Investigating Microstructural and Mechanical Behavior of DLP-Printed Nickel Microparticle Composites. *J. Compos. Sci.* **2024**, *8*, 247. [CrossRef]
14. Zach, T.F.; Dudescu, M.C. The Three-Dimensional Printing of Composites: A Review of the Finite Element/Finite Volume Modelling of the Process. *J. Compos. Sci.* **2024**, *8*, 146. [CrossRef]
15. Salgueiro, M.P.; Pereira, F.A.M.; Faria, C.L.; Pereira, E.B.; Almeida, J.A.P.P.; Campos, T.D.; Fakher, C.; Zille, A.; Nguyễn, Q.; Dourado, N. Numerical and Experimental Characterisation of Polylactic Acid (PLA) Processed by Additive Manufacturing (AM): Bending and Tensile Tests. *J. Compos. Sci.* **2024**, *8*, 55. [CrossRef]
16. Daly, M.; Tarfaoui, M.; Bouali, M.; Bendarma, A. Effects of Infill Density and Pattern on the Tensile Mechanical Behavior of 3D-Printed Glycolyzed Polyethylene Terephthalate Reinforced with Carbon-Fiber Composites by the FDM Process. *J. Compos. Sci.* **2024**, *8*, 115. [CrossRef]
17. Ogaili, A.A.F.; Basem, A.; Kadhim, M.S.; Al-Sharify, Z.T.; Jaber, A.A.; Njim, E.K.; Al-Haddad, L.A.; Hamzah, M.N.; Al-Ameen, E.S. The Effect of Chopped Carbon Fibers on the Mechanical Properties and Fracture Toughness of 3D-Printed PLA Parts: An Experimental and Simulation Study. *J. Compos. Sci.* **2024**, *8*, 273. [CrossRef]
18. Paulo, A.; Santos, J.; da Rocha, J.; Lima, R.; Ribeiro, J. Mechanical Properties of PLA Specimens Obtained by Additive Manufacturing Process Reinforced with Flax Fibers. *J. Compos. Sci.* **2023**, *7*, 27. [CrossRef]

19. Billings, C.; Siddique, R.; Sherwood, B.; Hall, J.; Liu, Y. Additive Manufacturing and Characterization of Sustainable Wood Fiber-Reinforced Green Composites. *J. Compos. Sci.* **2023**, *7*, 489. [CrossRef]
20. Adeniran, O.; Osa-uwagboe, N.; Cong, W.; Ramoni, M. Fabrication Temperature-Related Porosity Effects on the Mechanical Properties of Additively Manufactured CFRP Composites. *J. Compos. Sci.* **2023**, *7*, 12. [CrossRef]
21. Spitaels, L.; Dantinne, H.; Bossu, J.; Rivière-Lorphèvre, E.; Ducobu, F. A Systematic Approach to Determine the Cutting Parameters of AM Green Zirconia in Finish Milling. *J. Compos. Sci.* **2023**, *7*, 112. [CrossRef]
22. Sadeghi, B.; Cavaliere, P.; Sadeghian, B. Enhancing Strength and Toughness of Aluminum Laminated Composites through Hybrid Reinforcement Using Dispersion Engineering. *J. Compos. Sci.* **2023**, *7*, 332. [CrossRef]
23. Demarbaix, A.; Ochana, I.; Levrie, J.; Coutinho, I.; Cunha, S.S., Jr.; Moonens, M. Additively Manufactured Multifunctional Composite Parts with the Help of Coextrusion Continuous Carbon Fiber: Study of Feasibility to Print Self-Sensing without Doped Raw Material. *J. Compos. Sci.* **2023**, *7*, 355. [CrossRef]
24. Meißner, S.; Kafka, J.; Isermann, H.; Labisch, S.; Kesel, A.; Eberhardt, O.; Kuolt, H.; Scholz, S.; Kalisch, D.; Müller, S.; et al. Development and Evaluation of a Novel Method for Reinforcing Additively Manufactured Polymer Structures with Continuous Fiber Composites. *J. Compos. Sci.* **2024**, *8*, 272. [CrossRef]
25. Chen, Y.; Mai, Y.-W.; Ye, L. Perspectives for multiphase mechanical metamaterials. *Mater. Sci. Eng. R. Rep.* **2023**, *153*, 100725. [CrossRef]
26. Jiang, B.N.; Chen, Y.; Ye, L.; Dong, H.; Chang, L. Residual stress and warpage of additively manufactured SCF/PLA composite parts. *Adv. Manuf. Polym. Compos. Sci.* **2023**, *9*, 2171940. [CrossRef]
27. Kamaal, M.; Anas, M.; Rastogi, H.; Bhardwaj, N.; Rahaman, A. Effect of FDM Process Parameters on Mechanical Properties of 3D-Printed Carbon Fibre–PLA Composite. *Prog. Addit. Manuf.* **2021**, *6*, 63–69. [CrossRef]

Disclaimer/Publisher's Note: The statements, opinions and data contained in all publications are solely those of the individual author(s) and contributor(s) and not of MDPI and/or the editor(s). MDPI and/or the editor(s) disclaim responsibility for any injury to people or property resulting from any ideas, methods, instructions or products referred to in the content.

Review

The Three-Dimensional Printing of Composites: A Review of the Finite Element/Finite Volume Modelling of the Process

Theodor Florian Zach and Mircea Cristian Dudescu *

Department of Mechanical Engineering, Technical University of Cluj-Napoca, Blvd. Muncii No. 103-105, 400641 Cluj-Napoca, Romania; theodor.zach@campus.utcluj.ro
* Correspondence: mircea.dudescu@rezi.utcluj.ro

Abstract: Composite materials represent the evolution of material science and technology, maximizing the properties for high-end industry applications. The fields concerned include aerospace and defense, automotive, or naval industries. Additive manufacturing (AM) technologies are increasingly growing in market shares due to the elimination of shape barriers, a plethora of available materials, and the reduced costs. The AM technologies of composite materials combine the two growing trends in manufacturing, combining the advantages of both, with a specific enhancement being the elimination of the need for mold manufacturing for composites, or even post-curing treatments. The challenge of AM composites is to compete with their conventional counterparts. The aim of the current paper is to present the additive manufacturing process across different spectrums of finite element analyses (FEA). The first outcomes are building definition (support definition) and the optimization of deposition trajectories. In addition, the multi-physics of melting/solidification using computational fluid dynamics (CFD) are performed to predict the fiber orientation and extrusion profiles. The process modelling continues with the displacement/temperature distribution, which influences porosity, warping, and residual stresses that influence characteristics of the component. This leads to the tuning of the technological parameters, thus improving the manufacturing process.

Keywords: composite additive manufacturing; continuous fiber fabrication; direct ink writing; process finite element analysis

Citation: Zach, T.F.; Dudescu, M.C. The Three-Dimensional Printing of Composites: A Review of the Finite Element/Finite Volume Modelling of the Process. *J. Compos. Sci.* **2024**, *8*, 146. https://doi.org/10.3390/jcs8040146

Academic Editor: Yuan Chen

Received: 5 March 2024
Revised: 27 March 2024
Accepted: 8 April 2024
Published: 12 April 2024

Copyright: © 2024 by the authors. Licensee MDPI, Basel, Switzerland. This article is an open access article distributed under the terms and conditions of the Creative Commons Attribution (CC BY) license (https://creativecommons.org/licenses/by/4.0/).

1. Introduction

Additive manufacturing technologies represent the processes of incrementally adding material to obtain a pre-defined tridimensional shape [1]. The technologies have been developed continuously over the last 45 years. First, the commercially launched equipment (1987) was based on stereolithography (the photopolymerization of liquid resin, SLA) via the use of 3D Systems (Rock Hill, SC, USA), although the first patent was registered in 1977 [2–4]. The thermoplastics were first used in the industry for additive manufacturing in 1991, when fused deposition modelling (FDM)/fused filament fabrication (FFF) technology was commercially launched by the company Stratasys (Edina, MN, USA), gradually becoming a popular technology for daily users [5]. Subsequently, in 1992, the plastic and ceramic powders were laser sintered (selective laser sintering, SLS) [6–9]. Eventually, beginning from 1999 (EOS GmbH, Krailling, Germany), the metallic powders (titanium, aluminum, stainless steel, nickel-based alloys) were melted and welded together via guided lasers or electron beam fluxes under the generically named powder bed fusion (PBF), in particular selective laser melting (SLM-1999) or electron beam melting (EBM-2000), considering the energy source [10,11]. In the beginning, the additive manufacturing technologies satisfied the industrial needs of rapid prototyping, serving as design studies, mock-ups, rapid tooling for molds, and, eventually, fully functional components in sectors such as aerospace (fuel tanks, fuselage, or interior elements) [12,13], automotive (body panels, interior elements, or spare parts) [14,15], medical applications (prosthetics, dental, or bone implants) [16,17], or electronics (MEMS, LED, transistors, or batteries) [18,19]. The global market of additive manufacturing has an estimated annual growth of 18.41% from 2023 to 2032, and the market share has risen from 13.16 billion dollars (2022) to 109.52 billion dollars (2032) [20–22]. Considering the progress, high performance materials like composites have become of great interest being processed by AM technologies. In addition, the elimination of molds brings a significant reduction of costs. The first endeavors were performed by introducing short fibers in filaments for FDM/FFF such as ABS-fiberglass, 2001 [23] and ABS-carbon fiber (CF), 2003 [24], developing into reinforcement particles of different natures, including metallic, such as ABS-iron/copper [25], PLA-magnesium [26], and ABS-Stainless Steel [27]; ceramics, such as ABS-BaTiO$_3$ [28] and Nylon-Al$_2$O$_3$ [29]; or short synthetic fibers, such as PLA-CF [30], ABS-CF [31], Polycarbonate-CF [32], Nylon-CF [33], PP-glass fiber (GF) [34], PEEK-CF [35]. This has resulted in randomly oriented composites [36–40], such as in a layered extrusion, as shown in Figure 1a. Short fibered structures present improved mechanical characteristics than pure plastics, due to the reinforcement embedded into the component, as noticed in Figure 1b [41–44].

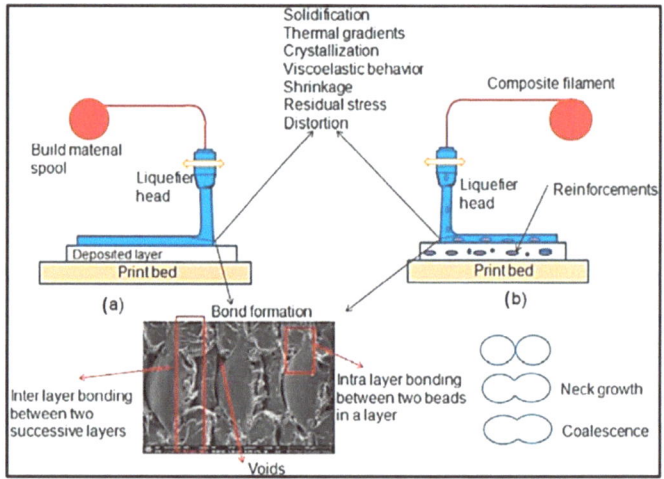

Figure 1. Principle of FDM with pure thermoplastics (**a**); reinforced thermoplastics (**b**) [38]. Reproduced with permission: Elsevier.

The additive manufacturing of continuous fiber composites appeared experimentally in 2012, then being based on SLA (continuous scaled manufacturing, CSM) using thermal curing thermosets [45–47]. The evolution that occurred in 2017 relied on a photo-curable matrix (direct ink writing, DIW), and developed commercially into Continuous Composite's equipment (Coeur d'Alene, ID, USA) [48]. The principle included the in-situ impregnation of fibers through a liquid resin reservoir, resulting in the composite being extruded through a nozzle and then cured either by a heat or light source, such as one may see in Figure 2. Reinforcements are mainly synthetic fibers such as CF, GF, or Aramid [49–51], with thermoset resins such as epoxy [52,53], photo-curable acrylic resins [54,55], phenolic resins [56], or frontal polymerization resins [57].

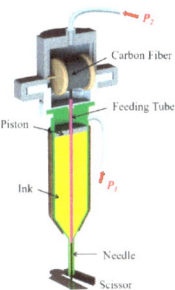

Figure 2. Principle of continuous fiber DIW with a thermoset matrix [58]. Reproduced with permission: Elsevier.

In parallel, in 2015, the continuous fiber fabrication technology (CFF), first commercially developed by Markforged (Waltham, MA, USA), resulted in continuous composite materials [59–61], either through dual deposition or in situ impregnation as shown in Figure 3a, utilizing a thermoplastic matrix. This could be completed using prepreg filaments, as presented in Figure 3b, resulting in a continuous fiber with a matrix interface, while the second extruder is deposed in parallel using the matrix between the fibers. Commercially available filaments include the nanocomposite matrix CF-Nylon (Onyx) with GF, Carbon, and Kevlar, or using Aramid fibers as reinforcements [62]. Alternatively, some filaments were experimentally developed, such as CF-PLA [63], Kevlar-PLA [64], Glass-PLA [65], Aramid-PLA [66], CF-ABS [67], CF-PA [68], CF-PEEK [69], Basalt-PLA, and Stainless Steel-ABS [70].

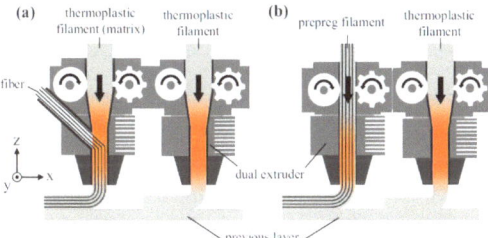

Figure 3. Principle of the in-situ impregnation of CFF (**a**); Prepreg (**b**) [71]. Under CC–BY Licensing 4.0.

2. Challenges and Key Factors in the 3D Printing of Composites

Being a relatively new technology, several key aspects are worth studying. The results of the manufacturing process must respect the designated quality criteria [72].

In the context of an engineering assembly, the quality criteria relate first and foremost to the functional role. Alternatively, the aspect of the part is influenced by the technological parameters. Lastly, the process performance is also influenced, allowing competitiveness with conventional composites. The parameters' influences are shown in Table 1. The slicing strategy enables a better coverage of the built volume, which is crucial for anisotropic structures like composites, resulting in an enhanced structural stability and deposed material in the designated stress areas. Voided volumes are inevitable in 3D printing due to the elemental deposition, resulting in a factor of embrittlement and a common cause of structural failure. The optimization of trajectories induces the continuity of fibers and their orientation, representing a key factor for the directional strength of the composite. The extrusion temperature leads the recrystallization process, resulting in dimensional deviations (warping) and induced residual stresses, thus making the component brittle. The nozzle diameter influences the fiber orientation and fiber volume content (proportional with strength) due to the flow of solid reinforcements into the molten matrix. The building plate temperature impacts the formation of the first layer and the interlaminar adhesion, leading to delamination failure. In less severe cases, the surface could be affected by wrinkles or sinkholes. The layer height directly influences the lead time of the process. Nevertheless, an increased layer height causes a coarser structure and a high-temperature gradient, which impacts the warping and residual stresses; the shrinkage may also cause the delamination of the consecutive layers, which can lead to the failure of the component and voids the formation. The aspect of the part could be affected when the layer height or the temperature is too high. Ultimately, the printing speed causes fibers to settle on a different orientation than expected due to a faster crystallization; indeed, some areas may cause voids. All in all, the combination of optimized parameters results in a performant process and component.

Table 1. Technological parameters for AM composites [73–79].

No.	Technological Parameters	Defects Induced in the Structure	Influenced Characteristic
1.	Slicing Strategy	Voids Creation/Fiber Interruptions/Fiber Orientation	Mechanical Strength
2.	Extrusion Temperature	Warping/Residual Stresses/Nozzle Clogging	Precision, Mechanical Strength
3.	Nozzle Diameter	Fiber Volume Content/Fiber Orientation	Mechanical Strength
4.	Printing-Bed Temperature	Delamination/Residual Stresses	Precision/Mechanical Strength/Surface Defects
5.	Layer Height	Warping, Surface Defects, Voids, Delamination, Residual Stresses	Precision, Mechanical Strength
6.	Printing Speed	Fiber Orientation/Void Formation	Mechanical Strength

3. Finite Element Simulations of Additive Manufacturing Processes for Composites

The current section aims to assess the types of finite element analyses performed on the processes presented before, as shown in Table 2. The aim of these analyses is to optimize AM processes for composite structures, tuning the technological parameters and predicting defects, as presented in the prior section. The global AM process for polymers can be seen as a series of physical phenomena that are potentially modelled via the finite element method (FEM) [80] or the finite volume method (FVM) [81]. Considering the random orientation of the fibers and the computational capacity, composites are widely modelled using a representative volume element (RVE) [82].

Table 2. Categorization of the analyses for composite 3D printing.

No.	Process Phase	Subject	FE Approach	Application
1.	Pre-Processing	Path Optimization	Topological Optimization, Fiber Direction Optimization	Trajectory Definition/Optimization
2.	AM Process	Extrusion	Computational Fluid Dynamics	Melting Simulation
				In-Nozzle Flow and Fiber Orientation
				Nozzle Clogging Simulation
		Deposition	Multi-Physics (Thermo-Mechanical)	First Layer Formation
		Solidification	Multi-Physics (Thermo-Mechanical)	Residual Stresses
				Dimensional Accuracy
				Curing Thermoset Composites
3.	Post-Processing	Defects and Post-processing Treatments	Multi-Physics (Thermo-Mechanical)	Internal Defects
				Void Formation
				Surface Roughness

The deposition process of AM composites can be divided into several areas of interest. The current section presents, via FEM or FVM, analytical models of the manufacturing phases in the technologies presented in Section 1, mainly CF/GF and thermoplastic and thermoset matrices. The phases of AM are categorized (Brenken et al., 2018) [36], as presented in Figure 4; the first area of interest is the nozzle, where the matrix undergoes a phase change. The solid fibers show a viscous flow into the matrix, resulting in fiber orientation. The unitary layer might be studied as a bonding between the current and previous raster, with the fibers having to flow on top due to the gravitational effect. Lastly, the solidification of the layer is studied with a focus on shrinkage (warping and residual stresses), the thermal trace, and, ultimately, the elastic behavior of the structure.

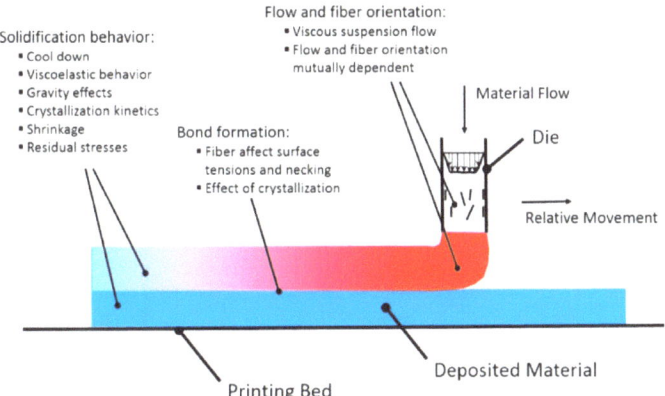

Figure 4. Physics of FFF processes [36]. Reproduced with permission: Elsevier.

3.1. Pre-Processing (Slicing and Trajectory) Optimization

Composite specimens are mainly formed on a 0° fiber orientation, though AM enables one to develop various deposition strategies, presenting opportunities for design alternatives via the use of techniques such as topology optimization (mass/stiffness-oriented redesign). The case of short carbon fibers (SCF) and the polyethylene terephthalate glycol (PETG) matrix (Shafighfard et al., 2021) [83], is based on a rectangular plate with a hole in the middle, as used in a common stress concentrator test, as shown in Figure 5. The method was developed using the Halpin–Tsai homogenization method for design optimization (LSC), where the boundaries of the raster are defined, see Figure 5a, and the different orientation areas are emphasized, as seen in Figure 5b.

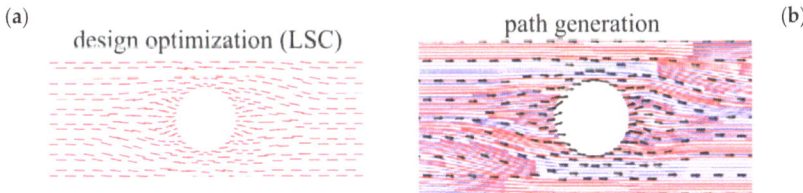

Figure 5. Design optimization for SCF/PETG specimen (**a**); path generation (**b**) [83]. Reproduced with permission: Elsevier.

The difference with unidirectional strands lies in their ability to prevent interruptions in the path near the stress concentrator [84]. The conventional building strategy (0°), as seen in Figure 6a, shows higher peak values around the stress concentrator as the load increases (10 MPa, 12.5 MPa, respectively; 15 MPa), whereas the curvilinear optimized path, presented in Figure 6b, presents lower values with a relatively evenly distributed strain due to the stress occurring along a continuous path, which, in other areas, unloads in a normal direction, as in the case of 0°, where tangential inter-fiber unloading can be observed.

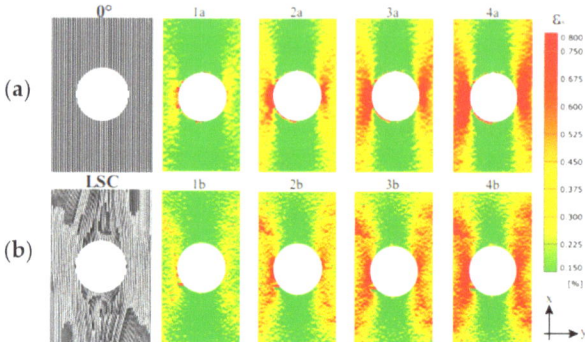

Figure 6. Tensile test for 0° orientation (**a**); path optimization SCF/PETG specimens (**b**) [83]. Reproduced with permission: Elsevier.

Further investigation has demonstrated continuous deposition paths, as one can see in Figure 7, which form the basis of the concept of a flow over a cylinder. The idea behind this concept originates from path definitions following curvilinear directions, as inspired by natural structures (such as branches on a tree trunk). In utilizing this method, the material avoids any interruptions in the deposition around the stress concentrator, even though the geometry is slightly altered.

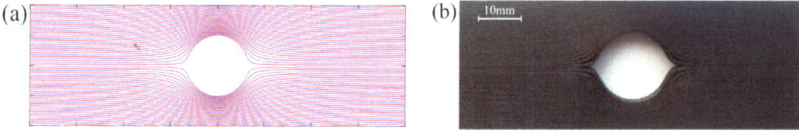

Figure 7. Continuous flow paths for PETG/SCF specimens: virtual (**a**); physical (**b**) [83]. Reproduced with permission: Elsevier.

The path optimization becomes of more relevance in the case of continuous fibers, which can assure the continuity of the fibers. The path topology optimization, in the case of continuous carbon fibers (CCF) and PA, (Chen et al., 2023) [85] develops models where the initial volume of the part was reduced, as visible in Figure 8.

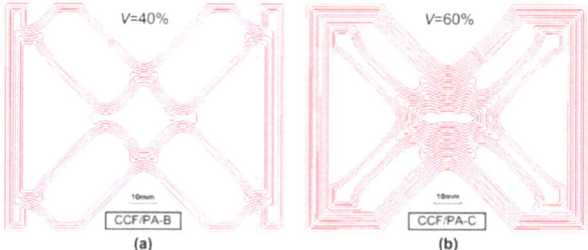

Figure 8. Trajectory optimization for CCF structures with 40% infill (**a**); 60% infill (**b**) [85]. Reproduced with permission: Elsevier.

Stresses were observed when the material was subjected to 1% effective strains, particularly those occurring in the intersectional areas for CCF-PA, as shown in Figure 9b, as the rest of the material was less vulnerable. On the contrary, for PA, the strain is distributed throughout the whole structure. This may lead to a decrease in the fiber content, eventually leading to a cost optimization.

Figure 9. von Mises stress diagram of structures made with different volume fractions for PA specimens (**a**); CCF-PA specimens (**b**) [85]. Reproduced with permission: Elsevier.

Due to the complex shapes, where continuous fibers can become difficult to lay on the trajectories and fill the surfaces, a combination of SCF/PA and CCF/PA (Chen et al., 2021) [86] is used toto manufacture a complex component. The application starts as a fixed-end plate featuring a hole, as one may see in Figure 10a. Following the topology optimization, the filling strategy was performed to prioritize continuous fiber paths (without interruptions), as one can see via the blue paths in Figure 10b; in the areas where continuous path generation was achieved, a SCF-PA composite filament was employed (as indicated by the red contours in Figure 10b).

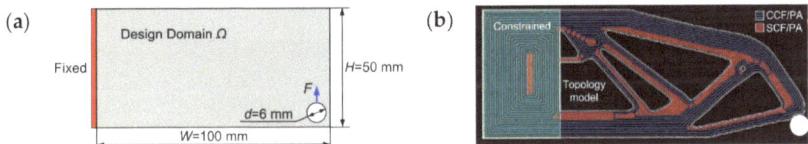

Figure 10. Definitions for optimization (**a**); paths of CCF/SCF in the optimized part (**b**) [86]. Reproduced with permission: Elsevier.

A comparison of the levels of stiffness between pure PA, SCF-PA, and CCF-PA is illustrated in Figure 11a, with the levels being almost two times higher for SCF, and twelve times higher for CCF. The peak load is five times higher for SCF, and sixteen times higher for CCF, as shown in Figure 11b.

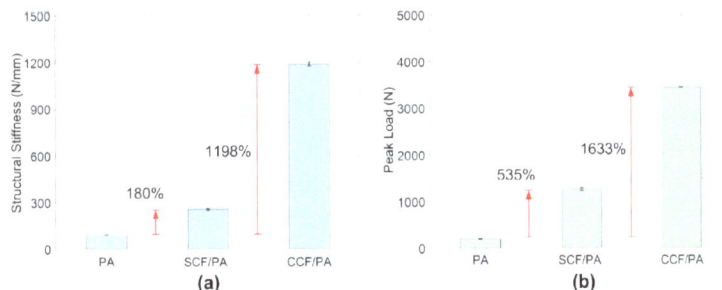

Figure 11. Structural stiffness of PA, SCF-PA, and CCF-PA specimens (**a**); peak load (**b**) [86]. Reproduced with permission: Elsevier.

The presentation of the path optimization of thermoset composites (DIW) (Qian et al., 2021) [87] is significant. It was built a porous SCF/Poly-carbosilane catalyst structure for the production of hydrogen via reforming methanol steam for fuel cells, as one can see in Figure 12a. It is meshed so the channels can be more refined, as shown in Figure 12b.

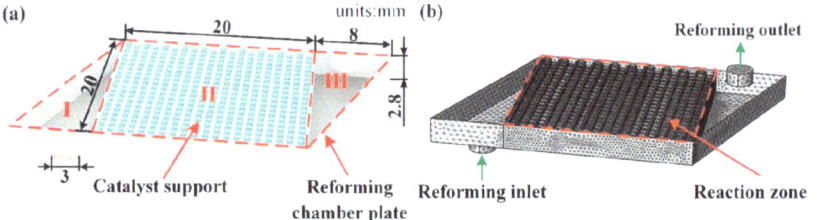

Figure 12. Methanol reactor geometry (**a**); discretization (**b**) [87]. Reproduced with permission: Elsevier.

The scope of this optimization mainly concerns the reformation of the methanol steams process, with the aim of slowing the flow of the methanol, as seen in Figure 13a. On the other hand, the hydrogen concentration is the highest at the end of the channels, as one can see in Figure 13b. This optimization leads to a better overall efficiency of the process, as well as the construction of small reactors for applications such as hydrogen fuel cells to be used in vehicles, which can be fueled using methanol, making storage easier and less hazardous than in liquid hydrogen tanks.

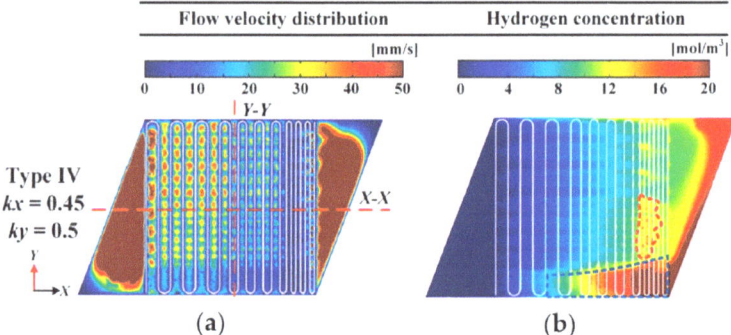

Figure 13. Reactor simulation for flow velocity (**a**); hydrogen concentration (**b**) [87]. Reproduced with permission: Elsevier.

3.2. Computational Fluid Dynamics of Additive Maufacturing Composites

Computational fluid dynamics represent the differential numerical modelling of the equations of energy transfer in a fluid tridimensional medium (Navier-Stocks) (Date Anil W 2005; Hu 2012) [88,89]. Xu et al. made a categorization of CFD domains in the context of 3D-printed composites (Xu et al., 2023) [90]. The process starts with a multi-phase flow (solid/liquid) into the nozzle, as shown in Figure 14a. In the extrusion process, the fibers are flowed into the molten matrix, resulting the fiber orientation, whereas, in the continuous fiber composites, this will result in fiber deformation, as seen in Figure 14b. Following the phase transformation (liquid/solid), the dimensional deflection is simulated until convergence is achieved.

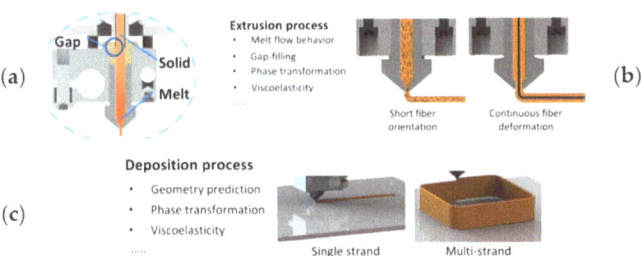

Figure 14. Computational fluid dynamics in the AM of composites: multi-phase transformation (**a**); extrusion process (**b**); deposition process (**c**) [90]. Reproduced with permission: Elsevier.

3.3. Melting Simulation of Additive Maufacturing Composites

A multi-phase transformation FVM model of PEEK AM (Wang et al., 2019) [91], which proved important for applications that entailed high temperatures/high strength (aerospace and defense applications) and biocompatibility. This model employed Ansys Fluent, utilizing the Cross-WLF viscosity law [92]. The geometry of the multi-stepped printing head demonstrates the gradient heating of the filament from the chamber temperature (68 °C) to the extrusion temperature within the nozzle (380 °C), and the filament displays convex-shape heating, beginning in the walls, as seen in Figure 15a. Before the extrusion, the filament starts the phase transformation from the heated walls, continuing to the tip of the nozzle, as seen Figure 15b. The multi-phase transformation temperature gradient is shown in Figure 15c, with a decreased viscosity of 14 Pa·s (liquid phase).

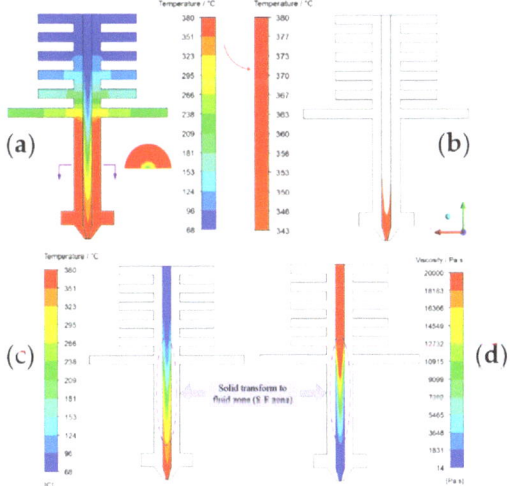

Figure 15. Temperature plots in the printing head for PEEK (**a**); temperature above melting in the nozzle area (343 °C) (**b**); temperature at phase change (**c**); viscosity at phase change (**d**) [91]. Reproduced with permission: Elsevier.

3.4. In-Nozzle Flow for Additive Manufacturing Composites (Fiber Orientation)

The second phenomenon of interest in CFD concerns the flow dynamics. In this case, an isothermal model of short-glass fibers in the molten ABS suspension (Yang et al., 2017) [93] who utilized the SPH-DEM (smoothed particle hydrodynamics–discrete element modelling) approach [94,95]. This model consists of solid short fiber (DEM) elements and fully fluid ABS DEM elements, as shown Figure 16a, where the moving wall pushes the composite through the nozzle at a certain flow rate. The transient model presented in Figure 16b depicts the flow velocities at two separate moments: firstly, when the first particles arrive at the diameter reduction (nozzle), where the flow rate increases, and secondly, when the particles passed through nozzles demonstrate free flow on the deposition bed, thus slowing the flow rate; however, the rate in the nozzle increases due to the gravity of the particles to be extruded.

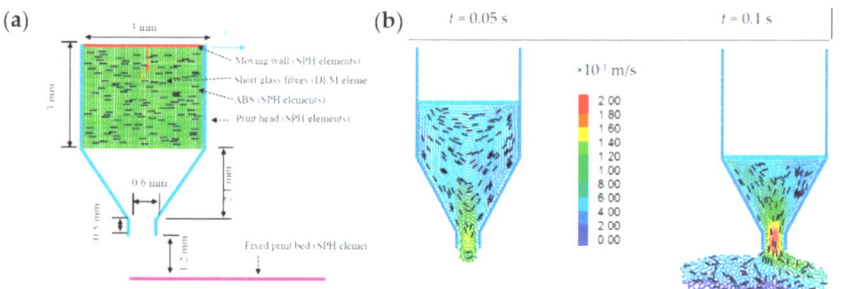

Figure 16. Model definition of GF-ABS AM (**a**); short GF/ABS flow simulation (**b**) [93]. Under CC–BY Licensing 4.0.

Concerning continuous fibers, a SPH-DEM model (Yang et al., 2017) [93] with similar approach to the case of short fibers, as seen in Figure 17a. The in-situ impregnation of continuous CF-Nylon is simplified due to the rigidity of the center. The flow simulation, visible in Figure 17b, presents a continuous carbon fiber flow with a high-speed flow velocity under the fiber, which can be attributed to the diameter reduction, as well as the pushing and higher viscosity of the solid fiber, which is continuously fed. The laying of the continuous fiber shows a non-linear placement that may eventually influence the material's isotropy.

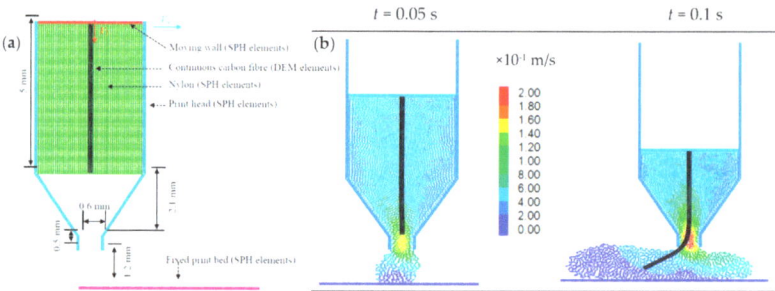

Figure 17. Model definition for the continuous fiber fabrication of Nylon-CF (**a**); flow map of the continuous fiber fabrication of Nylon-CF (**b**) [93]. Under CC–BY Licensing 4.0.

In the continuous fiber, stresses can be withstood due to the pressure in the area opposite to the deposition direction, as one may see in Figure 18; these stresses can have values up to 450 MPa, which, in some cases, may lead to fiber breakage.

The flow of the thermoset resin and short carbon fibers by DIW (Kanarska et al., 2019) [96], is inspired by the Low Mach Code with Particle Transport (LMC-PT) algorithm, based on the Distributed Large Multiplier (DLM) technique found in the SAMRAI software v. 4.3.0. The suspension of fibers into the resin was modeled after RVE. The suspension flow was simulated in the transient domain with a constant increase in viscosity as the diameter diminished, as seen in Figure 19a,b. The first section shows a relatively even fiber dispersion over the angle, with a small peak around 20°. The tendency in the second area is similar, with peaks between 0° and 20°; however, in the third part (close to the extrusion area), most fibers are oriented around 10°, as shown in Figure 19c. This results in the control of the directional behavior of the resulting composite.

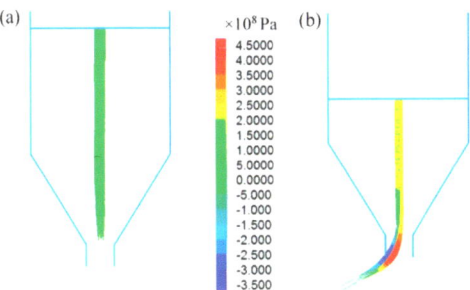

Figure 18. Tensile stress in the continuous fiber before deposition (**a**); after deposition (**b**) [93]. Under CC–BY Licensing 4.0.

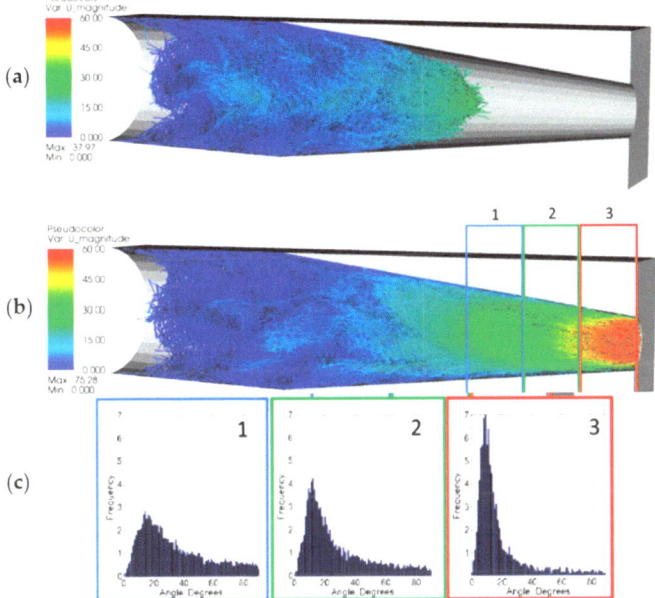

Figure 19. Fiber flow into the nozzle at 0.25 s (**a**); 0.55 s (**b**); graphics of orientation (**c**) [96]. Reproduced with permission: Elsevier.

The cylindrical nozzle geometry emphasizes the differences between short fibers of different diameters, as shown in Figure 20. In the first case (Figure 20a), there are 0.15 mm diameter carbon fibers. The flow is relatively constant in the central area, and decreases near the walls due to the adhesion. The orientation is more dispersed, which is due to the lower viscosity of the resulting composite material. As the fiber diameter increases, the flowing speed decreases, and the fiber orientation is closer to 0° (Figure 20b,c).

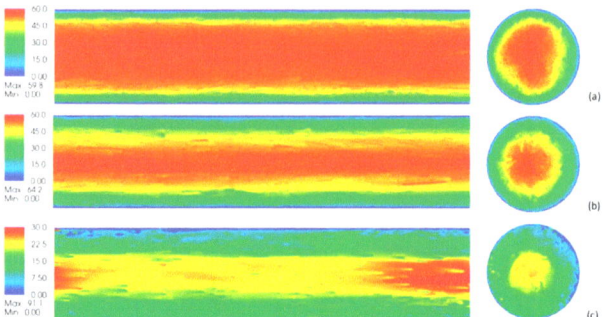

Figure 20. Fiber orientation and flow in the cylindrical nozzle with different fiber diameters: 0.15 mm (**a**); 0.3 mm (**b**); 0.5 mm (**c**) [96]. Reproduced with permission: Elsevier.

3.5. Extrusion Defects Simulation

A specific defect of composite AM, both in terms of thermoplastic and thermoset, is the clogging of the nozzle, which occurs due to the higher viscosity caused by the solid fibers in the nozzle. Due to this phenomenon, the fibers will adhere to the extruder's walls and the composite material will have less fiber volume content, thus resulting in less mechanical strength. Another downside is the turbulent flow that will occur inside the nozzle, which can lead to an unpredictable fiber orientation. This will affect the directional strength of the composite structure. Lastly, the process performance is affected by the frequency of nozzle changes resulting from abrasion and the increased failure rate due to any clogging that occurs during printing [77,79].

The FVM model (Zhang et al., 2021) [97] of nozzle extrusion (SCF-PA) via the SPH-DEM method (CF as DEM particles, SPH matrix) was performed using Ansys Fluent. During this process, various reinforcement volume fractions (V_f) were considered, including 13.34%, 20.01%, and 26.68%, as well as various raster widths (L), such as 0.12, 0.24, and 0.35 mm. The Figure 21 presents the filaments into the nozzle in different areas (I, II—beginning of diameter reduction, III—stabilization of extrusion diameter, IV—reduction to the extrusion diameter, V—extrusion area). The IV area of Figure 21, which is the most vulnerable area due to the diameter change. The initial results are coordination numbers (CN), which are a measure of the packing structure of the composite and the contact points between particles.

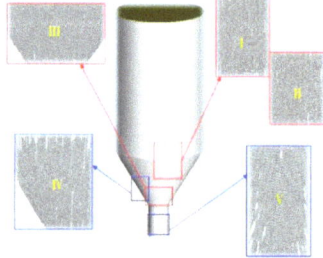

Figure 21. Definition of areas in the printing nozzle [97]. Under CC–BY Licensing 4.0.

In all cases of raster widths, there is a Gaussian distribution of median CN with higher peaks (probability density function) for V_f = 26.68%, as seen Figure 22a; this is due to more solid particles being in contact. Alternatively, the variation of singular values for particles has been studied, as one can see in Figure 22b–d. The small particles represent the matrix, whereas the small cylinders represent the contact between the matrix and the solid particles; as the volume reduces, the CN grows due to the space reduction.

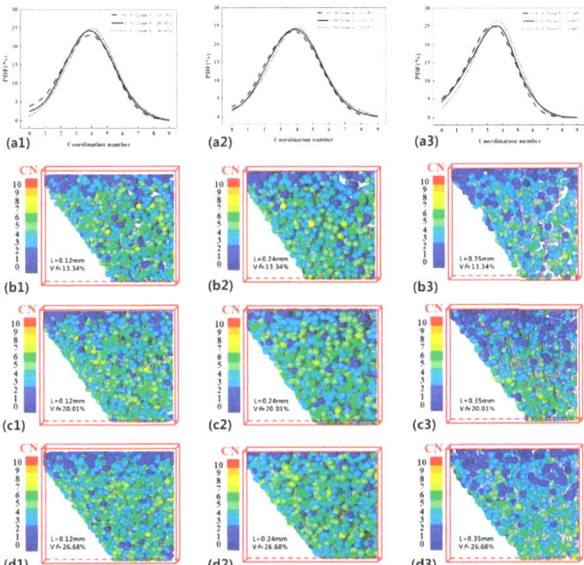

Figure 22. Probability density function (**a**); coordination number for SCF-PA for 0.12 mm raster width (**b**); 0.24 mm width (**c**); 0.35 mm width (**d**) [97]. Under CC−BY Licensing 4.0.

The flow analysis shown in Figure 23a emphasizes the construction of the fiber flow over time (bridging), where, initially, the front takes a convex shape (a laminar flow), and then the fibers tend to spread to the walls due to the adhesion of the fiber matrix. Ultimately, due to the feed rate, the process converges after 0.75% in a relatively constant tubular flow of the fibers, which is useful for the prediction of the fiber orientation during the process. The geometry of the nozzle influences the contact force of the composites at the wall of the nozzle in the IV area, as one may see in Figure 23b. In the conical area, the original design shows a high contact pressure, whereas, in the second design, the contact force is balanced. The last design shows a central constant contact force.

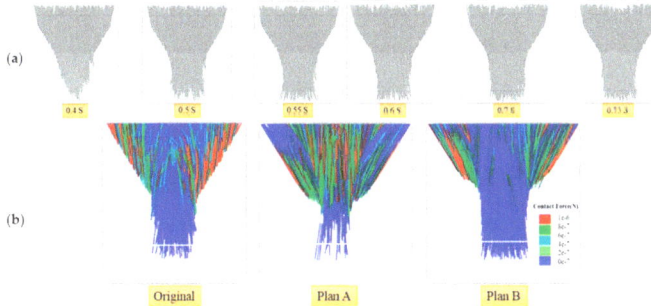

Figure 23. Fiber flow for L = 0.24 mm and V_f = 26.68% (**a**); contact forces in the nozzle (**b**) [97]. Under CC−BY Licensing 4.0.

3.6. Deposition Simulation (First Layer Simulation for Thermoplastic Composites)

The following phenomenon studied includes the assessment of the elemental deposition trace and the fiber orientation relative to the direction of deposition, both in the cases of thermoplastic and thermoset matrix AM composites. The deposition of the first layer of the 3D-printed structure was studied. The numerical method can be defined as a series of boundary conditions, as seen in Figure 24. In general, the nozzle is considered a slip-free heating source for simplification. The first layer in contact with the printing bed (temperature regulated) is reduced to a constant height and length area.

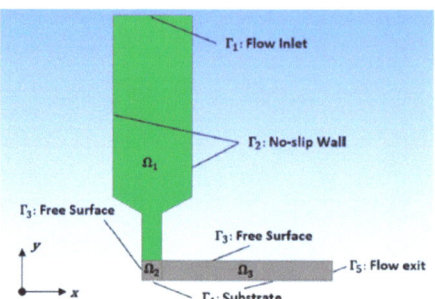

Figure 24. A 2D schematic representation of matrix flow of composite filaments [98]. Under CC–BY Licensing 4.0.

In the case of SCF, the fluid flow in the first layer FVM model (Kermani et al., 2023) [99] was developed in COMSOL Multiphysics, using the Advani–Tucker model [100], and comparing Newtonian and Carreau viscosity models [101]. Reinforcement fibers are considered directional vectors, and the fiber orientation model is user defined (Coefficient PDE model). The short fibers in the nozzle are considered to be randomly oriented.

Therefore, it can be remarked in Figure 25A that the orientation tensor of the fibers shifts abruptly in an area of interference between a 90° direction change and the deposed layer. The orientation of SCF can be observed in Figure 25B. The core shows the gradients concentrically, ranging from 40% to 0%, indicating the probability of the desired outcome to be higher than at the center of gravity. The outer layers show an almost certain orientation of 0°, On the transversal cross-section, as seen in Figure 25C, the gradual adhesion to the printing bed can be seen, with a core of 0–20% around the deposition orientation, whereas the outer surface that has been subjected to shrinking almost certainly displays the predicted direction.

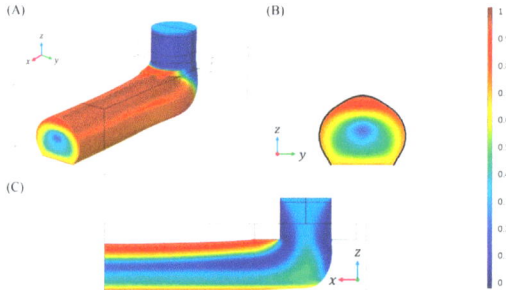

Figure 25. Fiber orientation of SCF on the X direction (**A**); view (**B**); and XZ view (**C**) [99]. Reproduced with permission: Elsevier.

The raster cross-section displays a flat base attached to the printing bed and a bell-shaped path trace, as seen in Figure 26. The strand profile tends to increase in height as the distance from the nozzle is made smaller, due to the compression reaction. In time, it could be remarked that, after 0.5 s, the strand height tends to display a constant profile. The experimental setup is presented in Figure 26a. This is defined by G (the gap between the nozzle and the printing bed) and D (the diameter of the nozzle). A reduced diameter nozzle forms a more rectangular strand, where the material in the middle tends to migrate to the corners of the strand due to the compression induced by the following strand, as visible in the yellow contour presented in Figure 26b. The third strand shape corresponds to the speed (V) to feed rate (U) ratio being decreased to 0.5, as seen in the red path in Figure 26b, resulting in a higher volume, an enlarged base, and an increased contact surface with the adjacent raster and the printing bed.

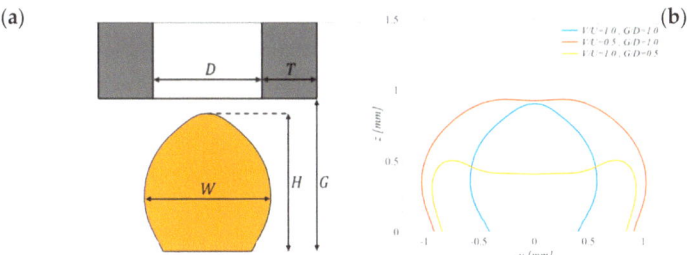

Figure 26. Geometric description (**a**); strand form (**b**) [99]. Reproduced with permission: Elsevier.

The difference between the first row of images in Figure 27 and the second is the probability of short fiber orientation on the X-axis (as discussed in the previous section) and in the Y-direction, resulting in a fiber orientation of 50% in the y-axis direction of the core. This means that the material can be considered orthotropic with a probability of 60%. The lower G/D shows a superior fiber orientation stability, with a relatively smaller "core"-oriented volume, as one can see in Figure 27C,D.

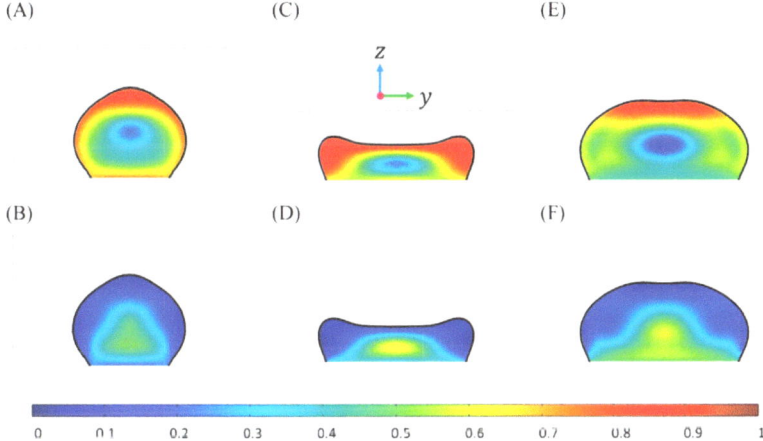

Figure 27. Printing layer shape and fiber orientation (X-axis; (**A**,**C**,**E**)), Y-axis (**B**,**D**,**F**) [99]. Reproduced with permission: Elsevier.

As discussed in terms of the single direction extrusion process, the more practical case of in-plane deposition and direction change has been analyzed (Zhang et al., 2023) [102]. The numerical model is a non-linear transient for CCF-PA, developed in ABAQUS. The normal strain, see Figure 28a, is the measure of fiber deformation, where the direction changes in trajectories at 30° show an increased compression at the angle change and shows tension at the deposition layer. The shear stress, see Figure 28b, shows the fiber's tendency to break, whereas the angled raster shows a significant risk of breakage. The contact pressure (nozzle-filament), see Figure 28c, has a similar profile due to the change of direction, where red circle 1 area is compressed in the nozzle, and the red circle 2 area is elongated, and the deposed raster tends to bond less to the previous material.

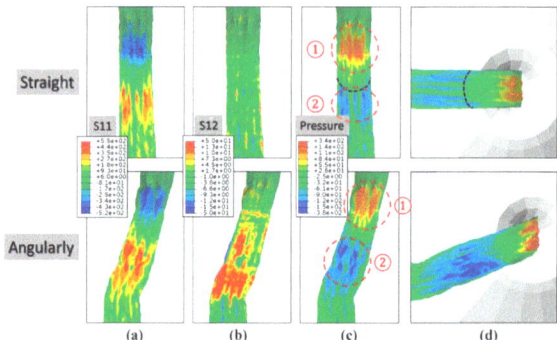

Figure 28. Straight and angular change direction in CCF: normal stress (**a**); shear stress (**b**); contact pressure (**c**); bottom view (**d**) [102]. Reproduced with permission: Elsevier.

The effect of normal stress when the angle is changing is shown in Figure 29, where, for the interval between 15–30°, the printing width and strain is constant in the changing direction, whereas, for angles between 45–75°, the width increases due to the high stress, and a bending of the inner fiber is represented by the ellipses dotted area in Figure 29. In the extreme case of 90°, the width becomes unrealistic, corresponding to a break into the fibers.

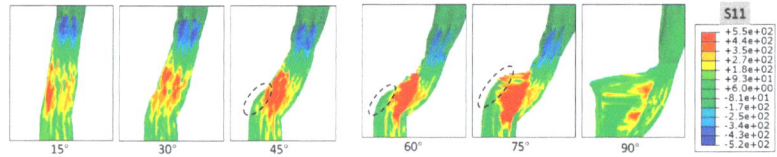

Figure 29. Angular variation effect on residual stress [102]. Reproduced with permission: Elsevier.

The variation of the curvature radii effects is shown in Figure 30. The shear stress is reduced when R = 20 mm, whereas, for radii of 5–10 mm, the wrinkling phenomenon occurs (dotted ellipse area), as the stress area increases in the turning area. In the case where R = 2.5 mm, the shear appears in both external and internal areas, which signalizes a potential fiber breakage, emphasized in the turning area as the dotted contour. The variation of the raster width displays an increase in the shear strain with the augmentation of the bundle, which then increases the possibility of fiber misalignment and breakage.

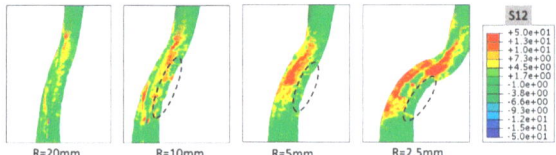

Figure 30. Radius variation effects on residual stress [102]. Reproduced with permission: Elsevier.

The previous model was developed for planar paths and curvilinear printing paths (K. Zhang et al., 2023) [103]. The model is a meso-scale RVE, where the volume of continuous filaments is visible in the cube (Figure 31) and the filaments oriented along the extrusion direction. There were two nozzles in this analysis as follows: a geometry cylindrical M-shaped nozzle (planar deposition), and a rotational R-shaped nozzle (multi-axis deposition). Firstly, the raster is compressed by the nozzle tip. The second phase includes the nozzle moving in a single linear direction, preceding the curvilinear trajectory of the nozzle (only planar for the M-shaped and Z-axis increments for the R-shape).

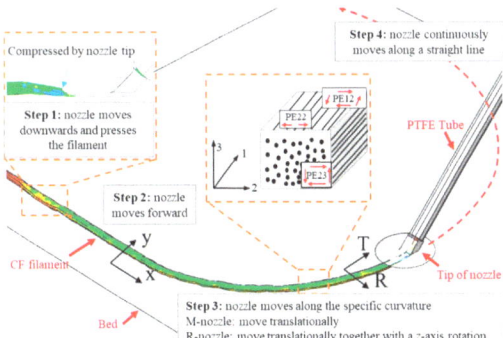

Figure 31. FEA setup process of CFF [103]. Under CC−BY Licensing 4.0.

The first step is rectilinear deposition (highlighted on the dotted contour), wherein transverse normal plastic strain occurs due to the compression of the nozzle, as shown in Figure 32. This indicates an uneven but smaller amplitude (maximum 0.19) distribution for the M-nozzle, as visible in Figure 32a, which is understandable due to the higher levels of compression resulting from the higher bending transition, whereas, in the case of the R-shaped nozzle, a wider angle (135°) assures a constant but higher residual strain (maximum 0.4), as one can see in Figure 32b. The bottom surfaces show the same tendency, with discontinuity occurring for the M-shaped nozzle due to the non-uniform contact pressure. In this way, a higher value means a higher warping tendency, but also a constant raster with better fiber alignment.

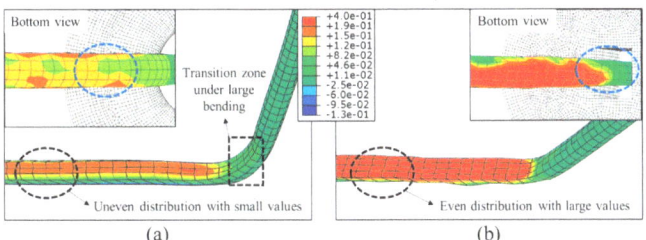

Figure 32. Transverse plastic strain: M-nozzle (**a**); R-nozzle (**b**) [103]. Under CC−BY Licensing 4.0.

In terms of the transverse shear strain (the measure of shearing fiber occurrences), it can be observed in Figure 33a that the more curvature in the transition area in the case of M-nozzle shows a higher concentration, with a 0.29 relative strain, and a punctual compression area (−0.9), displaying variations along the way, which can then cause fiber slitting and out-of-plane buckling. The R-nozzle displays a relatively gradual deposition area, where the material suffers a relative compression of −0.6, as one can see in Figure 33b.

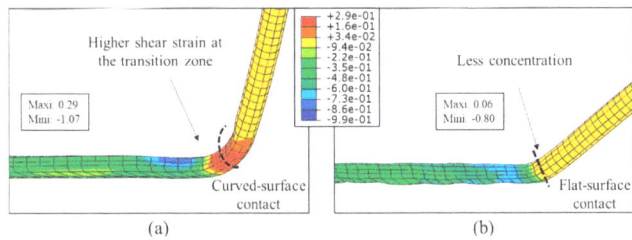

Figure 33. Transverse shear strain: M-nozzle (**a**); R-nozzle (**b**) [103]. Under CC−BY Licensing 4.0.

The second step is printing on a circular path using a constant 2.5 mm curvature for both geometries. The logarithmic strain was used as a measure to assess the filament torsion during the printing process. The small values indicate no torsion filaments, a phenomenon that increases before deposition in the M-nozzle across the entire filament, as one can see in Figure 34a, whereas, in the R-nozzle, the strain only manifests on the outer surface, as one can see in Figure 34b in the transversal section.

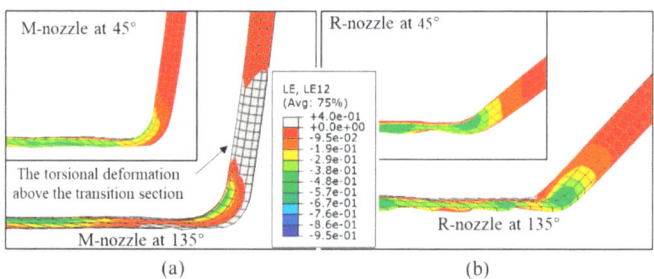

Figure 34. Logarithmic strain: M-nozzle (**a**); R-nozzle (**b**) [103]. Under CC–BY Licensing 4.0.

The curvilinear direction change (R = 10 mm between the trajectories) can be seen from above in Figure 35. The plastic strain tends to be smaller but inconstant (compression from the nozzle) for the planar M-shaped nozzle, with a gradual filament twisting of 45–60°; the fibers have a tendency to irregularly distribute in the middle of the directional change, making the structure with the M-shaped nozzle more vulnerable to breakage, as seen in Figure 35a. The R-shaped nozzle path, see Figure 35b, results in a more strained structure, which can be compensated via shrinkage, but this results in a regular distribution along the path, which results in a better performance of the resulting composite.

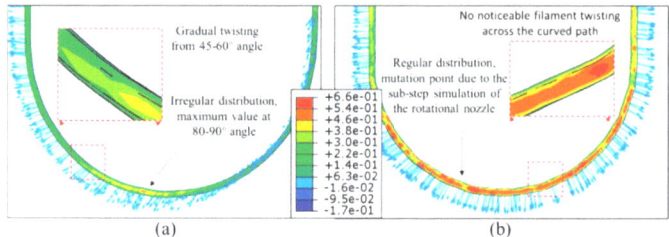

Figure 35. Normal plastic strain on the circular path (R = 10 mm): M-nozzle (**a**); and R-nozzle (**b**) [103]. Under CC–BY Licensing 4.0.

A tighter radial directional change (R = 2.5 mm) analysis was performed. The M-nozzle shows a significant strain at the 60–75° area (was occurred earlier and at an increased level when compared with the previous case), even if the fiber has a gradual tendency to twist (see Figure 36a). On the other hand, the R-nozzle demonstrated steep twisting around the 90° area (see Figure 36b), where the fiber has a sudden orientational change, and sudden folding occurs at 135°; this means that the curvature is considered too tight for this nozzle configuration.

The last step is performed on the frictional contact point between the filament and the nozzle in the dotted contour area, as seen in Figure 37, in a non-planar deposition, where the M-shaped nozzle displays a positive/negative split line at the middle when printing. The structure becomes widely compressed following this, as visible in Figure 37a. However, the R-shaped nozzle maintains a relatively constant split line throughout the process, which reduces the twisting and optimizes the fiber orientation along the deposition direction, as presented in Figure 37b.

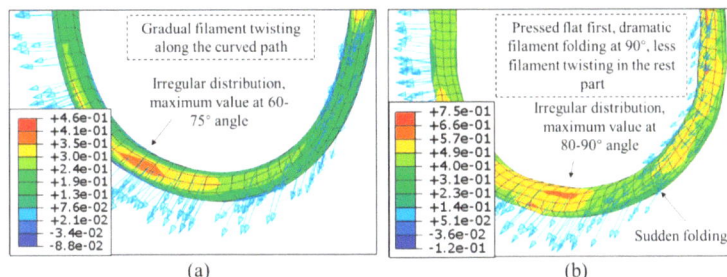

Figure 36. Normal strain on circular path (R = 2.5 mm) on: M-nozzle (**a**); R-nozzle (**b**) [103]. Under CC–BY Licensing 4.0.

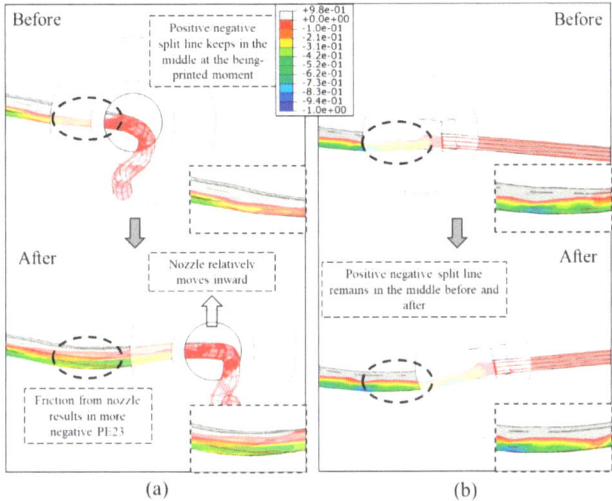

Figure 37. Strains on non-planar deposition: M-shaped nozzle (**a**); R-shaped nozzle (**b**) [103]. Under CC–BY Licensing 4.0.

3.7. Solidification of 3D-Printed Composites (Residual Stress and Dimensional Precision)

The solidification process is worth evaluating in terms of the temperature field distribution, the induced thermal deformation, or the residual stresses within the printed part. Also, reconstituting the material deposition is an FEA case in itself, which presents particularities for composite materials due to the resulting directional properties.

The case of short wood fibers and PLA is a thermo-mechanical numerical model of the deposition (Moryanova et al., 2023) [104], which was completed in parallel with the experimental model of the 20 × 20 × 20 mm cubic composite specimen. The first aspect studied was the warping of the 3D-printed composite with different layer heights (0.15 mm and 0.25 mm), as can be seen in Figure 38a. For a layer of 0.5 mm, the warping tendency is higher on the middle-top layers towards the barrier of 0.5 mm, showing a gradual decrease towards the building plate. The lower height (0.15 mm) displays an improved warping tendency, with the middle layers being the most deformed around 0.35–0.4 mm. The experimental results show coarser top edges with deposition defects for layers at higher heights, as visible in Figure 38b.

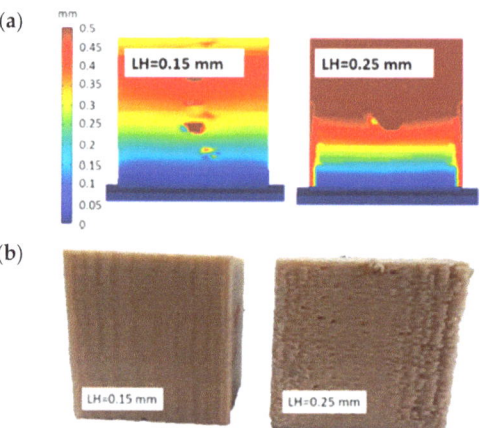

Figure 38. Residual displacements of the PLA–wood composite: simulation (**a**); experimental (**b**) [104]. Under CC−BY Licensing 4.0.

In the cooling process, deformation results in stresses due to the movement of the particles, adhesion, and the phase change of the matrix. As one can see in Figure 39, the lower the extrusion temperature, the lower the stress levels. The areas where the residual stresses are highest are on the boundaries of the part, where lower heights result in a thinner boundary stress of 30 MPa, whereas, for the increased layer heights, the boundary stresses meet 45 MPa on the lower area (0.2 mm), as well as on all the 0.25 mm boundaries of the cube.

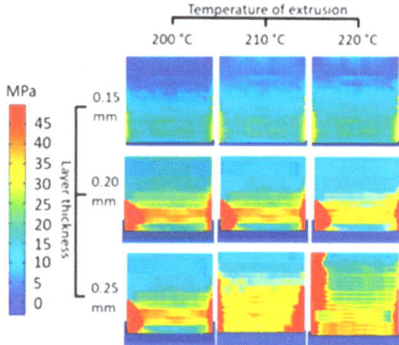

Figure 39. Residual stress on PLA–wood specimens [104]. Under CC−BY Licensing 4.0.

Lastly, the temperature field distribution was studied and found to be uniformly distributed throughout the printed cube (3 s after finishing the process) for all layer heights, as one can see in Figure 40. The boundary layers cool at a higher rate, resulting in, as remarked in the previous sections, more defects, and coarser structures.

The numerical model of deposition (Chnatios et al., 2023) [105] is a thermo-mechanical coupled simulation for CCF-PEEK. The magnitude of warping caused by the variation of process parameters was demonstrated. Due to the computational difficulty, a single-layer deflection was calculated, and it was found that warping increased continuously as the manufactured thickness grew; this was due to the reheating of the previously manufactured areas. The high-temperature PEEK composites leave a thermal trace, as one may notice in Figure 41. The higher the thickness, the more pronounced the thermal footprint; Figure 41a shows a lower strain than that shown in Figure 41b. The printing bed temperature (150 °C) will cause additional thermal deflection following printing when cooling to room temperature.

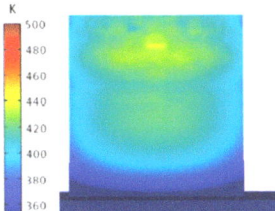

Figure 40. Temperature 3.5 s after completing the specimen [104]. Under CC–BY Licensing 4.0.

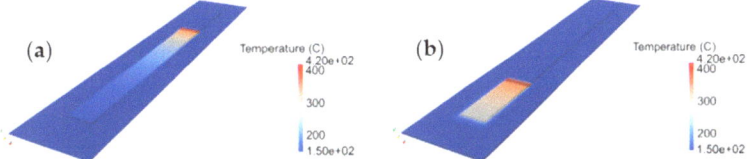

Figure 41. Temperature field for CCF-PEEK for different thicknesses: 0.14 mm/layer (**a**); 0.28 mm/layer (**b**) [105]. Reproduced with permission: Elsevier.

As the heat transfer exists between the previously deposed raster and the current layer, the deflection grows with the number of layers (Z-height), as well as with the width of the part (Y-axis), as one can see in Figure 42, where a change from 2.02 µm (layer 3, Figure 42a,b) to 3.8 µm (layer 5, Figure 42c,d) can be seen. The deflection can be observed on the edges of the raster, where the cooling occurs last, causing the material to warp (free surface on the edges). This is considered the most important factor in warping, as it is quite unavoidable in most applications.

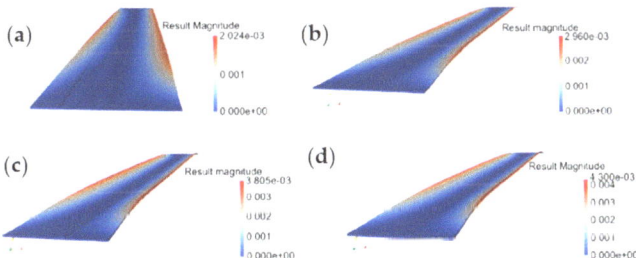

Figure 42. Raster deflection for CCF-PEEK for 11th trajectory (**a**); 15th trajectory (**b**); 23rd trajectory (**c**); 25th trajectory (**d**) [105]. Reproduced with permission: Elsevier.

The printing bed temperature can limit the warping, reducing from 9.048 µm for a standard temperature (150 °C, Figure 43b), to the insignificantly lower 0.904 µm (155 °C, Figure 43c), and again to 0.891 µm (160 °C, Figure 43d). The deflection has a reversed tendency for lower temperatures, which is even more pronounced at 0.917 µm (145 °C, Figure 43a). This is a result of a lower temperature variation between the raster and the printing bed.

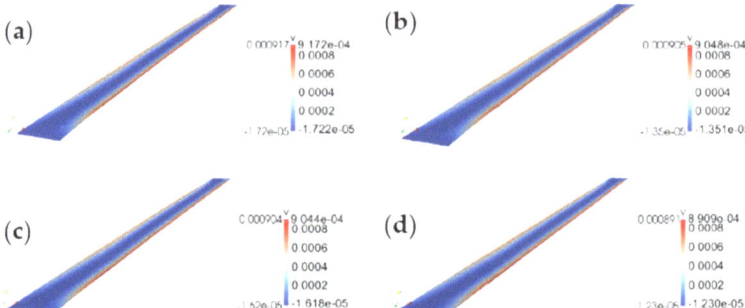

Figure 43. Printing bed temperature effects on warping for CCF-PEEK: 145 °C (**a**); 150 °C (**b**); 155 °C (**c**); 160 °C (**d**) [105]. Reproduced with permission: Elsevier.

In addition, the printing temperature variation was studied, as seen in Figure 44. The more significant warping was 1.52 μm (390 °C, Figure 44a), constantly decreasing to 1.425 μm (400 °C, Figure 44b) or to 1.323 μm (410 °C, Figure 44c). The standard temperature (420 °C, Figure 44d) shows the smallest deflection of 1.21 μm. This is understandable due to the viscosity variation of PEEK when suspended with CF.

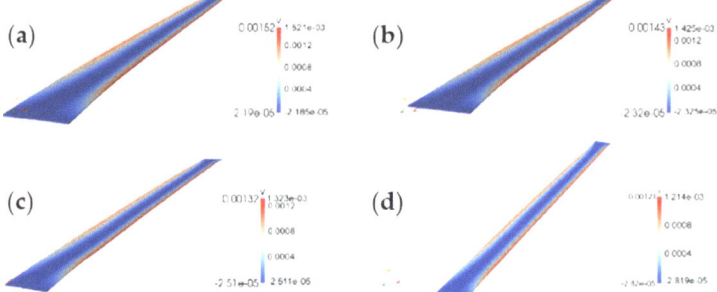

Figure 44. Deposition temperature effect on warping CCF-PEEK for $T_{deposition}$ = 390 °C (**a**); $T_{deposition}$ = 400 °C (**b**); $T_{deposition}$ = 410 °C (**c**); $T_{deposition}$ = 420 °C (**d**) [105]. Reproduced with permission: Elsevier.

The printing speed influences the warping, as shown in Figure 45. The lowest speed (v = 1 mm/s, Figure 45a) shows the smallest deflection of 1.172 μm. The increase observed in the speed results in a 1.21 μm deflection (Figure 45b,c) stabilized in the range of v = 3–5 mm/s. Exceeding the standard speed results (v = 3–5 mm/s, Figure 45c) following a steep increase in deflection to 1.362 μm can be explained via the faster deposition, higher heat transfer over time, and the optimal speed being a compromise between the precision and manufacturing time.

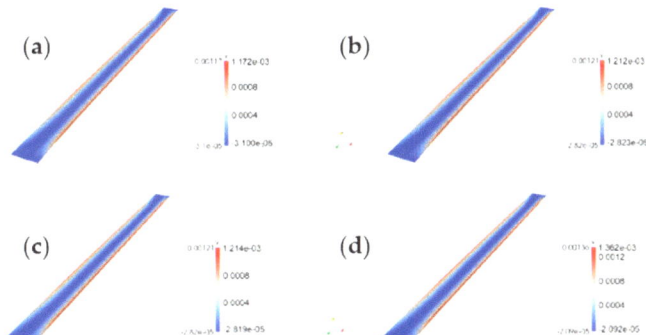

Figure 45. Printing speed effect on warping CCF-PEEK for υ = 1 mm/s (**a**); υ = 3 mm/s (**b**); υ = 5 mm/s (**c**); υ = 7 mm/s (**d**) [105]. Reproduced with permission: Elsevier.

3.8. Solidification of 3D-Printed Thermoset Composites

In the case of thermoset composites, the solidification process (curing) is performed via several sources of external energy or via the chemistry of the matrix. The thermal polymerization of thermoset matrix composites (Struzziero et al., 2023) [106], is focused on Epoxy-CF-based composites. The model was a 2D deposition layer, as seen in Figure 46a, where a convection surface can be found at the boundary of the deposed layer. This was developed in ABAQUS, utilizing a C++ general algorithm for the material properties based on the logic shown in Figure 46b.

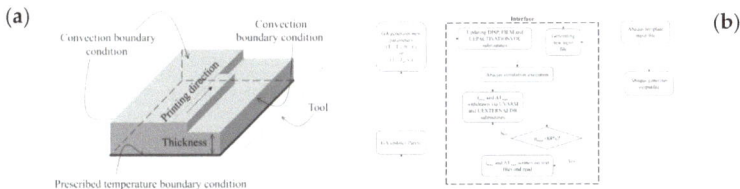

Figure 46. Process simulation setup for thermoset composites (**a**); pseudo-code of the algorithm used for the ABAQUS link (**b**) [106]. Under CC–BY Licensing 4.0.

The temperature map of the curing process is shown in Figure 47a, highlighting the tendency of curing around 183 °C, spreading concentrically throughout the raster. The model was developed in ABAQUS, using USDVLD to calculate the curing state in differential time HEVAL subroutines for the heat generated via exothermic curing, following the conditions displayed in Figure 47b. Based on the algorithm earlier described, the state of curing is shown in Figure 47c, where the previously deposited layers display a curing state of 89–92%. It can be observed that the material consumed most of its activation energy, maintaining a constant curing rate during the process, and thus guaranteeing a constant thickness.

A prospective material development encompasses the usage of frontal polymerization thermoset resin for eliminating the need of molds and, especially, of post-processing treatments. The model for the state of curing for frontal polymerization (Sharifi et al., 2023) [107], is focused on continuous carbon fiber PDCPD resin (promising improved impact and corrosion resistance) as used on RVE, as shown in Figure 48a, in 3D and as cross-section of red-dotted contour. The stochastic material is defined via a routine in Python, as described in Figure 48b. The RVE is defined as a bundle equation creation. By applying the Box–Muller algorithm [108], the matrix is given random properties, respecting the standard deviation from the mean actual values of the characteristics.

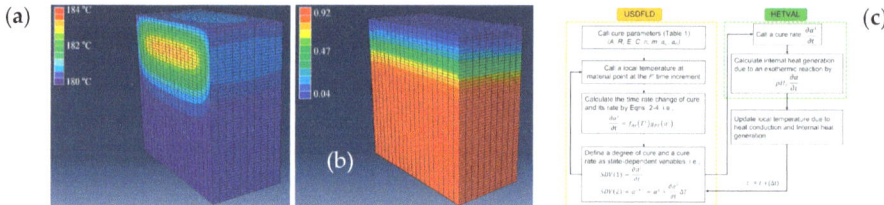

Figure 47. Contour of temperature distribution for thermoset composites (**a**); curing status distribution for thermoset composites (**b**); USDFLD and HETVAL coupling subroutines coupling (**c**) [106]. Under CC−BY Licensing 4.0.

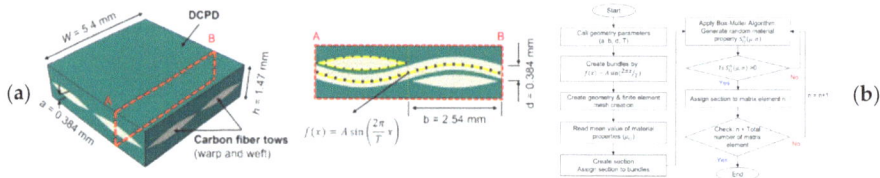

Figure 48. Meso-scale RVE frontal polymerization thermoset composite element (**a**); script for stochastic material setup (**b**) [107]. Reproduced with permission: Elsevier.

Three types of material configurations were created for this simulation. The deterministic material is shown in Figure 49a with no voids. The second material configuration is stochastic Figure 49b, where the spatial distribution of the properties is randomly generated in different color elements, which is more suited for liquid resin that is prone to impurities. The third model is the first configuration, with randomly distributed 3% voids (red elements) throughout the volume Figure 49c.

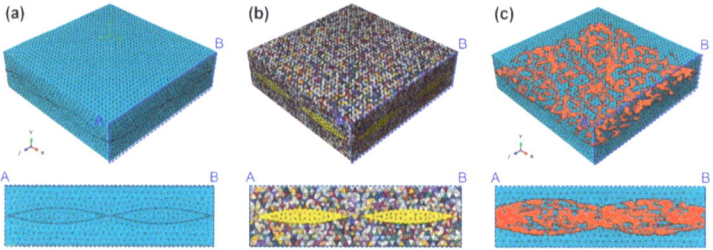

Figure 49. Material hypotheses in frontal polymerization with the following: deterministic properties (**a**); stochastic Properties (**b**); deterministic properties with 3% void (**c**) [107]. Reproduced with permission: Elsevier.

The state of curing commences on a normal direction of the printed composite, as one may see in Figure 50 in 3D, then in A-B cross-section. The cases consider various standard deviations from the mean cured properties (SD), where the curing occurs around the fibers. On the other hand, the fiber/matrix interface cures last (after 3 s), where the cure state decreases with the SD augmentation, converging at 100% for each SP following 5 s. The other case treated in Figure 50 is polymerization in the direction of printing, where the curing is slower around the fibers and in between the fibers, the maximum curing latency being for SD = 40% at 3 s.

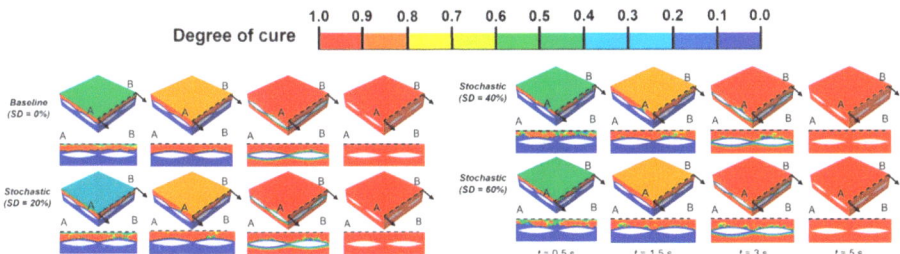

Figure 50. Curing simulation of frontal polymerization composites: XY plane heat source (**a**) [107]. Reproduced with permission: Elsevier.

3.9. Defect Simulations of Completed 3D-Printed Composites

The defects numerical model (Moryanova et al., 2023) [104], that are visible in the Figure 38b are revealed in detail in Figure 51. The first remarked defects are the splattering marks (cracks, holes, traces), caused by the difference in crystallization states during the cooling of the layers, which is due to the high extrusion temperature and the wood fibers in the resulting composite. Splatter traces of 0.5 mm magnitudes were observed both numerically and experimentally, as seen in Figure 51a. For a lower layer height, a more stable surface is visible in Figure 51b, which is due to a more uniform distribution of wood reinforcements (solid fibers are lighter than the matrix and flow above), and the similar size magnitudes of fibers and the PLA layer.

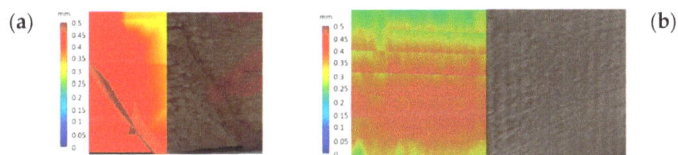

Figure 51. Defects caused by the thermal splatter simulation and experiment for a layer height of 0.25 mm (**a**) and a layer height of 0.15 mm (**b**) [104]. Under CC–BY Licensing 4.0.

Along the traces, in increased layer heights (0.25 mm), both numerical simulations and experimental observations demonstrated pultrusion into the outer surface due to twisting of the material (see Figure 52a). The quality of the interlayer seams and the sinkholes can be observed in Figure 52b, resulting from the time difference of the material deposition time in the cooling process. Ultimately, at higher extrusion temperatures (220 °C), a failure of the first layer was observed, which was due to the molten matrix and the considerable deposed mass, as one can see in Figure 52c.

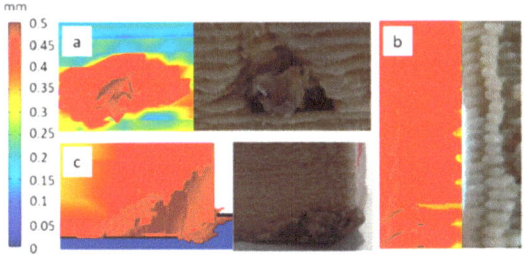

Figure 52. Defect simulation for cubical PLA–wood specimens for layer height = 0.25 mm: pultrusion (**a**); sinkholes (**b**); failure of lower edges (**c**) [104]. Under CC–BY Licensing 4.0.

3.10. Void Formation Simulation

As seen in the previous section, the cross section of the deposed layer is widely influenced by the process parameters, meaning the internal structure depends on each path. The process simulation of SCF-PEEK (Fu et al., 2023) [109], is done using ABAQUS with user-defined material (UMAT) subroutines [110]. Micro-scale RVE with periodicity was applied, as seen in Figure 53. Unitary deposed element is represented as a matrix cube, 5% SCF (black elements), modelled as cylinders inside the RVE, whereas voids are introduced in random volumes (blue particles). The orientation of the layers is defined as per the picture below, resulting in the tensile specimen.

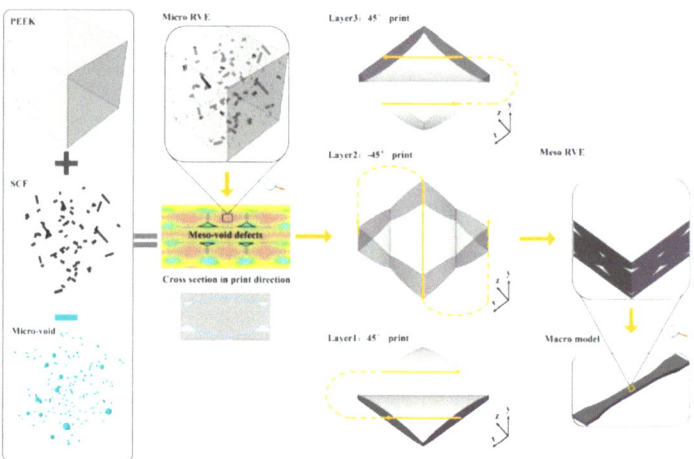

Figure 53. Multi-scale analysis of SCF-PEEK [109]. Reproduced with permission: Elsevier.

The crystallization is presented within the printed layers like intervals of 0–1 (ω_c), PEEK being a semi-crystalline matrix. Firstly, the deposed element is completely uncrystallized, as seen in Figure 54a. The crystallization process begins at the margins, where the matrix starts to cool, as visible in Figure 54b. When changing direction, as is visible in Figure 54c,d, the areas adjacent to the previous layer diminish their solidification state due to the interfacial heat transfer. As the process continues, the layer returns to a complete solid state as the distance grows from the currently deposed raster Figure 54e–g, and as voids appear between the printing tracks. In the end, the transfer to the following layer results in an insignificant change within the previous raster, as shown in Figure 54h. Once the process is over, the crystallization process is completed to almost 100%, with the crystallization between trajectories being over 85% due to the irregular heat flux from the adjacent deposed raster. Overall, it was calculated that the porosity simulation showed a peak value of 14.29% without any post-processing treatment.

The macro-scale model for CCF-PA (Zhilyaev et al., 2022) [111], developed in ANSYS with user-defined routines, where the phase transition is a Nakamura algorithm [112], and the mechanical properties are assigned based on experimental activities, as they are temperature dependent. The primary drawback of the continuous fiber composites is the level of porosity, which is exacerbated by the upright printed surfaces on the complex geometry, resulting in the highest void volumes of 8–10% in the support areas and at the overhanging hole areas, as visible in Figure 55a. The base area was subjected to higher porosities at points of sharp directional changes, as one can see in Figure 55b. The cumulative defects from the cases presented above result in a maximum of 24.1%, as visible in Figure 55c.

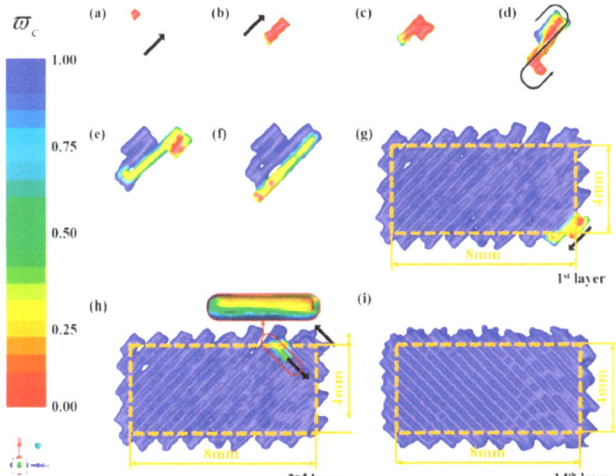

Figure 54. Solidification simulation for SCF-PEEK specimens: single element trajectory (**a**,**b**); single layer changing direction trajectory (**c**–**g**); next layer deposition (**h**,**i**) [109]. Reproduced with permission: Elsevier.

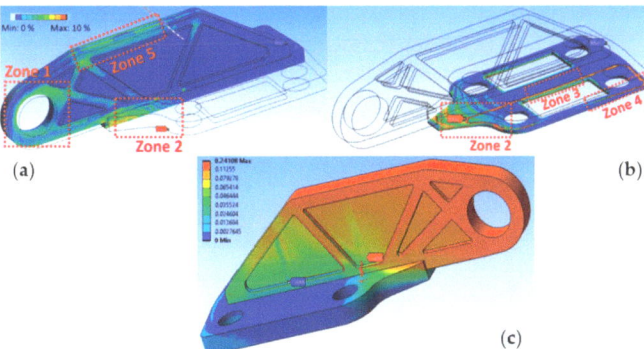

Figure 55. Porosity simulation on CCF-PA12 fixture: 90° area (**a**); base area (**b**); complete part (**c**) [111]. Under CC−BY Licensing 4.0.

3.11. Surface Roughness Simulation

The surface roughness is important, especially for the aspect of the component. In addition, the contact area with other parts in the assembly is significant. Due to this, the roughness before any post-processing treatment is of great concern, especially in the case of thermoset matrix composites manufactured via AM, due to amorphous polymerization. The roughness simulation for continuous GF and bisphenol photosensitive (Lorenz et al., 2022) [113]. The model was developed in TexGen software 3.13.1, with experimental material properties. The model was developed for two directions, with mean profiles and extremum values, as shown in Figure 56. Plotted profiles for both experimental and simulation results can be found below.

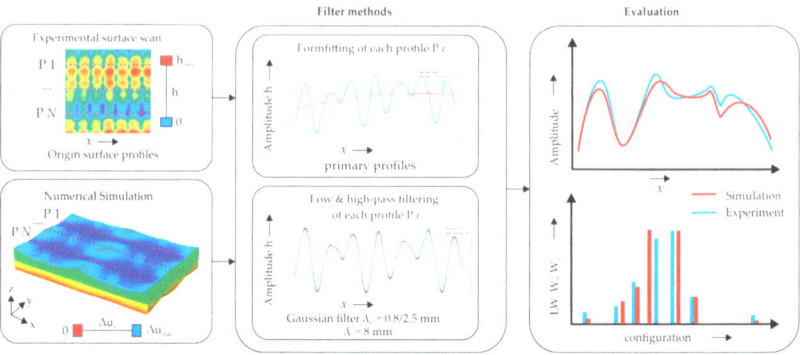

Figure 56. Surface profile numerical model for thermoset composites [113]. Under CC−BY Licensing 4.0.

3.12. Post-Processing Simulation

A post-processing treatment simulation (Grieder et al., 2022) [114], including hot pressing decreased the porosity from 20.28% to only 1.4% in the case of 240 °C/10 bars. The maximum void pressure was 1.6 MPa, as shown in Figure 57a, whereas the void content exists at the bottom of the bar Figure 57b, opposite the hot-pressing treatment. The fiber variation content was similar in profile to the porosity Figure 57c, ranging between 53–57%. Because it is used as a heat-resistant thermoplastic matrix, the elastic transverse modulus at heat deflection temperatures (175 °C) is studied and presented in Figure 57d, ranging between 3.84–6.47 MPa.

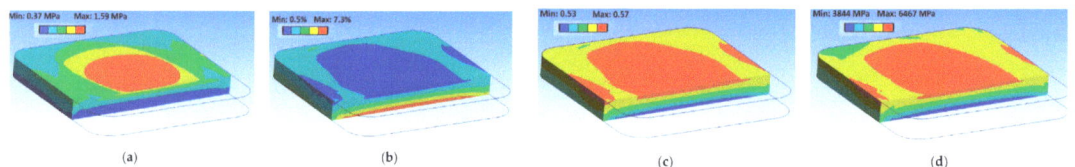

Figure 57. Simulation results for void pressure (**a**); porosity (**b**); fiber fraction (**c**); elastic modulus at the heat deflection temperature (2 MPa, 175 °C) (**d**) [114]. Under CC−BY Licensing 4.0.

4. Conclusions

To conclude, the AM processes are diverse in terms of the material range and the nature of the reinforcement, as presented in Section 1. Due to their similarities with common AM technologies such as FDM or SLA, these have been developing at an accelerated pace over the last 5–10 years, and are showing strong potential to become widely used in composite manufacturing.

As the processes are in the development phase and their cost is a significant factor, especially for continuous fibers composites, addressing potential defects arising from the technologies they derive from, as well as specific composite defects, the digital twinning of the processes is highly desirable for industrial applications to remain competitive with their conventional counterparts. FEM represents a technique used to model the physical phenomena. In the case of AM composites, as presented in Section 2, there are several aspects to be considered, which are relevant if the process parameters are to be digitally optimized, involving many disciplines of finite elements.

Author Contributions: For this review concept, T.F.Z.; Resources T.F.Z.; Original and Draft Preparation: T.F.Z.; Review and Editing M.C.D.; Supervision: M.C.D. All authors have read and agreed to the published version of the manuscript.

Funding: This research received no external funding.

Data Availability Statement: Not applicable.

Conflicts of Interest: The authors declare no conflicts of interest.

References

1. Gibson, I.; Rosen, D.; Stucker, B.; Khorasani, M. Introduction and Basic Principles. In *Additive Manufacturing Technologies*; Springer: Cham, Switzerland, 2021; pp. 1–19. [CrossRef]
2. Associates, W. *Wohlers Report 2022: 3D Printing and Additive Manufacturing Global State of the Industry*; Wohlers Associates: Fort Collins, CO, USA, 2022.
3. Huang, J.; Qin, Q.; Wang, J. A review of stereolithography: Processes and systems. *Processes* **2020**, *8*, 1138. [CrossRef]
4. Hull, C.W. Apparatus for Production of Three-Dimensional Objects by Stereolithography. U.S. Patent 45,753,30A, 1 December 1977.
5. Siemiński, P. Introduction to Fused Deposition Modeling. In *Handbooks in Advanced Manufacturing*; Elsevier: Amsterdam, The Netherlands, 2021; pp. 217–275. [CrossRef]
6. Agarwal, R. Chapter 5—Additive manufacturing, Materials, technologies, and applications. In *Additive Manufacturing: Advanced Materials and Design Techniques*, 1st ed.; Pandey, P.M., Singh, N.K., Singh, Y., Eds.; CRC Press: Boca Raton, FL, USA, 2023; pp. 77–97.
7. Yan, C.; Shi, Y.; Li, Z.; Wen, S.; Wei, Q. *Selective Laser Sintering Additive Manufacturing Technology*; Elsevier: New York, NY, USA, 2020. [CrossRef]
8. The Evolution of SLS: New Technologies, Materials and Applications. Autonomous Manufacturing. 2021. Available online: https://amfg.ai/2020/01/21/the-evolution-of-sls-new-technologies-materials-and-applications/ (accessed on 12 January 2024).
9. Prince, J.D. 3D Printing: An Industrial Revolution. *J. Electron. Resour. Med. Libr.* **2014**, *11*, 39–45. [CrossRef]
10. Huang, Y.; Leu, M.C.; Mazumder, J.; Donmez, A. Additive manufacturing: Current state, future potential, gaps and needs, and recommendations. *J. Manuf. Sci. Eng. Trans. ASME* **2015**, *137*, 014001. [CrossRef]
11. Sames, W.J.; List, F.A.; Pannala, S.; Dehoff, R.R.; Babu, S.S. The metallurgy and processing science of metal additive manufacturing. *Int. Mater. Rev.* **2016**, *61*, 315–360. [CrossRef]
12. Joshi, S.C.; Sheikh, A.A. 3D printing in aerospace and its long-term sustainability. *Virtual Phys. Prototyp.* **2015**, *10*, 175–185. [CrossRef]
13. Martinez, D.W.; Espino, M.T.; Cascolan, H.M.; Crisostomo, J.L.; Dizon, J.R.C. A Comprehensive Review on the Application of 3D Printing in the Aerospace Industry. *Key Eng. Mater.* **2022**, *913*, 27–34. [CrossRef]
14. Nichols, M.R. How does the automotive industry benefit from 3D metal printing? *Met. Powder Rep.* **2019**, *74*, 257–258. [CrossRef]
15. Rajak, D.K.; Pagar, D.D.; Behera, A.; Menezes, P.L. Role of Composite Materials in Automotive Sector: Potential Applications. In *Advances in Engine Tribology*; Kumar, V., Agarwal, A.K., Jena, A., Upadhyay, R.K., Eds.; Springer: Singapore, 2022; pp. 193–217. [CrossRef]
16. Yan, Q.; Dong, H.; Su, J.; Han, J.; Song, B.; Wei, Q.; Shi, Y. A Review of 3D Printing Technology for Medical Applications. *Engineering* **2018**, *4*, 729–742. [CrossRef]
17. Aimar, A.; Palermo, A.; Innocenti, B. The Role of 3D Printing in Medical Applications: A State of the Art. *J. Healthc. Eng.* **2019**, *2019*, 5340616. [CrossRef]
18. Nagarajan, B.; Hu, Z.; Song, X.; Zhai, W.; Wei, J. Development of Micro Selective Laser Melting: The State of the Art and Future Perspectives. *Engineering* **2019**, *5*, 702–720. [CrossRef]
19. Saengchairat, N.; Tran, T.; Chua, C.K. A review: Additive manufacturing for active electronic components. *Virtual Phys. Prototyp.* **2017**, *12*, 31–46. [CrossRef]
20. Inkwood Research. Global 3D Printing Market Growth_Global Opportunities. 2023. Available online: https://inkwoodresearch.com/reports/3d-printing-market/ (accessed on 12 January 2024).
21. Statista. Global 3D Printing Industry Market Size. 2021. Available online: https://www.statista.com/statistics/315386/global-market-for-3d-printers/ (accessed on 12 January 2024).
22. Report Linker. Additive Manufacturing Global Market Report 2023. 2023. Available online: https://www.globenewswire.com/news-release/2023/06/21/2692058/0/en/Additive-Manufacturing-Global-Market-Report-2023.html (accessed on 12 January 2024).
23. Zhong, W.; Li, F.; Zhang, Z.; Song, L.; Li, Z. Short Fiber Reinforced Composites for Fused Deposition Modeling. *Mater. Sci. Eng. A* **2001**, *301*, 125–130. [CrossRef]
24. Shofner, M.L.; Lozano, K.; Rodríguez-Macías, F.J.; Barrera, E.V. Nanofiber-Reinforced Polymers Prepared by Fused Deposition Modeling. *J. Appl. Polym. Sci.* **2003**, *89*, 3081–3090. [CrossRef]
25. Hwang, S.; Reyes, E.I.; Moon, K.; Rumpf, R.C.; Kim, N.S. Thermo-mechanical Characterization of Metal/Polymer Composite Filaments and Printing Parameter Study for Fused Deposition Modeling in the 3D Printing Process. *J. Electron. Mater.* **2015**, *44*, 771–777. [CrossRef]
26. Antoniac, I.; Popescu, D.; Zapciu, A.; Antoniac, A.; Miculescu, F.; Moldovan, H. Magnesium filled polylactic acid (PLA) material for filament based 3D printing. *Materials* **2019**, *12*, 719. [CrossRef]
27. Ryder, M.A.; Lados, D.A.; Iannacchione, G.S.; Peterson, A.M. Fabrication and properties of novel polymer-metal composites using fused deposition modeling. *Compos. Sci. Technol.* **2018**, *158*, 43–50. [CrossRef]

28. Kalsoom, U.; Nesterenko, P.N.; Paull, B. Recent developments in 3D printable composite materials. *RSC Adv.* **2016**, *6*, 60355–60371. [CrossRef]
29. Singh, R.; Bedi, P.; Fraternali, F.; Ahuja, I. Effect of single particle size, double particle size and triple particle size Al_2O_3 in Nylon-6 matrix on mechanical properties of feed stock filament for FDM. *Compos. B Eng.* **2016**, *106*, 20–27. [CrossRef]
30. Ferreira, R.T.L.; Amatte, I.C.; Dutra, T.A.; Bürger, D. Experimental characterization and micrography of 3D printed PLA and PLA reinforced with short carbon fibers. *Compos. B Eng.* **2017**, *124*, 88–100. [CrossRef]
31. Zhang, W.; Cotton, C.; Sun, J.; Heider, D.; Gu, B.; Sun, B.; Chou, T.-W. Interfacial bonding strength of short carbon fiber/acrylonitrile-butadiene-styrene composites fabricated by fused deposition modeling. *Compos. B Eng.* **2018**, *137*, 51–59. [CrossRef]
32. Gupta, A.; Hasanov, S.; Fidan, I. *Processing and Characterization of 3D-Printed Polymer Matrix Composites Reinforced with Discontinuous Fibers*; University of Texas at Austin: Austin, TX, USA, 2019.
33. Mohammadizadeh, M.; Gupta, A.; Fidan, I. Mechanical benchmarking of additively manufactured continuous and short carbon fiber reinforced nylon. *J. Compos. Mater.* **2021**, *55*, 3629–3638. [CrossRef]
34. Sodeifian, G.; Ghaseminejad, S.; Yousefi, A.A. Preparation of polypropylene/short glass fiber composite as Fused Deposition Modeling (FDM) filament. *Results Phys.* **2019**, *12*, 205–222. [CrossRef]
35. Khatri, B.; Lappe, K.; Habedank, M.; Mueller, T.; Megnin, C.; Hanemann, T. Fused deposition modeling of ABS-barium titanate composites: A simple route towards tailored dielectric devices. *Polymers* **2018**, *10*, 666. [CrossRef] [PubMed]
36. Brenken, B.; Barocio, E.; Favaloro, A.; Kunc, V.; Pipes, R.B. Fused filament fabrication of fiber-reinforced polymers: A review. *Addit. Manuf.* **2018**, *21*, 1–16. [CrossRef]
37. Dhavalikar, P.; Lan, Z.; Kar, R.; Salhadar, K. Biomedical Applications of Additive Manufacturing. In *Biomaterials Science*, 4th ed.; Wagner, W.R., Sakiyama-Elbert, S.E., Yaszemski, M.J., Eds.; Academic Press: New York, NY, USA, 2020; pp. 623–639.
38. Penumakala, P.; Santo, J.; Thomas, A. A critical review on the fused deposition modeling of thermoplastic polymer composites. *Compos. B Eng.* **2020**, *201*, 108336. [CrossRef]
39. Solomon, I.J.; Sevvel, P.; Gunasekaran, J. A review on the various processing parameters in FDM. *Mater. Today Proc.* **2020**, *37*, 509–514. [CrossRef]
40. Yin, J.; Lu, C.; Fu, J.; Huang, Y.; Zheng, Y. Interfacial bonding during multi-material fused deposition modeling (FDM) process due to inter-molecular diffusion. *Mater. Des.* **2018**, *150*, 104–112. [CrossRef]
41. Liu, Z.; Wang, Y.; Wu, B.; Cui, C.; Guo, Y.; Yan, C. A critical review of fused deposition modeling 3D printing technology in manufacturing polylactic acid parts. *Int. J. Adv. Manuf. Technol.* **2019**, *102*, 2877–2889. [CrossRef]
42. Tse, L.Y.L.; Kapila, S.; Barton, K. *Contoured 3D Printing of Fiber Reinforced Polymers*; University of Texas at Austin: Austin, TX, USA, 2016.
43. Blanco, I. The use of composite materials in 3D printing. *J. Compos. Sci.* **2020**, *4*, 42. [CrossRef]
44. Pervaiz, S.; Qureshi, T.A.; Kashwani, G.; Kannan, S. 3D Printing of Fiber-Reinforced Plastic Composites Using Fused Deposition Modeling: A Status Review. *Materials* **2021**, *14*, 4520. [CrossRef]
45. Tyller, K. Method and Apparatus for Continuous Composite Three-Dimensional Printing. US20140061974A1, 24 August 2013.
46. Pandelidi, C.; Bateman, S.; Piegert, S.; Hoehner, R.; Kelbassa, I.; Brandt, M. The technology of continuous fibre-reinforced polymers: A review on extrusion additive manufacturing methods. *Int. J. Adv. Manuf. Technol.* **2021**, *113*, 3057–3077. [CrossRef]
47. Tamez, M.B.A.; Taha, I. A review of additive manufacturing technologies and markets for thermosetting resins and their potential for carbon fiber integration. *Addit. Manuf.* **2021**, *37*, 101748. [CrossRef]
48. Continuous Composites. CF3D®. Available online: https://www.continuouscomposites.com/technology (accessed on 12 January 2024).
49. Struzziero, G.; Barbezat, M.; Skordos, A.A. Consolidation of continuous fibre reinforced composites in additive processes: A review. *Addit. Manuf.* **2021**, *48*, 102458. [CrossRef]
50. Beyene, S.D.; Ayalew, B.; Pilla, S. *Nonlinear Model Predictive Control of UV-Induced Thick Composite Manufacturing Process*; University of Texas at Austin: Austin, TX, USA, 2019.
51. Lewicki, J.P.; Rodriguez, J.N.; Zhu, C.; Worsley, M.A.; Wu, A.S.; Kanarska, Y.; Horn, J.D.; Duoss, E.B.; Ortega, J.M.; Elmer, W.; et al. 3D-Printing of Meso-structurally Ordered Carbon Fiber/Polymer Composites with Unprecedented Orthotropic Physical Properties. *Sci. Rep.* **2017**, *7*, 43401. [CrossRef]
52. Chandrasekaran, S.; Duoss, E.B.; Worsley, M.A.; Lewicki, J.P. 3D printing of high performance cyanate ester thermoset polymers. *J. Mater. Chem. A Mater.* **2018**, *6*, 853–858. [CrossRef]
53. Mahshid, R.; Isfahani, M.N.; Heidari-Rarani, M.; Mirkhalaf, M. Recent advances in development of additively manufactured thermosets and fiber reinforced thermosetting composites: Technologies, materials, and mechanical properties. *Compos. Part. A Appl. Sci. Manuf.* **2023**, *171*, 107584. [CrossRef]
54. Thakur, A.; Dong, X. Printing with 3D continuous carbon fiber multifunctional composites via UV-assisted coextrusion deposition. *Manuf. Lett.* **2020**, *24*, 1–5. [CrossRef]
55. Islam, Z.; Rahman, A.; Gibbon, L.; Hall, E.; Ulven, C.; Scala, J. Mechanical Characterization and Production of Complex Shapes Using Continuous Carbon Fiber Reinforced Thermoset Resin Based 3d Printing. *SSRN* **2023**. [CrossRef]
56. Xiao, H.; He, Q.; Duan, Y.; Wang, J.; Qi, Y.; Ming, Y.; Zhang, C.; Zhu, Y. Low-temperature 3D printing and curing process of continuous fiber-reinforced thermosetting polymer composites. *Polym. Compos.* **2023**, *44*, 2322–2330. [CrossRef]

57. Robertson, I.D.; Yourdkhani, M.; Centellas, P.J.; Aw, J.E.; Ivanoff, D.G.; Goli, E.; Lloyd, E.M.; Dean, L.M.; Sottos, N.R.; Geubelle, P.H.; et al. Rapid energy-efficient manufacturing of polymers and composites via frontal polymerization. *Nature* **2018**, *557*, 223–227. [CrossRef]
58. He, X.; Ding, Y.; Lei, Z.; Welch, S.; Zhang, W.; Dunn, M.; Yu, K. 3D printing of continuous fiber-reinforced thermoset composites. *Addit. Manuf.* **2021**, *40*, 101921. [CrossRef]
59. Matsuzaki, R.; Ueda, M.; Namiki, M.; Jeong, T.-K.; Asahara, H.; Horiguchi, K.; Nakamura, T.; Todoroki, A.; Hirano, Y. Three-dimensional printing of continuous-fiber composites by in-nozzle impregnation. *Sci. Rep.* **2016**, *6*, 23058. [CrossRef] [PubMed]
60. Kabir, S.M.F.; Mathur, K.; Seyam, A.F.M. A critical review on 3D printed continuous fiber-reinforced composites: History, mechanism, materials and properties. *Compos. Struct.* **2020**, *232*, 111476. [CrossRef]
61. Zhang, H.; Huang, T.; Jiang, Q.; He, L.; Bismarck, A.; Hu, Q. Recent progress of 3D printed continuous fiber reinforced polymer composites based on fused deposition modeling: A review. *J. Mater. Sci.* **2021**, *56*, 12999–13022. [CrossRef]
62. Markforged. Continuous Carbon Fiber—High Strength 3D Printing Material. Available online: https://markforged.com/materials/continuous-fibers (accessed on 18 January 2024).
63. Li, N.; Li, Y.; Liu, S. Rapid prototyping of continuous carbon fiber reinforced polylactic acid composites by 3D printing. *J. Mater. Process Technol.* **2016**, *238*, 218–225. [CrossRef]
64. Melenka, G.W.; Cheung, B.K.O.; Schofield, J.S.; Dawson, M.R.; Carey, J.P. Evaluation and prediction of the tensile properties of continuous fiber-reinforced 3D printed structures. *Compos. Struct.* **2016**, *153*, 866–875. [CrossRef]
65. Akhoundi, B.; Behravesh, A.H.; Bagheri Saed, A. Improving mechanical properties of continuous fiber-reinforced thermoplastic composites produced by FDM 3D printer. *J. Reinf. Plast. Compos.* **2019**, *38*, 99–116. [CrossRef]
66. Bettini, P.; Alitta, G.; Sala, G.; Di Landro, L. Fused Deposition Technique for Continuous Fiber Reinforced Thermoplastic. *J. Mater. Eng. Perform.* **2017**, *26*, 843–848. [CrossRef]
67. Mori, K.I.; Maeno, T.; Nakagawa, Y. Dieless forming of carbon fibre reinforced plastic parts using 3D printer. *Procedia Eng.* **2014**, *81*, 1595–1600. [CrossRef]
68. Araya-Calvo, M.; López-Gómez, I.; Chamberlain-Simon, N.; León-Salazar, J.L.; Guillén-Girón, T.; Corrales-Cordero, J.S.; Sánchez-Brenes, O. Evaluation of compressive and flexural properties of continuous fiber fabrication additive manufacturing technology. *Addit. Manuf.* **2018**, *22*, 157–164. [CrossRef]
69. Luo, M.; Tian, X.; Shang, J.; Zhu, W.; Li, D.; Qin, Y. Impregnation and interlayer bonding behaviours of 3D-printed continuous carbon-fiber-reinforced poly-ether-ether-ketone composites. *Compos. Part A Appl. Sci. Manuf.* **2019**, *121*, 130–138. [CrossRef]
70. Brooks, H.; Molony, S. Design and evaluation of additively manufactured parts with three dimensional continuous fibre reinforcement. *Mater. Des.* **2016**, *90*, 276–283. [CrossRef]
71. Heitkamp, T.; Kuschmitz, S.; Girnth, S.; Marx, J.-D.; Klawitter, G.; Waldt, N.; Vietor, T. Stress-adapted fiber orientation along the principal stress directions for continuous fiber-reinforced material extrusion. *Progress. Addit. Manuf.* **2023**, *8*, 541–559. [CrossRef]
72. Yeong, W.Y.; Goh, G.D. 3D Printing of Carbon Fiber Composite: The Future of Composite Industry? *Matter* **2020**, *2*, 1361–1363. [CrossRef]
73. Stamopoulos, A.G.; Glinz, J.; Senck, S. Assessment of the effects of the addition of continuous fiber filaments in PA 6/short fiber 3D-printed components using interrupted in-situ x-ray CT tensile testing. *Eng. Fail. Anal.* **2024**, *159*, 108121. [CrossRef]
74. Ding, S.; Zou, B.; Zhang, P.; Liu, Q.; Zhuang, Y.; Feng, Z.; Wang, F.; Wang, X. Layer thickness and path width setting in 3D printing of pre-impregnated continuous carbon, glass fibers and their hybrid composites. *Addit. Manuf.* **2024**, *83*, 104054. [CrossRef]
75. Almeida, J.H.S.; Jayaprakash, S.; Kolari, K.; Kuva, J.; Kukko, K.; Partanen, J. The role of printing parameters on the short beam strength of 3D-printed continuous carbon fibre reinforced epoxy-PETG composites. *Compos. Struct.* **2024**, *337*, 118034. [CrossRef]
76. Baechle-Clayton, M.; Loos, E.; Taheri, M.; Taheri, H. Failures and Flaws in Fused Deposition Modeling (FDM) Additively Manufactured Polymers and Composites. *J. Compos. Sci.* **2022**, *6*, 202. [CrossRef]
77. Akhoundi, B.; Behravesh, A.H.; Bagheri Saed, A. An innovative design approach in three-dimensional printing of continuous fiber–reinforced thermoplastic composites via fused deposition modeling process: In melt simultaneous impregnation. *Proc. Inst. Mech. Eng. B J. Eng. Manuf.* **2020**, *234*, 243–259. [CrossRef]
78. Lupone, F.; Padovano, E.; Venezia, C.; Badini, C. Experimental Characterization and Modeling of 3D Printed Continuous Carbon Fibers Composites with Different Fiber Orientation Produced by FFF Process. *Polymers* **2022**, *14*, 426. [CrossRef]
79. Croom, B.P.; Abbott, A.; Kemp, J.W.; Rueschhoff, L.; Smieska, L.; Woll, A.; Stoupin, S.; Koerner, H. Mechanics of nozzle clogging during direct ink writing of fiber-reinforced composites. *Addit. Manuf.* **2021**, *37*, 101701. [CrossRef]
80. Muftu, S. Introduction. In *Finite Element Method*; Academic Press: Cambridge, MA, USA, 2022; pp. 1–8. [CrossRef]
81. Zienkiewicz, O.C.; Taylor, R.L. *The Finite Element Method Volume 1: The Basis*; Wiley: Hoboken, NJ, USA, 2000; Volume 1.
82. Li, S.; Sitnikova, E. Representative volume elements and unit cells. In *Representative Volume Elements and Unit Cells: Concepts, Theory, Applications and Implementation*; Woodhead Publishing: Sawston, UK, 2020; pp. 67–77. [CrossRef]
83. Shafighfard, T.; Cender, T.A.; Demir, E. Additive manufacturing of compliance optimized variable stiffness composites through short fiber alignment along curvilinear paths. *Addit. Manuf.* **2021**, *37*, 101728. [CrossRef]

84. Li, N.; Link, G.; Wang, T.; Ramopoulos, V.; Neumaier, D.; Hofele, J.; Walter, M.; Jelonnek, J. Path-designed 3D printing for topological optimized continuous carbon fibre reinforced composite structures. *Compos. B Eng.* **2020**, *182*, 107612. [CrossRef]
85. Chen, Y.; Klingler, A.; Fu, K.; Ye, L. 3D printing and modelling of continuous carbon fibre reinforced composite grids with enhanced shear modulus. *Eng. Struct.* **2023**, *286*, 116165. [CrossRef]
86. Chen, Y.; Ye, L. Topological design for 3D-printing of carbon fibre reinforced composite structural parts. *Compos. Sci. Technol.* **2021**, *204*, 108644. [CrossRef]
87. Qian, S.; Liu, H.; Wang, Y.; Mei, D. Structural optimization of 3D printed SiC scaffold with gradient pore size distribution as catalyst support for methanol steam reforming. *Fuel* **2023**, *341*, 127612. [CrossRef]
88. Date, A.W. (Ed.) Introduction. In *Introduction to Computational Fluid Dynamics*; Cambridge University Press: Cambridge, UK, 2005; pp. 1–16. [CrossRef]
89. Hu, H.H. Chapter 10—Computational Fluid Dynamics. In *Fluid Mechanics*, 5th ed.; Kundu, P.K., Cohen, I.M., Dowling, D.R., Eds.; Academic Press: Boston, MA, USA, 2012; pp. 421–472. [CrossRef]
90. Xu, X.; Ren, H.; Chen, S.; Luo, X.; Zhao, F.; Xiong, Y. Review on melt flow simulations for thermoplastics and their fiber reinforced composites in fused deposition modeling. *J. Manuf. Process.* **2023**, *92*, 272–286. [CrossRef]
91. Wang, P.; Zou, B.; Xiao, H.; Ding, S.; Huang, C. Effects of printing parameters of fused deposition modeling on mechanical properties, surface quality, and microstructure of PEEK. *J. Mater. Process Technol.* **2019**, *271*, 62–74. [CrossRef]
92. Shi, X.Z.; Huang, M.; Zhao, Z.F.; Shen, C.Y. Nonlinear fitting technology of 7-parameter Cross-WLF viscosity model. *Adv. Mater. Res.* **2011**, *189–193*, 2103–2106. [CrossRef]
93. Yang, D.; Wu, K.; Wan, L.; Sheng, Y. A particle element approach for modelling the 3d printing process of fibre reinforced polymer composites. *J. Manuf. Mater. Process.* **2017**, *1*, 10. [CrossRef]
94. Sun, X.; Sakai, M.; Yamada, Y. Three-dimensional simulation of a solid-liquid flow by the DEM-SPH method. *J. Comput. Phys.* **2013**, *248*, 147–176. [CrossRef]
95. Monaghan, J.J. An Introduction to SPH. *Comput. Phys. Commun.* **1988**, *48*, 89–96. [CrossRef]
96. Kanarska, Y.; Duoss, E.B.; Lewicki, J.P.; Rodriguez, J.N.; Wu, A. Fiber motion in highly confined flows of carbon fiber and non-Newtonian polymer. *J. Nonnewton Fluid. Mech.* **2019**, *265*, 41–52. [CrossRef]
97. Zhang, L.; Zhang, H.; Wu, J.; An, X.; Yang, D. Fibre bridging and nozzle clogging in 3D printing of discontinuous carbon fibre-reinforced polymer composites: Coupled CFD-DEM modelling. *Int. J. Adv. Manuf. Technol.* **2021**, *117*, 3549–3562. [CrossRef]
98. Wang, Z.; Smith, D.E. A fully coupled simulation of planar deposition flow and fiber orientation in polymer composites additive manufacturing. *Materials* **2021**, *14*, 2596. [CrossRef]
99. Kermani, N.N.; Advani, S.G.; Férec, J. Orientation Predictions of Fibers Within 3D Printed Strand in Material Extrusion of Polymer Composites. *Addit. Manuf.* **2023**, *77*, 103781. [CrossRef]
100. Advani, S.G.; Tucker, C.L. The Use of Tensors to Describe and Predict Fiber Orientation in Short Fiber Composites. *J. Rheol.* **1987**, *31*, 751–784. [CrossRef]
101. Carreau, P.J.; De Kee, D.C.R.; Chhabra, R.P. Rheology of Polymeric Systems. In *Rheology of Polymeric Systems*; Carl Hanser Verlag GmbH & Co. KG: Munich, Germany, 2021. [CrossRef]
102. Zhang, H.; Chen, J.; Yang, D. Fibre misalignment and breakage in 3D printing of continuous carbon fibre reinforced thermoplastic composites. *Addit. Manuf.* **2021**, *38*, 101775. [CrossRef]
103. Zhang, K.; Zhang, H.; Wu, J.; Chen, J.; Yang, D. Improved fibre placement in filament-based 3D printing of continuous carbon fibre reinforced thermoplastic composites. *Compos. Part A Appl. Sci. Manuf.* **2023**, *168*, 107454. [CrossRef]
104. Morvayová, A.; Contuzzi, N.; Casalino, G. Defects and residual stresses finite element prediction of FDM 3D printed wood/PLA biocomposite. *Int. J. Adv. Manuf. Technol.* **2023**, *129*, 2281–2293. [CrossRef]
105. Ghnatios, C.; Fayazbakhsh, K. Warping estimation of continuous fiber-reinforced composites made by robotic 3D printing. *Addit. Manuf.* **2022**, *55*, 102796. [CrossRef]
106. Struzziero, G.; Barbezat, M.; Skordos, A.A. Assessment of the benefits of 3D printing of advanced thermosetting composites using process simulation and numerical optimisation. *Addit. Manuf.* **2023**, *63*, 103417. [CrossRef]
107. Sharifi, A.M.; Kwon, D.J.; Shah, S.Z.H.; Lee, J. Modeling of frontal polymerization of carbon fiber and dicyclopentadiene woven composites with stochastic material uncertainty. *Compos. Struct.* **2023**, *326*, 117582. [CrossRef]
108. Ross, S.M. Generating continuous random variables. In *Simulation*; Academic Press: Cambridge, MA, USA, 2023. [CrossRef]
109. Fu, Y.T.; Li, J.; Li, Y.Q.; Fu, S.Y.; Guo, F.L. Full-process multi-scale morphological and mechanical analyses of 3D printed short carbon fiber reinforced polyetheretherketone composites. *Compos. Sci. Technol.* **2023**, *236*, 109999. [CrossRef]
110. Gao, C.Y. FE Realization of a Thermo-Visco-Plastic Constitutive Model using VUMAT in ABAQUS/Explicit Program. In *Computational Mechanics*; Springer: Berlin/Heidelberg, Germany, 2007. [CrossRef]
111. Zhilyaev, I.; Grieder, S.; Küng, M.; Brauner, C.; Akermann, M.; Bosshard, J.; Inderkum, P.; Francisco, J.; Eichenhofer, M. Experimental and numerical analysis of the consolidation process for additive manufactured continuous carbon fiber-reinforced polyamide 12 composites. *Front. Mater.* **2022**, *9*, 1068261. [CrossRef]
112. Nakamura, K.; Katayama, K.; Amano, T. Some aspects of nonisothermal crystallization of polymers. II. Consideration of the isokinetic condition. *J. Appl. Polym. Sci.* **1973**, *17*, 1031–1041. [CrossRef]

113. Lorenz, N.; Gröger, B.; Müller-Pabel, M.; Gerritzen, J.; Müller, J.; Wang, A.; Fischer, K.; Gude, M.; Hopmann, C. Development and verification of a cure-dependent visco-thermo-elastic simulation model for predicting the process-induced surface waviness of continuous fiber reinforced thermosets. *J. Compos. Mater.* **2023**, *57*, 1105–1120. [CrossRef]
114. Grieder, S.; Zhilyaev, I.; Küng, M.; Brauner, C.; Akermann, M.; Bosshard, J.; Inderkum, P.; Francisco, J.; Willemin, Y.; Eichenhofer, M. Consolidation of Additive Manufactured Continuous Carbon Fiber Reinforced Polyamide 12 Composites and the Development of Process-Related Numerical Simulation Methods. *Polymers* **2022**, *14*, 3429. [CrossRef]

Disclaimer/Publisher's Note: The statements, opinions and data contained in all publications are solely those of the individual author(s) and contributor(s) and not of MDPI and/or the editor(s). MDPI and/or the editor(s) disclaim responsibility for any injury to people or property resulting from any ideas, methods, instructions or products referred to in the content.

Article

Optimising Additive Manufacturing to Produce PLA Sandwich Structures by Varying Cell Type and Infill: Effect on Flexural Properties

Gabriele Marabello, Mohamed Chairi and Guido Di Bella *

Department of Engineering, University of Messina, Contrada di Dio, 98166 Messina, Italy; gabriele.marabello@studenti.unime.it (G.M.); mohamed.chairi@unime.it (M.C.)
* Correspondence: guido.dibella@unime.it

Abstract: The objective of this research is to optimize additive manufacturing processes, specifically Fused Filament Fabrication (FFF) techniques, to produce sandwich structures. Mono-material specimens made of polylactic acid (PLA) were produced, where both the skin and core were fabricated in a single print. To optimize the process, variations were made in both the base cell geometry of the core (Tri-Hexagon and Gyroid) and the core infill (5%, 25%, 50%, and 75%), evaluating their effects on static three-point bending behavior. Optical microscopy was employed to assess both the structure generated by additive manufacturing and the fracture modes. The findings reveal that increasing the infill, and thus the core density, enhances the mechanical properties of the structure, although the improvement is such that samples with 50% infill already demonstrate excellent performance. The difference between hexagonal and Gyroid structures is not significant. Based on microscopic analyses, it is believed that the evolution of 3D printers, from open to closed chamber designs, could significantly improve the deposition of the various layers.

Keywords: additive manufacturing; fused filament fabrication; sandwich

Citation: Marabello, G.; Chairi, M.; Di Bella, G. Optimising Additive Manufacturing to Produce PLA Sandwich Structures by Varying Cell Type and Infill: Effect on Flexural Properties. *J. Compos. Sci.* **2024**, *8*, 360. https://doi.org/10.3390/jcs8090360

Academic Editor: Yuan Chen

Received: 17 August 2024
Revised: 5 September 2024
Accepted: 12 September 2024
Published: 14 September 2024

Copyright: © 2024 by the authors. Licensee MDPI, Basel, Switzerland. This article is an open access article distributed under the terms and conditions of the Creative Commons Attribution (CC BY) license (https://creativecommons.org/licenses/by/4.0/).

1. Introduction

Additive Manufacturing (AM) [1] plays a crucial role in industrial production [2]. Commonly known as 3D printing, AM enables the creation of complex and innovative structures using innovative materials, depositing layer by layer, and eliminating many of the restrictions associated with traditional manufacturing methods [3]. This manufacturing paradigm represents a significant turning point in design and manufacturing processes [4].

A sandwich panel, also known as a sandwich structure, consists of two strong layers called skins or faces, separated and solidly connected by a central element known as a core. This configuration gives the panel considerable structural stability compared to the individual components. The core, which is usually made of a lightweight, low-strength material, is designed to keep the skins, which are made of high-quality, thin-gauge materials, separate [5,6]. While the skins distribute the loads in the plane, the presence of the core significantly increases the flexural stiffness of the panel, influencing the distance of the faces from the midplane [7]. This concept can be compared to the structure of an I-section beam, where the web contributes to increasing the bending stiffness in the same direction. Increasing the distance between the skins leads to a significant improvement in stiffness without a significant increase in weight. For these reasons, the use of sandwich panels has become increasingly common in the aerospace industry over the past forty years. A common example of a sandwich panel is made of a cardboard material, where the outer layers are flat and separated by a layer of corrugated cardboard. Skins are commonly made from high mechanical strength materials such as fiberglass, carbon or Kevlar composites, or from thin sheets of aluminum or steel. As for the core, structures with honeycomb cells (honeycomb) [8], foams or other materials are used. Honeycomb cells [9] can be made in several ways, for example by processing thin sheets of aluminum or by forming cells of aramid fibers in a thermosetting resin matrix. However, sandwich panels with honeycomb cores may have some buckling issues. To solve these problems, it is essential to correctly size the cell thickness and cell area. Foams, on the other hand, are cellular materials obtained by dispersing a gas in a solid plastic material. They can be open or closed cell, flexible, semi-rigid or rigid, and made of thermoplastic or thermosetting materials. While foams offer excellent thermal and acoustic insulation, vibration damping, and impact resistance properties, it is important to note that their mechanical properties are inferior to honeycomb cores.

This research aims to explore in depth the advanced potential of additive manufacturing (AM) techniques in the production of sandwich structures, with a particular focus on the creation of complex geometries obtainable exclusively using additive manufacturing. The main objective is to significantly modulate the physical properties of the resulting sandwich structure by intervening on the core geometry, such as the kind of cell, and a process parameter, such as the infill, and keeping the type of material used constant. The material used in this study is polylactic acid (PLA) [10,11].

The intricate structures under study are triply periodic minimal surfaces (TPMS) [12–14], known for their ability to provide lightness and strength. In particular, the investigation focuses on the lightening of the panels through the adoption of these complex geometries [15,16], and on the detailed analysis of their failure behavior. The TPMS surfaces [17], characterized by a continuous configuration without intersections, enable the creation of structures with high mechanical resistance and low mass, ideal for advanced applications.

The TPMS [18–20] geometry under investigation will be compared to a two-dimensional (2D) filling pattern. To achieve these objectives, the research examines in detail the different types of geometries, varying the infill and, consequently, the density of the core. This approach allows us to better understand how these variables influence the mechanical and physical properties of the sandwich panel by optimizing the process of additive manufacturing. Three-point bending tests are carried out to evaluate the performance of the panels with different geometric configurations and filling densities. Three-point bending tests are crucial for determining the mechanical behavior of the panel, offering important data on the failure behavior of the panels.

The ultimate goal is to develop sandwich panels with optimal physical properties for specific applications [21], using additive manufacturing to overcome the limitations of traditional production techniques [22]. These panels could find application in sectors such as aerospace, automotive and biomedical [23], where the combination of lightness and resistance is particularly advantageous [24].

This study aims to analyze complex geometries for the creation of the core of sandwich structures [25], focusing on geometries not yet used in the production of such panels [26,27]. In particular, we chose to examine a 2D structure, derived from the honeycomb structure [28], known as "Tri-Hexagonal", and a 3D structure, i.e., formed by an elementary cell that repeats itself unchanged in the three directions of space [29], in order to evaluate the impact of the third dimension on the repeatability of the elementary cells [30]. In this specific case, a structure known as a triply periodic minimal surface (TPMS), or a "Gyroid", selected for its intrinsic physical properties, will be analyzed. The 2D structure, characterized by hexagons connected by triangles, offers high stability in the X and Y directions and moderate stability in the Z direction.

This study introduces an innovative approach to optimizing the mechanical properties of 3D-printed PLA sandwich structures by exploring both traditional (Tri-Hexagon) and novel (Gyroid) cell geometries. By systematically varying the infill density, the research offers new insights into how these geometric configurations can be fine-tuned to achieve specific performance goals. The ability to manipulate these parameters through additive manufacturing represents a significant advancement over conventional manufacturing methods, which are often limited in their capacity to produce complex, customized structures with such precision.

Among the main developments that this research activity can have, there are both design and production evolutions. It is plausible to hypothesize that the use and testing of other types of filling, be they two-dimensional or three-dimensional geometries, could enrich the literature with significant examples of the application of additive manufacturing in the field of sandwich structures. The testing of lattice geometries, known as "lattice structures" [31], which are easily achievable using AM, is also envisaged. Furthermore, the possible change in technology must be considered, moving from the current FDM used in this work, to other types of additive manufacturing [32], which could broaden the horizons of this research. Finally, the production of specimens via multi-material FDM printing is planned, with the aim of creating composite sandwich structures in which it will be possible to use different materials [33], both for the skins and for the core of the panel, including composite materials containing polymers [34] and fibers [35–37], short or long, made of materials such as carbon [38,39], Kevlar or vegetable fibers [40–42].

2. Materials and Methods

In the present work, innovative sandwich panel specimens were generated and produced using additive manufacturing techniques with consequent mechanical characterization through three-point bending tests. The core of these sandwich structures was modeled with the help of two different types of geometries: Tri-Hexagonal (2D) and Gyroid (3D). These geometries, or patterns, were generated with four different filling percentages: 5%, 25%, 50%, and 75%. The filling percentage, in the merit of additive manufacturing technologies [43], is also defined through the term "infill". This parameter indicates the percentage of fullness present inside the component in relation to the void, meaning that an infill equal to 0% indicates an empty component while 100% filling indicates a completely full component.

The research methodology follows established standards to ensure consistency and reliability. Specifically, the flexural tests were conducted in accordance with the ASTM D790 standard, which outlines the procedures for testing the flexural properties of plastics. The optimization process was structured around this standard, with the objective function being the maximization of flexural strength relative to the material used. Constraints included the selected infill densities and cell geometries, which were chosen based on their relevance to real-world applications. This adherence to standard testing protocols ensures that the findings can be compared with other studies and applied in practical scenarios.

In the next sections, the design and manufacturing process will be discussed, with particular attention to the geometries used and the related geometric parameters, such as their filling density.

2.1. Design of Specimens

We proceeded by examining the process of producing the specimens; specifically, it is noted that additive manufacturing technologies are based on a process known as "design-driven manufacturing". The procedure begins with the three-dimensional design of the component using three-dimensional design software, such as CAD version 24.2 (Computer Aided Design). In the context of this study, parametric modeling was performed using Autodesk Inventor software 2022. For the creation of the specimens, reference was made to the D790 standard, through which the appropriate dimensions with which to create the specimens were chosen; finally, the specimens were weighed, and the relative weights were reported in Table 1. The dimensions of the bending test specimens are 122.9 mm × 25.5 mm × 6.4 mm. It is evident that the difference in weight between the two cells is low.

Table 1. Mean weight of each kind of sample.

	5%	25%	50%	75%
Trihexagonal	8.32 g	11.87 g	15.93 g	19.91 g
Gyroid	8.47 g	11.96 g	16.06 g	20.21 g
	+1.80%	+0.75%	+0.82%	+1.51%

Once the CAD modeling was complete, the file was exported in a format called STL (Stereo Lithography Interface Format). This format represents a solid whose surface is discretized into triangles. Essentially, the STL file contains the X, Y, and Z coordinates for each of the three vertices of each triangle, along with a vector describing the orientation of the surface normal of the triangle in question. The STL file was imported into the slicing software. A slicing software is an application used mainly in the field of 3D printing. This type of software is designed to convert digital 3D models into specific instructions, called "slices", that a 3D printer can understand and use to create a physical object layer by layer. The slicing process involves dividing the 3D model into a series of thin horizontal layers, determining the paths that the printer will take to deposit the material and create each layer. This software allows the customization of various printing parameters, such as infill density, print speed, and other settings, to achieve optimal results based on your specific project needs. Some examples of slicing software include Cura, Simplify3D 3.1.0, Slic3r 1.3.0, and PrusaSlicer 2.8.0. UltiMaker Cura software version 5.7.2 was used in this study. Within this software it was possible to modify the cell configuration and the infill. For the purposes of this study, two different types of filling configuration were taken into consideration:

- The Tri-Hexagon pattern (Figure 1a) is a distinctive geometric pattern used in 3D printing and various engineering applications due to its excellent mechanical characteristics. Geometrically, this pattern combines regular hexagons with equilateral triangles that fill the spaces between the hexagons. Regular hexagons are six-sided polygons with internal angles of 120 degrees, and in a two-dimensional grid, each hexagon is surrounded by other hexagons, creating a tessellated arrangement with no gaps. Between each pair of adjacent hexagons, there are equilateral triangles, all of which have three equal sides and internal angles of 60 degrees. Each equilateral triangle fits perfectly into the spaces between three adjacent hexagons. The combination of hexagons and triangles creates a periodic structure that repeats itself infinitely in all directions of the plane, giving the pattern a high degree of geometric symmetry. The Tri-Hexagon pattern can be visualized as a mesh whose nodes are the vertices of the hexagons and triangles, contributing to the uniform distribution of the applied forces across the structure. The combination of hexagons and triangles creates a network that evenly distributes the mechanical forces. The hexagons provide stability and compressive strength, while the triangles reinforce the structure and prevent deformation. Hexagons, known to be one of the most efficient shapes for stress distribution, together with triangles, contribute to high stiffness, which is essential to resist deformation under load. The Tri-Hexagon pattern enables efficient use of material and reduces the overall mass of the structure without compromising strength. This is particularly useful in 3D printing where the aim is to reduce weight while maintaining mechanical properties. The geometric configuration of the Tri-Hexagon pattern takes advantage of the optimal arrangement of hexagons and triangles to maximize mechanical properties, giving a unique combination of lightness, strength and stiffness, ideal for advanced applications in 3D printing and engineering. Its excellent mechanical characteristics arise from the ability to evenly distribute forces and use the material efficiently, making it a preferred choice for structures requiring a balance between strength and lightness.
- The TPMS (Triply Periodic Minimal Surface) [44–46] Gyroid structure (Figure 1b) is an advanced geometric model used in 3D printing and various fields of engineering due to its extraordinary mechanical and physical properties. Geometrically, a Gyroid surface is one of the minimal triply periodic surfaces, characterized by a continuous

and intersection-free configuration that repeats three-dimensionally in space [47]. This surface is mathematically defined and has a constant mean curvature of zero, meaning that every point on the surface is subjected to uniformly distributed tensile forces. The Gyroid is composed of a network of sinusoidal channels [48] that intertwine in three spatial directions and form a highly symmetric and periodic structure. The complexity of the Gyroid lies in its ability to divide space into two interconnected but non-overlapping regions [49,50], creating a three-dimensional lattice [51,52] that offers high mechanical resistance and great lightness. This configuration allows a uniform distribution of forces across the entire structure and makes it particularly resistant to both compression and tension. The Gyroid configuration maximizes the strength-to-weight ratio through efficient material distribution. This geometric model allows you to create components with an optimized internal structure, reducing the amount of material needed without compromising overall strength [53]. The Gyroid surface is particularly effective at absorbing and dispersing energy [54,55], making it ideal for applications requiring high shock absorption and superior mechanical strength. In 3D printing, the Gyroid structure is used as an infill to improve the mechanical properties [56,57] of the printed objects. Thanks to its unique geometric configuration, the Gyroid offers an optimal balance between rigidity and flexibility, making it suitable for a wide range of applications, from aerospace to biomedical. 3D printing allows the fabrication of Gyroid structures with a high degree of precision, fully exploiting the potential of this minimal surface to create advanced components with superior mechanical and physical properties. Like some other triply periodic minimal surfaces, the Gyroid surface can be approximated trigonometrically by a short equation (Equation (1)):

$$\sin x \cdot \cos y + \sin y \cdot \cos z + \sin z \cdot \cos x = 0, \tag{1}$$

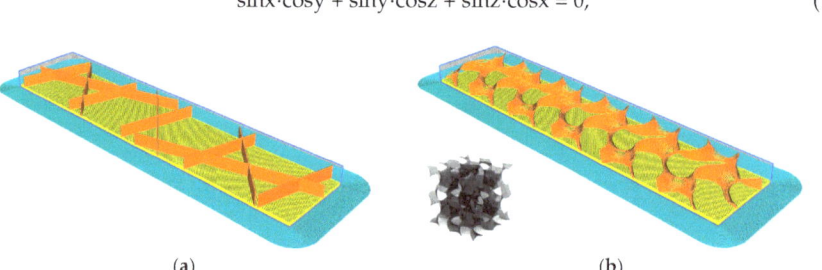

Figure 1. Filling configurations (examples with 5% infill): (**a**) Tri-Hexagon pattern; (**b**) TPMS (Triply Periodic Minimal Surface) Gyroid structure.

After slicing the model in the appropriate software, the file is transferred to the 3D printer, which then creates the component using the various additive manufacturing technologies available. The test specimen is positioned on the printing surface so that the smallest dimension, i.e., the thickness, aligned along the Z-axis of the machine, and the horizontal layers at a 90° to the testing machine [58,59]. After production, the manufactured component can be subjected to post-production treatments, if necessary, which may include chemical and thermal treatments as well as finishing processes using machine tools. In the subject of this study, no post-production treatments were implemented.

2.2. Production of Specimens Using Additive Manufacturing Techniques

In this study, the additive manufacturing technology called "Fused Deposition Modeling" (FDM) [60] was used, with an "Artillery Sidewinder X2" 3D printer. Table 2 shows the data relating to the printing parameters of the specimens [61] and the photo of the 3D printer.

Table 2. Process parameters.

Process Parameter	Value
Printing temperature	190 °C
Bed temperature	60 °C
Filling percentage	5%/25%/50%/75%
Layer height	0.2 mm
Material	Polylactic acid (PLA)

Below are the main features of the Sidewinder X2:

- Model: Artillery Sidewinder X2
- Build Volume: 300 × 300 × 400 mm
- Extruder Type: Direct Drive
- Auto-bed Leveling: Yes
- Heat Bed Type: AC heat bed
- Nozzle Type: Volcano
- Build Speed: 60 mm/s–150 mm/s
- Z-axis Design: Synchronized Dual Z System

This machine is equipped with a "Titan" extruder equipped with a "Volcano" nozzle that can reach a maximum extrusion temperature of 240 °C. A temperature of 190 °C is used to produce the specimens. The bed is heated with alternative current and can reach a temperature of 130 °C. In this study, a temperature of 60 °C was set to prevent the printed material from detaching from the printing bed.

Particularly, Figure 2a shows a Tri-Hexagon sample with an infill of 50% after production. The combination of triangles and hexagons can be seen. Figure 2b, on the other hand, shows the additive manufacturing process of the Gyroid sample with 50% infill. In this case, the core appears more complex, and the tridimensional structure can be seen.

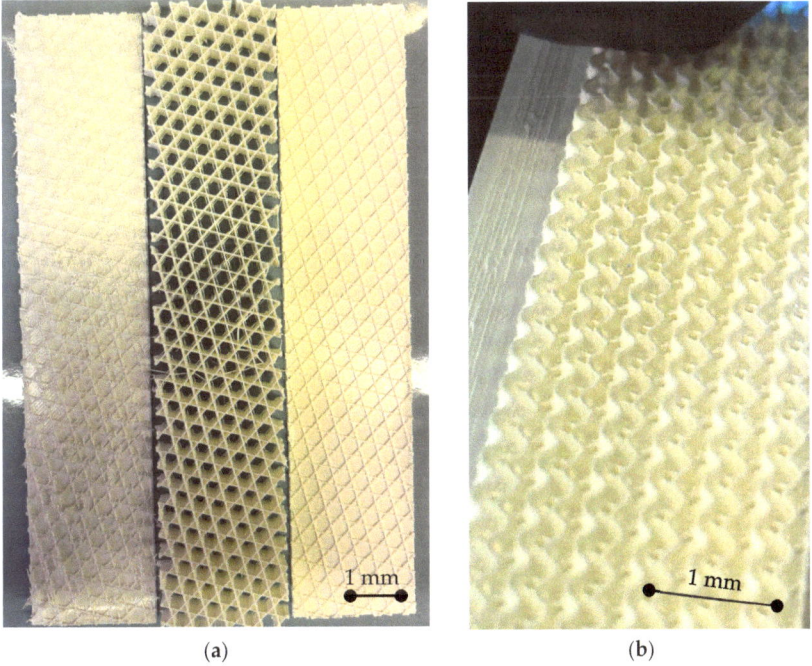

Figure 2. (**a**) Core of a Tri-Hexagon 50% sample after manufacturing process; (**b**) Additive manufacturing of a Gyroid 50% sample.

2.3. Flexural Tests

To carry out the mechanical characterization of the specimens, a "ZwichRoell" testing machine equipped with a 2.5 kN load cell was used, in accordance with ASTM D790 [62].

To carry out the bending tests, forty specimens were produced using AM, twenty of which had a Tri-Hexagon filling pattern and twenty with Gyroid geometry, i.e., five for each type. Each class of specimens, representing one of the two geometric patterns, was divided and then produced with four different filling densities, as previously mentioned. Five specimens were therefore produced for each filling density.

The tests were performed with a strain rate of 0.1 mm/min.

To analyze the samples and mainly to evaluate the failure modes, a Hirox Digital Microscope KH 8700 (Hirox, Tokyo, Japan) was used.

3. Results and Discussion

Figure 3 reports the typical load-displacement curves by varying the infill and the cell configuration (i.e., Tri-Hexagon and Gyroid). It is possible to observe that:

- For both the cell configurations, an infill equal to 5% leads to poor structural properties. In this case, the cell walls were unable to sustain the applied loads during flexural testing due to significant structural instability (Figure 4). In both types of structures, the thin-walled architecture lacks the necessary stiffness and strength at low density, making it highly susceptible to buckling under compressive stresses [63]. As the load increases, these thin walls initially deform elastically. However, due to their slenderness and geometric complexity, they rapidly reach a critical buckling threshold. Once this threshold is surpassed, the cell walls undergo localized collapse or crumpling, resulting in a significant reduction in load-bearing capacity and ultimately leading to structural failure. The unique geometries of both Tri-Hexagonal and Gyroid cells, while advantageous for weight reduction, limit their ability to evenly distribute and manage stresses, especially under bending loads. This makes them particularly prone to instability-related failures at low density. This phenomenon is illustrated in Figure 3, where the failure mode of a Tri-Hexagon 5% sample is reported.
- On increasing the infill, the load-displacement curve exhibits a characteristic response typical of higher-density sandwich structures under flexural testing [64]. At the beginning of the test, the load increases linearly with the displacement. This linear region indicates that the structure deforms elastically, which means that the material is returning to its original shape when the load is removed. The slope of this region reflects the stiffness of the structure, which is higher in this case due to the increased density of the material. As the load continues to increase, the curve reaches a peak value indicating the maximum load the structure can withstand before significant plastic deformation occurs. At this point, the cell walls of the structure begin to yield, and permanent deformation occurs. Following the peak load, cracks begin to form in the lower skin of the sandwich structure, particularly towards the point of contact with the punch, where the tensile stresses are highest (Figure 5). These tensile stresses cause micro-cracks to initiate, which subsequently propagate through the material. As these cracks grow, they lead to a reduction in the load-bearing capacity of the structure, which is reflected in the downward slope of the curve. The propagation of these cracks leads to a change in the slope of the load-displacement curve, indicating a loss of stiffness and the onset of material failure. As the cracks continue to grow and coalesce, the structure can no longer support the applied load, eventually leading to the failure of the sandwich panel. This phase is characterized by a gradual decrease in load, even if the displacement continues to increase. For higher-density Tri-Hexagon and Gyroid structures, the increased material density enhances the stiffness and strength, allowing the structure to withstand greater loads and exhibit more ductile behavior before failure. However, with continued loading, the tensile stresses in the lower skin lead to the formation and propagation of cracks, which ultimately govern the failure process. These cracks are a critical factor in the

observed change in slope and subsequent reduction in load-bearing capacity, leading to the eventual rupture of the structure.
- As the density of the Tri-Hexagon and Gyroid structures increases, the initial slope of the load-displacement curve becomes steeper, indicating an increase in stiffness. This is due to the reduction of voids within the structure as the density increases. With fewer voids, the material has a more continuous and solid network, allowing it to resist deformation more effectively under applied loads. This enhanced structural integrity requires a greater force to achieve the same displacement, reflecting the increased stiffness.
- The maximum load that the structures can withstand before they fail also increases with density, as indicated by the higher peak values in the load-displacement curves. The reduction in voids means that more material is available to distribute and bear the applied loads, leading to an enhanced load-bearing capacity. Denser structures are therefore able to sustain higher loads before significant plastic deformation and eventual failure occurs.

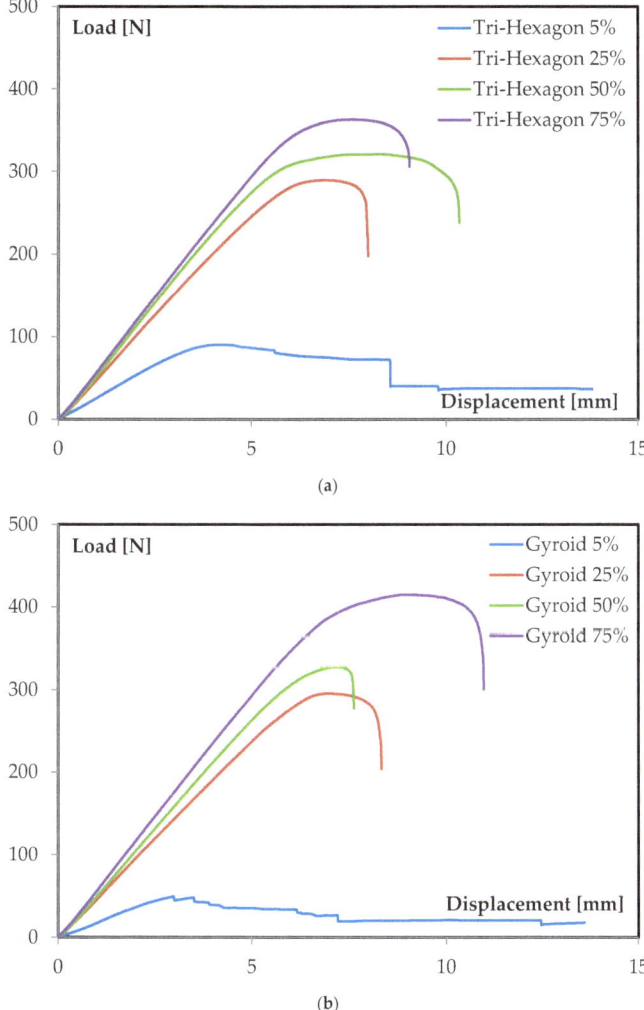

Figure 3. Typical load-displacement curves by varying the infill: (**a**) Tri-Hexagon; (**b**) Gyroid.

Figure 4. Failure mode of a Tri-Hexagon 5% structure.

The failure modes observed during flexural testing, as shown in Figures 4 and 5, reveal the structural response of the 3D-printed sandwich panels under load. In particular, the Tri-Hexagon structure (Figure 4) demonstrates buckling as the primary failure mode at low infill densities. This buckling occurs due to the collapse of the thin cell walls under compressive stress, highlighting the structural limitations when material distribution is insufficient. As the load increases, the lack of sufficient stiffness leads to significant deformation and eventual failure.

In Figure 5, which shows a Tri-Hexagon structure with higher infill, a different failure mechanism is observed. Here, the structure shows more ductile behavior, with cracks initiating in the lower skin and propagating through the core. This shift from buckling to crack propagation reflects the increased material density, which enhances the structure's ability to distribute loads more evenly. However, even with this improvement, the failure still originates at stress concentration points, such as the area near the punch in the three-point bending test. This analysis of failure modes during testing underscores the importance of balancing infill density and geometry to optimize structural performance under load.

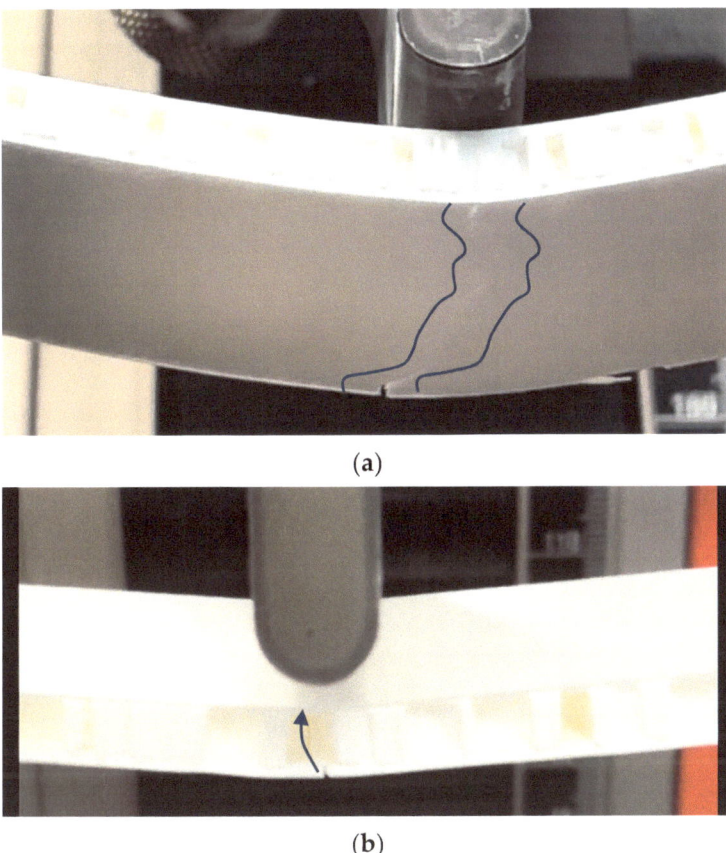

Figure 5. Failure mode of a Tri-Hexagon 50% structure: (**a**) cracks on the bottom skin; (**b**) propagation on the thickness.

Figure 6 compares the maximum loads by varying the infill and the cell configuration. The graph shows the following:

- At 5% density, the Tri-Hexagon structure exhibits a significantly higher maximum load compared to the Gyroid structure. This suggests that the Tri-Hexagon configuration has a better load-bearing capacity at lower densities, i.e., the 2D architecture allows for better resistance to the load. The standard deviation is also higher for the Tri-Hexagon structure, indicating greater variability in its performance. The Gyroid structure, while having a lower maximum load, shows less variability, which might imply more consistent performance but at a lower strength.
- At 25% density, the Gyroid structure surpasses the Tri-Hexagon in terms of maximum load capacity. This indicates that the Gyroid structure may have better strength or load distribution at this density. The standard deviation for the Gyroid is considerably lower, suggesting more uniform performance across all samples compared to the Tri-Hexagon. This could mean that the Gyroid structure is more reliable and consistent at this density, while the Tri-Hexagon has more variability in its load-bearing capacity.
- At 50% density, both structures show similar maximum load-bearing capacities, with the Tri-Hexagon slightly outperforming the Gyroid. The standard deviations at this density are relatively low for both, but the Tri-Hexagon has a slightly lower deviation, suggesting a slightly more consistent performance at this density. The similar

load capacities indicate that at mid-range density, both structures have comparable performance, but the Tri-Hexagon may offer a slight edge in terms of consistency.
- At 75% density, the Gyroid structure clearly outperforms the Tri-Hexagon in terms of maximum load capacity. This suggests that the Gyroid structure's design becomes more advantageous at higher densities, due to more effective load distribution and structural efficiency. The Gyroid also shows a much lower standard deviation, indicating a very consistent performance across all samples at this density. In contrast, the Tri-Hexagon structure, while still strong, shows slightly more variability in its performance.

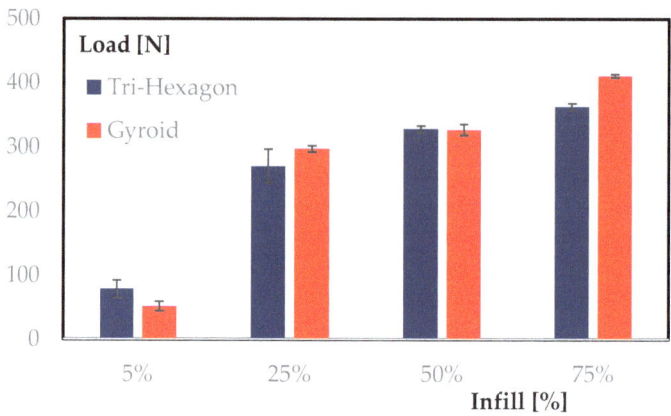

Figure 6. Comparison in maximum load between Tri-Hexagon and Gyroid structures.

To verify if the cell configuration is a significant parameter and, consequently, that it influences the flexural performance of the sample, a variance analysis was performed using MINITAB software 22.1.0, by considering the following factors:
- Cell configuration with 2 levels (i.e., Tri-Hexagon and Gyroid);
- Infill with 4 levels (i.e., 5%, 25%, 50%, and 75%).

In applying ANOVA, several assumptions were made, including the normality of residuals, homoscedasticity (equal variance), and independence of observations. These assumptions are fundamental to ensure that the ANOVA results are valid:
- Normality of residuals refers to the assumption that the differences between the observed and predicted values (i.e., residuals) follow a normal distribution. This ensures that the statistical tests used in ANOVA are appropriate for the data.
- Homoscedasticity (or equal variance) means that the variability in the response variable is consistent across all levels of the independent variables. This is important because unequal variances can lead to inaccurate estimates of the factor effects.
- Independence of observations implies that the data points are not related to each other. For example, the outcome of one observation should not influence another. Violations of this assumption can result in misleading conclusions.

To validate these assumptions, a detailed residual analysis was performed. As shown in Figure 7, the residuals are symmetrically distributed around zero, indicating normality. The assumption of homoscedasticity is confirmed by the consistent spread of residuals across the predicted values. No significant autocorrelation patterns were observed, affirming the independence of the data. These verifications indicate that the ANOVA results are reliable and that the significant effects observed are valid under the conditions of the study.

Table 3 summarizes the main results of the analysis of variance (ANOVA). The main objective was to determine whether the effects of the investigated factors on load were significant. DF are the degrees of freedom, used to calculate the mean square (MS).

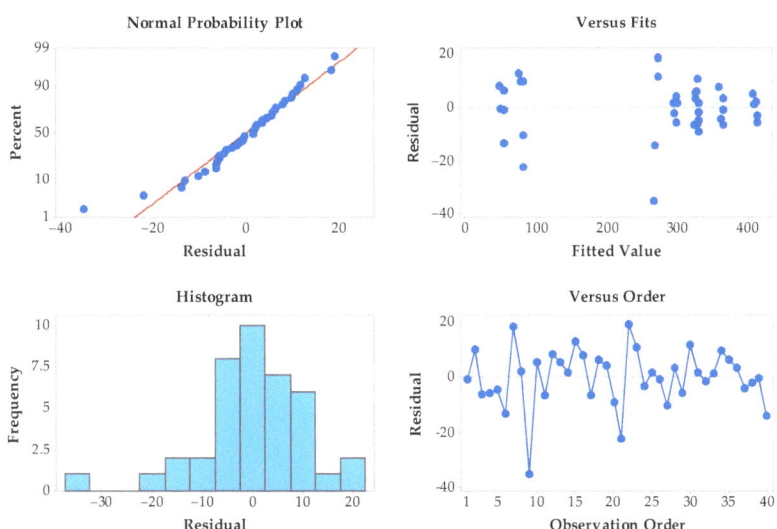

Figure 7. Residual plots for maximum load.

Table 3. ANOVA: Analysis of Variance for maximum load.

Source		DF	Adj SS	Adj MS	F-Value	*p*-Value
Model		11	599,647	54,513	367.46	0.000
Blocks		4	217	54	0.37	0.831
Linear		4	591,334	147,833	996.51	0.000
	Cell	1	1418	1418	9.56	0.004
	Infill	3	589,916	196,639	1325.50	0.000
2-Way Interactions		3	8097	2699	18,19	0.000
	Cell*Infill	3	8097	2699	18,19	0.000
Error		28	4154	148		
Total		39	603,801			
		S = 12.1799	R-sq = 99.31%	R-sq(adj) = 99.04%	R-sq(pred) = 98.60%		

In general, they measure how much "independent" information is available to calculate each sum of squares (SS). This latter, also called the sum of the squared deviations, measures the total variability in the data, which is made up of: (i) the sum of squares for each of the two factors, which measures how much the means of the levels differ within each factor; (ii) the sum of squares for the interaction, which measures how much the effects of one factor depend on the level of the other factor; and (iii) the sum of squares for the error, which measures the variability that remains after the factors and the interaction have been taken into account. MS is simply the sum of squares (SS) divided by the degrees of freedom. The mean squared error is an estimate of the variance in the data that remains after accounting for the mean differences. F is used to determine the *p*-value (*p*), which defines whether the effect for a term is significant: i.e., if *p* is less than or equal to a selected level (e.g., 0.05), the effect for the term is significant.

From the table it is possible draw the following considerations:

- the individual effect of all the factors on the load is significant. The different infills induce a change in the level of maximum load that the sandwich can withstand, whereas the different cell structures induce different types of crack propagation, and consequently different behaviors depending on the kind infill by creating competition between the two cells.
- the interaction between the factors is also significant. This is evident in Figure 8, where the lines or intersects between them or are not parallel.

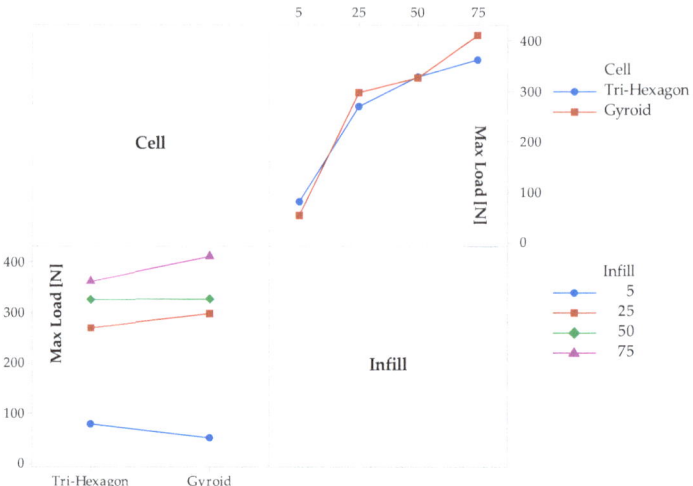

Figure 8. Interaction plot for maximum load.

Figure 9 summarizes the results for all the combinations investigated.

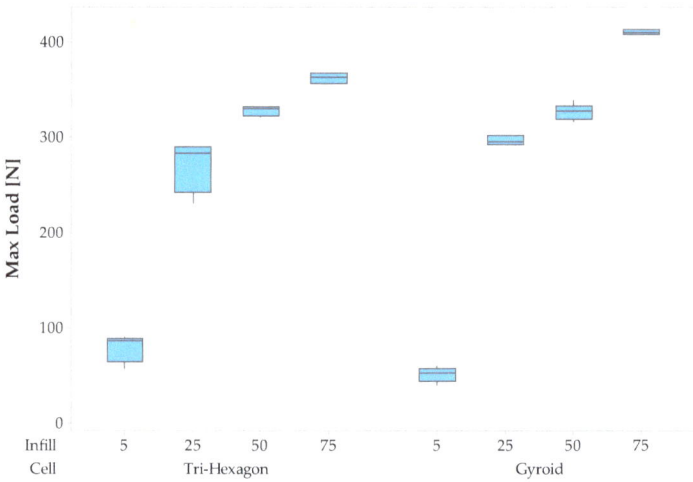

Figure 9. Interaction plot for maximum load.

While the statistical significance of factors such as core geometry and infill density is evident from the ANOVA, translating these findings into practical applications is crucial for advancing the field of additive manufacturing. In real-world applications, such as aerospace or automotive components, the trade-offs between mechanical performance and material efficiency become particularly important. The observed differences between geometries, where the Tri-Hexagon structure shows superior performance at lower densities, and the Gyroid structure, which excels at higher densities, suggest that the choice of structure should be application-specific. In situations where weight reduction is a priority, but mechanical integrity cannot be compromised, the Tri-Hexagon structure may be preferable at lower infills. Conversely, for applications requiring high structural efficiency, the Gyroid structure becomes more advantageous at higher infills. These trade-offs can guide the selection of the most appropriate configuration for specific needs, balancing factors such as material cost, production time, and mechanical performance.

Finally, optical microscope analysis shows the typical failure modes that occur in the two structures by varying the infill.

Figure 10 presents the analysis of a sample characterized by a Gyroid cell and an infill of 5%. For this structure, it is evident that the buckling phenomenon typical of the Try-Hexagon structure is coupled with an interlayer crack that affects the whole sample by causing its premature fracture. This phenomenon is due to the presence of several defects generated by the manufacturing process, the presence of which are promoted by the low infill. This explains the poorer behavior than with the Tri-Hexagon structure, where the 2D basis architecture involves only a fracture for buckling, as previously observed.

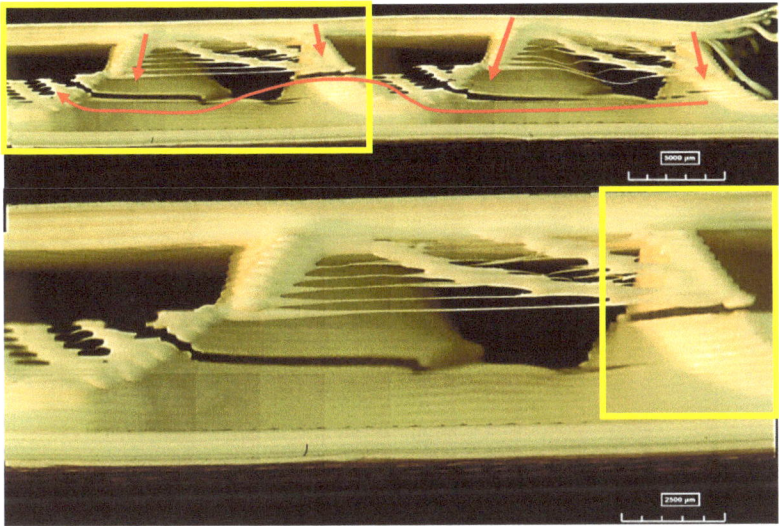

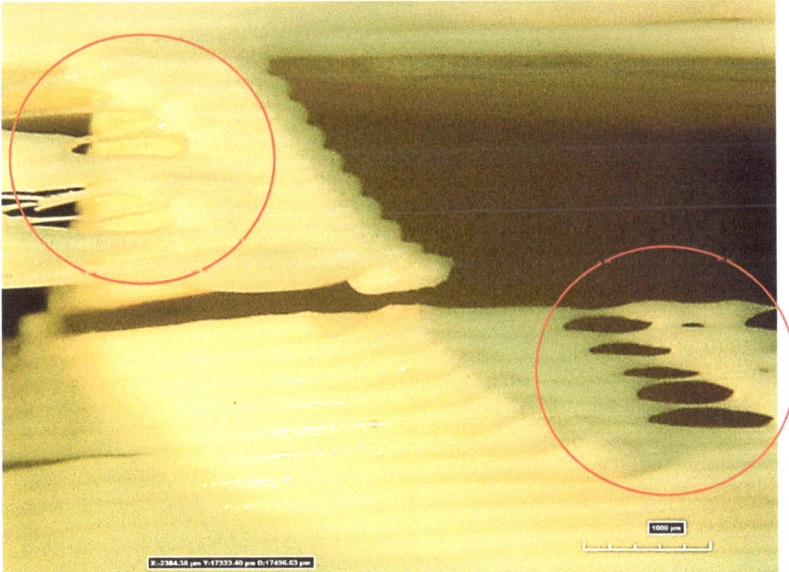

Figure 10. Optical analysis of a Gyroid 5% sample.

For higher infills, the structures of the samples are more compact and, consequently, these defects are not obvious. The crack starts in the lower skin near the pin, then it propagates in the core and reaches the upper skin, leading to the failure of the sample, as can be observed in Figure 11 for a Gyroid 25% sample.

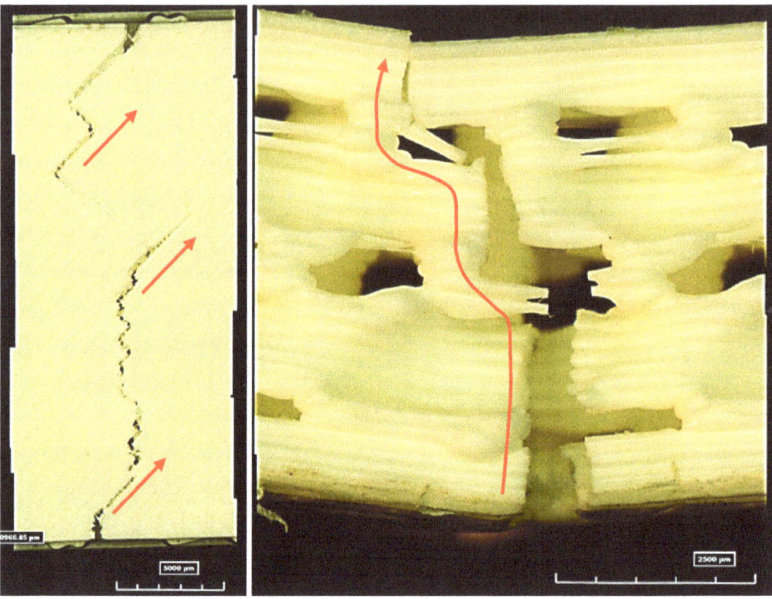

Figure 11. Optical analysis of a Gyroid 25% sample.

It is evident that:
- in the lower skin, the cracks mainly follow the direction of filament deposition;
- after the crack opens, this propagates into the core until it reaches the upper skin.

The detailed microscopy analysis presented in Figures 10 and 11 provides further insights into the failure mechanisms observed in the Gyroid structures at different infill densities. Figure 10 shows a Gyroid structure with 5% infill, where both buckling and interlayer cracking are evident. The presence of interlayer cracks suggests that the 3D printing process may have introduced defects, such as poor layer adhesion and voids, particularly at low infill levels. These defects, combined with the complex geometry of the Gyroid structure, lead to premature failure. The microscopy images highlight the limitations of current printing technology, and the challenges associated with achieving uniform material deposition, especially at low densities.

In contrast, Figure 11 illustrates the failure mode of a Gyroid structure with 25% infill. At this higher density, the structure exhibits a more compact and robust form, resulting in a different failure behavior. Here, the primary mode of failure is crack propagation, which begins in the lower skin and progresses through the core until it reaches the upper skin. This behavior is consistent with the findings from the flexural tests, where the increased material density leads to improved load distribution and a more ductile response. The direction of crack propagation aligns with the filament deposition pattern, indicating that even at higher densities, the printing process plays a crucial role in determining the failure characteristics of the structure. These observations emphasize the need for further optimization of both the printing process and the design of infill patterns to minimize defects and enhance the overall mechanical performance of 3D-printed sandwich structures.

The preferred direction of crack propagation that characterizes the Gyroid samples is not evident in the Tri-Hexagon sandwiches. Here, the crack follows a different trend corresponding to the cell walls and also after it has propagated into the core, as observed in Figure 12. This consideration explains the significance of the cell configuration.

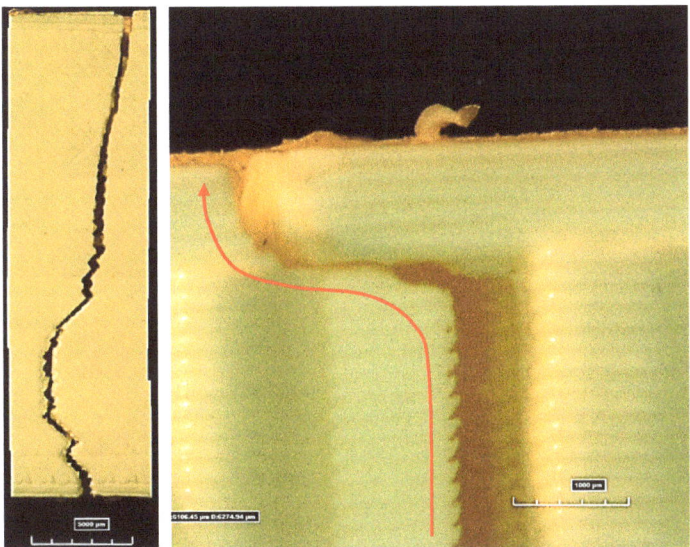

Figure 12. Optical analysis of a Tri-Hexagon 50% sample.

This optical analysis highlights the limitations of the technology used in this study. In fact, in addition to the defects generated by the low infill, the 3D printer caused both poor deposition of the filament, because the different layers are not perfectly distributed along the z-axis, and burning with a consequent deformation of the first layer, as can be seen in Figure 13 [65]. This can be overcome by using next-generation 3D printers, characterized by a closed chamber and a more rigid structure, as evidenced in Figure 14, where a sample of PLA, the subject of a future study, is realized by a Bambu Lab X1 Carbon.

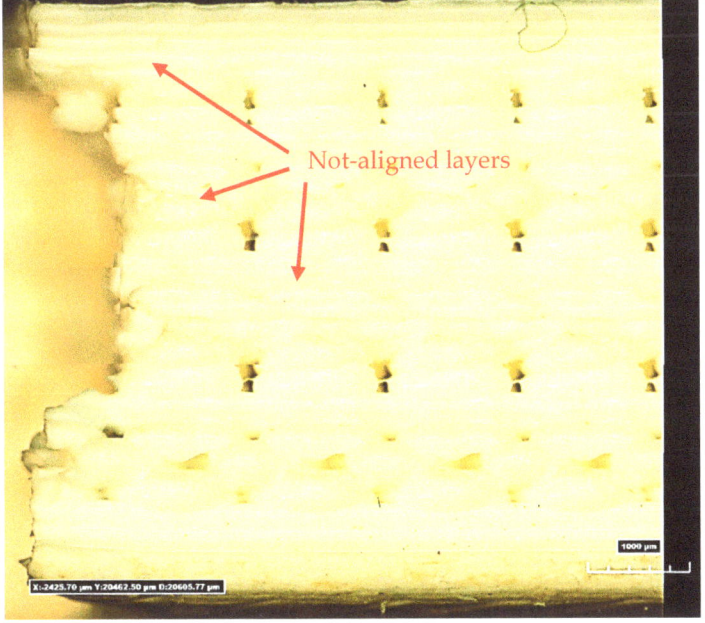

Figure 13. *Cont.*

Figure 13. Defects induced by technology.

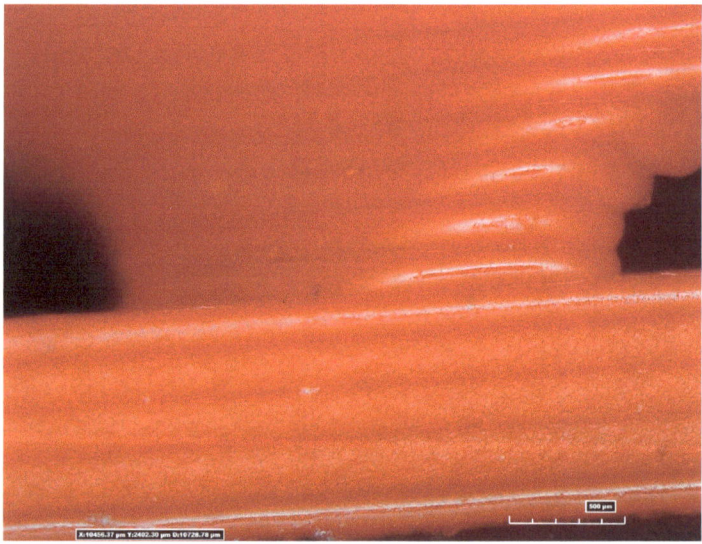

Figure 14. Example of sample in PLA generated on a next-generation 3D printer.

4. Conclusions

In this study, the additive manufacturing of PLA sandwich structures was optimized by varying the type of cell (i.e., the typical Tri-Hexagon and the new TPMS Gyroid) and infill (i.e., 5%, 25%, 50%, and 75%). The effects on flexural tests were investigated. It can be concluded that:

- For low infill (i.e., 5%), the architecture of the core for both cells does not resist, leading to premature failure. In the Tri-Hexagon samples, this occurs due to buckling, while in the Gyroid samples, buckling is coupled with interlayer fractures. This highlights the structural limitations of low-density infills, where the material distribution is insufficient to provide the necessary stiffness and load-bearing capacity.

- For high infill, the architectures are more resistant, and both cells exhibit good responses to bending loads. At these higher densities, cracks initiate at the lower skin and propagate through the core until reaching the upper skin. The difference between the two cells lies in the crack propagation direction: in Gyroid samples, the crack follows the direction of the filament deposition, while in the Tri-Hexagon samples, it follows the cell walls. This behavior underscores the importance of infill density in enhancing structural integrity, with both geometries showing improved performance as material density increases.

This behavior is confirmed by the analysis of variance (ANOVA), which demonstrates that both factors (i.e., cell geometry and infill) have a statistically significant impact on flexural behavior. For example, the Gyroid structure at 75% infill exhibited a maximum load-bearing capacity, outperforming the Tri-Hexagon structure. These findings suggest that, depending on the application, specific combinations of geometry and infill density can be tailored to optimize performance.

Moreover, the optical analysis reveals not only the differences in failure modes but also the defects introduced by the manufacturing process, such as unaligned layers, voids, and burns. These defects, particularly in low-density structures, further compromise mechanical performance.

Looking forward, future research should explore advanced printing technologies, such as multi-material printing or closed-chamber systems, to further minimize defects and improve the structural performance of 3D-printed sandwich structures. Additionally, expanding this research to other materials, including composites or polymers reinforced with fibers, could open new possibilities for developing high-performance structures with even greater versatility. By connecting these findings to the broader literature, ongoing advancements in additive manufacturing will continue to offer innovative solutions for complex engineering challenges.

Author Contributions: Conceptualization, G.D.B. and G.M.; methodology, G.D.B. and G.M.; software, G.M. and M.C.; validation, G.D.B., G.M. and M.C.; formal analysis, G.M. and M.C.; investigation, G.D.B., G.M. and M.C.; data curation, G.D.B., G.M. and M.C.; writing—original draft preparation, G.M.; writing—review and editing, G.D.B., G.M. and M.C.; supervision, G.D.B.; funding acquisition, G.D.B. All authors have read and agreed to the published version of the manuscript.

Funding: This research received no external funding.

Data Availability Statement: No new data were created or analyzed in this study. Data sharing is not applicable to this article.

Conflicts of Interest: The authors declare no conflicts of interest.

References

1. Ngo, T.D.; Kashani, A.; Imbalzano, G.; Nguyen, K.T.Q.; Hui, D. Additive Manufacturing (3D Printing): A Review of Materials, Methods, Applications and Challenges. *Compos. B Eng.* **2018**, *143*, 172–196. [CrossRef]
2. Shahrubudin, N.; Lee, T.C.; Ramlan, R. An Overview on 3D Printing Technology: Technological, Materials, and Applications. *Procedia Manuf.* **2019**, *35*, 1286–1296. [CrossRef]
3. DebRoy, T.; Wei, H.L.; Zuback, J.S.; Mukherjee, T.; Elmer, J.W.; Milewski, J.O.; Beese, A.M.; Wilson-Heid, A.; De, A.; Zhang, W. Additive Manufacturing of Metallic Components—Process, Structure and Properties. *Prog. Mater. Sci.* **2018**, *92*, 112–224. [CrossRef]
4. Tofail, S.A.M.; Koumoulos, E.P.; Bandyopadhyay, A.; Bose, S.; O'Donoghue, L.; Charitidis, C. Additive Manufacturing: Scientific and Technological Challenges, Market Uptake and Opportunities. *Mater. Today* **2018**, *21*, 22–37. [CrossRef]
5. Xu, F.; Liu, Z.; Selvaraj, R.; Ahsan, M. Bending and Eigenfrequency Analysis of Glass Fiber Reinforced Honeycomb Sandwich Shell Panels Containing Graphene-Decorated with Graphene Quantum Dots. *Eng. Struct.* **2023**, *294*, 116685. [CrossRef]
6. Birman, V.; Kardomateas, G.A. Review of Current Trends in Research and Applications of Sandwich Structures. *Compos. B Eng.* **2018**, *142*, 221–240. [CrossRef]
7. Moradi, A.; Ansari, R.; Hassanzadeh-Aghdam, M.K.; Jang, S.H. The Effect of Fuzzy Fiber-Reinforced Composite Skin on Bending Stiffness and Debonding Growth between Skin and Polymeric Foam Core of Sandwich Structures. *Eur. J. Mech. A/Solids* **2024**, *103*, 105182. [CrossRef]

8. Nian, Y.; Wan, S.; Li, X.; Su, Q.; Li, M. How Does Bio-Inspired Graded Honeycomb Filler Affect Energy Absorption Characteristics? *Thin-Walled Struct.* **2019**, *144*, 106269. [CrossRef]
9. Wang, H.; Chen, J.; Wang, C.; Guo, G.; An, Q.; Ming, W.; Chen, M. Burr Formation Mechanism and Morphological Transformation Critical Conditions in Grinding of Nickel-Based Superalloy Honeycomb Cores. *Chin. J. Aeronaut.* **2023**, *36*, 434–451. [CrossRef]
10. Chacón, J.M.; Caminero, M.A.; García-Plaza, E.; Núñez, P.J. Additive Manufacturing of PLA Structures Using Fused Deposition Modelling: Effect of Process Parameters on Mechanical Properties and Their Optimal Selection. *Mater. Des.* **2017**, *124*, 143–157. [CrossRef]
11. Ligon, S.C.; Liska, R.; Stampfl, J.; Gurr, M.; Mülhaupt, R. Polymers for 3D Printing and Customized Additive Manufacturing. *Chem. Rev.* **2017**, *117*, 10212–10290. [CrossRef] [PubMed]
12. Thomas, N.; Sreedhar, N.; Al-Ketan, O.; Rowshan, R.; Abu Al-Rub, R.K.; Arafat, H. 3D Printed Triply Periodic Minimal Surfaces as Spacers for Enhanced Heat and Mass Transfer in Membrane Distillation. *Desalination* **2018**, *443*, 256–271. [CrossRef]
13. Yuan, L.; Ding, S.; Wen, C. Additive Manufacturing Technology for Porous Metal Implant Applications and Triple Minimal Surface Structures: A Review. *Bioact. Mater.* **2019**, *4*, 56–70. [CrossRef] [PubMed]
14. Al-Ketan, O.; Abu Al-Rub, R.K. Multifunctional Mechanical Metamaterials Based on Triply Periodic Minimal Surface Lattices. *Adv. Eng. Mater.* **2019**, *21*, e56088. [CrossRef]
15. Al-Ketan, O.; Al-Rub, R.K.A.; Rowshan, R. Mechanical Properties of a New Type of Architected Interpenetrating Phase Composite Materials. *Adv. Mater. Technol.* **2017**, *2*, 1600235. [CrossRef]
16. Benedetti, M.; du Plessis, A.; Ritchie, R.O.; Dallago, M.; Razavi, N.; Berto, F. Architected Cellular Materials: A Review on Their Mechanical Properties towards Fatigue-Tolerant Design and Fabrication. *Mater. Sci. Eng. R Rep.* **2021**, *144*, 100606. [CrossRef]
17. Attarzadeh, R.; Rovira, M.; Duwig, C. Design Analysis of the "Schwartz D" Based Heat Exchanger: A Numerical Study. *Int. J. Heat. Mass. Transf.* **2021**, *177*, 121415. [CrossRef]
18. Krishnan, K.; Lee, D.-W.; Al Teneji, M.; Abu Al-Rub, R.K. Effective Stiffness, Strength, Buckling and Anisotropy of Foams Based on Nine Unique Triple Periodic Minimal Surfaces. *Int. J. Solids Struct.* **2022**, *238*, 111418. [CrossRef]
19. Iyer, J.; Moore, T.; Nguyen, D.; Roy, P.; Stolaroff, J. Heat Transfer and Pressure Drop Characteristics of Heat Exchangers Based on Triply Periodic Minimal and Periodic Nodal Surfaces. *Appl. Therm. Eng.* **2022**, *209*, 118192. [CrossRef]
20. Qureshi, Z.A.; Elnajjar, E.; Al-Ketan, O.; Al-Rub, R.A.; Al-Omari, S.B. Heat Transfer Performance of a Finned Metal Foam-Phase Change Material (FMF-PCM) System Incorporating Triply Periodic Minimal Surfaces (TPMS). *Int. J. Heat. Mass. Transf.* **2021**, *170*, 121001. [CrossRef]
21. Maskery, I.; Aremu, A.O.; Parry, L.; Wildman, R.D.; Tuck, C.J.; Ashcroft, I.A. Effective Design and Simulation of Surface-Based Lattice Structures Featuring Volume Fraction and Cell Type Grading. *Mater. Des.* **2018**, *155*, 220–232. [CrossRef]
22. Buchanan, C.; Gardner, L. Metal 3D Printing in Construction: A Review of Methods, Research, Applications, Opportunities and Challenges. *Eng. Struct.* **2019**, *180*, 332–348. [CrossRef]
23. Li, W.; Yu, G.; Yu, Z. Bioinspired Heat Exchangers Based on Triply Periodic Minimal Surfaces for Supercritical CO2 Cycles. *Appl. Therm. Eng.* **2020**, *179*, 115686. [CrossRef]
24. Attaran, M. The Rise of 3-D Printing: The Advantages of Additive Manufacturing over Traditional Manufacturing. *Bus. Horiz.* **2017**, *60*, 677–688. [CrossRef]
25. Li, T.; Wang, L. Bending Behavior of Sandwich Composite Structures with Tunable 3D-Printed Core Materials. *Compos. Struct.* **2017**, *175*, 46–57. [CrossRef]
26. Iranmanesh, N.; Yazdani Sarvestani, H.; Ashrafi, B.; Hojjati, M. Beyond Honeycombs: Core Topology's Role in 3D-Printed Sandwich Panels. *Mater. Today Commun.* **2023**, *37*, 107548. [CrossRef]
27. Wang, Y.; Liu, F.; Zhang, X.; Zhang, K.; Wang, X.; Gan, D.; Yang, B. Cell-Size Graded Sandwich Enhances Additive Manufacturing Fidelity and Energy Absorption. *Int. J. Mech. Sci.* **2021**, *211*, 106798. [CrossRef]
28. Lubombo, C.; Huneault, M.A. Effect of Infill Patterns on the Mechanical Performance of Lightweight 3D-Printed Cellular PLA Parts. *Mater. Today Commun.* **2018**, *17*, 214–228. [CrossRef]
29. Nian, Y.; Wan, S.; Avcar, M.; Yue, R.; Li, M. 3D Printing Functionally Graded Metamaterial Structure: Design, Fabrication, Reinforcement, Optimization. *Int. J. Mech. Sci.* **2023**, *258*, 108580. [CrossRef]
30. Sugiyama, K.; Matsuzaki, R.; Ueda, M.; Todoroki, A.; Hirano, Y. 3D Printing of Composite Sandwich Structures Using Continuous Carbon Fiber and Fiber Tension. *Compos. Part. A Appl. Sci. Manuf.* **2018**, *113*, 114–121. [CrossRef]
31. Nian, Y.; Wan, S.; Wang, X.; Zhou, P.; Avcar, M.; Li, M. Study on Crashworthiness of Nature-Inspired Functionally Graded Lattice Metamaterials for Bridge Pier Protection against Ship Collision. *Eng. Struct.* **2023**, *277*, 115404. [CrossRef]
32. Dilag, J.; Chen, T.; Li, S.; Bateman, S.A. Design and Direct Additive Manufacturing of Three-Dimensional Surface Micro-Structures Using Material Jetting Technologies. *Addit. Manuf.* **2019**, *27*, 167–174. [CrossRef]
33. Lee, J.Y.; An, J.; Chua, C.K. Fundamentals and Applications of 3D Printing for Novel Materials. *Appl. Mater. Today* **2017**, *7*, 120–133. [CrossRef]
34. Dizon, J.R.C.; Espera, A.H.; Chen, Q.; Advincula, R.C. Mechanical Characterization of 3D-Printed Polymers. *Addit. Manuf.* **2018**, *20*, 44–67. [CrossRef]
35. Wang, X.; Jiang, M.; Zhou, Z.; Gou, J.; Hui, D. 3D Printing of Polymer Matrix Composites: A Review and Prospective. *Compos. B Eng.* **2017**, *110*, 442–458. [CrossRef]
36. Guo, H.; Lv, R.; Bai, S. Recent Advances on 3D Printing Graphene-Based Composites. *Nano Mater. Sci.* **2019**, *1*, 101–115. [CrossRef]

37. Marabello, G.; Borsellino, C.; Di Bella, G. Carbon Fiber 3D Printing: Technologies and Performance—A Brief Review. *Materials* **2023**, *16*, 7311. [CrossRef] [PubMed]
38. Wang, Z.; Yao, G. Nonlinear Vibration and Stability of Sandwich Functionally Graded Porous Plates Reinforced with Graphene Platelets in Subsonic Flow on Elastic Foundation. *Thin-Walled Struct.* **2024**, *194*, 111327. [CrossRef]
39. Sukia, I.; Morales, U.; Esnaola, A.; Aurrekoetxea, J.; Erice, B. Low-Velocity Impact Performance of Integrally-3D Printed Continuous Carbon Fibre Composite Sandwich Panels. *Mater. Lett.* **2024**, *354*, 135374. [CrossRef]
40. Penumakala, P.K.; Santo, J.; Thomas, A. A Critical Review on the Fused Deposition Modeling of Thermoplastic Polymer Composites. *Compos. B Eng.* **2020**, *201*, 108336. [CrossRef]
41. Ayrilmis, N. Effect of Layer Thickness on Surface Properties of 3D Printed Materials Produced from Wood Flour/PLA Filament. *Polym. Test.* **2018**, *71*, 163–166. [CrossRef]
42. Peng, X.; Zhang, M.; Guo, Z.; Sang, L.; Hou, W. Investigation of Processing Parameters on Tensile Performance for FDM-Printed Carbon Fiber Reinforced Polyamide 6 Composites. *Compos. Commun.* **2020**, *22*, 100478. [CrossRef]
43. Yao, T.; Deng, Z.; Zhang, K.; Li, S. A Method to Predict the Ultimate Tensile Strength of 3D Printing Polylactic Acid (PLA) Materials with Different Printing Orientations. *Compos. B Eng.* **2019**, *163*, 393–402. [CrossRef]
44. Feng, J.; Liu, B.; Lin, Z.; Fu, J. Isotropic Porous Structure Design Methods Based on Triply Periodic Minimal Surfaces. *Mater. Des.* **2021**, *210*, 110050. [CrossRef]
45. Jones, A.; Leary, M.; Bateman, S.; Easton, M. TPMS Designer: A Tool for Generating and Analyzing Triply Periodic Minimal Surfaces[Formula Presented]. *Softw. Impacts* **2021**, *10*, 100167. [CrossRef]
46. Feng, J.; Fu, J.; Yao, X.; He, Y. Triply Periodic Minimal Surface (TPMS) Porous Structures: From Multi-Scale Design, Precise Additive Manufacturing to Multidisciplinary Applications. *Int. J. Extrem. Manuf.* **2022**, *4*, 022001. [CrossRef]
47. Dixit, T.; Al-Hajri, E.; Paul, M.C.; Nithiarasu, P.; Kumar, S. High Performance, Microarchitected, Compact Heat Exchanger Enabled by 3D Printing. *Appl. Therm. Eng.* **2022**, *210*, 118339. [CrossRef]
48. Qureshi, Z.A.; Al-Omari, S.A.B.; Elnajjar, E.; Al-Ketan, O.; Al-Rub, R.A. Using Triply Periodic Minimal Surfaces (TPMS)-Based Metal Foams Structures as Skeleton for Metal-Foam-PCM Composites for Thermal Energy Storage and Energy Management Applications. *Int. Commun. Heat. Mass. Transf.* **2021**, *124*, 105265. [CrossRef]
49. Attarzadeh, R.; Attarzadeh-Niaki, S.H.; Duwig, C. Multi-Objective Optimization of TPMS-Based Heat Exchangers for Low-Temperature Waste Heat Recovery. *Appl. Therm. Eng.* **2022**, *212*, 118448. [CrossRef]
50. Qureshi, Z.A.; Al-Omari, S.A.B.; Elnajjar, E.; Al-Ketan, O.; Abu Al-Rub, R. Nature-Inspired Triply Periodic Minimal Surface-Based Structures in Sheet and Solid Configurations for Performance Enhancement of a Low-Thermal-Conductivity Phase-Change Material for Latent-Heat Thermal-Energy-Storage Applications. *Int. J. Therm. Sci.* **2022**, *173*, 107361. [CrossRef]
51. Nian, Y.; Wan, S.; Zhou, P.; Wang, X.; Santiago, R.; Li, M. Energy Absorption Characteristics of Functionally Graded Polymer-Based Lattice Structures Filled Aluminum Tubes under Transverse Impact Loading. *Mater. Des.* **2021**, *209*, 110011. [CrossRef]
52. Xie, C.; Wang, D.; Zong, L.; Kong, D. Crashworthiness Analysis and Multi-Objective Optimization of Spatial Lattice Structure under Dynamic Compression. *Int. J. Impact Eng.* **2023**, *180*, 104713. [CrossRef]
53. Clarke, D.A.; Dolamore, F.; Fee, C.J.; Galvosas, P.; Holland, D.J. Investigation of Flow through Triply Periodic Minimal Surface-Structured Porous Media Using MRI and CFD. *Chem. Eng. Sci.* **2021**, *231*, 116264. [CrossRef]
54. Qureshi, Z.A.; Al Omari, S.A.B.; Elnajjar, E.; Mahmoud, F.; Al-Ketan, O.; Al-Rub, R.A. Thermal Characterization of 3D-Printed Lattices Based on Triply Periodic Minimal Surfaces Embedded with Organic Phase Change Material. *Case Stud. Therm. Eng.* **2021**, *27*, 101315. [CrossRef]
55. Mirabolghasemi, A.; Akbarzadeh, A.H.; Rodrigue, D.; Therriault, D. Thermal Conductivity of Architected Cellular Metamaterials. *Acta Mater.* **2019**, *174*, 61–80. [CrossRef]
56. Qureshi, Z.A.; Addin Burhan Al-Omari, S.; Elnajjar, E.; Al-Ketan, O.; Al-Rub, R.A. On the Effect of Porosity and Functional Grading of 3D Printable Triply Periodic Minimal Surface (TPMS) Based Architected Lattices Embedded with a Phase Change Material. *Int. J. Heat. Mass. Transf.* **2022**, *183*, 122111. [CrossRef]
57. Lu, Y.; Zhao, W.; Cui, Z.; Zhu, H.; Wu, C. The Anisotropic Elastic Behavior of the Widely-Used Triply-Periodic Minimal Surface Based Scaffolds. *J. Mech. Behav. Biomed. Mater.* **2019**, *99*, 56–65. [CrossRef]
58. Dudescu, C.; Racz, L. Effects of Raster Orientation, Infill Rate and Infill Pattern on the Mechanical Properties of 3D Printed Materials. *ACTA Univ. Cibiniensis* **2017**, *69*, 23–30. [CrossRef]
59. Shanmugam, V.; Das, O.; Babu, K.; Marimuthu, U.; Veerasimman, A.; Johnson, D.J.; Neisiany, R.E.; Hedenqvist, M.S.; Ramakrishna, S.; Berto, F. Fatigue Behaviour of FDM-3D Printed Polymers, Polymeric Composites and Architected Cellular Materials. *Int. J. Fatigue* **2021**, *143*, 106007. [CrossRef]
60. Singh, S.; Singh, G.; Prakash, C.; Ramakrishna, S. Current Status and Future Directions of Fused Filament Fabrication. *J. Manuf. Process* **2020**, *55*, 288–306. [CrossRef]
61. Popescu, D.; Zapciu, A.; Amza, C.; Baciu, F.; Marinescu, R. FDM Process Parameters Influence over the Mechanical Properties of Polymer Specimens: A Review. *Polym. Test.* **2018**, *69*, 157–166. [CrossRef]
62. *ASTM D790-17*; Standard Test Methods for Flexural Properties of Unreinforced and Reinforced Plastics and Electrical Insulating Materials. ASTM International: West Conshohocken, PA, USA, 2017.
63. Daynes, S. High Stiffness Topology Optimised Lattice Structures with Increased Toughness by Porosity Constraints. *Mater. Des.* **2023**, *232*, 112183. [CrossRef]

64. Epasto, G.; Rizzo, D.; Landolfi, L.; Detry, A.L.H.S.; Papa, I.; Squillace, A. Design of Monomaterial Sandwich Structures Made with Foam Additive Manufacturing. *J. Manuf. Process* **2024**, *121*, 323–332. [CrossRef]
65. Kim, S.; Kim, E.-H.; Lee, W.; Sim, M.; Kim, I.; Noh, J.; Kim, J.-H.; Lee, S.; Park, I.; Su, P.-C.; et al. Real-Time in-Process Control Methods of Process Parameters for Additive Manufacturing. *J. Manuf. Syst.* **2024**, *74*, 1067–1090. [CrossRef]

Disclaimer/Publisher's Note: The statements, opinions and data contained in all publications are solely those of the individual author(s) and contributor(s) and not of MDPI and/or the editor(s). MDPI and/or the editor(s) disclaim responsibility for any injury to people or property resulting from any ideas, methods, instructions or products referred to in the content.

Article

The Effect of Chopped Carbon Fibers on the Mechanical Properties and Fracture Toughness of 3D-Printed PLA Parts: An Experimental and Simulation Study

Ahmed Ali Farhan Ogaili [1], Ali Basem [2], Mohammed Salman Kadhim [3], Zainab T. Al-Sharify [4,5], Alaa Abdulhady Jaber [6,*], Emad Kadum Njim [7], Luttfi A. Al-Haddad [8], Mohsin Noori Hamzah [6] and Ehsan S. Al-Ameen [1]

1. Mechanical Engineering Department, College of Engineering, Mustansiriyah University, Baghdad 10052, Iraq; ahmed_ogaili@uomustansiriyah.edu.iq (A.A.F.O.); apehsanameen@uomustansiriyah.edu.iq (E.S.A.-A.)
2. Air Conditioning Engineering Department, Faculty of Engineering, Warith Al-Anbiyaa University, Karbala 56001, Iraq; ali.basem@uowa.edu.iq
3. Applied Science Department, University of Technology, Baghdad 10066, Iraq; 100358@uotechnology.edu.iq
4. Environmental Engineering Department, Al Hikma University College, Baghdad 10052, Iraq; zta011@alumni.bham.ac.uk
5. Chemical Engineering Department, Birmingham University, Birmingham B15 2TT, UK
6. Mechanical Engineering Department, University of Technology-Iraq, Baghdad 10066, Iraq; mohsin.n.hamzah@uotechnology.edu.iq
7. Ministry of Industry and Minerals, State Company for Rubber and Tires Industries, Baghdad 10052, Iraq; emad.njim@gmail.com
8. Training and Workshops Center, University of Technology-Iraq, Baghdad 19006, Iraq; luttfi.a.alhaddad@uotechnology.edu.iq
* Correspondence: alaa.a.jaber@uotechnology.edu.iq

Abstract: The incorporation of fiber reinforcements into polymer matrices has emerged as an effective strategy to enhance the mechanical properties of composites. This study investigated the tensile and fracture behavior of 3D-printed polylactic acid (PLA) composites reinforced with chopped carbon fibers (CCFs) through experimental characterization and finite element analysis (FEA). Composite samples with varying CCF orientations (0°, 0°/90°, +45°/−45°, and 0°/+45°/−45°/90°) were fabricated via fused filament fabrication (FFF) and subjected to tensile and single-edge notched bend (SENB) tests. The experimental results revealed a significant improvement in tensile strength, elastic modulus, and fracture toughness compared to unreinforced PLA. The 0°/+45°/90° orientation exhibited a 3.6% increase in tensile strength, while the +45°/−45° orientation displayed a 29.9% enhancement in elastic modulus and a 29.9% improvement in fracture toughness (259.12 MPa) relative to neat PLA (199.34 MPa$\sqrt{m}$). An inverse correlation between tensile strength and fracture toughness was observed, attributed to mechanisms such as crack deflection, fiber bridging, and fiber pull-out facilitated by multi-directional fiber orientations. FEA simulations incorporating a transversely isotropic material model and the J-integral approach were conducted using Abaqus, accurately predicting fracture toughness trends with a maximum discrepancy of 8% compared to experimental data. Fractographic analysis elucidated the strengthening mechanisms, highlighting the potential of tailoring CCF orientation to optimize mechanical performance for structural applications.

Keywords: AM; 3D printing; PLA/CF; experimental tests; FEM

Citation: Ogaili, A.A.F.; Basem, A.; Kadhim, M.S.; Al-Sharify, Z.T.; Jaber, A.A.; Njim, E.K.; Al-Haddad, L.A.; Hamzah, M.N.; Al-Ameen, E.S. The Effect of Chopped Carbon Fibers on the Mechanical Properties and Fracture Toughness of 3D-Printed PLA Parts: An Experimental and Simulation Study. *J. Compos. Sci.* **2024**, *8*, 273. https://doi.org/10.3390/jcs8070273

Academic Editor: Yuan Chen

Received: 20 June 2024
Revised: 9 July 2024
Accepted: 12 July 2024
Published: 15 July 2024

Copyright: © 2024 by the authors. Licensee MDPI, Basel, Switzerland. This article is an open access article distributed under the terms and conditions of the Creative Commons Attribution (CC BY) license (https://creativecommons.org/licenses/by/4.0/).

1. Introduction

Additive manufacturing (AM), more commonly referred to as three-dimensional (3D) printing, is a disruptive technology that allows the manufacture of complex geometries with a level of precision and accuracy that has potential to be improved [1,2]. Additive manufacturing (AM) has revolutionized the production of complex geometries with unprecedented precision and accuracy [3,4]. One of the key advantages of 3D printing, especially fused filament fabrication (FFF), is its ability to significantly enhance mechanical properties through the incorporation of reinforcing materials [5,6]. Among the many materials applied in additive manufacturing, polylactic acid (PLA) is well liked, not only for its biodegradability and renewability, but also because it is biocompatible [7,8]. PLA is a kind of thermoplastic polyester; it is derived from renewable sources such as corn starch or sugarcane and is an environmentally friendly type of plastic [5]. However, this resin often has relatively inferior mechanical properties, such as brittleness and poor impact strength, which results in limitations to its applications in many areas where high mechanical performance is needed.

Compared to traditional manufacturing methods, which often involve subtractive processes leading to material wastage, 3D printing is highly efficient and environmentally friendly. The integration of carbon fibers into PLA (polylactic acid) matrices has been particularly noteworthy. Carbon fibers enhance the mechanical properties of PLA by providing superior tensile strength, fracture toughness, and durability. This is attributed to carbon fibers' high strength-to-weight ratio, excellent stiffness, and thermal stability, making the resulting PLA–carbon fiber composites suitable for high-performance engineering applications. These enhancements are crucial for expanding the applicability of PLA in load-bearing and structural components—applications where pure PLA would traditionally fall short due to its inherent mechanical limitations.

Among the various AM techniques, fused filament fabrication (FFF) or fused deposition modeling (FDM)—a popular technique—has the advantages of easy use, low cost, and processing of a wide range of thermoplastic materials [5,6] Additive manufacturing (AM) has now been termed the new era of manufacturing, mainly because it allows increasing complexity in the production of substances across a diverse range of industries, such as aerospace, automobile, biomedical, and customer products [7,9,10]. The utilization of FDM has been pursued in AM strategies because of its cost-effectiveness, ease of use, and flexibility [11,12]. MAKINGPLA is advantageous; however, its applicability in load-bearing and structural components has been restricted by its intrinsic mechanical deficiencies, such as low strength, stiffness, and fracture toughness [13].

Recent advancements in 3D printing have facilitated the exploration of various composite materials to enhance mechanical properties. Significant research has been carried out to enhance the mechanical performance of PLA by the addition of reinforcing materials, in particular, carbon fibers (CFs). The addition of reinforcing materials, especially carbon fibers, has been used to enhancing PLA's mechanical properties and fracture toughness for components designed for 3D printing. Numerous research studies have been conducted on the effect of both chopped and continuous carbon fibers on the properties of PLA matrices. Knowledge of the effects of chopped carbon fibers on the properties of 3D-PLA parts can be used to develop new advanced composites for application as AM products with tailor-made properties [14].

One study addressed the circular economy by converting waste polypropylene (PP) and carbon fibers into upcycled composite materials suitable for additive manufacturing. This research optimized material extrusion and 3D printing to overcome adhesion and warpage issues, examining the effects of various carbon fiber weight fractions on filament properties and printability. The results highlighted the successful fabrication of fiber-reinforced filaments despite some reduction in mechanical properties due to thermal processes during production [15]. Another study explored the micromechanics of short carbon fiber-reinforced thermoplastics (sCFRTP) produced via 3D printing. This investigation analyzed the effects of different printing parameters on tensile properties and internal structure through tensile testing, X-ray computed tomography, and theoretical analysis. The study found that fiber length, void volume, and fiber orientation significantly impacted mechanical properties, with optimal parameters enhancing Young's moduli and tensile strengths [16]. Additionally, research on 3D-printed continuous fiber-reinforced PA6 composites examined their degradation in salt water. Using a Markforged® Mark Two 3D printer, the study fabricated samples reinforced with carbon, glass, and Kevlar fibers. Mechanical tests post-ageing in saltwater revealed significant reductions in tensile and flexural strengths, with carbon fiber-reinforced samples showing the least degradation [17]. One of the most promising routes towards the enhancement of the strength, stiffness, and fracture toughness of PLA-based composites is reinforcing them with chopped carbon fibers. As a result, there is a need for further research to assess the impact of chopped carbon fiber reinforcement on both the mechanical properties and the fracture toughness of 3D-printed PLA parts [11,12]. Much research effort has already been devoted to developing strategies for improving the mechanical performance of PLA, such as the addition of different reinforcing fillers. Vălean et al. [18] conducted static and fatigue trials of 3D-printed PLA and PLA short carbon fiber-reinforced parts. They found that the addition of short carbon fibers resulted in a significant improvement in the tensile strength, fracture toughness, and fatigue life of the specimens. Iragi et al. [19] researched the ply and interlaminar behaviors of 3D-printed continuous carbon fiber-reinforced thermoplastic laminates, relating these to processing conditions and microstructure. Their study revealed the importance of process parameter optimization and showed good potential regarding the targeted mechanical properties.

Li et al. [20] studied a binding layer of a carbon fiber/PLA in FDM samples, and tensile strength and elastic modulus improvements were recorded. Another study [21] designed and cohesively tested structurally alternate-layered polymer composites using PLA material and carbon fiber-reinforced PLA. The studies showed the intended applications of the composites and the good prospects of using them as carbon fiber-reinforced materials. It has been reported that the lower melt flow rate (MFR) of PLA/CCF samples enhanced interfilament adhesion, which significantly improved the mechanical properties of the composite [22]. The strong adhesion forces between the filaments lead to enhanced mechanical performance. Additionally, the homogeneous orientation of the CCFs within the PLA/CCF matrix facilitates effective load transfer and absorption from external forces, thereby further enhancing the mechanical properties [23]. Moreover, the uniformly dispersed carbon fibers act as heterogeneous nucleation sites, reducing the crystallization temperature of PLA and increasing its degree of crystallinity. These changes in crystallization behavior resulted in improved mechanical properties of PLA/CCF composites compared to pure PLA samples [24]. In addition to short carbon fibers, continuous carbon fibers have been considered for use with PLA matrices in additive manufacturing applications. Wang et al. [25] studied the preparation of continuous carbon fiber-reinforced PLA prepreg filaments, while Li et al. [26] studied the mechanical performances of continuous carbon fiber-reinforced PLA composites printed in vacuum environments.

Naranjo-Lozada et al. [27] compared the tensile properties and failure behavior of chopped and continuous carbon fiber composites manufactured by additive manufacturing. Their study showed the better mechanical performance of the continuous fiber composites in comparison to their chopped counterparts. Yadav et al. [28] studied the flexural strength and surface profiling of carbon-based PLA parts developed by additive manufacturing. The results showed that carbon fiber reinforcement has the potential for achieving improved flexural performance of 3D-printed PLA components. However, further research will be required in order to fully optimize processing conditions, fiber content, and material compositions with the intention of fully exploiting the advantages of reinforced composites for a wide range of applications in additive manufacturing. Few researchers have published their work due to the influence of several factors on the mechanical properties of PLA in 3D printing. The present study gives an account of the conspicuous factors that have been identified for altering the mechanical properties of PLA in 3D printing. Similarly, Kumar Patro et al. [29] worked on the mechanical properties of 3D-printed sandwich structures developed using PLA and acrylonitrile butadiene styrene (ABS). The conclusion derived from this study was that the infill pattern, layer thickness, and raster angle had a significant effect on the mechanical properties of the sandwich structures developed using this infill. The samples developed with a hexagonal infill pattern and a $0°$ raster angle had the highest tensile strengths and tensile moduli. Zwawi [30] evaluated the integrity of a structure and PLA 3D-printed eye grab hooks' fracture prediction with different cross-sections, and it was concluded that the mechanical behavior of the hooks was significantly influenced by the cross-sectional shape and size. More specifically, hooks with a rectangular cross-section showed better strength together with fracture toughness compared to those featuring a circular cross-section. Wu et al. [31] studied the influence of layer thickness, raster angle, deformation temperature, and recovery temperature of shape memory on 3D-printed PLA samples. The study found that it was mainly the angle of the raster and the deformation temperature that influenced the shape memory effect. It was shown that samples achieved the most significant shape memory effect at a deformation temperature of 60 °C with a change in raster angle of 0°. Kartikeyan et al. [5] found that the mechanical properties of specimens fabricated by FDM 3D printing were affected by the relation between layer thickness, raster angle, and build orientation. The tensile strength and modulus results indicated that the best performance was achieved by a specimen with a 0° raster angle and a build orientation in parallel with the loading direction. Sajjadi et al. [32] also presented a study on 3D-printed high-impact polystyrene fracture properties with and without raster angles based on essential work of the fracture method.

Each research work showed that the raster angle exerts a remarkable influence on the fracture toughness of HIPS specimens, whereas specimens with a 0° raster angle exhibit higher fracture toughness. Ayatollahi et al. [33] studied the in-plane raster angle effect on the tensile and fracture strengths of 3D-printed PLA specimens. Therefore, the raster angle dominated the mechanical behavior of the specimens. It has been indicated that the tensile and fracture strengths are higher for specimens with a 0° raster angle. Another experimental investigation [34] evaluated a 3D-printed ABS specimen under tension–tear loading and the influence of raster orientation on fracture behavior. The results showed that raster orientation noticeably contributed to the fracture toughness and flexibility of the specimen. Therefore, the specimen with a raster angle of 90° showed larger values of fracture toughness and ductility. Gongabadi et al. [35] studied the influence of raster angle, build orientation, and infill density on the elasticity of 3D-printed parts. The specimens for the experiments were modelled using finite elements with microstructural modeling and homogenization techniques.

The raster angle, the build orientation, and the infill density were found to be the principal factors that affected the elastic response of specimens. The results obtained indicated that the specimens with a 0° raster angle, parallel orientation to the loading direction, and high infill density presented the highest elastic modulus values [36]. Studies examined the effects of printing parameters on the mechanical properties of 3D-printed polymer composites/structures. Khan et al. [26] found for FFF-printed PLA-PETG-ABS composites that infill density and raster angle significantly impacted the tensile strength/modulus, with 100% infill and a 0° raster angle being optimal. Karimi et al. [36] analyzed the effect of the layer angle as well as the ambient temperature on the mechanical and fracture properties of unidirectional 3D-printed PLA materials. The study concluded that both the layer angle and the ambient temperature had major effects on the mechanical and fracture behavior of the specimens. Samples made with a 0° layer angle and tested at room temperature showed the highest tensile strengths and fracture toughness. These studies indicate that the optimal values of these factors depend on the specific application of interest and the desired properties.

The novelty of this work lies in its comprehensive approach to evaluating the mechanical properties and fracture toughness of 3D-printed PLA composites reinforced with chopped carbon fibers. Unlike previous studies that primarily focused on continuous carbon fibers, this research investigates the effects of chopped fibers, which offer distinct advantages in terms of processing and material behavior. This study aims to fill the gap in the literature by providing detailed insights into the performance enhancements achievable through the use of chopped carbon fibers in PLA matrices, thereby advancing the potential applications of 3D-printed composites in engineering fields.

The motivation for studying PLA composites with chopped carbon fibers stems from the need to enhance the mechanical properties of PLA for broader engineering applications.

PLA's biodegradability and ease of processing make it an attractive material; however, its inherent mechanical limitations restrict its use in high-performance applications. By incorporating chopped carbon fibers, we aim to leverage their high strength-to-weight ratio and excellent stiffness to create a composite material that combines the environmental benefits of PLA with the superior mechanical properties of carbon fibers. This study not only provides a sustainable solution but also extends the application range of 3D-printed PLA composites.

This study aimed to investigate the effect of chopped carbon fibers on the mechanical properties and fracture toughness of PLA 3D-printed parts. Specifically, the objectives were as follows:

1. To fabricate PLA-CF composite samples with varying CCF orientations ($0°$, $0°/90°$, $+45°/-45°$, and $0°/+45°/-45°/90°$) using the FFF technique.
2. To experimentally evaluate the tensile properties and fracture toughness (via the single-edge notched bend (SENB) method) of the composite samples.
3. To look at how in-plane raster orientation affects the tensile and fracture strengths of PLA parts made with the FDM method. (The paper starts with a theoretical background of the stress field around the crack tip.)
4. To develop and validate a finite element analysis (FEA) model using Abaqus to predict the fracture toughness of the composite samples.
5. To analyze the effects of CF orientation on the mechanical properties and fracture toughness of the PLA-CF composites.
6. To provide insights into the strengthening and toughening mechanisms of PLA-CF composites and their potential applications.

This study aims to contribute to the understanding of the mechanical behavior of PLA-CF composites fabricated via FFF, particularly the effects of CF orientation on fracture toughness. By employing both experimental and simulation approaches, this research seeks to provide a comprehensive analysis of the mechanical performance of these composites and their potential for structural applications.

The rest of this paper is organized as follows. Section 2 defines the materials and experimental methods used. These include the PLA and carbon fiber materials, details of the composite fabrication procedure through the FFF process, the mechanical testing methodologies (tensile and fracture toughness), and the FEA modeling approach. In Section 3, the obtained results, such as the tensile properties, the values of fracture toughness, the fractographic analysis, and the mechanisms of strengthening and toughening, along with a comparison with earlier studies, are discussed. Section 4 summarizes the major conclusions and practical implications, as well as future research directions.

2. Theoretical Background

The theoretical framework underpinning fracture mechanics analysis is essential for interpreting the experimental findings and comprehending the fracture behavior of the composite samples. In the realm of linear elastic fracture mechanics (LEFM), the stress field near the crack tip in a linear elastic material subjected to mode I loading can be defined by the following equation [37]:

$$\sigma_{rr} = \frac{K_I}{\sqrt{2\pi r}} \left[\frac{5}{4}\cos\left(\frac{\theta}{2}\right) - \frac{1}{4}\cos\left(\frac{3\theta}{2}\right) \right] + \ldots$$

$$\sigma_{\theta\theta} = \frac{K_I}{\sqrt{2\pi r}} \left[\frac{3}{4}\cos\left(\frac{\theta}{2}\right) + \frac{1}{4}\cos\left(\frac{3\theta}{2}\right) \right] + \ldots \quad (1)$$

$$\sigma_{r\theta} = \frac{K_I}{\sqrt{2\pi r}} \left[\frac{1}{4}\sin\left(\frac{\theta}{2}\right) + \frac{1}{4}\sin\left(\frac{3\theta}{2}\right) \right] + \ldots$$

where σ_{ij} represents the stress tensor components, r and θ are the polar coordinates centered at the crack tip, K_I is the mode I stress intensity factor that quantifies the magnitude of stress around the crack tip, and $f_{ij}(\theta)$ is a dimensionless function of the angle θ. It is important to note that while Equation (1) is accurate for isotropic materials, it can still serve as a good approximation for linear elastic stresses near the crack tip for materials exhibiting slight anisotropic behavior [38]. According to the principles of LEFM, mode I fracture occurs when the stress intensity factor, K_I, reaches its critical value, K_{IC}, which is defined as the material's fracture toughness [37]. However, when the amount of plastic deformation around the crack tip is not negligible, the assumption of linear elasticity breaks down and an alternative characterizing parameter, J, is employed. The J-integral, proposed by Rice [39], represents the rate of change in the net potential energy for non-linear elastic solids and serves as a parameter quantifying the singularity strength at the crack tip in elastic–plastic fracture mechanics.

The J-integral is defined as follows:

$$J = \int \Gamma \left(W dy - T_i \, \partial u_i / \partial x \, \partial s \right) \qquad (2)$$

where W is the strain energy density, T_i is the traction vector, u_i is the displacement vector, and Γ is a counterclockwise integration path surrounding the crack tip [37,38]. The J-integral provides a measure of the energy release rate associated with crack growth and is particularly useful for materials exhibiting significant plastic deformation at the crack tip. In the context of this study, both the stress intensity factor (KI) and the J-integral will be employed to characterize the fracture behavior of the PLA-CF composite samples, depending on the level of plastic deformation observed. The theoretical background provided by LEFM and elastic–plastic fracture mechanics will guide the interpretation of experimental data and facilitate a comprehensive understanding of the fracture toughness of the composites under investigation.

3. Materials and Methods

3.1. Materials

In this work, the following components were utilized: PLA granules and short carbon fibers (CCFs) of polyacrylonitrile (PAN), which were all acquired commercially from Dongguan ANT Plastic Technology Co., Ltd., Dongguan, China. The CCFs had diameters in the range of 40–60 µm and lengths in the range of 15–20 µm, as recommended for better reinforcement of the polymer composites [40]. Before mixing the polymer components, the carbon fibers' surfaces were treated to improve their interfacial bonding with the PLA matrix. The CCFs were placed in a solution of 20 wt% H_2SO_4 so that oxygen-attached functional groups could develop on the surface, enhancing wettability and achieving greater surface rugosity for better interfacial bonding [41]. In addition, to improve the performance properties of PLA and the PLA/CCF filaments, the CCFs were coated with a KH570 saline coupling agent (γ-aminopropyltriethoxysilane). Silane coupling agents are widely used to enhance interfacial adhesion between reinforcing fibers and polymeric matrices. The processing was performed using a TY-7004 single-screw extrusion machine obtained from Dongguan Tienyu Machinery Co., Ltd., China. The extrusion was performed at a temperature of 180 °C and a screw speed of 200 revolutions per minute. These parameters were set based on general recommendations for PLA processing [42]. The quantity of carbon fiber in the PLA/CCF filaments was 15 wt%—a common value in chopped fiber-reinforced PLA composites. The filaments were extruded to a diameter of 1.75 mm—a standard size for FDM printers [24]. To ensure precise control of the filament diameter, a diameter measurement system was incorporated into the extruder. This system continuously monitored the filament during production, automatically adjusting the extrusion parameters to maintain a consistent diameter of 1.75 mm. Real-time feedback and adjustments were made to correct any deviations, ensuring uniform filament dimensions essential for high-quality 3D printing. Figure 1 schematically describes the reinforcement mechanism of the PLA composite samples with the incorporation of CCF.

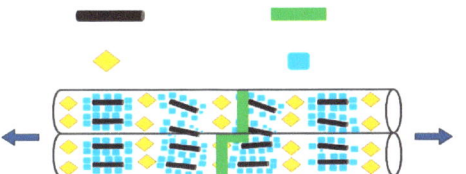

Figure 1. Reinforcement mechanism of carbon fibers in PLA.

3.2. Composite Fabrication

The composite samples were fabricated using the fused deposition modeling (FDM) technique, also referred to as fused filament fabrication (FFF), with a Creality Ender 3 Pro 3D printer (Creality, Shenzhen, China). Initially, a 3D model was created in SolidWorks software2017, which was then converted into an STL format file. This STL file was processed by slicing software specific to (AM) technology, which divided the model into layers and generated a file containing the necessary information for each layer, including toolpath coordinates and extrusion rates. The slicing data were subsequently translated into machine instructions using G-code.

During the manufacturing process, the filament was heated to a printing temperature of 200 °C and extruded through a nozzle with a diameter of 0.4 mm to print each layer of the object. After each layer was printed, the nozzle moved vertically according to a predetermined layer height to begin printing the next layer. The slicing software defined the filament trajectory to fill the product geometry and create a shell with a specific raster angle pattern, which was achieved by alternating the orientation of successive layers. To address the challenge of printing with a high fiber content using a 0.4 mm nozzle, a heated chamber was utilized, and the filament was pre-heated to a molten state before extrusion in the production process, as shown in Figure 2. The platform temperature was maintained at 50 °C to mitigate residual stress, facilitating smoother extrusion and preventing nozzle clogging, thus enabling the successful fabrication of PLA/CCF samples with enhanced mechanical properties.

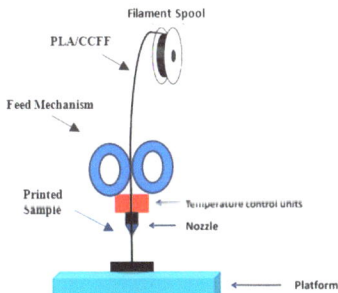

Figure 2. The production process for the samples.

The following sample groups were used to investigate the influence of fiber orientation, the composite samples being fabricated with four different fiber orientations: $0°$, $0°/90°$, $+45°/-45°$, and $0°/+45°/-45°/90°$. These orientations were achieved by controlling the filament deposition pathways during the printing process via the slicing software. The $0°$ orientation indicates that all layers were printed with the fibers aligned along the primary loading direction, while the $0°/90°$ orientation alternated between $0°$ and $90°$ fiber orientations in successive layers. The $+45°/-45°$ orientation alternated between $+45°$ and $-45°$ fiber orientations, and the $0°/+45°/-45°/90°$ orientation combined all three orientations in a specific sequence [34]. For each orientation group, a sufficient number of samples were printed to ensure statistical significance in the subsequent mechanical testing.

The critical printing parameters employed for fabricating the PLA-CCF composite samples via the FDM process are summarized in Table 1. These parameters, including printing speed, nozzle diameter, layer thickness, raster angle, and print temperature, were carefully selected based on prior optimization studies to ensure consistent and high-quality printing. Table 2 outlines the different composite sample groups produced, each characterized by a distinct fiber orientation configuration ($0°$, $0°/90°$, $+45°/-45°$, and $0°/+45°/-45°/90°$). The dimensions of the printed specimens, tailored for subsequent mechanical testing and fracture toughness evaluation, are also specified in Table 2.

Table 1. Technical parameters for manufacturing polylactic acid carbon fiber-reinforced samples.

Parameters	Value
Printing speed (mm/s)	60
Nozzle diameter (mm)	0.4
Layer thickness (mm)	0.2
Supplied speed (mm/s)	60
Print temperature (°C)	200
Infill ratio	100%

Table 2. Composite sample groups with different fiber orientations.

Sample Group	Raster Orientation	Description
A_1	$0°$	All layers are printed with fibers aligned along the primary loading direction
A_2	$0°/90°$	Alternating layers with fibers oriented at $0°$ and $90°$
A_3	$+45°/-45°$	Alternating layers with fibers oriented at $+45°$ and $-45°$
A_4	$0°/+45°/-45°/90°$	Layers with fibers oriented at $0°$, $+45°$, and $90°$ in a specific sequence

The comprehensive experimental workflow, encompassing sample fabrication and the various testing procedures, is illustrated in Figure 3. This diagram provides a step-by-step overview of the entire process, starting from the preparation of the 3D printer to the loading of the PLA/CCF filament. The 3D printing process, including the manufacturing of samples with different orientations of the fibers, was followed by sample preparation steps, including cutting samples into the required shape and size and making notches on the samples for fracture toughness tests. Other stages included performing tensile tests according to ASTM standards, from which load–displacement data could be extracted, and the evaluation of fracture toughness using the SENB method. The final steps included analysis of the data, where the acquired mechanical properties were interpreted, the fracture toughness values were calculated, and the effect of fiber orientation on the composite's performance was investigated.

3.2.1. Tensile Testing

Uniaxial tensile tests were conducted using an Instron 5980 system, operating at a crosshead speed of 1 mm/min in accordance with ASTM D638 procedures [43–45]. Dogbone specimens, prepared following the ASTM D638 Type IV standard, were subjected to tension testing. The average maximum tensile strength, elastic modulus, and elongation at break were calculated based on results obtained from three specimens per group [46]. This approach ensured consistency with established testing methodologies, thereby facilitating reliable comparison and reproducibility of the results. The use of the ASTM D638 standard, despite its primary application to unreinforced plastics, is justified by its widespread recognition and detailed procedural guidelines, which are applicable to composite materials as well.

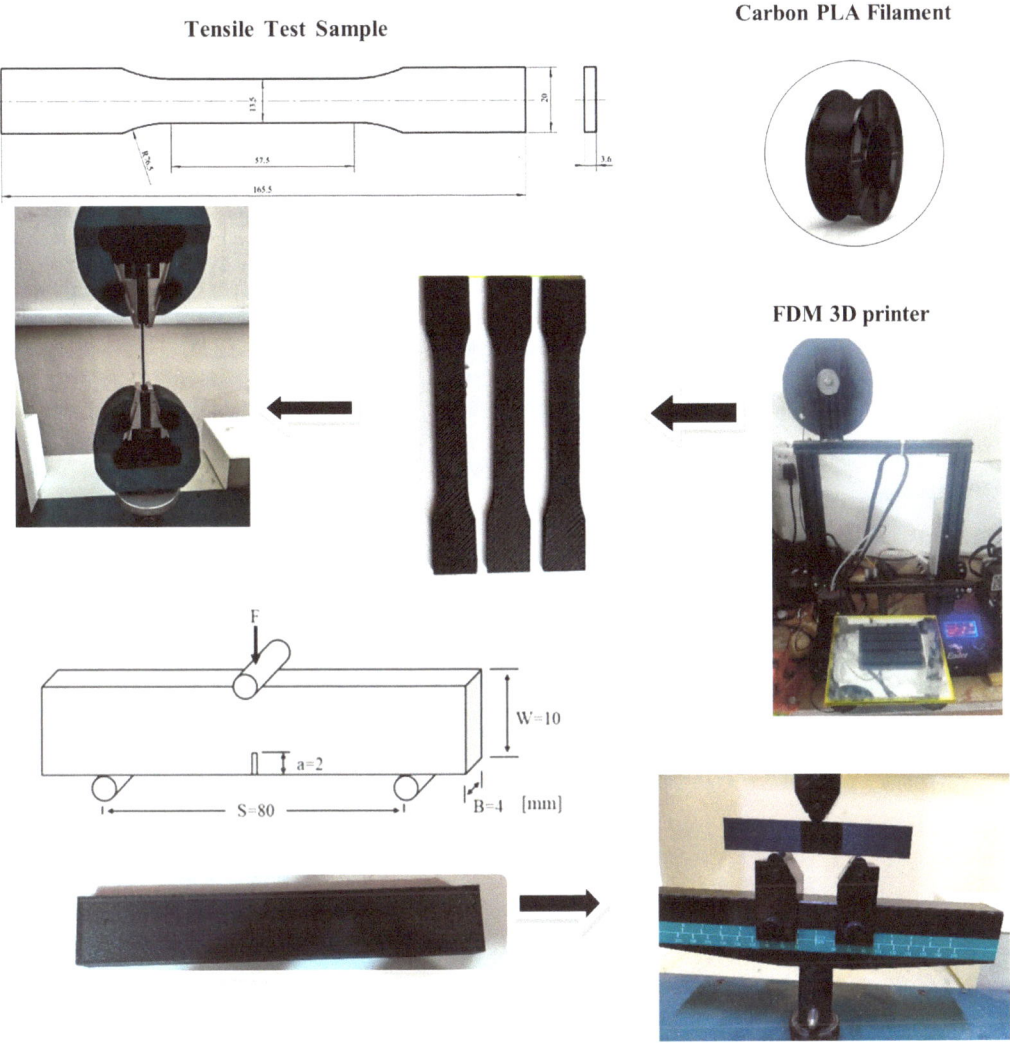

Figure 3. Sample fabrication and testing process for the experiment.

3.2.2. Fracture Testing for Mode I

The fracture toughness of the SENB samples was evaluated under three-point bending on an Instron 5980 system. The samples were loaded at a rate of 0.5 mm/min until failure [47,48]. The K_{IC} critical stress intensity factor was determined on the basis of the maximum load measured for each raster pattern. The K_{IC} could be obtained using the equation below.

$$K_{IC} = \frac{P_{max}}{BW^{1/2}} f\left(\frac{a}{W}\right)$$

where P_{max} is the maximum load recorded during the test (N), B is the thickness of the specimen (mm), W is the width of the specimen (mm), and a is the initial crack length (mm). The function $f\left(\frac{a}{W}\right)$ is a dimensionless geometry factor that depends on the ratio of the crack length to the specimen width and can be determined from standard tables or empirical equations.

4. Finite Element Analysis

The finite element analysis (FEA) was performed using the commercial software Abaqus (version 2017; Dassault Systèmes, Providence, RI, USA) to predict the fracture toughness values of the PLA-CCF composite samples. A three-dimensional (3D) model of the single-edge notched bend (SENB) specimen geometry was created based on the experimental dimensions as shown in Figure 4 and material defined from specified in Table 2. The SENB specimen was discretized using a structured mesh of hexahedral elements with a higher mesh density in the crack-tip region to capture the stress singularity accurately. Convergence studies were performed to ensure that the mesh size did not significantly influence the fracture toughness predictions. The PLA-CCF composite was modelled as a transversely isotropic material, with the material properties defined using the elastic constants and strengths obtained from experimental characterization and data reported in the literature [43,44]. The carbon fibers were assumed to be aligned in the specified orientations ($0°$, $0°/90°$, $+45°/-45°$, and $0°/+45°/-45°/90°$) within the PLA matrix.

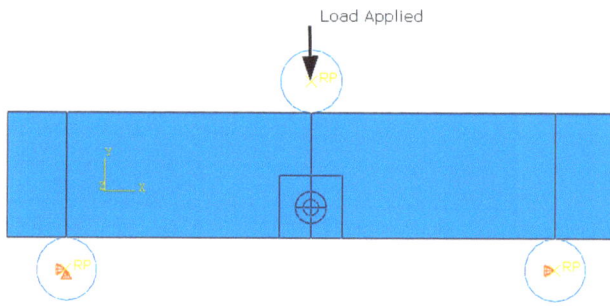

Figure 4. The applied load and the boundary condition.

The boundary conditions were applied to replicate the three-point bend configuration used in the experimental SENB tests. The bottom surface of the specimen was constrained in the vertical direction, while the loading was applied through a prescribed displacement on the top surface at the midspan location.

The crack was modelled as a seam crack with an initial crack length equal to the notch length introduced during sample preparation. The crack tip was defined as a focused circular region with a radius of 2 mm, consistent with the experimental observations.

A mesh sensitivity study was conducted to quantify the degree of discretization required for the convergence of the finite element solutions [49,50]. This was achieved by systematically varying the number of elements in the SENB model and evaluating the response. Models were generated with total element counts ranging from 2000 to 16,000 two-dimensional 8-noded quadrilateral elements (CPS8) as shown in Figure 5. All other parameters, including geometry, material properties, loads, and boundary conditions, were kept constant.

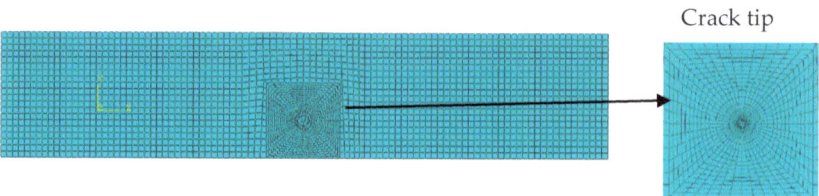

Figure 5. The generated mesh.

Prediction of Fracture Toughness Values

The fracture toughness predictions were based on the J-integral approach, which is suitable for materials exhibiting significant plastic deformation at the crack tip. The J-integral was evaluated along a contour integral path surrounding the crack tip, and the critical value of the J-integral at fracture initiation was determined. The critical J-integral value (Jc) was then used to calculate the fracture toughness (K_{IC}) using the following equation [37]:

$$K_{IC} = \sqrt{E * Jc} \tag{3}$$

where E is the effective elastic modulus of the PLA-CCF composite, which accounts for the anisotropic behavior of the material.

The FEA simulations were performed for each fiber orientation group, and the predicted fracture toughness values were compared with the experimental results. The influence of fiber orientation on fracture toughness was analyzed, and the mechanisms governing the fracture behavior of the PLA-CCF composites were investigated. Additionally, the stress and strain distributions within the composite samples were examined to gain insights into the failure mechanisms and the role of fiber reinforcement in enhancing fracture toughness.

5. Results and Discussion

5.1. Tensile Properties

The tensile stress–strain curves obtained from the uniaxial tensile tests on the PLA-CCF composite samples are presented in Figure 6. The curves exhibit a typical elastoplastic behavior, with an initial linear elastic region followed by a non-linear plastic deformation region until failure. The results indicate a significant improvement in tensile properties with the incorporation of chopped carbon fibers (CCFs) into the PLA matrix, corroborating findings from previous studies [9,10].

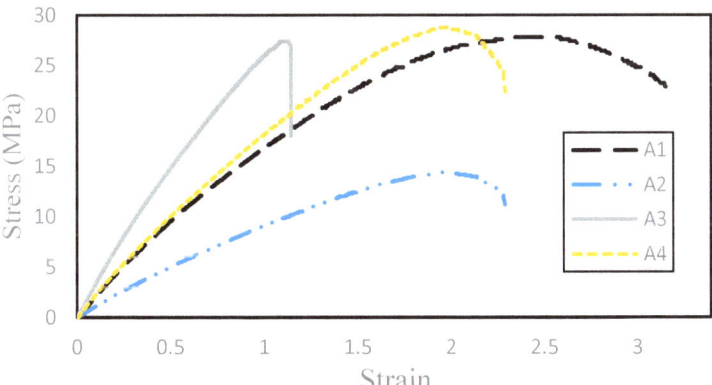

Figure 6. The stress–strain curves for the samples.

The tensile strength and elastic modulus values of the PLA-CCF composites for different fiber orientations are presented in Figure 6. As evident observed from the Figure 7, the incorporation of chopped carbon fibers (CCFs) into the PLA matrix resulted in significant improvements in tensile properties compared to unreinforced PLA. The 0° fiber orientation (sample A1) exhibited a tensile strength of 27.8 MPa, benefiting from the fibers being aligned along the loading direction, which effectively transferred the applied loads. Similarly, sample A3, another 0°-oriented sample, showed a high tensile strength of 27.4 MPa.

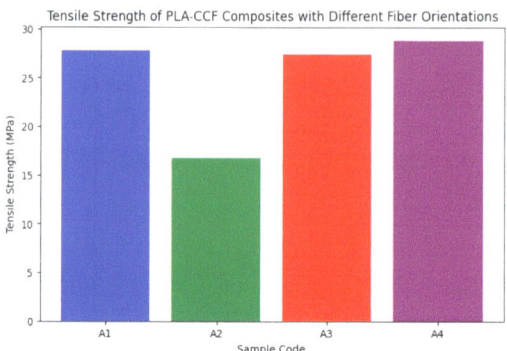

Figure 7. Tensile strength of PLA-CCF composites with different fiber orientations.

In contrast, the 0°/90° orientation (sample A2) displayed a lower tensile strength of 16.786 MPa, suggesting that while fibers oriented perpendicularly to the loading direction can still contribute to load bearing, their efficiency is reduced compared to unidirectional alignment. Sample A4, with a tensile strength of 28.8 MPa, outperformed the other orientations, indicating that the specific distribution and arrangement of fibers in this sample facilitated superior mechanical performance.

The variation in tensile strengths among the different fiber orientations highlights the anisotropic nature of the composite material. The fibers' ability to transfer and absorb applied loads effectively depends significantly on their orientation within the matrix. This anisotropic behavior is crucial in designing composite materials for specific engineering applications where directional strength and stiffness are paramount.

In summary, the study underscores the importance of fiber orientation in enhancing the mechanical properties of PLA-CCF composites. Aligning fibers along the primary load direction maximizes tensile strength and modulus, making these composites more suitable for high-performance applications requiring specific mechanical characteristics.

5.2. Fracture Toughness Values from SENB Testing

The fracture toughness values obtained from the single-edge notched bend (SENB) tests are presented in Figure 8, and Table 3 shows the load–displacement curves of the samples. The results demonstrate that the incorporation of chopped carbon fibers (CCFs) significantly enhanced the fracture toughness of the PLA composites compared to unreinforced PLA.

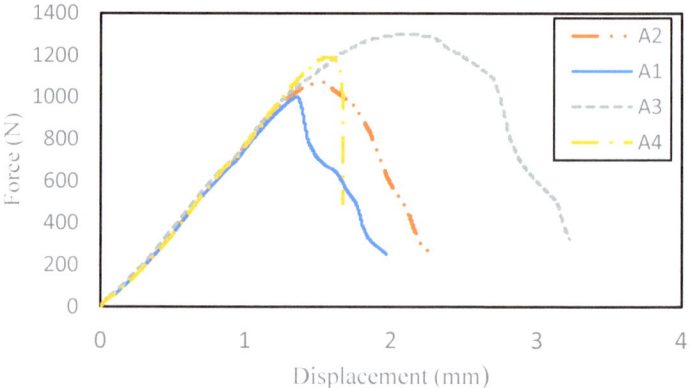

Figure 8. The SENB load–displacement results for the samples.

Table 3. Fracture loads and fracture toughness (K_{IC}) values for different SENB specimens.

Sample	Max. Fracture Loads (N)	K_{IC} (MPa $\sqrt{m}$)
A1	1000	199.34
A2	1072	213.28
A3	1300	259.12
A4	1190	237.21

The load–displacement curves (Figure 8) and the corresponding fracture toughness values (Table 3) reveal that the $0°/+45°/-45°/90°$ fiber orientation exhibited the highest fracture toughness, followed by the $+45°/-45°$ orientation. This trend can be attributed to the combined effects of fiber bridging, crack deflection, and fiber pull-out mechanisms, which enhance energy dissipation and crack growth resistance.

The significant improvement in fracture toughness with the incorporation of CCF underscores the effectiveness of fiber reinforcement in enhancing the mechanical properties of PLA composites can be observed that in Figure 9.

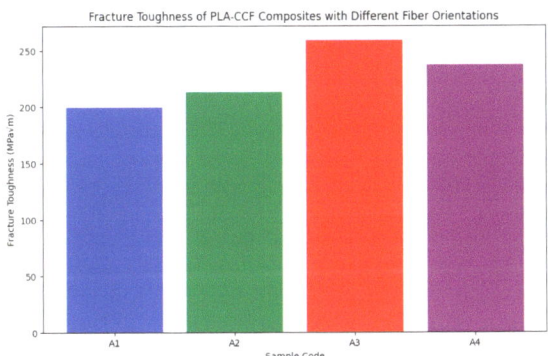

Figure 9. Fracture toughness of different samples.

The chopped carbon fibers provide a high strength-to-weight ratio, excellent stiffness, and thermal stability, which contribute to their ability to improve fracture toughness. These properties allow the fibers to effectively transfer and absorb applied loads, preventing crack propagation and increasing the material's resistance to fracture [51].

The orientation of the fibers played a crucial role; among the different fiber orientations, the samples with a $0°/+45°/-45°/90°$ orientation exhibited the highest fracture toughness values, followed by the $+45°/-45°$ orientation. This behavior can be attributed to the combined effects of fiber bridging, crack deflection, and fiber pull-out mechanisms that contribute to enhanced energy dissipation and crack growth resistance. Similar observations were made by Ayatollahi et al. [34], who reported that the fracture strengths of 3D-printed PLA specimens were influenced by the raster angle (fiber orientation), with the highest fracture strengths being observed for specimens with a $0°/90°$ raster angle. The fracture loads obtained from the SENB tests, along with the corresponding fracture toughness values, are shown in Figure 6. The fracture loads follow a similar trend to the fracture toughness values, with the highest loads observed for the $0°/+45°/90°$ and $+45°/-45°$ orientations.

5.3. Analysis of Effects of Fiber Orientation on Properties

The variations in tensile performance and fracture toughness can be explained by the different fiber orientations and their effect on load transfer and failure mechanisms within the composite. In the $0°$ orientation, the fibers are aligned along the loading direction to effectively transfer the applied load, increasing the composite's strength and rigidity. However, this leads to a limitation in ductility and fracture toughness due to the lack of fiber bridging and crack deflection mechanisms. In contrast, the +45°/−45° and 0°/+45°/90° orientations distribute the fibers in multiple directions, promoting more effective crack deflection, fiber bridging, and fiber pull-out mechanisms. These mechanisms contribute to increased energy dissipation and crack growth resistance, leading to improved fracture toughness. According to the results reported in Table 3 the best mechanical properties, including tensile strength, elongation at break, and fracture toughness, were obtained by the specimens oriented in the +45°/−45° printing direction.

Similar results were reported by Oviedo et al. [52] and Jap et al. [53] for the tensile behavior and fatigue performance of parts produced by the FDM technique with PLA. Oviedo et al. showed that the strain to failure of parts additively manufactured from PLA with a +45°/−45° printing direction was about 50% higher than those additively manufactured with a 0°/90° printing direction. On the other hand, dealing with fatigue performance, also in the study [53] reported that, for a constant stress amplitude, the number of cycles to failure for samples with a +45°/−45° raster orientation was more than twice that of the 0°/90° raster orientation. The fact that the tensile properties and the fracture toughness values were found to be anisotropic shows the need to include fiber orientation in such analyses and, in order to assure the best performance of PLA-CCF composites in different applications, an optimization of the single directivity orientation.

5.4. Microscopic Examination of Fracture Surfaces

Scanning electron microscopy (SEM) analysis was conducted to investigate the fracture surfaces of the PLA-CCF composite samples and gain insights into the failure mechanisms. Figure 10 shows an SEM micrograph of the chopped carbon fibers dispersed within the PLA matrix. The image was captured at a magnification of 891×, with an accelerating voltage of 20.00 kV and a working distance of 2.768 mm. The scale bar indicates a length of 50 micrometers (μm). Several carbon fibers with varying diameters can be observed, indicated as D1 (21.6 μm), D2 (9.36 μm), D3 (8.15 μm), and D4 (18.09 μm). The distribution and orientation of the chopped carbon fibers within the PLA matrix play a crucial role in determining the composite's mechanical properties, including tensile strength, modulus, and toughness. In this micrograph, the fibers appear randomly oriented and non-uniformly distributed, which is typical for short-fiber-reinforced composites. The dispersion and orientation of the fibers can significantly influence the composite's performance, as aligned fibers generally provide better mechanical properties along the alignment direction. In comparison, random dispersion can lead to more isotropic properties.

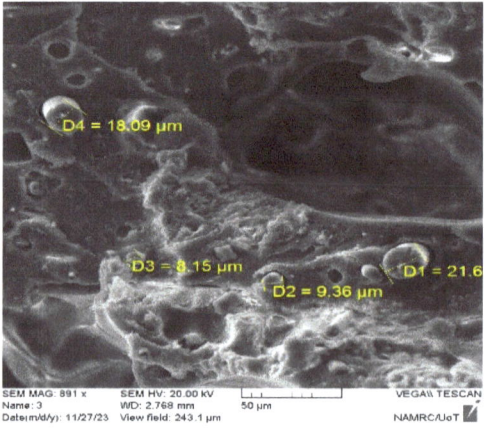

Figure 10. SEM of PLA reinforced by CCF.

The distribution of the chopped carbon fibers in the PLA matrix is a critical factor in determining the composite material's mechanical properties, such as tensile strength, modulus, and toughness. In this image, the fibers appear to be randomly oriented and not uniformly distributed, which is typical for short-fiber-reinforced composites. The dispersion and orientation of the fibers can significantly affect the composite's performance, as aligned fibers generally provide better mechanical properties in the direction of alignment, while random dispersion can lead to more isotropic properties.

Figure 11 presents an SEM micrograph of the fracture surface of a tensile specimen from Group 1, which has a 0° fiber orientation. The image reveals slight air gaps between the filaments—an inherent feature of 3D-printed specimens. Notably, the bonding lines between the filaments do not contribute significantly to energy dissipation during fracture, as evidenced by the presence of only minor damage zones in these regions.

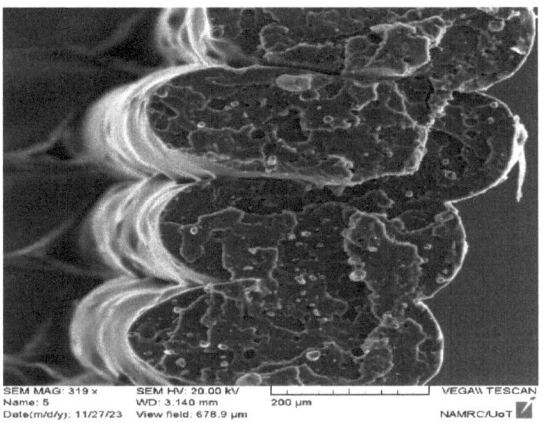

Figure 11. Fracture surface of samples.

The SEM analysis provided valuable information on the fracture mechanisms of the 3D-printed materials and their effect on mechanical properties and performance. Hence, the importance of considering fiber orientation and other intrinsic features of 3D-printed materials in designing and analyzing components for a specific application is brought into perspective. The micrograph of the fracture surface suggests that the adhesion between filaments is less perfect, which will influence the mechanical properties of the material. These small damage zones in the bonding lines suggest that the strength of the bonding is very weak in relation to dissipating energy during fracture; hence, crack propagation may result along these lines. The fiber orientation can affect the fracture mechanism of the specimen. A +45/−45°-oriented specimen will have different fracture behavior to a 0/90-oriented specimen. The differences between the SEM micrographs accentuate how the fiber orientation influences the mechanical properties and performance of a 3D-printed material. These results are in good agreement with those obtained by [54,55], in which it was found that the bonding in 3D-printed PLA samples did not make a high contribution to the fracture energy. In general, SEM imagery has offered invaluable data in comprehending the fracture mechanisms of 3D-printed materials. Their influence on mechanical properties and performance as well as fiber orientation and other intrinsic features of 3D-printed materials have to be considered in designing and analyzing components for specific applications.

5.5. Finite Element Analysis Results

FEA was performed for each fiber orientation group, and the predicted fracture toughness values were compared with the experimental results. The association between the FEA results and the experimental data was mostly satisfactory, and the model represented the trends and fracture toughness for the various fiber orientations [56,57]. The data from the FEA, together with the outcomes of the experiment, gave an insight into the toughening and strengthening mechanisms of the PLA-CCF composites.

Stress Distribution and Crack-Tip Fields

The FEA model enabled the visualization and analysis of the stress distribution and crack-tip fields within the SENB specimens. Figure 12 shows a contour plot of the von Mises stress distribution for sample A1, representing the 0° fiber orientation. For the A1 sample, the highest stress was at the crack tip, as expected. The distribution of stress ahead of the crack tip showed a signature pattern because of the crack-tip singularity, and the maximum occurred at the crack and then decayed radially outwards. Crack-tip stress fields were determined using the J-integral approach, which is adequate for materials undergoing large plastic deformation at the tip of the crack.

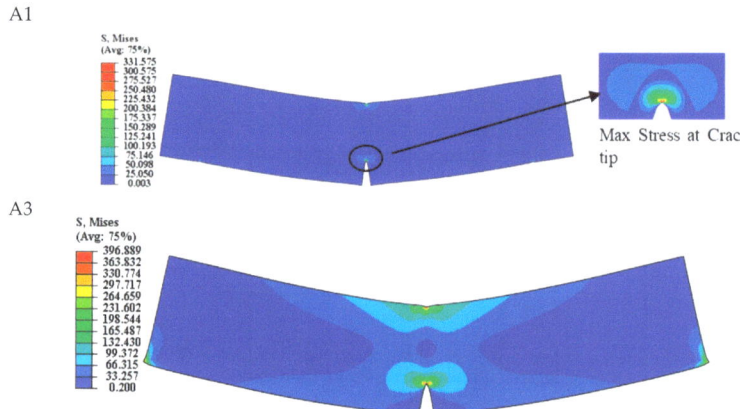

Figure 12. Von Mises stress distribution and crack-tip fields obtained from FEA simulations for the A1 and A3 samples.

Using a contour integral path surrounding the crack tip, the J-integral was evaluated, and the critical value of the J-integral at fracture initiation (Jc) was determined. The critical value of the J-integral (Jc) was used to calculate the value of the fracture toughness (K_{IC}) through the following expression: $K_{IC} = \sqrt{E * J_C}$, where E is the effective elastic modulus of the material undergoing fracture and takes into account the anisotropic behavior of the material. It is obvious that the FEA simulations proved that ghost elements exist, as did the prevention of cracking and the closely related stress distributions and crack-tip fields responsible for the fracture behavior of the PLA-CCF composites to a great extent. The predictions of the FEA model were in very good agreement with the experimental data, accurately capturing the trends and relative differences in fracture toughness associated with the several fiber orientations. The J-integral approach, combined with the transversely isotropic material model and adequate boundary conditions, allowed reliable fracture behavior predictions of the PLA-CCF composites. In general, the results of this investigation reveal the high potential for PLA-CCF composites as really high-performance materials with very good feasibility for obtaining tailored mechanical properties, especially fracture toughness, by the proper selection of fiber orientations. These findings enrich the research on the development of advanced composite materials intended for several structural and functional applications.

6. Conclusions

The present study documents a systematic investigation into the influence of chopped carbon fiber (CCF) reinforcement and fiber orientation on the tensile strength and fracture toughness of 3D-printed PLA. The key achievements of this study include the following:

- Enhanced tensile strength:
  - The incorporation of CCF reinforcement resulted in a significant enhancement in tensile strength compared to unreinforced PLA (27.8 MPa).
  - The 0°/+45°/90° fiber orientation (sample A4) exhibited the highest tensile strength of 28.8 MPa, representing a 3.6% improvement over the unreinforced baseline.
  - Conversely, the 0°/90° orientation (sample A2) displayed a 39.6% decrease in tensile strength (16.786 MPa), highlighting the profound impact of fiber orientation on tensile properties.
- Improved fracture toughness:
  - The fracture toughness data revealed a trend where multi-directional fiber orientations outperformed unidirectional orientations.
  - The +45°/−45° orientation (sample A3) yielded the maximum fracture toughness of 259.12 MPa$\sqrt{m}$—a remarkable 29.9% increase compared to unreinforced PLA (199.34 MPa$\sqrt{m}$).
  - The 0°/+45°/90° orientation (sample A4) exhibited an 18.9% improvement in fracture toughness (237.21 MPa$\sqrt{m}$), while the 0°/90° orientation (sample A2) showed a modest 7.0% increase (213.28 MPa$\sqrt{m}$).
- Critical role of fiber orientation:
  - The study underscores the critical role played by fiber orientation in tailoring the mechanical properties of CCF-reinforced PLA composites.
  - Unidirectional fiber orientations, such as 0°, enhance tensile strength through efficient load transfer.
  - Multi-directional orientations facilitate improved fracture toughness by promoting mechanisms like crack deflection, fiber bridging, and fiber pull-out.
  - The observed inverse correlation between tensile strength and fracture toughness highlights the importance of judiciously selecting the fiber orientation to meet specific mechanical performance requirements.
- Optimizing composite architectures:
  - Composite designers can leverage the trade-off between tensile strength and fracture toughness to optimize the desired combination of strength, stiffness, and fracture resistance by strategically tailoring composite architectures.
- Future research directions:
  - Future research endeavors should focus on further exploring the underlying mechanisms governing the observed trends, potentially through in-depth microstructural and fractographic analyses.
  Investigating the effect of varying fiber content and aspect ratios could provide valuable insights for optimizing the mechanical performance of CCF-reinforced PLA composites.

This study contributes to the rapidly evolving field of additive manufacturing by demonstrating the potential of CCF reinforcement and strategic fiber orientation in enhancing the mechanical properties of 3D printed PLA composites. The findings pave the way for the development of high-performance, sustainable, and lightweight composite materials for a wide range of structural and functional applications, aligning with the principles of eco-friendly and resource-efficient manufacturing.

Author Contributions: Conceptualization, A.A.F.O., M.S.K. and E.K.N.; Methodology, A.A.F.O.; Software, A.A.F.O. and M.S.K.; Validation, A.A.J., E.K.N., L.A.A.-H., M.N.H. and E.S.A.-A.; Resources, L.A.A.-H. and M.N.H.; Writing—original draft, A.A.F.O.; Writing—review and editing, A.A.J.; Visualization, A.B., Z.T.A.-S. and E.S.A.-A.; Project administration, A.A.J.; Funding acquisition, A.B. and Z.T.A.-S. All authors have read and agreed to the published version of the manuscript.

Funding: This research received no external funding.

Data Availability Statement: The datasets generated and/or analyzed during the current study are available from the corresponding author upon reasonable request.

Acknowledgments: The authors wish to thank Mustansiriyah University (College of Engineering) and the University of Technology, Iraq, for the use of facilities in their labs. Thanks also to both Warith Al-Anbiyaa University, Iraq, and Al Hikma University College, Baghdad, Iraq.

Conflicts of Interest: The authors have no conflicts of interest to declare. All co-authors have seen and agree with the contents of this manuscript, and there are no financial interests to report.

References

1. Ngo, T.D.; Kashani, A.; Imbalzano, G.; Nguyen, K.T.Q.; Hui, D. Additive Manufacturing (3D Printing): A Review of Materials, Methods, Applications and Challenges. *Compos. Part B Eng.* **2018**, *143*, 172–196. [CrossRef]
2. Murariu, M.; Dubois, P. PLA composites: From production to properties. *Adv. Drug Deliv. Rev.* **2016**, *107*, 17–46. [CrossRef]
3. Gao, W.; Zhang, Y.; Ramanujan, D.; Ramani, K.; Chen, Y.; Williams, C.B.; Wang, C.C.L.; Shin, Y.C.; Zhang, S.; Zavattieri, P.D. The status, challenges, and future of additive manufacturing in engineering. *Comput.-Aided Des.* **2015**, *69*, 65–89. [CrossRef]
4. Thompson, M.K.; Moroni, G.; Vaneker, T.; Fadel, G.; Campbell, R.I.; Gibson, I.; Bernard, A.; Schulz, J.; Graf, P.; Ahuja, B.; et al. Design for Additive Manufacturing: Trends, opportunities, considerations, and constraints. *CIRP Ann.* **2016**, *65*, 737–760. [CrossRef]
5. Rajpurohit, S.R.; Dave, H.K. Impact strength of 3D printed PLA using open source FFF-based 3D printer. *Prog. Addit. Manuf.* **2021**, *6*, 119–131. [CrossRef]
6. Kartikeyan, B.; Ponshanmugakumar, A.; Saravanan, G.; BharathGanesh, S.; Hemamalini, V. Experimental and theoretical analysis of FDM AM PLA mechanical properties. *Mater. Today Proc.* **2023**. [CrossRef]
7. Caminero, M.Á.; Chacón, J.M.; García-Plaza, E.; Núñez, P.J.; Reverte, J.M.; Becar, J.P. Additive Manufacturing of PLA-Based Composites Using Fused Filament Fabrication: Effect of Graphene Nanoplatelet Reinforcement on Mechanical Properties, Dimensional Accuracy and Texture. *Polymers* **2019**, *11*, 799. [CrossRef] [PubMed]
8. Erdaş, M.U.; Yildiz, B.S.; Yildiz, A.R. Crash performance of a novel bio-inspired energy absorber produced by additive manufacturing using PLA and ABS materials. *Mater. Test.* **2024**, *66*, 696–704. [CrossRef]
9. Ogaili, A.A.F.; Jaber, A.A.; Hamzah, M.N. Wind turbine blades fault diagnosis based on vibration dataset analysis. *Data Brief* **2023**, *49*, 109414. [CrossRef]
10. Farah, S.; Anderson, D.G.; Langer, R. Physical and mechanical properties of PLA, and their functions in widespread applications—A comprehensive review. *Adv. Drug Deliv. Rev.* **2016**, *107*, 367–392. [CrossRef]
11. Mohamed, O.A.; Masood, S.H.; Bhowmik, J.L. Optimization of fused deposition modeling process parameters: A review of current research and future prospects. *Adv. Manuf.* **2015**, *3*, 42–53. [CrossRef]
12. Gardan, J. Additive manufacturing technologies for polymer composites: State-of-the-art and future trends. In *Structure and Properties of Additive Manufactured Polymer Components*; Friedrich, K., Walter, R., Soutis, C., Advani, S.G., Fiedler, B., Eds.; Woodhead Publishing: Cambridge, UK, 2020; pp. 3–15. [CrossRef]
13. Koç, O.O.; Meram, A.; Çetin, M.E.; Öztürk, S. Acoustic properties of ABS and PLA parts produced by additive manufacturing using different printing parameters. *Mater. Test.* **2024**, *66*, 705–714. [CrossRef]
14. Erdaş, M.U.; Yıldız, B.S.; Yıldız, A.R. Experimental analysis of the effects of different production directions on the mechanical characteristics of ABS, PLA, and PETG materials produced by FDM. *Mater. Test.* **2024**, *66*, 198–206. [CrossRef]
15. Ghabezi, P.; Sam-Daliri, O.; Flanagan, T.; Walls, M.; Harrison, N.M.; Ghabezi, P.; Sam-Daliri, O.; Flanagan, T.; Walls, M.; Harrison, N.M. Circular economy innovation: A deep investigation on 3D printing of industrial waste polypropylene and carbon fibre composites. *Resour. Conserv. Recycl.* **2024**, *206*, 107667. [CrossRef]
16. Shirasu, K.; Yamaguchi, Y.; Hoshikawa, Y.; Kikugawa, G.; Tohmyoh, H.; Okabe, T. Micromechanics study of short carbon fiber-reinforced thermoplastics fabricated via 3D printing using design of experiments. *Mater. Sci. Eng. A* **2024**, *891*, 145971. [CrossRef]
17. Ghabezi, P.; Flanagan, T.; Walls, M.; Harrison, N.M. Degradation characteristics of 3D printed continuous fibre-reinforced PA6/chopped fibre composites in simulated saltwater. *Prog. Addit. Manuf.* **2024**, *9*, 1–14. [CrossRef]
18. Vălean, E.; Foti, P.; Razavi, S.M.J.; Berto, F.; Marșavina, L. Static and fatigue behavior of 3D printed PLA and PLA reinforced with short carbon fibers. *J. Mech. Sci. Technol.* **2023**, *37*, 5555–5559. [CrossRef]
19. Iragi, M.; Pascual-González, C.; Esnaola, A.; Lopes, C.; Aretxabaleta, L. Ply and interlaminar behaviours of 3D printed continuous carbon fibre-reinforced thermoplastic laminates; effects of processing conditions and microstructure. *Addit. Manuf.* **2019**, *30*, 100884. [CrossRef]
20. Li, Y.; Gao, S.; Dong, R.; Ding, X.; Duan, X. Additive Manufacturing of PLA and CF/PLA Binding Layer Specimens via Fused Deposition Modeling. *J. Mater. Eng. Perform.* **2018**, *27*, 492–500. [CrossRef]
21. Thirugnanasamabandam, A.; Subramaniyan, M.; Prabhu, B.; Ramachandran, K. Development and comprehensive investigation on PLA / carbon fiber reinforced PLA based structurally alternate layered polymer composites. *J. Ind. Eng. Chem.* **2024**, *136*, 248–257. [CrossRef]

22. Nanni, A.; Parisi, M.; Colonna, M.; Messori, M. Thermo-Mechanical and Morphological Properties of Polymer Composites Reinforced by Natural Fibers Derived from Wet Blue Leather Wastes: A Comparative Study. *Polymers* **2021**, *13*, 1837. [CrossRef] [PubMed]
23. Ramesh, P.; Prasad, B.D.; Narayana, K.L. Effect of MMT Clay on Mechanical, Thermal and Barrier Properties of Treated Aloevera Fiber/PLA-Hybrid Biocomposites. *Silicon* **2020**, *12*, 1751–1760. [CrossRef]
24. Cao, M.; Cui, T.; Yue, Y.; Li, C.; Guo, X.; Jia, X.; Wang, B. Preparation and Characterization for the Thermal Stability and Mechanical Property of PLA and PLA/CF Samples Built by FFF Approach. *Materials* **2023**, *16*, 5023. [CrossRef] [PubMed]
25. Wang, Q.; Zhang, Q.; Kang, Y.; Wang, Y.; Liu, J. An investigation of preparation of continuous carbon fiber reinforced PLA prepreg filament. *Compos. Commun.* **2023**, *39*, 101530. [CrossRef]
26. Li, H.; Liu, B.; Ge, L.; Chen, Y.; Zheng, H.; Fang, D. Mechanical performances of continuous carbon fiber reinforced PLA composites printed in vacuum. *Compos. Part B Eng.* **2021**, *225*, 109277. [CrossRef]
27. Naranjo-Lozada, J.; Ahuett-Garza, H.; Orta-Castañón, P.; Verbeeten, W.M.; Sáiz-González, D. Tensile properties and failure behavior of chopped and continuous carbon fiber composites produced by additive manufacturing. *Addit. Manuf.* **2019**, *26*, 227–241. [CrossRef]
28. Yadav, P.; Sahai, A.; Sharma, R.S. Flexural Strength and Surface Profiling of Carbon-Based PLA Parts by Additive Manufacturing. *J. Inst. Eng. India Ser. C* **2021**, *102*, 1–11. [CrossRef]
29. Kumar Patro, P.; Kandregula, S.; Suhail Khan, M.; Das, S. Investigation of mechanical properties of 3D printed sandwich structures using PLA and ABS. *Mater. Proc.* **2023**. [CrossRef]
30. Zwawi, M. Structure Integrity and Fracture Prediction of PLA 3D-Printed Eye Grab Hooks with Different Cross Sections. *Procedia Struct. Integr.* **2022**, *37*, 1057–1064. [CrossRef]
31. Wu, W.Z.; Ye, W.L.; Wu, Z.C.; Geng, P.; Wang, Y.L.; Zhao, J. Influence of Layer Thickness, Raster Angle, Deformation Temperature and Recovery Temperature on the Shape-Memory Effect of 3D-Printed Polylactic Acid Samples. *Materials* **2017**, *10*, 970. [CrossRef]
32. Sajjadi, S.A.; Ghasemi, F.A.; Rajaee, P.; Fasihi, M. Evaluation of fracture properties of 3D printed high impact polystyrene according to essential work of fracture: Effect of raster angle. *Addit. Manuf.* **2022**, *59*, 103191. [CrossRef]
33. Ayatollahi, M.R.; Nabavi-Kivi, A.; Bahrami, B.; Yahya, M.Y.; Khosravani, M.R. The influence of in-plane raster angle on tensile and fracture strengths of 3D-printed PLA specimens. *Eng. Fract. Mech.* **2020**, *237*, 107225. [CrossRef]
34. Nabavi-Kivi, A.; Ayatollahi, M.R.; Razavi, N. Investigating the effect of raster orientation on fracture behavior of 3D-printed ABS specimens under tension-tear loading. *Eur. J. Mech. A/Solids* **2023**, *99*, 104944. [CrossRef]
35. Gonabadi, H.; Chen, Y.; Yadav, A.; Bull, S. Investigation of the effect of raster angle, build orientation, and infill density on the elastic response of 3D printed parts using finite element microstructural modeling and homogenization techniques. *Int. J. Adv. Manuf. Technol.* **2022**, *118*, 1485–1510. [CrossRef]
36. Karimi, H.R.; Khedri, E.; Nazemzadeh, N.; Mohamadi, R. Effect of layer angle and ambient temperature on the mechanical and fracture characteristics of unidirectional 3D printed PLA material. *Mater. Today Commun.* **2023**, *35*, 106174. [CrossRef]
37. Anderson, T.L.; Anderson, T.L. *Fracture Mechanics: Fundamentals and Applications*; CRC Press: Boca Raton, FL, USA, 2005.
38. Sih, G.C.; Paris, P.C.; Irwin, G.R. On cracks in rectilinearly anisotropic bodies. *Int. J. Fract.* **1965**, *1*, 189–203. [CrossRef]
39. Rice, J.R. A Path Independent Integral and the Approximate Analysis of Strain Concentration by Notches and Cracks. *J. Appl. Mech.* **1968**, *35*, 379–386. [CrossRef]
40. Ning, F.; Cong, W.; Qiu, J.; Wei, J.; Wang, S. Additive manufacturing of carbon fiber reinforced thermoplastic composites using fused deposition modeling. *Compos. Part B Eng.* **2015**, *80*, 369–378. [CrossRef]
41. Hu, C.; Hau, W.N.J.; Chen, W.; Qin, Q.-H. The fabrication of long carbon fiber reinforced polylactic acid composites via fused deposition modelling: Experimental analysis and machine learning. *J. Compos. Mater.* **2020**, *55*, 1459–1472. [CrossRef]
42. Arrigo, R.; Bartoli, M.; Malucelli, G. Poly(lactic Acid)–Biochar Biocomposites: Effect of Processing and Filler Content on Rheological, Thermal, and Mechanical Properties. *Polymers* **2020**, *12*, 892. [CrossRef]
43. Abdulla, F.A.; Hamid, K.L.; Ogaili, A.A.F.; Abdulrazzaq, M.A. Experimental study of Wear Rate Behavior for Composite Materials under Hygrothermal Effect. *IOP Conf. Ser. Mater. Sci. Eng.* **2020**, *928*, 022009. [CrossRef]
44. Ogaili, A.A.F.; Al-Ameen, E.S.; Kadhim, M.S.; Mustafa, M.N. Evaluation of mechanical and electrical properties of GFRP composite strengthened with hybrid nanomaterial fillers. *AIMS Mater. Sci.* **2020**, *7*, 93–102. [CrossRef]
45. Ogaili, A.A.F.; Abdulla, F.A.; Al-Sabbagh, M.N.M.; Waheeb, R.R. Prediction of Mechanical, Thermal and Electrical Properties of Wool/Glass Fiber based Hybrid Composites. In *IOP Conference Series: Materials Science and Engineering*; IOP Publishing: Bristol, UK, 2020; p. 022004.
46. Ogaili, A.A.F.; Jaber, A.A.; Hamzah, M.N. A methodological approach for detecting multiple faults in wind turbine blades based on vibration signals and machine learning. *Curved Layer. Struct.* **2023**, *10*, 20220214. [CrossRef]
47. Al-Ameen, E.S.; Abdulhameed, J.J.; Abdulla, F.A.; Ogaili, A.A.F.; Al-Sabbagh, M.N.M. Strength characteristics of pol-yester filled with recycled GFRP waste. *J. Mech. Eng. Res. Dev.* **2020**, *43*, 178–185.
48. Al-Ameen, E.S.; Al-Sabbagh, M.N.M.; Ogaili, A.A.F.; Kurji, A. Role of pre-stressing on anti-penetration properties for Kevlar/Epoxy composite plates. *Int. J. Nanoelectron. Mater.* **2022**, *15*, 293–302.
49. Al-Ameen, E.S.; Abdulla, F.A.; Ogaili, A.A.F. Effect of Nano TiO_2 on Static Fracture Toughness of Fiberglass/Epoxy Composite Materials in Hot Climate regions. *IOP Conf. Ser. Mater. Sci. Eng.* **2020**, *870*, 012170. [CrossRef]

50. Mohammed, S.A.; Al-Haddad, L.A.; Alawee, W.H.; Dhahad, H.A.; Jaber, A.A.; Al-Haddad, S.A. Forecasting the productivity of a solar distiller enhanced with an inclined absorber plate using stochastic gradient descent in artificial neural networks. *Multiscale Multidiscip. Model. Exp. Des.* **2023**, *7*, 1–11. [CrossRef]
51. Akasheh, F.; Aglan, H. Fracture toughness enhancement of carbon fiber–reinforced polymer composites utilizing additive manufacturing fabrication. *J. Elastomers Plast.* **2018**, *51*, 698–711. [CrossRef]
52. Oviedo, A.; Puente, A.; Bernal, C.; Pérez, E. Mechanical evaluation of polymeric filaments and their corresponding 3D printed samples. *Polym. Test.* **2020**, *88*, 106561. [CrossRef]
53. Jap, N.S.; Pearce, G.M.; Hellier, A.K.; Russell, N.; Parr, W.C.; Walsh, W.R. The effect of raster orientation on the static and fatigue properties of filament deposited ABS polymer. *Int. J. Fatigue* **2019**, *124*, 328–337. [CrossRef]
54. Khosravani, M.R.; Reinicke, T. Effects of printing parameters on the fracture toughness of 3D-printed polymer parts. *Procedia Struct. Integr.* **2023**, *47*, 454–459. [CrossRef]
55. Vălean, E.; Foti, P.; Berto, F.; Marșavina, L. Static and fatigue behavior of 3D printed smooth and notched PLA and short carbon fibers reinforced PLA. *Theor. Appl. Fract. Mech.* **2024**, *131*, 104417. [CrossRef]
56. Ogaili, A.A.F.; Hamzah, M.N.; Jaber, A.A. Integration of Machine Learning (ML) and Finite Element Analysis (FEA) for Predicting the Failure Modes of a Small Horizontal Composite Blade. *Int. J. Renew. Energy Res.* **2022**, *12*, 2168–2179. [CrossRef]
57. Ogaili, A.A.; Hamzah, M.N.; Jaber, A.A. Free Vibration Analysis of a Wind Turbine Blade Made of Composite Materials. In Proceedings of the International Middle Eastern Simulation and Modeling Conference, Baghdad, Iraq, 27–29 June 2022; pp. 27–29.

Disclaimer/Publisher's Note: The statements, opinions and data contained in all publications are solely those of the individual author(s) and contributor(s) and not of MDPI and/or the editor(s). MDPI and/or the editor(s) disclaim responsibility for any injury to people or property resulting from any ideas, methods, instructions or products referred to in the content.

Article

Development and Evaluation of a Novel Method for Reinforcing Additively Manufactured Polymer Structures with Continuous Fiber Composites

Sven Meißner [1,*], Jiri Kafka [1], Hannah Isermann [2], Susanna Labisch [2], Antonia Kesel [2], Oliver Eberhardt [3], Harald Kuolt [3], Sebastian Scholz [1], Daniel Kalisch [1], Sascha Müller [4], Axel Spickenheuer [5] and Lothar Kroll [4]

1. Fraunhofer Plastics Technology Center Oberlausitz, Fraunhofer Institute for Machine Tools and Forming Technology IWU, Theodor-Koerner-Allee 6, 02763 Zittau, Germany; jiri.kafka@iwu.fraunhofer.de (J.K.); sebastian.scholz@iwu.fraunhofer.de (S.S.); daniel.kalisch@iwu.fraunhofer.de (D.K.)
2. Biomimetics-Innovation-Centre (B-I-C), University of Applied Sciences Bremen, Neustadtswall 30, 28199 Bremen, Germany; hannah.isermann@gmx.de (H.I.); slabisch@bionik.hs-bremen.de (S.L.); akesel@bionik.hs-bremen.de (A.K.)
3. J. Schmalz GmbH, Johannes-Schmalz-Str. 1, 72293 Glatten, Germany; harald.kuolt@schmalz.de (H.K.)
4. Department of Lightweight Structures and Polymer Technology, Faculty of Mechanical Engineering, Chemnitz University of Technology, 09111 Chemnitz, Germany; sascha.mueller@mb.tu-chemnitz.de (S.M.); lothar.kroll@mb.tu-chemnitz.de (L.K.)
5. Mechanics and Composite Materials Department, Leibniz-Institut für Polymerforschung Dresden e. V., Hohe Str. 6, 01069 Dresden, Germany; spickenheuer@ipfdd.de
* Correspondence: sven.meissner@iwu.fraunhofer.de

Citation: Meißner, S.; Kafka, J.; Isermann, H.; Labisch, S.; Kesel, A.; Eberhardt, O.; Kuolt, H.; Scholz, S.; Kalisch, D.; Müller, S.; et al. Development and Evaluation of a Novel Method for Reinforcing Additively Manufactured Polymer Structures with Continuous Fiber Composites. *J. Compos. Sci.* **2024**, *8*, 272. https://doi.org/10.3390/jcs8070272

Academic Editor: Yuan Chen

Received: 24 June 2024
Revised: 8 July 2024
Accepted: 12 July 2024
Published: 14 July 2024

Copyright: © 2024 by the authors. Licensee MDPI, Basel, Switzerland. This article is an open access article distributed under the terms and conditions of the Creative Commons Attribution (CC BY) license (https://creativecommons.org/licenses/by/4.0/).

Abstract: Additively manufactured polymer structures often exhibit strong anisotropies due to their layered composition. Although existing methods in additive manufacturing (AM) for improving the mechanical properties are available, they usually do not eliminate the high degree of structural anisotropy. Existing methods for continuous fiber (cF) reinforcement in AM can significantly increase the mechanical properties in the strand direction, but often do not improve the interlaminar strength between the layers. In addition, it is mostly not possible to deposit cFs three-dimensionally and curved (variable–axial) and, thus, in a path that is suitable for the load case requirements. There is a need for AM methods and design approaches that enable cF reinforcements in a variable–axial way, independently of the AM mounting direction. Therefore, a novel two-stage method is proposed in which the process steps of AM and cF integration are decoupled from each other. This study presents the development and validation of the method. It was first investigated at the specimen level, where a significant improvement in the mechanical properties was achieved compared to unreinforced polymer structures. The Young's modulus and tensile strength were increased by factors of 9.1 and 2.7, respectively. In addition, the design guidelines were derived based on sample structures, and the feasibility of the method was demonstrated on complex cantilevers.

Keywords: additive manufacturing; pultrusion; continuous fiber; structural optimization; lightweight design; fused filament fabrication; fiber-reinforced polymer; thermoplastic polymer; thermoset polymer

1. Introduction

AM has gained significant importance in recent years and has now found broad applications in various industries such as mechanical engineering, the automotive industry, aerospace industry, medical industry, and consumer goods production. Compared to conventional manufacturing processes, AM offers the ability to produce components quickly and individually. The resulting flexibility in production allows companies to quickly respond to customer demands and market developments. Additionally, AM enables the realization of complex geometries and shapes that are difficult or even impossible to produce with other manufacturing methods. Furthermore, AM allows for a more efficient utilization of resources and materials [1]. These advantages have led to the development of numerous AM technologies, which are now utilized throughout the entire value chain, including upstream and downstream processes [2].

While AM offers many advantages, there are deficiencies in the mechanical properties of additively manufactured polymer components. In the AM process, the material is deposited layer by layer, which can lead to a weaker interface between individual layers. This leads to an attenuation of the structure in the mounting direction [3]. As a result, anisotropies can occur in additively manufactured components, whereas conventional polymer processing/manufacturing methods (such as unreinforced injection molding) typically achieve nearly isotropic structures. Therefore, additively manufactured structures generally cannot exhibit the same mechanical properties as components produced with the same material using conventional manufacturing processes.

Intensive research has been and is being conducted to improve the mechanical properties of additively manufactured polymer components. Specifically tailored materials were developed for AM, which exhibit a higher strength and durability, for example, by influencing the rheology through the compounding with additives [4–7]. Efforts also been made to optimize the AM processes to achieve better interfaces between the layers, such as through temperature control of the build chamber and build platform, as well as adjustments to the extrusion temperature or layer thickness [8–11]. Post-processing steps such as annealing are also considered to enhance the mechanical properties of the components [12]. Overall, addressing the deficiencies in the mechanical properties is an important area of research to further optimize AM and expand its application scope.

To overcome the deficiencies of a pure, unreinforced polymer in terms of its low mechanical properties, short fibers are often added to the polymer for structural reinforcement. For example, in Selective Laser Sintering (SLS), short fibers are mixed with the polymer powder [13–16], while in Fused Filament Fabrication (FFF), short fibers are added to the filament or granulate [16–19]. Short fibers offer the advantage of being relatively easy to process, increasing the mechanical properties in the direction of the fiber alignment, and significantly reducing part warping and distortion. However, compounding with short fibers does not contribute to an improvement in layer adhesion and therefore does not lead to a significant increase in the mechanical properties of additively manufactured polymer structures in the mounting direction.

The mechanical deficiencies prevent the realization of some applications, particularly mechanically highly stressed structures [20]. Therefore, enhancing the mechanical properties can help to expand the range of applications for AM to further increase the already high potential of AM. An improvement in the mechanical properties can be achieved, particularly through the integration of cFs. AM processes utilizing cFs can achieve comparable tensile strengths to conventionally manufactured fiber-reinforced polymer (FRP) structures [21]. Continuous carbon fibers are predestined, as they offer an exceedingly high Young's modulus as well as specific tensile strength compared to glass or Kevlar fibers, for example.

Figure 1 illustrates AM methods in which cFs are processed for component reinforcement. There is a possible classification according to the time and place of contact between the fibers and the melt [22]. The filament extrusion process uses cF-reinforced thermoplastic pre-impregnated filaments, known as towpregs. During extrusion, the towpreg is heated in the print head, the matrix is melted, and then, it is deposited layer by layer [23,24]. At the in situ impregnation process, dry cFs and a polymer filament are introduced separately into the print head. By heating, the polymer is melted, the cFs are impregnated, and then, they are deposited on the build platform [12,24]. In the dual extrusion process, a cF towpreg and a thermoplastic filament are deposited on the build platform through separate print heads [12,24]. The in situ co-extrusion process is characterized by the separate supply of the polymer filament and towpreg to the print head. In the print head, the towpreg is impregnated with the polymer matrix [25]. In situ consolidation is a scaled-down version of an Automated Tape Laying (ATL) process for thermoplastics. Pre-made towpregs or prepreg tapes are consolidated during and after deposition by applying heat [26]. At inline impregnation, dry fibers are pre-impregnated outside the print head, then transported into the print head and subsequently deposited [22].

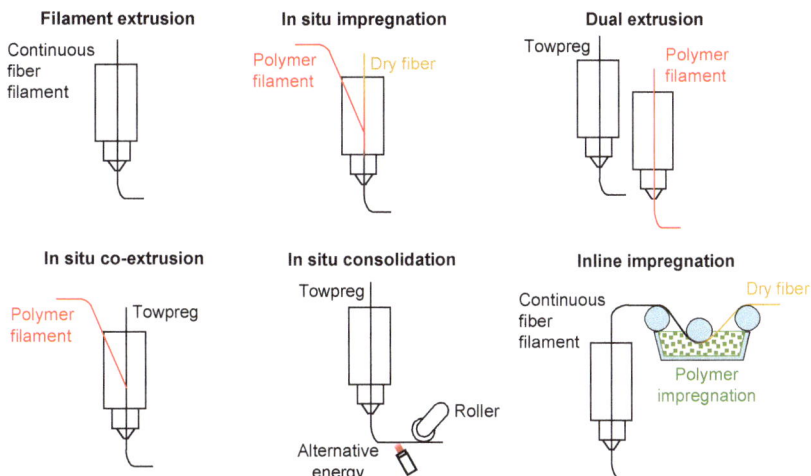

Figure 1. Technologies for AM with cF reinforcement, inspired by [22].

The described cF-AM methods deposit the fibers during the AM process mainly parallel to an even building plate. The in-plane deposition does not allow for the orientation of reinforcing material in a three-dimensional direction. Additionally, the deficiency of the layer interface strength in the mounting direction could not be improved either. There is a need for AM methods and design approaches that enable cF composite reinforcements in a three-dimensional curved (variable–axial) way, independently of the AM mounting direction. A variable–axial orientation of the reinforcement fibers offers the advantage that mechanical loads result in a more uniform distribution of stresses among the reinforcement fibers, while the matrix material is subjected to lower stresses. The high specific stiffness and strength of the fiber materials can consequently be optimally utilized. A significant reduction in material usage and an increase in lightweight construction are demonstrated by existing FRP processes that enable variable–axial fiber orientations (e.g., Tailored Fiber Placement), in contrast to conventional processes with multi-axial layering [27]. In AM, two-stage manufacturing processes have the potential to meet this demand by decoupling the integration of a continuous fiber-reinforced polymer (cFRP) from the AM process. One example is composite structures, where thermoplastic pre-impregnated unidirectional tapes are melted onto a polymer basic structure fabricated using an extrusion-based AM process [28,29].

The objective of this research is to develop and evaluate a novel two-stage method for reinforcing additively manufactured polymer structures with a cFRP. The approach decouples the process steps AM and cF implementation by inserting cFRP structures into component-integrated channels of an additively manufactured polymer's basic structure. The orientation of the integrated cFRP structures into the additively manufactured polymer structure corresponds to the principal stress lines of the load-bearing structure. The basic component design, which is initially based on classical topology optimization, is supplemented by a load-adapted, variable–axial orientation of the reinforcing fibers within the additively manufactured polymer structure.

The approach is mostly unexplored and distinguished by a high degree of novelty. This study presents the investigations on the feasibility of the method, as well as on the material characterization and the development of general design guidelines. In addition, a demonstrative implementation is intended. For the design of the demonstrators, a numerical tool will be developed to optimize the arrangement and orientation of the component-integrated cFRP. The demonstrators will be manufactured and analyzed, and potential improvements for the novel method will be derived.

2. Materials and Methods

The research program for the development and validation of the novel method is shown in Figure 2. The investigations start at the specimen level. The level of geometric complexity of the sample structures is subsequently increased, and finally, a proof of concept is provided by means of demonstrators.

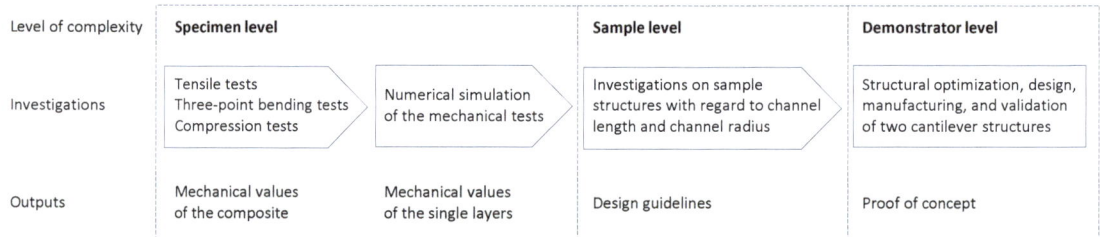

Figure 2. Research program for the development and validation of the novel method.

2.1. Manufacturing Method

The novel method was realized by using the manufacturing steps shown in Figure 3. First, a basic polymer structure is additively manufactured. In the basic structure, component-integrated channels are included. cF bundles are impregnated with a polymer matrix by passing them through a resin bath filled with liquid matrix material. The impregnated cF bundles are pulled through the component-integrated channels of the additively manufactured basic structure by using a traction rope. This pultrusion process enables a high fiber-volume ratio (FVR) to be realized. The matrix of the impregnated cF bundles cross-links within the component and forms a solid FRP with the additively manufactured basic structure.

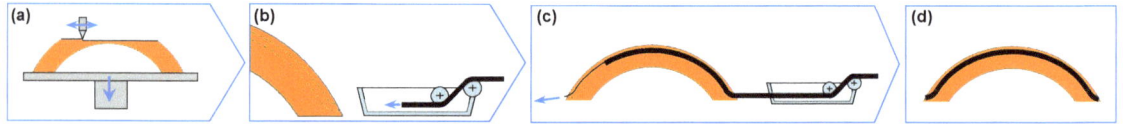

Figure 3. Process principle of the novel method for reinforcing additively manufactured polymer structures with cFRP. (**a**) AM of polymer basic structure. (**b**) cF impregnation by pulling a cF bundle through a resin bath. (**c**) Pulling impregnated cF bundles through component-integrated channels of the additively manufactured basic structure. (**d**) Cross-linking of the matrix.

The described process was implemented on a laboratory scale and applied to produce each of the specimens, sample components, and demonstrator structures described in this study. In this laboratory process, the fiber bundles were manually pulled through the resin bath and the additively manufactured channel structures using a wire.

2.2. Materials and Process Parameters

The objective of the study is to demonstrate the general feasibility of the two-stage method described. To this purpose, material components are selected, which are versatile and already established in AM as well as in cFRP fabrication and offer a good processability in the respective sub-processes. In addition, the focus in this phase of research with a low technology readiness level is on the use of materials that are available to a wide range of users.

Fused deposition modeling (FDM) by the 3D printing machine Stratasys Fortus 900mc, utilizing acrylonitrile butadiene styrene (ABS-M30-black from the company Stratasys,

Eden Prairie, MN, USA) was used to fabricate the basic polymer structures. All samples were printed with a layer height of 0.254 mm. The temperature of the installation chamber was 95 °C, and the nozzle temperature was 320 °C. The CAD files were exported in STL format and processed using the slicer software Insight 18.6 using single contour and +45°/−45° solid rasters, which are typical default settings.

cFs of type HTS40 from the company Toho Tenax, Chiyoda, Japan, were used as reinforcing fibers. The epoxy resin L + hardener EPH 161 with a viscosity of 560 ± 100 mPa·s from the company R&G, Waldenbuch, Germany, was selected for impregnating the cFs. The fiber impregnation and manually pultrusion process were performed at 20 °C. After the pultrusion, all samples were stored at 20 °C for 24 h, then treated at 60 °C for 15 h and stored again at 20 °C for at least 48 h.

Sufficient adhesion between the ABS structure and the cF-reinforced epoxy can be assumed [30] for the proof of concept according to the objective of the study. A detailed adaptation of the single material components is not the focus of the study. Further investigations regarding material modifications or substitutions should be expanded in further studies.

2.3. Specimen Geometry and Mechanical Tests

Tensile, compression, and flexure specimens were designed in accordance with the current polymer testing standards with production-related adaptations (Figure 4). In each case, the channel shape was selected in a drop shape to eliminate the need for an internal supporting structure in AM. Reinforced specimens with a channel diameter of 5 mm each were manufactured, and the cFRPs were inserted with FVRs φ_{FRP} of 41% and 57%, respectively. The unreinforced specimens had the same outer contour as the reinforced specimens, but the cross-section was closed (without channel). The mechanical tests were carried out with the Zwick/Roell Inspekt 100 kN universal testing device at 23 ± 2 °C and 50 ± 10% rH. The tensile tests were performed in accordance with the ISO 527-4 standard [31]. The speed of testing was 1 mm/min up to 0.3% and 10 mm/min after 0.3% elongation. The three-point bending tests were carried out according to the EN ISO 14125 standard [32]. The span was 80 mm, and the speed of testing 1 mm/min. The compression tests were in accordance with the EN ISO 14126 standard [33] method 1. The speed of testing was 1 mm/min. The material constants listed in Table 1 were extracted from the mechanical tests.

Figure 4. Reinforced specimens for determining mechanical properties consisting of additively manufactured basic structure (ABS) and cross-sectional geometry in the measuring area with drop-shaped channel for integration of cFRP (cF + epoxy). (**a**) Tensile specimen based on ISO 527-2 standard [34] type 1B. (**b**) Three-point bending specimen based on EN ISO 14125 standard. (**c**) Compression specimen based on EN ISO 14126 standard.

Table 1. Determined material constants for unreinforced specimens and reinforced specimens considering two different FVRs.

Tensile test	E:	Young's modulus	σ_m:	Tensile strength at initial failure
Three-point bending test	E_r:	Flexural modulus	σ_f:	Flexural strength at initial failure
Compression test	E_c:	Chord modulus	σ_c:	Compressive strength at initial failure

2.4. Application Scenario for Demonstrator Implementation

The application scenario is selected by the fact that it exhibits a highly loaded supporting structure and requires a high degree of lightweight construction as well as a significant need for customization. Horizontal cantilevers of an ergonomic handling system (Figure 5) [35] are selected for this purpose.

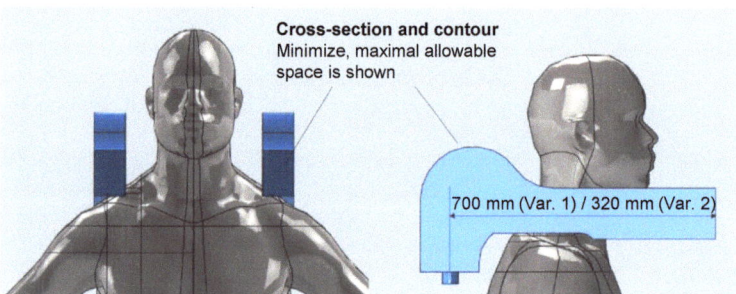

Figure 5. Maximal design space and placement for the "horizontal cantilever", that is, the main part of the planned exoskeleton.

2.5. Numerical Tool for Structure Optimization

The numerical optimization (in the context of the study, the term "optimization" is considered to describe the search for the optimum as well as the process of approximation based on the guideline VDI 6220 Sheet 2: Biomimetics–Biomimetic design methodology–Products and processes) was based on the finite element analysis (FEA) program MARC from Hexagon, release 2023.1, and used the interfaces offered there for data input and output. The control program was programmed in Python, release 3.8. The functionality of this control program BOT (Biomimetic Optimization Tool) is described below.

Two different bio-inspired optimization methods were used for the numerical component optimization, which are particularly effective in terms of lightweight construction optimization [36].

2.5.1. Topology Optimization via SKO

The "Soft Kill Option (SKO)" method was used for topology optimization [37]. In the SKO method, the Young's modulus of each element is varied based on a comparison of the local stress with the reference stress; see Equation (1). The optimization process is divided into several iterations with a linear increase in the reference stress σ_{ref}, each of which consists of several sub-iterations with a constant reference stress. The number of iterations and sub-iterations is determined in advance by the user.

The Young's modulus E of the iteration step $i + 1$ is determined using the following rule:

$$E_{i+1} = E_{max} \cdot \left(\frac{E_i}{E_{max}} + c \cdot \left(\frac{\sigma_i}{\sigma_{ref}} - 1 \right) \right). \tag{1}$$

Here, c is a constant factor, and σ_i is the von Mises yield criterion in the respective element. This update rule is normalized with the maximum Young's modulus E_{max} following Equation (2):

$$E_{i+1} = E_i + k \cdot (\sigma_i - \sigma_{ref}), \tag{2}$$

with

$$k = c \cdot \frac{E_{max}}{\sigma_{ref}}. \tag{3}$$

As suggested by Baumgartner et al. [38], the normalization provides independence from the magnitude of Young's modulus and the load, so that the result is independent of the units used.

To reduce the computing costs, Young's modulus was cut to the range $E_{max}/1000 \leq E \leq E_{max}$ and rounded to 10 discrete, equidistant elastic modulus gradations. The design proposal resulting from the optimization is made visible at the end by hiding all elements with the low Young's modulus $E = E_{max}/1000$.

Preliminary tests also showed that the usability of the optimization results can be improved by homogenizing the stresses of the individual elements using a moving average. This was achieved using a nearest neighbors algorithm by averaging the stress for each element with the stresses of the k-nearest neighbor elements. The number of neighbor elements determines the resulting structure width of the optimization results.

2.5.2. Fiber Orientation Optimization via CAIO

The second biomimetic optimization method implemented here is the "Computer-Aided Internal Optimization (CAIO)" method, which is used for load-appropriate fiber orientation in the component [39].

The CAIO method, like the SKO method, is an iterative optimization method that is carried out using the finite element analysis. The direction-dependent or orthotropic properties of each element are realigned in each iteration based on the local stress tensor in the major principal direction. The major principal direction is the eigenvector associated with the maximum absolute eigenvalue of the stress tensor of each element.

An implemented further development of the CAIO is the determination of tensile and compressive areas for appropriate material allocation. In each iteration, the tensile and compressive areas were identified in the fiber material, and the respective material parameters were assigned for the next iteration.

In addition to the separate topology and fiber orientation, the two optimization methods (SKO and CAIO) were programmed as a combined method based on the program blocks of both methods (Figure 6). In each iteration step, the updates of both methods are performed on the same model. This requires minor adaptations in the SKO method: To account for the orthotropic material properties during SKO, only Young's modulus in fiber direction E_x was used in the update rule (Section 2.5.1, Equation (1)). As in the pure SKO method, 10 discrete materials were used. As material properties, the Young's moduli in all directions $i \in \{x, y, z\}$ were interpolated towards $E_i = \min(E_x, E_y, E_z)/1000$, meaning that the material with the minimum Young's modulus has isotropic material properties and, thus, saves computing time during the CAIO part, as no fiber realignment is performed in this regions. Despite the orthotropic material properties, the von Mises stress was chosen as the characteristic stress σ for the material variation during SKO:

$$\sigma = \sqrt{\frac{1}{2}\left[(\sigma_I - \sigma_{II})^2 + (\sigma_{II} - \sigma_{III})^2 + (\sigma_{III} - \sigma_I)^2\right]} \qquad (4)$$

with the principal stresses σ_I, σ_{II}, and σ_{III}.

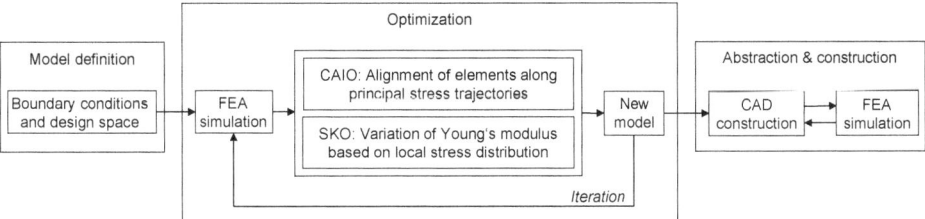

Figure 6. Workflow for the numerical tool used. Presetting of calculation steps and manual selection of a realizable model/geometry.

Element alignment is carried out according to the unmodified CAIO method, which also takes the various compressive and tensile properties of the material into account and assigns the material accordingly.

2.5.3. Component Design

The first step in creating the finite element model was to define a geometry with the maximum permissible installation space in a common CAD system (here: Autodesk Inventor Professional 2021). The geometry was loaded into the preprocessor and meshed there with four-node linear isoparametric tetrahedron elements using linear interpolation functions, resulting in 212,404 elements and 40,641 nodes (Figure 7). The material properties were applied according to the results of material characterization (Section 3.2.5) using the anisotropic mechanical values of the cFRP for tension and compression. The boundary conditions were applied with point loads on 78 nodes on the tip of the cantilever, resulting in a total load of −257.75 N in the x- and y-directions, respectively, and a fixation at the base.

Figure 7. FEA model (here: for the 700 mm long component) defining the construction space with applied boundary conditions; in red, the fixed displacements; and in blue, the applied point loads.

The load shall be transferred to the cantilever arm via a cable pull and a deflection roller. Hence, it was applied in negative x- and y-directions (Figure 7). Besides the 700 mm cantilever, a short component of 320 mm was also realized. Both cantilevers have the same materials and loading.

Finally, the created FEA models were exported for the finite element simulation and optimization with BOT.

2.5.4. FEA Validation of the CAD Construction

In order to validate the generated CAD models, the numerical tool for fiber orientation was extended so that it could implement a specified fiber orientation along the fiber-carrying channels. This has the advantage that compressive- and tension-dominated areas can still be identified, and the corresponding elements can be assigned the compressive or tensile modulus. The material property of the surrounding channel forming matrix was chosen according to the experimentally determined mechanical values in Section 3.2.5. Both demonstrator models, the long version (700 mm) and the short one (320 mm), were loaded in the same way.

3. Results

3.1. Theoretical Considerations on Tensile Behavior

A forecast is made to determine what mechanical properties could theoretically be expected. To calculate the effective FVR and effective Young's modulus, all material components (additively manufactured polymer, cF, and matrix phase) must be considered. The effective FVRs (Table 2) of the specimens (Figure 4) in the measuring range are calculated by

$$\varphi_{eff} = \varphi_{FRP} \cdot A_{FRP} / A_{eff}. \qquad (5)$$

The Young's modulus of the cFRP (consisting of cF and matrix phase) is derived by

$$E_{FRP} = \varphi_{FRP} \cdot E_{cF} + (1 - \varphi_{FRP}) \cdot E_{Epoxy}. \qquad (6)$$

The theoretically expected effective Young's modulus of the total composite (consisting of the additively manufactured polymer and cFRP phase) in the measuring range is calculated by

$$E_{theor} = \frac{1}{A_{eff}} \cdot \left(E_{FRP} \cdot A_{FRP} + E_{ABS} \cdot \left(A_{eff} \cdot A_{FRP} \right) \right). \qquad (7)$$

Table 2. Determination of the effective FVR φ_{eff} of the respective specimen in the measuring range (contains additively manufactured polymer and cFRP phase) according to Equation (5), with A_{FRP}—cross-sectional area of drop-shaped channel, A_{eff}—overall cross-sectional area of the respective specimen in the measuring range, and φ_{FRP}—FVR of the integrated cFRP (contains cF and matrix phase).

Specimen	A_{FRP} [mm²]	A_{eff} [mm²]	φ_{FRP}	φ_{eff}	φ_{FRP}	φ_{eff}
Tension		96.0		9.5%		13.2%
Flexure	22.2	105.0	41%	8.7%	57%	12.0%
Compression		140.0		6.5%		9.0%

The theoretical preliminary considerations in Table 3 show that an increase in stiffness by a factor of 10.5 or 14.1 can be expected compared to an unreinforced additively manufactured polymer.

Table 3. Determination of the theoretically expected Young's moduli E_{theor} of the total composite (contains additively manufactured polymer and cFRP phase) according to Equations (6) and (7), with A_{FRP}—cross-sectional area of drop-shaped channel, A_{eff}—overall cross-sectional area of the respective specimen in the measuring range, E_{ABS}—Young's Modulus of additively manufactured polymer, E_{cF}—Young's Modulus of reinforcing fibers, E_{Epoxy}—Young's Modulus of matrix, φ_{FRP}—FVR of integrated cFRP, φ_{eff}—effective FVR of respective specimen in the measuring range, E_{FRP}—Young's Modulus of the integrated cFRP (contains cF and matrix phase).

A_{FRP} [mm²]	A_{eff} [mm²]	E_{ABS} [MPa]	E_{cF} [MPa]	E_{Epoxy} [MPa]	φ_{FRP}	φ_{eff}	E_{FRP} [MPa]	E_{theor} [MPa]
22.2	96.0	2400 acc. datasheet ABS-M30 stratasys	240,000 acc. datasheet HTS40 Toho Tenax	4300 acc. datasheet resin L + EPH161 R&G	41%	9.5%	100,937	25,174
					57%	13.2%	138,649	33,891

3.2. Mechanical Tests

3.2.1. Tensile Tests

The failure behavior of the reinforced specimens basically works according to the following principle: The force is applied or constrained on the additively manufactured basic structure, which has a lower stiffness than the cFRP. With an increasing load, the basic structure is elongated to a degree that induces high interlaminar shear stresses at the interface between the cFRP and the additively manufactured channel wall, which leads to the failure of the interface. Subsequently, the complete load is transferred to the basic structure. The failure of the basic structure occurs immediately after the failure of the interface. The damage pattern is basically identical for all reinforced specimens. The specimens fail either near the top or bottom restraint (Figure 8).

To investigate the failure behavior, an FEA model was created to simulate the tensile test (Figure 9). The FEA results, shown in Figure 10 and summarized in Figure 11, demonstrate that the contact force between the cFRP and the additively manufactured basic structure is at its maximum in the region of the restraint. In the region of maximum shear stress, the interface fails, and an initial crack is formed. With further progress, the crack expands, and the principal stress in the additively manufactured basic structure increases. As soon as the permissible stress in the additively manufactured basic structure is exceeded, complete failure occurs.

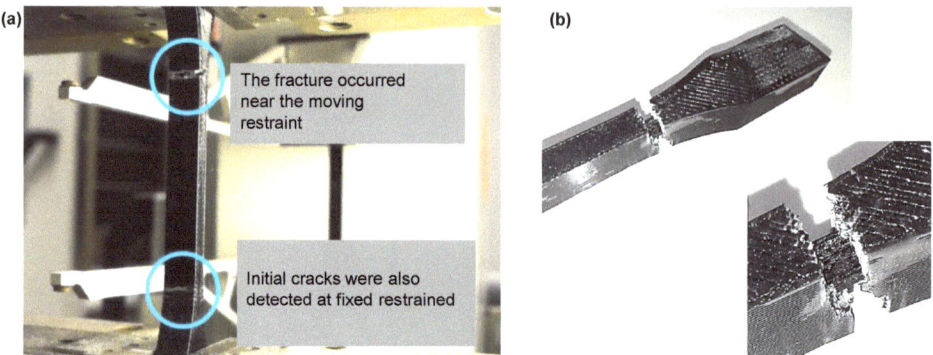

Figure 8. Restraining of specimen on tensile testing machine (**a**) and typical damage pattern (interface failure) on tensile specimens (**b**).

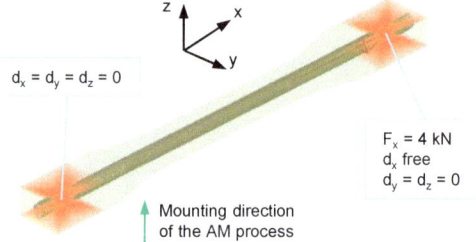

Figure 9. FEA model of a reinforced specimen with boundary conditions for the simulation of the tensile test. The model consists of an additively manufactured polymer basic structure and cFRP. The contact force is considered as failure criterion.

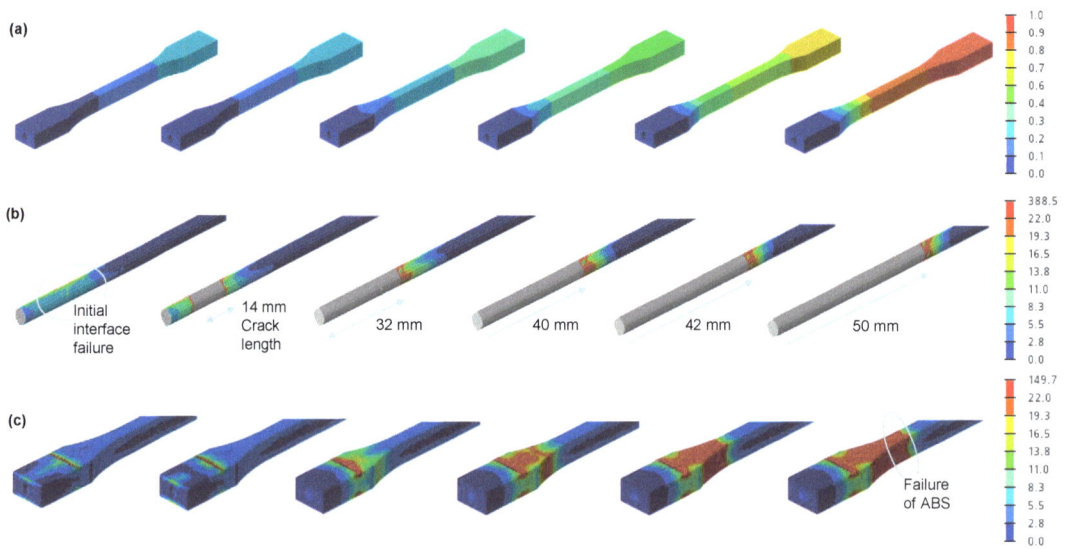

Figure 10. FEA results for (**a**) displacement in the x-direction [mm], (**b**) contact traction [N], and (**c**) principal major stress of additively manufactured basic structure [MPa] show initial crack and crack spreading at interface between cFRP and ABS.

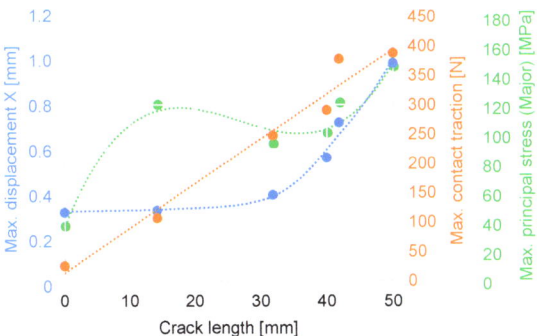

Figure 11. Summary of the FEA results according to the crack length.

A representative stress–strain curve of an unreinforced and a reinforced specimen is shown in Figure 12. The cFRP results in a significant increase in stiffness and strength. A detailed analysis of the reinforced specimen shows a drop in the stress–strain curve even before the maximum stress is reached (successive failure). This drop in the stress–strain curve indicates the initial failure. Parts of the matrix system detach from the channel wall, and the load is transferred to still-intact interface regions between the channel wall and the cFRP. Subsequently, the structure fails completely.

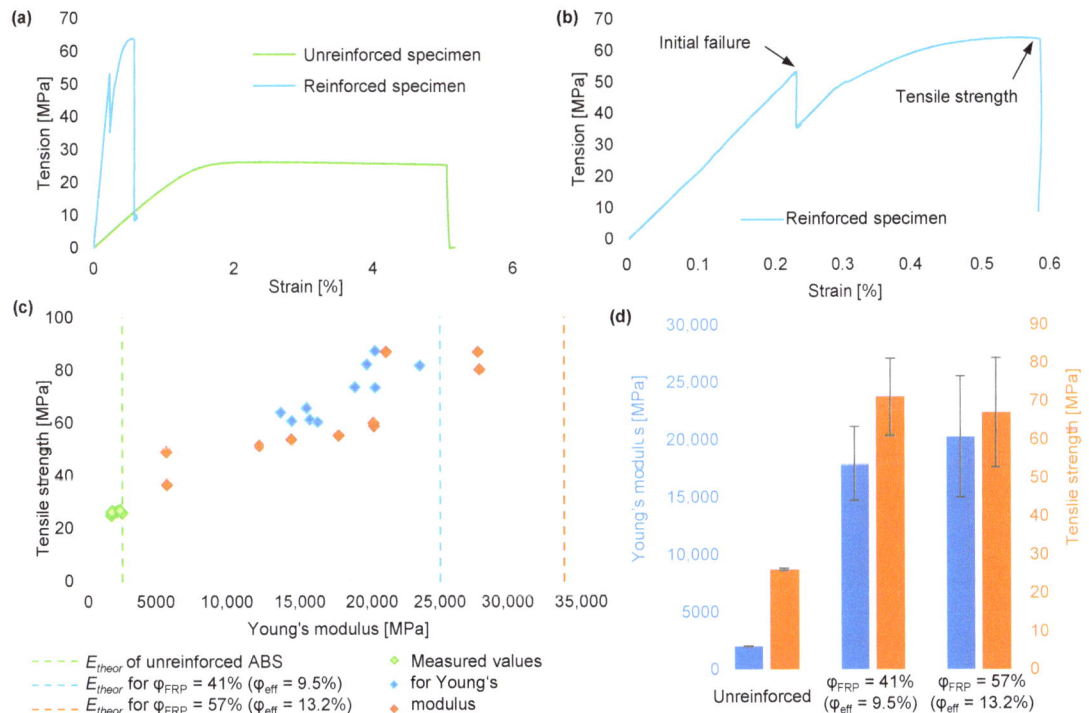

Figure 12. Evaluation of tensile tests. (**a**) Exemplary stress–strain curves on unreinforced and reinforced specimens. (**b**) Detailed stress–strain curve of exemplary reinforced specimen. (**c**) Individual measured and theoretically expected Young's moduli and tensile strengths. (**d**) Summary of the experimentally determined values with coefficients of variation.

Figure 12 and Table 4 show the results of the tensile tests. The Young's moduli of the unreinforced specimens determined experimentally are lower than those predicted according to the data sheet, which is a consequence of air inclusions due to the layer-by-layer deposition during the AM process (Figure 13). Consequently, the Young's moduli of the reinforced specimens are also lower than the previously theoretically determined values, with additional air inclusions in the matrix of the cFRP showing stiffness-reducing effects. Nevertheless, it is confirmed that the reinforced specimens exhibit significantly higher mechanical properties than the unreinforced specimens. The stress at initial failure of the reinforced specimens is 2.7 times higher than the tensile strength of the unreinforced specimens, and Young's modulus was increased by a factor of 9.1.

Table 4. Overview of the experimentally determined material properties of the unreinforced and reinforced additively manufactured structures with φ_{FRP}—FVR of cFRP, φ_{eff}—effective FVR, E—Young's modulus, σ_m—tensile strength at initial failure, E_r—flexural modulus, σ_f—flexural strength at initial failure, E_c—chord modulus, σ_c—compressive strength at initial failure, s—standard deviation, and V—coefficient of variation.

			Reinforced Specimens			
	Unreinforced Specimens		φ_{FRP} = 41% (φ_{eff} = 9.5%)		φ_{FRP} = 57% (φ_{eff} = 13.2%)	
	Material constant					
Tension	E / s / V	σ_m / s / V	E / s / V	σ_m / s / V	E / s / V	σ_m / s / V
Flexure	E_r / s / V	σ_f / s / V	E_r / s / V	σ_f / s / V	E_r / s / V	σ_f / s / V
Compression	E_c / s / V	σ_c / s / V	E_c / s / V	σ_c / s / V	E_c / s / V	σ_c / s / V
	Experimental results [MPa]					
Tension	1966 / 22 / 1.1%	26 / 0.3 / 1.2%	17,884 / 3217 / 18.0%	71 / 10.1 / 14.2%	20,212 / 5263 / 26.0%	67 / 14.3 / 21.3%
Flexure	1550 / 27 / 1.7%	50 / 0.4 / 0.8%	5300 / 196 / 3.7%	180 / 26 / 14.4%	5800 / 345 / 5.9%	200 / 33 / 16.5%
Compression	1750 / 33 / 1.9%	46 / 0.5 / 1.1%	8842 / 2380 / 26.9%	125 / 13 / 10.4%	- / - / -	- / - / -
	Normalized mean values [-] *					
Tension	1	1	9.1	2.7	10.3	2.6
Flexure	1	1	3.4	3.6	3.7	4.0
Compression	1	1	5.1	2.7	-	-

* The normalized mean values E_{norm} and σ_{norm} are calculated according to Equation (8) or Equation (9).

The tensile strengths and Young's moduli of the reinforced specimens obtained in the tensile tests show a higher scatter than the unreinforced specimens, and the scatter of the test results is larger for an FVR of 57% than for an FVR of 41%. Figure 13 shows micrographs of one specimen with low mechanical properties and one specimen with high mechanical properties. Force transfer from the additively manufactured basic structure to the cFRP occurs via the matrix covering the channel wall. The degree of matrix covering the channel wall is of immense importance for a high-performance interface between the cFRP and the additively manufactured basic structure. Thus, the impregnation quality is more important than the FVR because the interface between the channel wall and the fibers is the weakest point in the material composite.

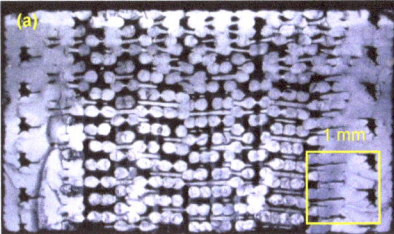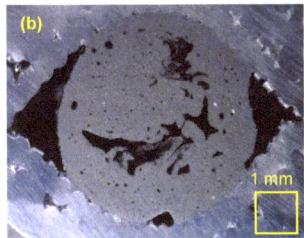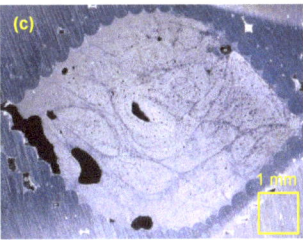

Figure 13. Micrographs of specimen (air inclusions in black). (**a**) Porosity due to layer-by-layer deposition of the polymer melt. (**b**) Reinforced specimen with low mechanical properties due to large air inclusions, especially at the channel wall. (**c**) Reinforced specimen with high mechanical properties due to good impregnation quality and good accumulation of the cFRP on channel wall.

3.2.2. Three-Point Bending Tests

The boundary conditions on the testing machine are shown in Figure 14. All specimens fail at the point of the maximum bending moment, which is located at the load application point. The specimens fail due to excessive deflection. In this case, the fibers carry almost the entire load. On the tensile-stressed side of the specimen, this is stretched to such an extent that a fiber fracture results, followed by a fracture of the additively manufactured basic structure. At the point of load application, an impression of the compression die (upper part of frame) was detected. Furthermore, as in the tensile tests, successive failure occurs. However, the reasons for this failure are not to be found in the failure of the interface, but in that of the fibers themselves.

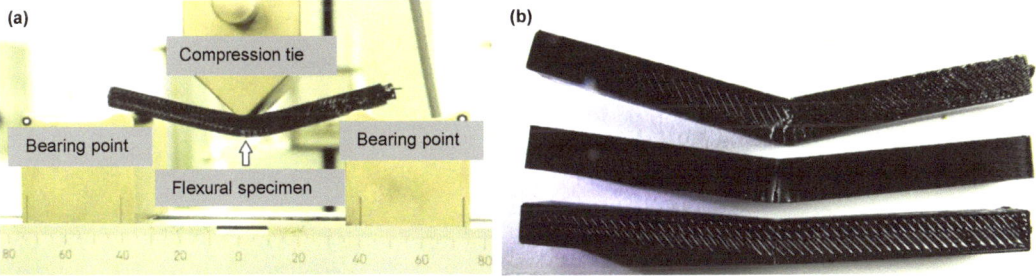

Figure 14. Three-point bending test of reinforced specimen (**a**) and details of tested specimens (**b**).

Figure 15 shows a representative stress–strain curve of an unreinforced and a reinforced specimen. In the case of the reinforced specimen, the stress increases approximately linearly until the initial failure is reached, and the stress drops abruptly to a lower level. The load is then transferred to fibers that are still intact and then fail completely (successive failure).

The individual results of the three-point bending tests are shown in Figure 15 and Table 4. Obviously, the cFRP leads to a significant increase in stiffness and strength. Compared with the unreinforced specimens, the flexural modulus of elasticity and the flexural strength were increased by factors of 5.1 and 2.7, respectively. As with the tensile tests, the results of the three-point bending tests scatter more with a high FVR than with a low FVR. Figure 16 shows two micrographs of a three-point bending specimen at two cross-sections. Air inclusion can be seen to occur at different locations. Coincidentally, if an air inclusion is located in the area of greatest stress, the performance of the composite structure will be negatively affected.

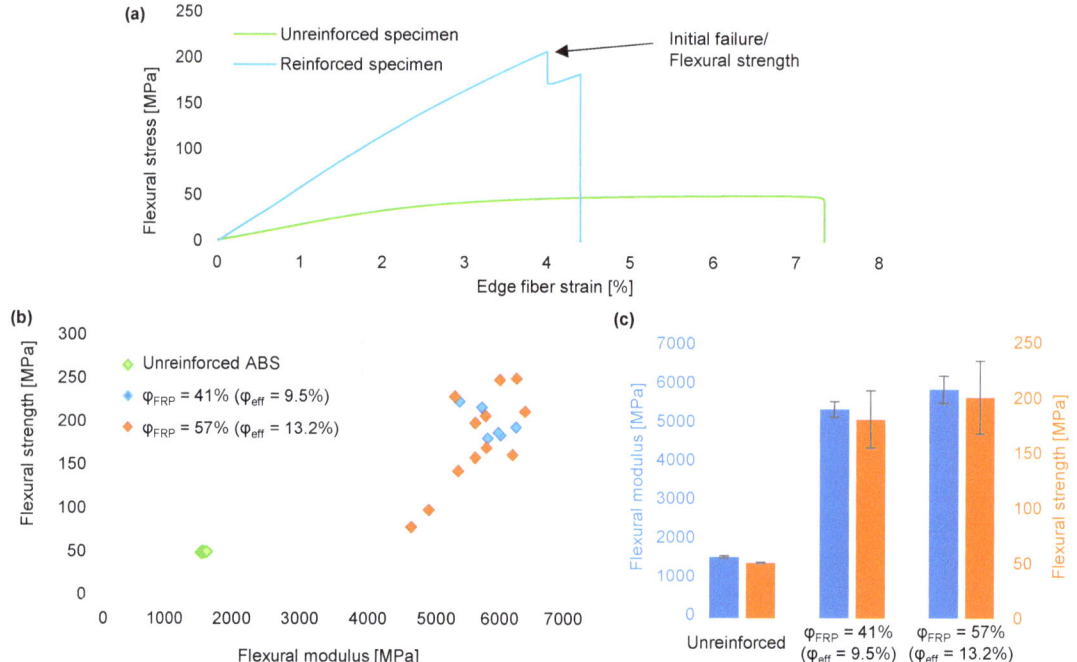

Figure 15. Evaluation of three-point bending tests. (**a**) Exemplary stress–strain curves on unreinforced and reinforced specimens. (**b**) Measured flexural moduli and flexural strengths. (**c**) Summary of the experimentally determined values with coefficients of variation.

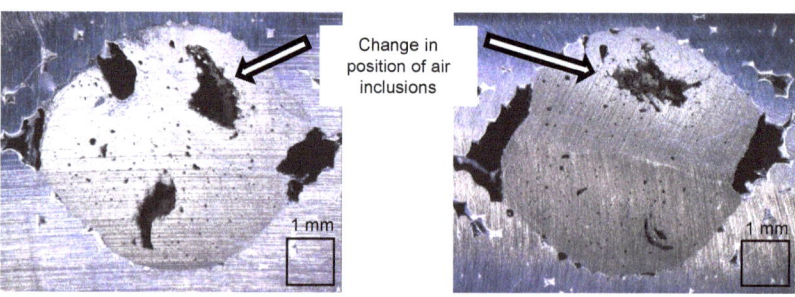

Figure 16. Micrographs of a flexure specimen at two cross-sections (air inclusions in black) show that air inclusions can occur at different locations along the specimen.

3.2.3. Compression Tests

Representative compressive stress–strain curves of an unreinforced or a reinforced specimen and a summary of the results of the compression tests are shown in Figure 17. While the unreinforced specimens showed significant compression, the reinforced specimens did not show any external damage. In the case of the reinforced specimen, the stress increases approximately linearly until the initial failure is reached, and the stress drops abruptly to a lower level. The introduction of the cFRP results in a factor of 3.4 in the chord modulus and a factor of 3.6 in the compressive strength. The mean values are summarized in Table 4.

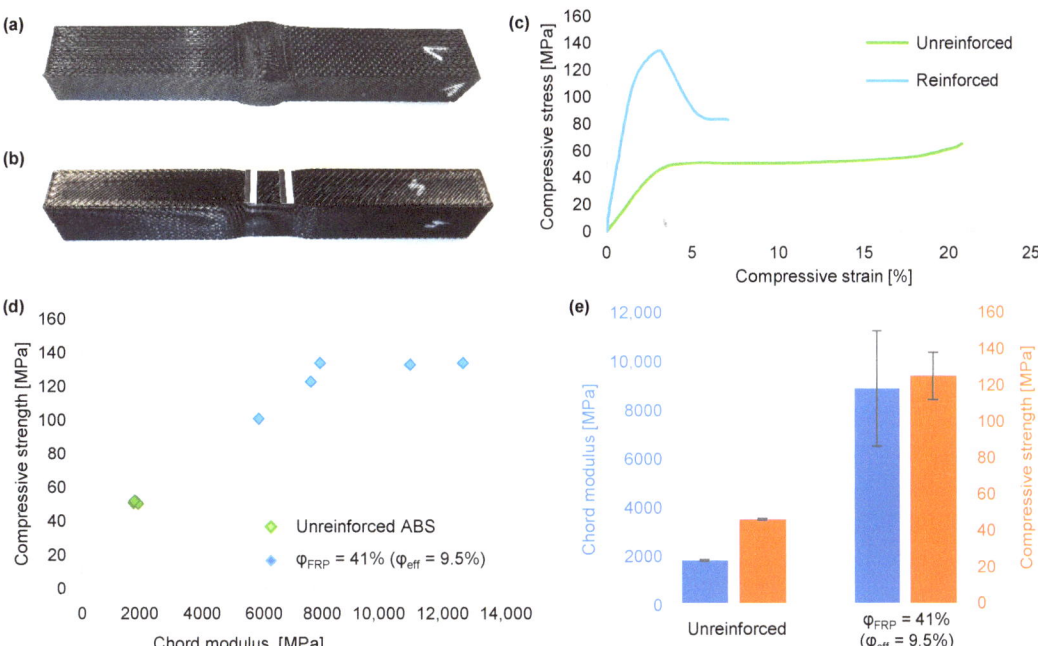

Figure 17. Unreinforced (**a**) and reinforced specimen (**b**) after compression test and evaluation of compression tests. (**c**) Exemplary stress–strain curves on unreinforced and reinforced specimens. (**d**) Chord moduli and compressive strengths. (**e**) Summary of the experimentally determined values with coefficients of variation.

3.2.4. Summary of the Mechanical Tests

Table 4 provides an overview of the experimentally determined material properties of the cFRP-reinforced and unreinforced additively manufactured polymer structures. In addition, the normalized mean values

$$E_{norm} = \frac{E_{reinforced}}{E_{unreinforced}} \tag{8}$$

and

$$\sigma_{norm} = \frac{\sigma_{reinforced}}{\sigma_{unreinforced}} \tag{9}$$

are provided, which indicate the stiffness and strength-influencing effect of the cFRP structure integrated into the additively manufactured basic structure (Figure 18). The mechanical tests for tension, compression, and flexure have shown that the mechanical properties of additively manufactured polymer structures are significantly increased by the integration of cFRP by means of the proposed method.

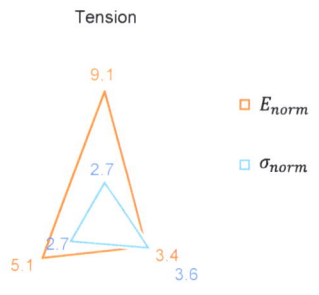

Figure 18. Normalized mean values of the mechanical tests at φ_{FRP} = 41% (φ_{eff} = 9.5%) demonstrate a significant increase in the mechanical properties, and coefficients of variation reveal the scattering of the measured values.

However, it should also be noted that the previously theoretically determined moduli (Table 3) are higher in comparison to the experimental results. These deviations can be caused by production-related effects (Figure 13), for example, air inclusions in the additively manufactured basic structure due to layer-by-layer deposition and in the integrated cFRP. Particularly, the interface between the cFRP and the additively manufactured basic structure at the channel wall offers potential for optimization. Further investigations must be carried out in further studies to utilize the full potential of the method presented.

3.2.5. Characteristic Mechanical Values of the Single Layers

Knowledge of the mechanical properties of the individual material components is essential for the dimensioning of composite structures. For this purpose, the properties of the unreinforced ABS determined by the shown experimental investigations can be used. However, the properties of the integrated cFRP cannot be measured directly. For the verification of the previously theoretically determined moduli (Table 3), the mechanical tests carried out were simulated using FEA using Ansys Workbench 2023. The contact between the additively manufactured basic structure and the cFRP at the channel wall is defined as bonded. It became clear that under the idealized assumption of the theoretical moduli, a strain occurs in the measuring range of the test specimens that is too low compared to the values measured in experiments. The deviation can be caused by production-related air inclusions in the cFRP or at the interface between the cFRP and the additively manufactured basic structure. Therefore, an adaption of the cFRP modulus was carried out in the FEA model, performed in a manual iteration process.

Figure 19 shows the corresponding FEA models and results of the tensile and compression tests, respectively, and a comparison with the measured strain. After iterative adjustment of the material parameters of the cFRP, a high level of correlation between the simulation and measurement results is achieved. In addition, the cFRP exhibits a bimodular behavior, i.e., the moduli under compressive and tensile stress significantly deviate from each other. Such a bimodular behavior can occur especially in fiber-reinforced materials with soft matrix [40,41]. During a flexure load, both tensile and compressive stresses occur. Therefore, a material model according to Ogden's 3rd order [42] was created for the FEA simulation of the flexural test, with which a bimodular behavior is replicated. The previously iteratively determined Young's modulus and chord modulus were transferred to the material model accordingly. A comparison of the numerical and experimental results of the three-point bending test shows a satisfactory correlation (Figure 19).

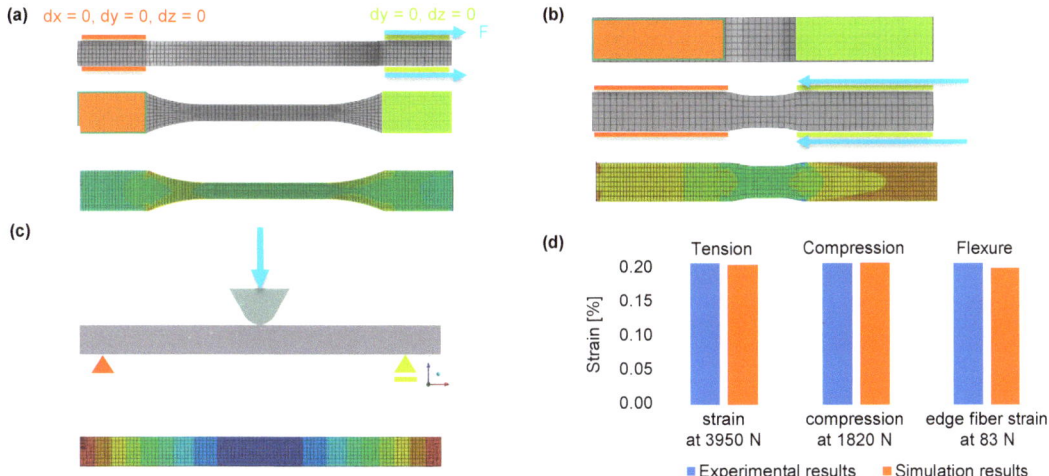

Figure 19. FEA model and simulation results (strain) with iteratively determined Young's modulus and chord modulus under tensile (**a**), compressive (**b**), and flexural loads (**c**). The comparison between the measured and simulated values shows a good correlation (**d**).

Table 5 summarizes the direction-dependent mechanical properties. The properties of the ABS are approximately quasi-isotropic. The properties of the cFRP are orthotropic and depend on the type of load (tension or compression).

Table 5. Overview of the iteratively determined material properties of unreinforced ABS and cFRP.

	Tension Properties [MPa]			**ABS**	**1800 (All Directions)**		
	X	Y	Z	cFRP	X	Y	Z
X	E_x	G_{xy}	G_{xz}	X	80,000	9000	9000
Y	-	E_y	G_{yz}	Y	-	6000	8000
Z	-	-	E_z	Z	-	-	6000
	Compression Properties [MPa]			**ABS**	**1800 (All Directions)**		
	X	Y	Z	cFRP	X	Y	Z
X	E_x	G_{xy}	G_{xz}	X	45,000	9000	9000
Y	-	E_y	G_{yz}	Y	-	6000	8000
Z	-	-	E_z	Z	-	-	6000
	Poisson Constant [-]			**ABS**	**0.35 (All Directions)**		
	X	Y	Z	cFRP	X	Y	Z
X	-	ν_{xy}	ν_{xz}	X	-	0.2	0.2
Y	-	-	ν_{yz}	Y	-	-	0.35
Z	-	-	-	Z	-	-	-

3.3. General Design Guidelines

The mechanical tests confirmed that the two-step manufacturing method developed significantly increases the mechanical properties of additively manufactured polymer structures. The next objective is a demonstrative implementation of the manufacturing method. For this purpose, general design guidelines must be developed. For the preparation of design guidelines, it is necessary to investigate which restrictions exist for the design of the component-integrated channels. This includes the design of the cross-sectional area and the arrangement of the channels. Central issues are that the basic structure can be produced by means of AM and that the impregnated cF bundles can be pulled through the channels.

In AM using the FDM process, it should be noted that overhangs >45° can only be realized with an additional support structure. This is usually washed out with a specific solvent. The support structure inside the channel cannot be washed out, because the solvent does not circulate sufficiently strongly. Therefore, when using the FDM process, a drop-shaped channel geometry (Figure 20) is preferred. This has an overhang that is smaller than 45° and can be realized without a support structure. In addition, the minimum wall thickness should not be less than 2 mm. Experiments with a lower wall thickness have shown that a fracture in the additively manufactured basic structure near a channel radius can occur when the impregnated fibers are pulled through the channel.

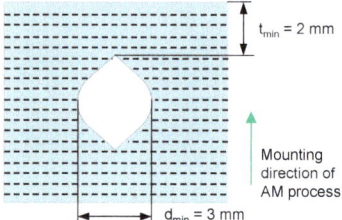

Figure 20. Design guideline for the channel cross-section in drop shape to avoid overhangs greater than 45° and a minimum wall thickness of 2 mm.

The channel length is a limiting design element in the manufacturing method developed. As the impregnated cF bundles are drawn in, friction occurs between the impregnated cF bundles and the channel wall of the additively manufactured basic structure. The frictional force to exceed increases with length of the channel. Tests were carried out with different channel lengths, and the traction force was measured. The test specimen and the results are shown in Figure 21. A maximum length of 800 mm is planned for the intended demonstrator, which was considered feasible for the experimental investigations carried out.

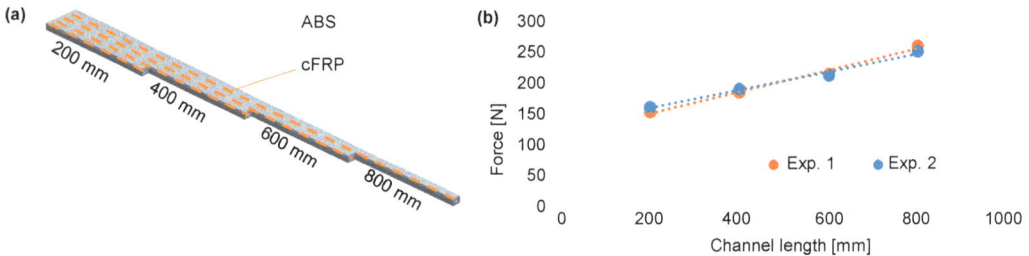

Figure 21. CAD model of specimens with different channel lengths (**a**) and traction force as a function of channel length (**b**).

The advantage of the presented method lies in the variable–axial design of the path of the cFRP. For this purpose, deflections of the fiber path must be realized. It was investigated whether a strong channel curvature, i.e., a small channel radius, has a negative influence on the impregnation quality and the interfacing with the component wall. For this purpose, a sample part was made with channels having different radii (Figure 22) and reinforced with cFRPs accordingly. Subsequently, samples were taken at the radius inlet and outlet, and quality control was carried out by means of micrographs. It is shown that at the minimum radius of 12.5 mm after channel curvature, the impregnation quality is similar to that before channel curvature. This channel radius is sufficient for the intended demonstrator.

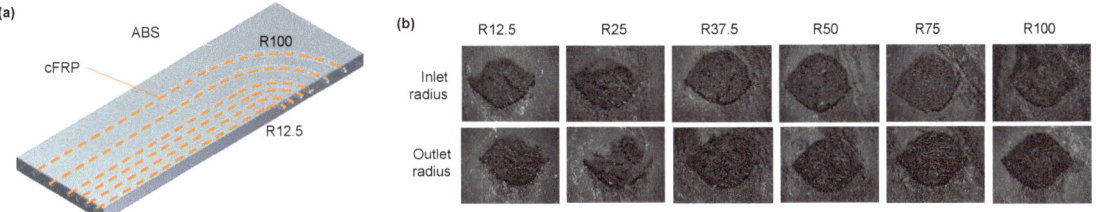

Figure 22. CAD model of the specimens with different channel radii (**a**) and micrographs for qualitative assessment of impregnation quality at radius inlet and outlet for different channel radii (**b**).

3.4. Structure Optimization and Design of Demonstrator Cantilevers

3.4.1. Topology Uptimization and Uiber Urientation

The progress of the combined optimization methods SKO and CAIO is shown in Figure 23. Visible is the topology optimization, where the elements with the minimum Young's modulus E_{min} = 6 MPa are hidden.

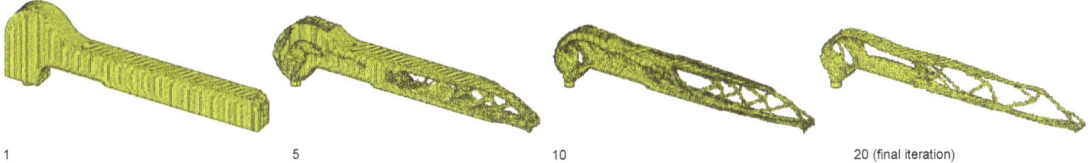

Figure 23. Development of topology optimization after selected iterations; the elements with the minimal Young's modulus E_{min} are hidden. The numbers given indicate the iteration number.

In addition to the consideration of orthotropic material properties, the integration of the CAIO approach allows for the element orientation to be adjusted in each iteration step. The compressive and tensile properties of the material are also taken into account and assigned accordingly in each case (Figure 24). This is necessary because the fiber-reinforced material used here has different tensile and compressive properties.

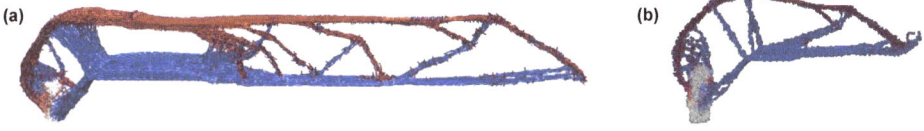

Figure 24. Optimization result of the combined fiber and topology optimization used to create the demonstrators. (**a**) Long version with 700 mm; (**b**) short version with 320 mm. The main stress directions are visualized in color (tension fibers in red and compression fibers in blue; visualization with Paraview, open source).

The development of a short (320 mm) and a long (700 mm) demonstrator shows different arrangements of supporting structures in the components.

3.4.2. Development of the Design Concept for the Fiber Channels

One special challenge within the method presented here is that there is currently no automated implementation option for all model requirements. In particular, the alignment of the fiber channels according to the determined design specifications must be emphasized here. Not only the minimum radius of 12.5 mm and the maximum length of 800 mm had to be taken into account. It was also necessary to minimize the number of radii in every single channel and to accept that the reinforced paths could not meet in a node, like in a space frame truss, but had to pass next to each other.

Consequently, the design proposal determined by the topology and fiber optimization must be abstracted and converted "manually", here, in CAD, into a wireframe model first (Figure 25). The cross-section for the fiber channels is created by thickening the wireframe model into a drop-shape form (Figure 26) and surrounded with matrix material (Figure 27). By this robust procedure, a basic geometry of the optimization results can be generated in a few steps and relatively short time and finally equipped with appropriate interfaces. For the professional use of the numerical tool, corresponding routines have to be further developed and implemented in the software packages.

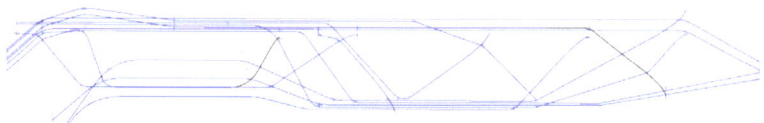

Figure 25. Creation of the CAD model for the demonstrator by generating a "wire model" consisting of lines and deflections.

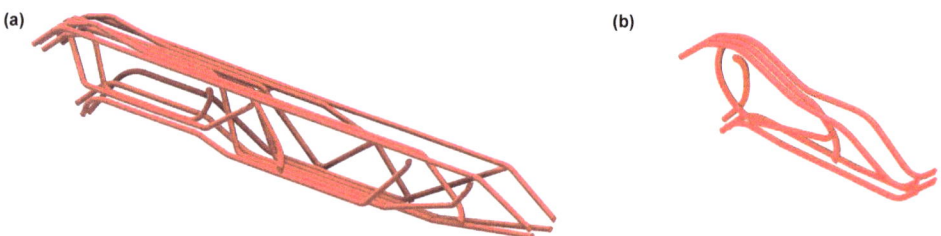

Figure 26. Realized fiber channels of the 700 mm demonstrator (**a**) as well as the 320 mm demonstrator (**b**); the channel diameter is 6.6 mm in each case.

Figure 27. Final CAD model of the 700 mm demonstrator (**a**) and the 320 mm demonstrator (**b**). Light red: input and output of the fiber channels into the construction.

3.4.3. Development and Preparation for Production of the Optimized Lightweight Structures

The final conception of the fiber channels (Figure 26) results in 13 fiber channels for the long version of the demonstrator; the short demonstrator is realized with 7 channels. Their channel diameters are 6.6 mm each. The cross-sectional shape of the fiber channels is approximately teardrop shaped to avoid overhangs greater than 45°. This means that no support structures are required within the channels during additive manufacturing. The matrix material is deposited around the fiber channels in the following process step (Figure 27).

Using the material components for the demonstrators as an example (ABS matrix: 1.05 g/cm^3, cFRP (FVR 41%): 1.4 g/cm^3), the mass of the 700 mm long model that can be extracted from the model simulation would be approx. 1164 g, and that of the 320 mm model would be approx. 349 g.

3.4.4. Numerical Validation of the Generated CAD Model

The numerical validation of the models takes into account both the matrix component and the fiber channels, i.e., their changed material characteristics compared to the properties of the matrix material (see Sections 2.5.3 and 2.5.4). This allows for the compression- and tension-dominated regions within the channels to be calculated correctly.

Both models, the long version and the short one, are calculated and loaded in the same way with respect to both the material characteristics and the load cases (Figure 28). Table 6 provides the values from the validation for both demonstrator variants.

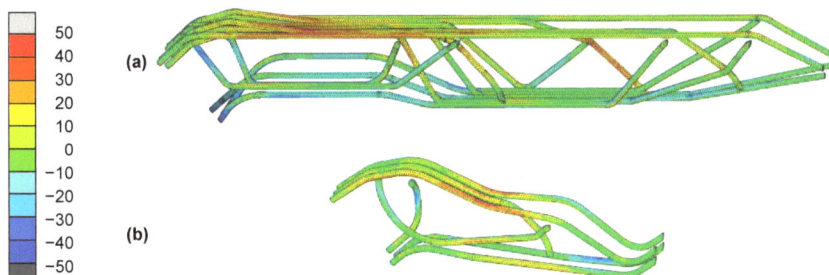

Figure 28. Major principal value of stress [MPa] in the fiber channels of the 700 mm demonstrator (**a**) and the 320 mm demonstrator (**b**). The surrounding ABS matrix is hidden to make the fiber channels visible.

Table 6. Results of the FEA validation.

	700 mm Demonstrator	320 mm Demonstrator
Max. tensile stress [MPa]	48	49
Max. compressive stress [MPa]	−66	−48
Max. total deformation [mm]	6.4	2.3
Max. lowering [mm]	6.2	1.5

The relevant yield strength for the fiber material is 57.9 MPa for tension and −127.8 MPa for compression. The major principal stress in the fiber channels does not exceed this limit, as shown in Figure 28, but due to the manual construction, not all areas are stressed evenly.

The analysis of the von Mises yield criterion in the demonstrator matrix shows that the stress is only slightly increased in the bearing area (Figure 29). The stresses determined in the FEA simulation are, in any case, below the yield strength of the material. These results indicate that the main load is carried by the fiber channels as intended.

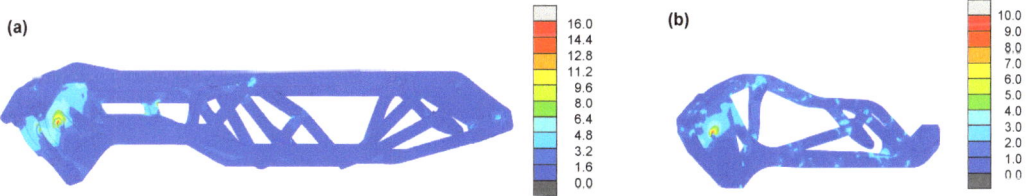

Figure 29. Von Mises yield criterion in the matrix of (**a**) the 700 mm demonstrator and (**b**) the 320 mm demonstrator (matrix material characteristics: isotropic, E = 1800 MPa, and ν = 0.35). The color reference is given in the unit MPa for both cases. To make the stress in the surrounding ABS matrix visible, the fiber channels are hidden. The large blue areas indicate that the matrix must endure little stress; only the bearing area shows slightly increased stress values.

3.5. Manufacturing and Validation of Demonstrator Cantilevers

The fabricated demonstrators are shown in Figure 30. In addition, a transparent structure was made to visualize the cF orientation. The demonstrator structure "long horizontal cantilever" was subjected to computed tomography (CT) to analyze the quality of the inserted cFRP (Figure 31). This revealed that air inclusions systematically occur in the channel radii. When the impregnated cFs are pulled in, they contact the inner channel radius, so that there is no connection to the component wall of the additively manufactured basic structure at the outer channel radius. The experimental investigations carried out at the specimen level have shown that the interface between the cFRP and the additively manufactured basic structure is decisive for the load-bearing capacity of the overall structure. Consequently, in further studies, the focus should be on eliminating the systematic faults by taking suitable measures.

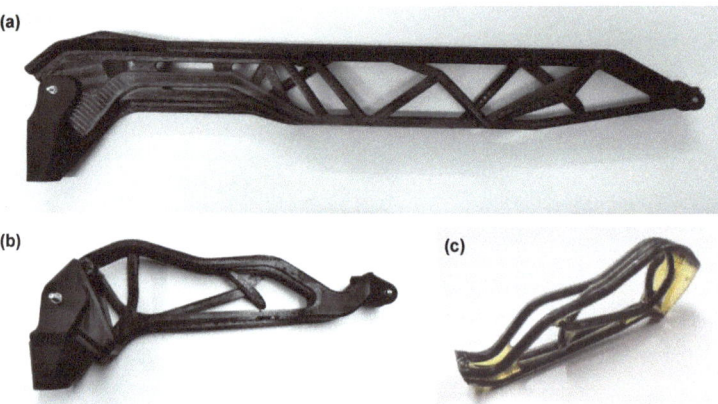

Figure 30. Demonstration structures manufactured using novel AM-cFRP method. (**a**) General view of long cantilever. (**b**) General view of short cantilever. (**c**) Short cantilever in transparent design shows fiber orientation.

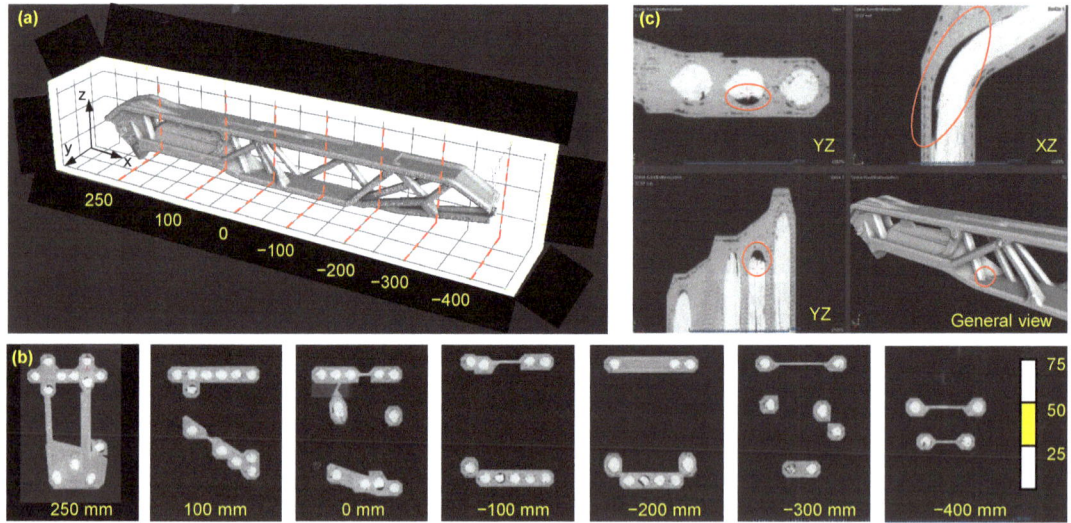

Figure 31. CT analysis of demonstrator cantilever. (**a**) General view with YZ cutting planes. (**b**) CT images on selected YZ section views. (**c**) Representative CT image at an exemplary deflection point with marked air inclusion in channel radius.

The developed cantilevers were implemented and validated in a functional demonstrator of an exoskeleton (Figure 32). For this purpose, corresponding interfaces of the cantilever were designed to enable a connection with the existing exoskeleton.

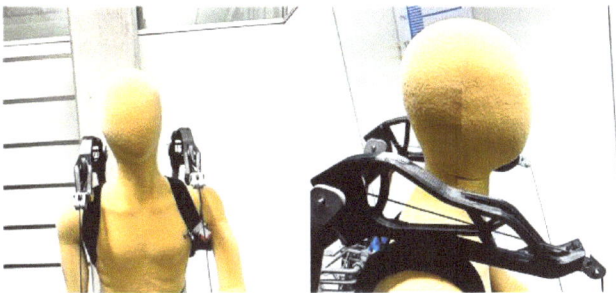

Figure 32. Integration of the developed (short) cantilevers into an existing test exoskeleton.

The cantilevers realized using the novel method were compared with existing unreinforced additively manufactured cantilevers. The newly developed load-bearing structures are significantly superior to the previously used ones. While the deflections of the unreinforced cantilevers were too large and thus unsuitable for the intended application, the deflections of the cantilevers have been considerably reduced to a suitable value by using the novel method. The results of the demonstrator validation proved the feasibility of the novel method for reinforcing additively manufactured polymer structures with component-integrated cFRPs.

4. Discussion

Compared to established cF-AM methods, the main advantage of the developed method is primarily the possibility of arranging reinforcing fibers variably–axially along the principal stress lines and independently from the AM mounting direction. The studies carried out have clearly demonstrated the feasibility of the method.

To assess the potential of the method, the results of the study were benchmarked against conventional FRP technologies, short-fiber AM and cF-AM methods (Figure 33). Especially compared to other cF-AM methods, the novel method is still in a lower performance range in terms of tensile strength in the fiber direction. A current overview regarding the mechanical properties of various cF-AM methods is provided by Safari et al. [21]. Studies presenting cF-AM methods with high tensile strength are, for example, [43–53].

The interface of the cFRP structure at the channel wall of the additively manufactured basic structure has emerged as a central, significant influencing factor. From the specimen and demonstrator validation, there is potential for optimizing the method in further studies. The following hypotheses are derived from the findings of the study:

(1) The interlaminar strength is significantly influenced by the amount of surface area covered with matrix material on the component-integrated channel wall. An adapted channel geometry with an increased surface will consequently result in an improvement in interlaminar strength.

(2) The interlaminar strength is essentially determined by the adhesive interaction between the matrix material and the polymer used for AM of the basic structure. A targeted modification or substitution of the material components will therefore result in an increase in the interlaminar strength.

(3) The matrix material must ideally cover the channel surface completely to create the conditions for a maximum number of adhesion points in the bonding zone. Consequently, the increase in impregnation quality will be associated with an improvement in the interlaminar strength.

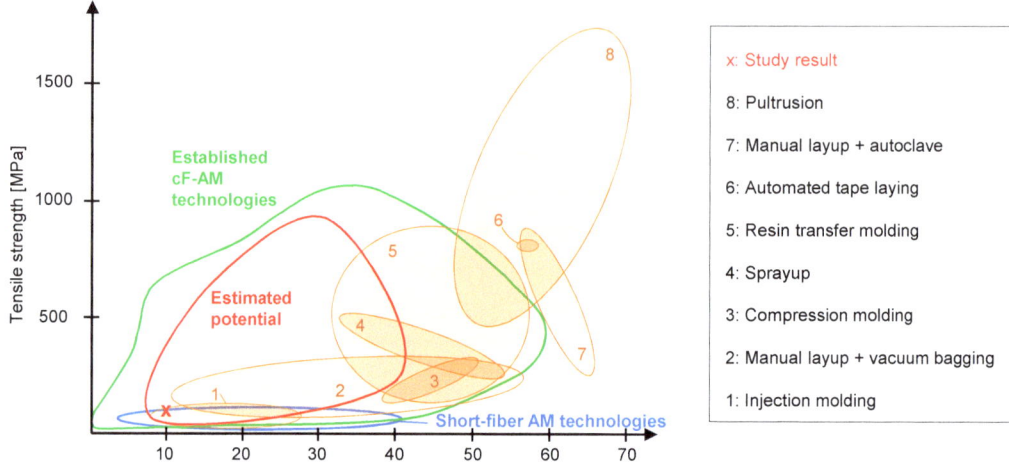

Figure 33. FVR and tensile strengths achievable with conventional FRP processes, short-fiber-reinforced AM, and established cF-reinforced AM methods (inspired by [21,23]). Assessment of the study results shows that the novel method is currently still at a lower performance level. However, a significant increase is expected due to an improvement in interlaminar strength.

5. Conclusions

A state-of-the-art analysis showed that processes integrating cFs in AM are very efficient in terms of their mechanical properties and can be comparable to established FRP processes. However, based on the evaluation of existing cF-AM methods, it was also determined that cF integration during the AM process can partially exclude applications as soon as an in-plane arrangement of the cF is not sufficient:

(1) The deficit of existing cF-AM methods regarding the high degree of structural anisotropy could be solved with the presented method, in which the cF integration is decoupled from the AM process.
(2) The method allows for a variable–axial arrangement of cFs independently of the AM mounting direction.
(3) The investigations at the specimen level evidenced a significant increase in mechanical properties regarding its tensile, compressive, and flexural properties compared with unreinforced specimens.
(4) Design restrictions in the execution of the component-integrated channels regarding the channel geometry, length, and radius were demonstrated based on sample structures. Design guidelines were derived.
(5) The proof of concept was provided by means of demonstrators.
(6) The numerical tool, which combines two methods of structural optimization, is well suited for the design of additively manufactured polymer structures with variable–axial cF reinforcement.
(7) As a result of the load-adjusted integration of cFRP in additively manufactured polymer structures, a significantly higher load-bearing capacity, an increased degree of lightweight construction, and a reduced material usage were realized compared with unreinforced additively manufactured polymer structures.
(8) The method is predestined to realize geometrically complex lightweight structures in single-part and small-series production.
(9) Compared to other cf-AM methods, the novel method is still in a lower performance range in terms of tensile strength in the fiber direction. Promising approaches to further improve the mechanical properties and reproducibility were presented. These

relate to the modification of the material composition, channel geometry, and impregnation quality.

Author Contributions: Conceptualization, S.M. (Sven Meißner); methodology—testing, S.M. (Sven Meißner) and J.K.; methodology—optimization, H.I. and S.L.; methodology—application scenario, O.E. and H.K.; validation, S.M. (Sven Meißner), H.I. and O.E.; investigation, J.K. and H.I.; writing—original draft preparation, S.M. (Sven Meißner); writing—review and editing, H.I., S.L., A.K., S.S., S.M. (Sascha Müller), A.S. and L.K.; visualization, D.K.; project administration, S.M. (Sven Meißner), A.K. and H.K.; funding acquisition, S.M. (Sven Meißner), A.K. and H.K. All authors have read and agreed to the published version of the manuscript.

Funding: This research was funded by the German Federal Ministry of Education and Research (BMBF) as part of the project "AM-Reinforce", grant numbers 02P20E160, 02P20E161, and 02P20E162.

Data Availability Statement: Data are contained within the article.

Conflicts of Interest: Oliver Eberhardt and Harald Kuolt were employed by the company J. Schmalz GmbH. The remaining authors declare that the research was conducted in the absence of any commercial or financial relationships that could be construed as a potential conflict of interest.

References

1. Wohlers, T.T.; Campbell, I.; Diegel, O.; Huff, R.; Kowen, J. *Wohlers Report 2022: 3D Printing and Additive Manufacturing Global State of the Industry*; Wohlers Associates: Fort Collins, CO, USA, 2022; ISBN 9780991333295.
2. Rehnberg, M.; Ponte, S. From smiling to smirking? 3D printing, upgrading and the restructuring of global value chains. *Glob. Netw.* **2018**, *18*, 57–80. [CrossRef]
3. Kuncius, T.; Rimašauskas, M.; Rimašauskienė, R. Interlayer Adhesion Analysis of 3D-Printed Continuous Carbon Fibre-Reinforced Composites. *Polymers* **2021**, *13*, 1653. [CrossRef] [PubMed]
4. Kristiawan, R.B.; Imaduddin, F.; Ariawan, D.; Ubaidillah; Arifin, Z. A review on the fused deposition modeling (FDM) 3D printing: Filament processing, materials, and printing parameters. *Open Eng.* **2021**, *11*, 639–649. [CrossRef]
5. Syrlybayev, D.; Zharylkassyn, B.; Seisekulova, A.; Akhmetov, M.; Perveen, A.; Talamona, D. Optimisation of Strength Properties of FDM Printed Parts—A Critical Review. *Polymers* **2021**, *13*, 1587. [CrossRef] [PubMed]
6. Thumsorn, S.; Prasong, W.; Kurose, T.; Ishigami, A.; Kobayashi, Y.; Ito, H. Rheological Behavior and Dynamic Mechanical Properties for Interpretation of Layer Adhesion in FDM 3D Printing. *Polymers* **2022**, *14*, 2721. [CrossRef] [PubMed]
7. Wach, R.A.; Wolszczak, P.; Adamus-Wlodarczyk, A. Enhancement of Mechanical Properties of FDM-PLA Parts via Thermal Annealing. *Macromol. Mater. Eng.* **2018**, *303*, 1800169. [CrossRef]
8. Spoerk, M.; Gonzalez-Gutierrez, J.; Sapkota, J.; Schuschnigg, S.; Holzer, C. Effect of the printing bed temperature on the adhesion of parts produced by fused filament fabrication. *Plast. Rubber Compos.* **2018**, *47*, 17–24. [CrossRef]
9. Popescu, D.; Zapciu, A.; Amza, C.; Baciu, F.; Marinescu, R. FDM process parameters influence over the mechanical properties of polymer specimens: A review. *Polym. Test.* **2018**, *69*, 157–166. [CrossRef]
10. Mohamed, O.A.; Masood, S.H.; Bhowmik, J.L. Optimization of fused deposition modeling process parameters: A review of current research and future prospects. *Adv. Manuf.* **2015**, *3*, 42–53. [CrossRef]
11. Li, H.; Wang, T.; Sun, J.; Yu, Z. The effect of process parameters in fused deposition modelling on bonding degree and mechanical properties. *Rapid Prototyp. J.* **2018**, *24*, 80–92. [CrossRef]
12. Wickramasinghe, S.; Do, T.; Tran, P. FDM-Based 3D Printing of Polymer and Associated Composite: A Review on Mechanical Properties, Defects and Treatments. *Polymers* **2020**, *12*, 1529. [CrossRef] [PubMed]
13. Chen, H.; Zhu, W.; Tang, H.; Yan, W. Oriented structure of short fiber reinforced polymer composites processed by selective laser sintering: The role of powder-spreading process. *Int. J. Mach. Tools Manuf.* **2021**, *163*, 103703. [CrossRef]
14. van de Werken, N.; Tekinalp, H.; Khanbolouki, P.; Ozcan, S.; Williams, A.; Tehrani, M. Additively manufactured carbon fiber-reinforced composites: State of the art and perspective. *Addit. Manuf.* **2020**, *31*, 100962. [CrossRef]
15. Salazar, A.; Rico, A.; Rodríguez, J.; Segurado Escudero, J.; Seltzer, R.; La Martin de Escalera Cutillas, F. Fatigue crack growth of SLS polyamide 12: Effect of reinforcement and temperature. *Compos. Part B Eng.* **2014**, *59*, 285–292. [CrossRef]
16. Badini, C.; Padovano, E.; de Camillis, R.; Lambertini, V.G.; Pietroluongo, M. Preferred orientation of chopped fibers in polymer-based composites processed by selective laser sintering and fused deposition modeling: Effects on mechanical properties. *J. Appl. Polym. Sci.* **2020**, *137*, 49152. [CrossRef]
17. Dickson, A.N.; Abourayana, H.M.; Dowling, D.P. 3D Printing of Fibre-Reinforced Thermoplastic Composites Using Fused Filament Fabrication—A Review. *Polymers* **2020**, *12*, 2188. [CrossRef] [PubMed]
18. Ahmadifar, M.; Benfriha, K.; Shirinbayan, M.; Tcharkhtchi, A. Additive Manufacturing of Polymer-Based Composites Using Fused Filament Fabrication (FFF): A Review. *Appl. Compos. Mater.* **2021**, *28*, 1335–1380. [CrossRef]
19. Pratama, J.; Cahyono, S.I.; Suyitno, S.; Muflikhun, M.A.; Salim, U.A.; Mahardika, M.; Arifvianto, B. A Review on Reinforcement Methods for Polymeric Materials Processed Using Fused Filament Fabrication (FFF). *Polymers* **2021**, *13*, 4022. [CrossRef] [PubMed]

20. Sculpteo—The State of 3D Printing 2021. Available online: https://www.sculpteo.com/en/ebooks/state-of-3d-printing-report-2021/ (accessed on 11 August 2023).
21. Safari, F.; Kami, A.; Abedini, V. 3D printing of continuous fiber reinforced composites: A review of the processing, pre- and post-processing effects on mechanical properties. *Polym. Polym. Compos.* **2022**, *30*, 096739112210987. [CrossRef]
22. Mashayekhi, F.; Bardon, J.; Berthé, V.; Perrin, H.; Westermann, S.; Addiego, F. Fused Filament Fabrication of Polymers and Continuous Fiber-Reinforced Polymer Composites: Advances in Structure Optimization and Health Monitoring. *Polymers* **2021**, *13*, 789. [CrossRef] [PubMed]
23. Goh, G.D.; Yap, Y.L.; Agarwala, S.; Yeong, W.Y. Recent Progress in Additive Manufacturing of Fiber Reinforced Polymer Composite. *Adv. Mater. Technol.* **2019**, *4*, 1800271. [CrossRef]
24. Pandelidi, C.; Bateman, S.; Piegert, S.; Hoehner, R.; Kelbassa, I.; Brandt, M. The technology of continuous fibre-reinforced polymers: A review on extrusion additive manufacturing methods. *Int. J. Adv. Manuf. Technol.* **2021**, *113*, 3057–3077. [CrossRef]
25. Zhuo, P.; Li, S.; Ashcroft, I.A.; Jones, A.I. Material extrusion additive manufacturing of continuous fibre reinforced polymer matrix composites: A review and outlook. *Compos. Part B Eng.* **2021**, *224*, 109143. [CrossRef]
26. Struzziero, G.; Barbezat, M.; Skordos, A.A. Consolidation of continuous fibre reinforced composites in additive processes: A review. *Addit. Manuf.* **2021**, *48*, 102458. [CrossRef]
27. Spickenheuer, A. Zur Fertigungsgerechten Auslegung von Faser-Kunststoff-Verbundbauteilen für den Extremen Leichtbau auf Basis des Variabelaxialen Fadenablageverfahrens Tailored Fiber Placement. Ph.D. Thesis, Technische Universität Dresden, Dresden, Germany, 2014.
28. Holzinger, M.; Blase, J.; Reinhardt, A.; Kroll, L. New additive manufacturing technology for fibre-reinforced plastics in skeleton structure. *J. Reinf. Plast. Compos.* **2018**, *37*, 1246–1254. [CrossRef]
29. Hirsch, P.; Scholz, S.; Borowitza, B.; Vyhnal, M.; Schlimper, R.; Zscheyge, M.; Kotera, O.; Stipkova, M.; Scholz, S. Processing and Analysis of Hybrid Fiber-Reinforced Polyamide Composite Structures Made by Fused Granular Fabrication and Automated Tape Laying. *J. Manuf. Mater. Process.* **2024**, *8*, 25. [CrossRef]
30. Habenicht, G. *Kleben*; Springer: Berlin/Heidelberg, Germany, 2009; ISBN 978-3-540-85264-3.
31. ISO 527-4:2023; Plastics—Determination of Tensile Properties—Part 4: Test Conditions for Isotropic and Orthotropic Fibre-Reinforced Plastic Composites. International Organization for Standardization (ISO): Geneva, Switzerland, 2023.
32. ISO 14125:1998; Fibre-Reinforced Plastic Composites—Determination of Flexural Properties. International Organization for Standardization (ISO): Geneva, Switzerland, 1998.
33. ISO 14126:2023; Fibre-Reinforced Plastic Composites—Determination of Compressive Properties In the in-Plane Direction. International Organization for Standardization (ISO): Geneva, Switzerland, 2023.
34. DIN EN ISO 527-2-2012; Plastics—Determination of Tensile Properties—Part 2: Test Conditions for Moulding and Extrusion Plastics. Beuth Verlag GmbH: Berlin, Germany, 2012.
35. J. Schmalz GmbH. 2023. Available online: https://www.schmalz.com/de-de/ (accessed on 12 December 2023).
36. Hamm, C. *Evolution of Lightweight Structures*; Springer: Dordrecht, The Netherlands, 2015; ISBN 978-94-017-9397-1.
37. Mattheck, C. Engineering Components grow like trees. *Mater. Werkst.* **1990**, *21*, 143–168. [CrossRef]
38. Baumgartner, A.; Harzheim, L.; Mattheck, C. SKO (soft kill option): The biological way to find an optimum structure topology. *Int. J. Fatigue* **1992**, *14*, 387–393. [CrossRef]
39. Kriechbaum, R. Ein Verfahren zur Optimierung der Faserverläufe in Verbundwerkstoffen durch Minimierung der Schubspannungen nach Vorbildern der Natur. Ph.D. Thesis, Universität Karlsruhe, Karlsruhe, Germany, 1994.
40. Zhang, L.; Zhang, H.W.; Wu, J.; Yan, B. A stabilized complementarity formulation for nonlinear analysis of 3D bimodular materials. *Acta Mech. Sin.* **2016**, *32*, 481–490. [CrossRef]
41. Sacco, E.; Reddy, J.N. A Constitutive Model for Bimodular Materials with an Application to Plate Bending. *J. Appl. Mech.* **1992**, *59*, 220–221. [CrossRef]
42. Ogden, R.W.; Ogden, R.W. *Non-Linear Elastic Deformations*; 1. Publ., Unabridged and Corr. Republ.; Dover Publications: Mineola, NY, USA, 1997; ISBN 0486696480.
43. Ueda, M.; Kishimoto, S.; Yamawaki, M.; Matsuzaki, R.; Todoroki, A.; Hirano, Y.; Le Duigou, A. 3D compaction printing of a continuous carbon fiber reinforced thermoplastic. *Compos. Part A Appl. Sci. Manuf.* **2020**, *137*, 105985. [CrossRef]
44. Giannakis, E.; Koidis, C.; Kyratsis, P.; Tzetzis, D. Static and fatigue properties of 3d printed continuous continuous carbon fiber nylon composites. *Int. J. Mod. Manuf. Technol.* **2019**, *XI*, 69–76.
45. Hao, W.; Liu, Y.; Zhou, H.; Chen, H.; Fang, D. Preparation and characterization of 3D printed continuous carbon fiber reinforced thermosetting composites. *Polym. Test.* **2018**, *65*, 29–34. [CrossRef]
46. Hou, Z.; Tian, X.; Zheng, Z.; Zhang, J.; Zhe, L.; Li, D.; Malakhov, A.V.; Polilov, A.N. A constitutive model for 3D printed continuous fiber reinforced composite structures with variable fiber content. *Compos. Part B Eng.* **2020**, *189*, 107893. [CrossRef]
47. Iragi, M.; Pascual-González, C.; Esnaola, A.; Lopes, C.S.; Aretxabaleta, L. Ply and interlaminar behaviours of 3D printed continuous carbon fibre-reinforced thermoplastic laminates: effects of processing conditions and microstructure. *Addit. Manuf.* **2019**, *30*, 100884. [CrossRef]
48. Pyl, L.; Kalteremidou, K.-A.; van Hemelrijck, D. Exploration of specimen geometry and tab configuration for tensile testing exploiting the potential of 3D printing freeform shape continuous carbon fibre-reinforced nylon matrix composites. *Polym. Test.* **2018**, *71*, 318–328. [CrossRef]

49. Todoroki, A.; Oasada, T.; Mizutani, Y.; Suzuki, Y.; Ueda, M.; Matsuzaki, R.; Hirano, Y. Tensile property evaluations of 3D printed continuous carbon fiber reinforced thermoplastic composites. *Adv. Compos. Mater.* **2020**, *29*, 147–162. [CrossRef]
50. Zhang, J.; Zhou, Z.; Zhang, F.; Tan, Y.; Tu, Y.; Yang, B. Performance of 3D-Printed Continuous-Carbon-Fiber-Reinforced Plastics with Pressure. *Materials* **2020**, *13*, 471. [CrossRef]
51. Goh, G.D.; Dikshit, V.; Nagalingam, A.P.; Goh, G.L.; Agarwala, S.; Sing, S.L.; Wei, J.; Yeong, W.Y. Characterization of mechanical properties and fracture mode of additively manufactured carbon fiber and glass fiber reinforced thermoplastics. *Mater. Des.* **2018**, *137*, 79–89. [CrossRef]
52. Ghebretinsae, F.; Mikkelsen, O.; Akessa, A.D. Strength analysis of 3D printed carbon fibre reinforced thermoplastic using experimental and numerical methods. *IOP Conf. Ser. Mater. Sci. Eng.* **2019**, *700*, 012024. [CrossRef]
53. Dutra, T.A.; Ferreira, R.T.L.; Resende, H.B.; Guimarães, A. Mechanical characterization and asymptotic homogenization of 3D-printed continuous carbon fiber-reinforced thermoplastic. *J. Braz. Soc. Mech. Sci. Eng.* **2019**, *41*, 133. [CrossRef]

Disclaimer/Publisher's Note: The statements, opinions and data contained in all publications are solely those of the individual author(s) and contributor(s) and not of MDPI and/or the editor(s). MDPI and/or the editor(s) disclaim responsibility for any injury to people or property resulting from any ideas, methods, instructions or products referred to in the content.

Article

Investigating Microstructural and Mechanical Behavior of DLP-Printed Nickel Microparticle Composites

Benny Susanto [1,†], Vishnu Vijay Kumar [2,†], Leonard Sean [3,†], Murni Handayani [4], Farid Triawan [5], Yosephin Dewiani Rahmayanti [4], Haris Ardianto [3,6] and Muhammad Akhsin Muflikhun [3,*]

1. PLN Puslitbang, Jl. Duren Tiga Raya No.102, Pancoran, Kota Jakarta Selatan, Jakarta 12760, Indonesia
2. International Institute of Aerospace Engineering and Management, Jain Deemed-to-be-University, JGI Global Campus, Bangalore 562112, India
3. Department of Mechanical and Industrial Engineering, Universitas Gadjah Mada, Jl. Grafika No. 2, Yogyakarta 55281, Indonesia
4. Research Center for Nanotechnology Systems, National Research and Innovation Agency (BRIN), Puspiptek Area, Tangerang Selatan 15314, Indonesia
5. Department of Mechanical Engineering, Sampoerna University, Jl. Raya Pasar Minggu No.Kav. 16, Kec. Pancoran, Jakarta 12780, Indonesia
6. Department of Aerospace Engineering, Sekolah Tinggi Teknologi Kedirgantaraan, Jl. Parangtritis km. 4,5, Yogyakarta 55281, Indonesia
* Correspondence: akhsin.muflikhun@ugm.ac.id
† These authors contributed equally to this work.

Abstract: The study investigates the fabrication and analysis of nickel microparticle-reinforced composites fabricated using the digital light processing (DLP) technique. A slurry is prepared by incorporating Ni-micro particles into a resin vat; it is thoroughly mixed to achieve homogeneity. Turbidity fluctuations are observed, initially peaking at 50% within the first two minutes of mixing and then stabilizing at 30% after 15–60 min. FTIR spectroscopy with varying Ni wt.% is performed to study the alterations in the composite material's molecular structure and bonding environment. Spectrophotometric analysis revealed distinctive transmittance signatures at specific wavelengths, particularly within the visible light spectrum, with a notable peak at 532 nm. The effects of printing orientation in the X, Y, and Z axes were also studied. Mechanical properties were computed using tensile strength, surface roughness, and hardness. The results indicate substantial enhancements in the tensile properties, with notable increases of 75.5% in the ultimate tensile strength and 160% in the maximum strain. Minimal alterations in surface roughness and hardness suggest favorable printability. Microscopic examination revealed characteristic fracture patterns in the particulate composite at different values for the wt.% of nickel. The findings demonstrate the potential of DLP-fabricated Ni-reinforced composites for applications demanding enhanced mechanical performance while maintaining favorable printability, paving the way for further exploration in this domain.

Keywords: particulate composites; additive manufacturing; nickel microparticles; digital light processing

Citation: Susanto, B.; Kumar, V.V.; Sean, L.; Handayani, M.; Triawan, F.; Rahmayanti, Y.D.; Ardianto, H.; Muflikhun, M.A. Investigating Microstructural and Mechanical Behavior of DLP-Printed Nickel Microparticle Composites. *J. Compos. Sci.* **2024**, *8*, 247. https://doi.org/10.3390/jcs8070247

Academic Editor: Francesco Tornabene

Received: 30 April 2024
Revised: 25 May 2024
Accepted: 28 May 2024
Published: 29 June 2024

Copyright: © 2024 by the authors. Licensee MDPI, Basel, Switzerland. This article is an open access article distributed under the terms and conditions of the Creative Commons Attribution (CC BY) license (https://creativecommons.org/licenses/by/4.0/).

1. Introduction

Composite materials are used in every field of engineering, like aerospace, construction, communication, military, ocean structure, and various other high-performance applications, owing to their high specific strength and modulus, increased design flexibility, desirable thermal expansion characteristics, good resistance to fatigue and corrosion, and economic efficiency [1–3]. Additive manufacturing (AM) has redefined the design process and the manufacturing of materials, components, and products in different sectors [4,5]. Among the many AM processes, digital light processing (DLP) printing has attracted significant interest, as it is capable of generating complex geometries with high accuracy and resolution. AM technology using photosensitive resin was first developed to cure resins using a precisely controlled laser involving DLP. Studies have shown that it is possible to develop a method for printing large models using a small DLP printer [6]. Introducing microparticles during AM processes is an important area for research because of the possibility of numerous property improvements such as mechanical, chemical, electrical, biological, and thermal characteristics of the printed components [7–9]. Though the introduction of microparticles has shown improvements in properties, the refractive index of the printing material strongly influences the curability and curing time of the printing photopolymers.

Micro- and nanoparticle-incorporated composites have shown superior properties [10,11]. A study on incorporating silicon nanoparticles like SiO_2, montmorillonite, and attapulgite found that printing was impossible with more than 10% filler concentration [12]. Nanographite dispersion with stereolithography has shown improvement in properties in the resulting composite [13]. The dispersion of nanoparticles in the base matrix is critical for maintaining homogeneous material characteristics. Agglomeration and sedimentation may occur if the nanoparticle surfaces do not adhere properly to the base matrix [14]. Nickel microparticles are widely employed in various applications due to their unique features, which include high strength, great corrosion resistance, and strong thermal conductivity. Adding nickel (Ni) microparticles to composite materials can improve their mechanical and functional performance. Adding a small quantity of Ni to iron can produce a corrosion-resistant Ni-based stainless steel [15]. Ni is used as a surface coating in electroplating to avoid substrate corrosion [16]. Ni plating has shown excellent biofouling reduction in seawater [17]. In recent years, the lithium battery industry has been inextricably linked with Ni; adding Ni to the battery can boost energy density while simultaneously lowering costs [18]. However, the characteristics of nickel microparticle composite materials created with DLP printing have not been thoroughly investigated.

DLP printing has produced various materials, including polymers, ceramics, and metals. A study on incorporating silver nanoparticles was performed with 3D conductive structures created by integrating silver nitrate into a photocurable oligomer in the presence of appropriate photoinitiators and subjecting them to a digital light system [19]. DLP-based manufacturing of zirconia scaffolds with 2–20% hydroxyapatite composites showed comparable mechanical strength and good cell proliferation and differentiation [20]. DLP-printed nanocomposite samples were reinforced with copper and magnetite nanoparticles and carbon nanofibers aligning and condensing conductive nanoparticles to produce embedded electronic components [21]. DLP printing incorporating metal microparticles remains an area of limited exploration, with only a few studies on this topic.

While some studies have explored the use of metal microparticles with DLP, the application of nickel microparticles as composite materials in DLP printing is still in its early stages. This study investigates the properties of nickel microparticle-reinforced composites fabricated using digital light processing (DLP). The results of this study provide insights into the possible uses of nickel microparticle composite materials created using DLP printing, as well as help develop novel materials for various sectors.

2. Materials and Methods

A photopolymer ANYCUBIC-plant-based UV resin procured from Indonesia was employed for DLP printing using a DLP ANYCUBIC Photon Ultra Printer. The properties of the resin are listed in Table A1 and the specification of the 3D printer is given in Table A2, both in the Appendix section. The nickel microparticle size varies from 5–60 μm, as observed through a digital microscope, as depicted in Figure 1. The printing parameters are given in Table 1. Initially, the resin mixture with nickel powder was prepared by combining 200 mL of resin with the requisite weight percentage in a beaker using a digital Taffware scale. Stirring was conducted with an IKA RW 20 mechanical stirrer at 288 RPM for five minutes to prevent gas bubble formation.

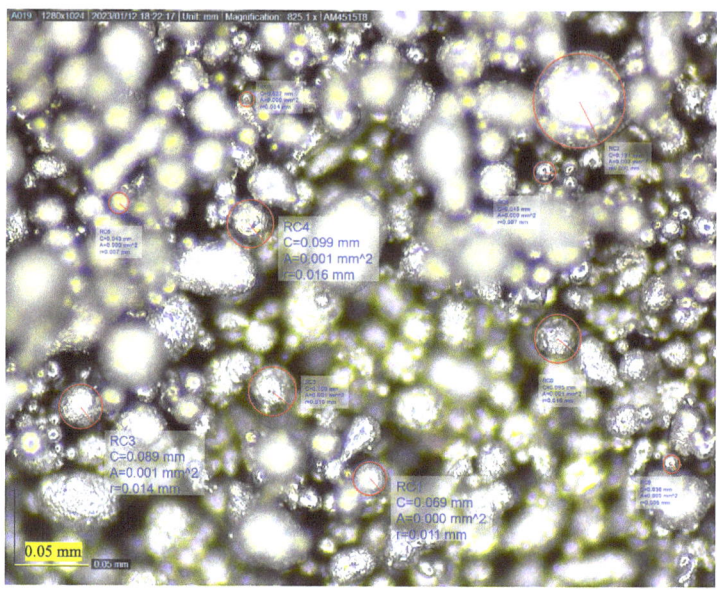

Figure 1. Digital microscope image showing the structure and morphology of the nickel microparticles.

Table 1. Printing parameters.

Printing Parameters	Values	Unit
Exposure time	2	s
Lifting platform height	5	mm
Lifting speed	120	mm/min
Bottom exposure time	35	s
Layer height	0.05	mm
Retracting speed	120	mm/min

The turbidity absorbance was plotted on a mixture of uncured micro-nickel resin with the mixing ratio that demonstrated the highest tensile characteristics while guaranteeing complete mixing. The material was scanned between 320–1100 nm at 0, 2, 4, 6, 8, 10, 15, 20, 25, 30, 40, 50, and 60 min. The results were then compared to those of a solution containing only photoresin. The mixture was then poured into the 3D DLP resin machine reservoir for printing. Subsequently, .stl format files were executed on the DLP printer, with the printing duration determined by the specimen's height along the z-axis. The printed specimens were rinsed in an alcohol solution, followed by curing for 24 h under ambient conditions. The testing involves determining the orientation of the tensile specimen, as categorized into horizontal and vertical printing, as shown in Figure 2.

Tests were conducted under two conditions: without additional particles (0 wt.%) and with 1 wt.% nickel microparticles. Orientation is considered optimal if it exhibits high tensile strength, maximum tensile load, high break elongation, and high break strain, while being generally repeatable. The research methodology is depicted in Figure A1 in Appendix A. The turbidity assessment is performed through spectrophotometric analysis of diverse values of wt.% of nickel microparticles integrated within the resin. FTIR spectroscopy of the resin samples with various nickel concentrations, ranging from 0 wt.% to 8 wt.%., and composed of Ni microparticles, are analyzed. The morphology and chemical composition of the specimens were characterized using various analytical techniques. Prior to analysis, the samples were sputter-coated with a thin film of gold to render them conductive for electron microscopy. Thermo-scientific equipment for scanning electron microscopy (SEM) was employed to examine the surface topography and microstructural features of the specimens. Energy dispersive X-ray spectroscopy (EDS) was performed in conjunction with SEM to obtain localized elemental compositions at the interface regions and neighboring areas. EDS mapping enabled the visualization of the spatial distribution of different elements within the samples, facilitating a comprehensive understanding of their elemental makeup and identifying the specific materials present at various locations.

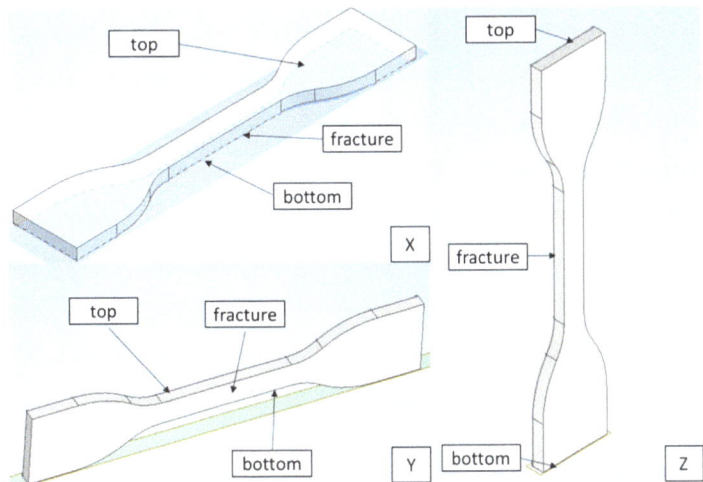

Figure 2. Schematic representation of different printing orientations of the tensile testing specimens: X, Y, and Z.

This study investigates the optimal printing orientation for 3D parts and the ideal nickel wt.% for enhanced performance. The investigation also evaluates the mechanical characteristics of various samples, including tensile strength, surface roughness, and material hardness, as a function of the increasing nickel content in the specimens. Microscopic observations elucidate the failure mechanisms occurring in the composite. The test results we obtained are examined and compared with findings in the literature to reach meaningful conclusions. A dedicated model is employed for conducting surface roughness with standard ISO 2021 [22] Shore D hardness, as per standard ASTM D2240-15 [23], and tensile testing as per ASTM D638 [23] type IV standard. Once the orientation is selected, the investigation determines the appropriate nickel ratio in the particle composite. Tensile tests were performed on photoresins with various nickel mass ratios, ranging from 2% to 8%. The collected data were evaluated to identify the optimal nickel wt.% ratio, which exhibits superior mechanical characteristics, including high tensile strength and maximum tensile load, break elongation, and break strain. The samples were then inspected using an AM4515T8 Dino-lite Edge DINOLITE digital microscope, Taiwan.

3. Results and Discussion

3.1. Spectrophotometric Analysis

Spectrophotometric analysis revealed distinctive transmittance signatures at specific wavelengths, particularly within the visible light spectrum, with a notable peak at 532 nm for all samples except the control, as shown in Figure 3a. This characteristic absorption pattern indicates the unique properties of the resin, which fully absorbs most of the UV light (324–444 nm) and visible light (660–694 nm) associated with the green color of the photoresin. The absorbance spectrum, which is complementary to the transmittance spectrum, provides information about the ability of the resin–micro-Ni mixture to absorb light at different wavelengths. The presence of nickel microparticles initially resulted in high absorbance across the spectrum, as shown in Figure 3b. The analysis further revealed a time-dependent trend in the transmittance and absorbance of the resin–nickel micropowder solution, as represented in Figure 3c.

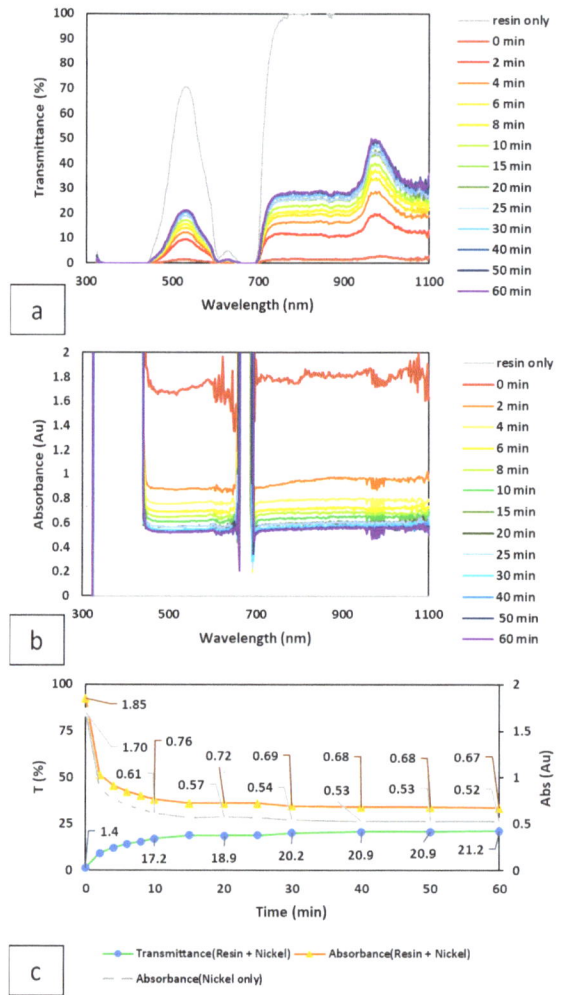

Figure 3. UV-Vis spectrophotometry analysis of the resin–micro-Ni mixture showing (**a**) transmittance spectrum; (**b**) absorbance spectrum; and (**c**) transmission and absorbance of the mixture in the wavelength of 532 nm as a factor of time.

Initially, the presence of nickel microparticles results in high absorbance across the spectrum, gradually transitioning to higher transmittance over time, and corresponding to a decrease in absorbance. The literature suggests a linear correlation between turbidity and absorbance, particularly at a wavelength peak of 532 nm. Examination of the absorbance solely attributed to nickel microparticles revealed a rapid reduction in light absorbance within the initial 2 min, followed by stabilization between 15 and 60 min. This indicates a turbidity reduction of approximately 50% within the first 2 min, stabilizing at approximately 30% thereafter. Previous research highlights the rapid sedimentation of resin–microparticle mixtures, with nanosized particles demonstrating superior suspension stability compared to microparticles [24]. Nonetheless, microparticle suspension stability can still be achieved within a shorter timeframe, typically under 12 h. Interestingly, the observed decrease in light absorbance over time may imply the presence of self-cleaning capabilities in the resin–nickel micropowder composite. As the microparticles settle, the resin surface becomes more transparent, possibly improving the material's capacity to collect and use light energy for a variety of purposes. Further research into the long-term stability and optical performance of this composite system might yield useful insights into its practical uses. Nevertheless, microparticle suspension stability may be attained in a shorter timescale, usually less than 12 h.

3.2. Fourier Transform Infrared (FT-IR) Spectroscopy

The spectral analysis provides insights into the chemical structure and composition variations resulting from the incorporation of nickel microparticles into the resin matrix. Figure 4 depicts the FT-IR spectra of the resin samples with various nickel concentrations ranging from 0 wt.% to 8 wt.%. The FT-IR spectrum of the resin without nickel exhibited distinct peaks at 1722, 1637, and 1110 cm^{-1}, corresponding to C=O, C=C, and C=O bonds, respectively. Upon adding 1, 2, and 4 wt.% nickel, the resin–Ni micro composites displayed similar peaks at 2360, 1670, and 1558 cm^{-1}, indicating consistent chemical compositions. Specifically, the peak at 2360 cm^{-1} is associated with C≡N, while the bands at 1670 and 1558 cm^{-1} correspond to C=C and C=N bonds, respectively. Conversely, resin–Ni micro composites containing 6 and 8 wt.% nickel exhibited additional peaks at 1724 and 1647 cm^{-1}, attributed to C=O and C=C/C=N bonds, respectively, in addition to the characteristic peaks observed in the lower nickel concentration composites. This suggests that the higher concentrations of nickel induce more significant changes in the chemical bonding and molecular structure of the resin. The inference from the FT-IR study suggested that the addition of nickel microparticles to the resin affects the chemical composition, resulting in changes in the FT-IR spectra. New peaks develop, and peak locations vary, indicating changes in the composite material's molecular structure and bonding environment. This information is critical for understanding how the resin matrix interacts with nickel particles, which can affect the composite material's overall characteristics and performance.

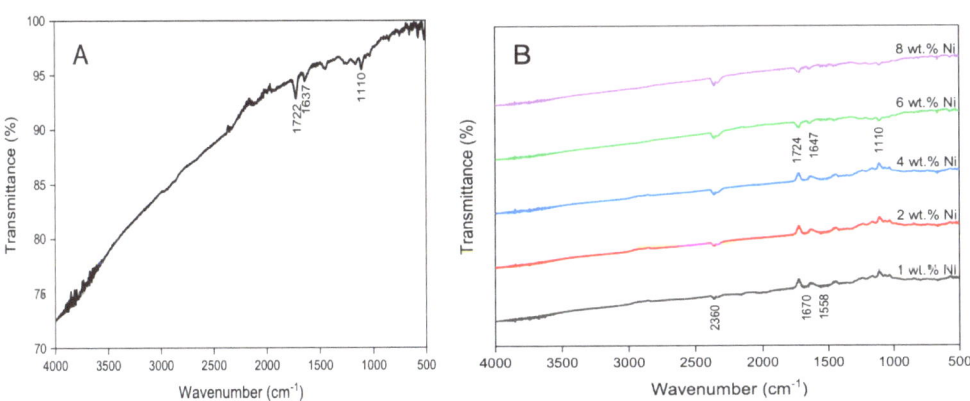

Figure 4. FT–IR spectra of 0 wt.% Ni (**A**) and (**B**) resin–Ni micro composites with different Ni wt.%.

3.3. Tensile Testing

Figure 5 depicts the various specimens after tensile testing. Figure 6 shows the stress–strain curve for the composites with different Ni wt.%. The composite containing 6 wt.% nickel has the capacity to endure the highest maximum load, along with the highest maximum tensile stress, elongation, and strain. The control sample registers the lowest values, followed by notable increases in the 1% and 2% nickel mass fractions. Within the 4% nickel mass fraction mixture, enhancements in the mechanical properties are discernible solely in the maximum load and maximum tensile stress thresholds. Finally, as compared to the 6% counterpart, the 8% nickel mass fraction combination has a greater mechanical characteristic. This trend is consistent with the findings that have reported that the mechanical performance of polymer composites reinforced with ceramic microparticles often determines an optimal filler content, beyond which the properties start to decline due to agglomeration and poor particle dispersion [24,25].

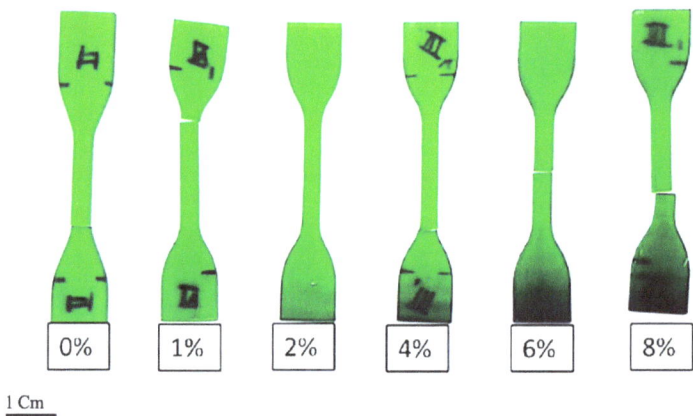

Figure 5. Tensile specimens post-testing.

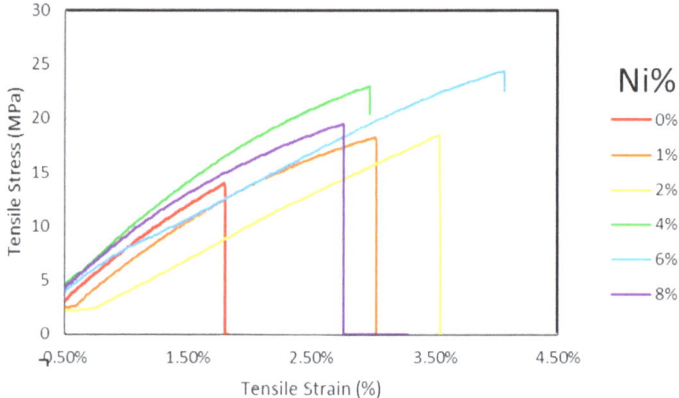

Figure 6. Stress–strain curves from tensile testing of the resin–Ni composites with different Ni wt.%.

The test outcomes were evaluated for three distinct orientations, namely, X, Y, and Z, each with a mass fraction of 0%, as depicted in Figure 7; the orientations follow the configuration shown in Figure 2. These orientations exhibited varying maximum tensile loads, ultimate tensile strengths, break elongations, and break strains. Specimen X demonstrated the highest tensile properties across all parameters, albeit with a noticeable deviation along the X-axis compared to the Z orientation. Conversely, the Y orientation displayed the lowest load-bearing capacity and tensile strength among the three orientations. Notably, the X orientation proved to be the most effective, yielding relatively high values of tensile properties while maintaining repeatability. Further testing with 1 wt.% nickel confirmed the superiority of printing in the Z orientation, as illustrated in Figure 8. The Z orientation exhibited the highest values across all the tested mechanical parameters, with minimal deviations [26,27].

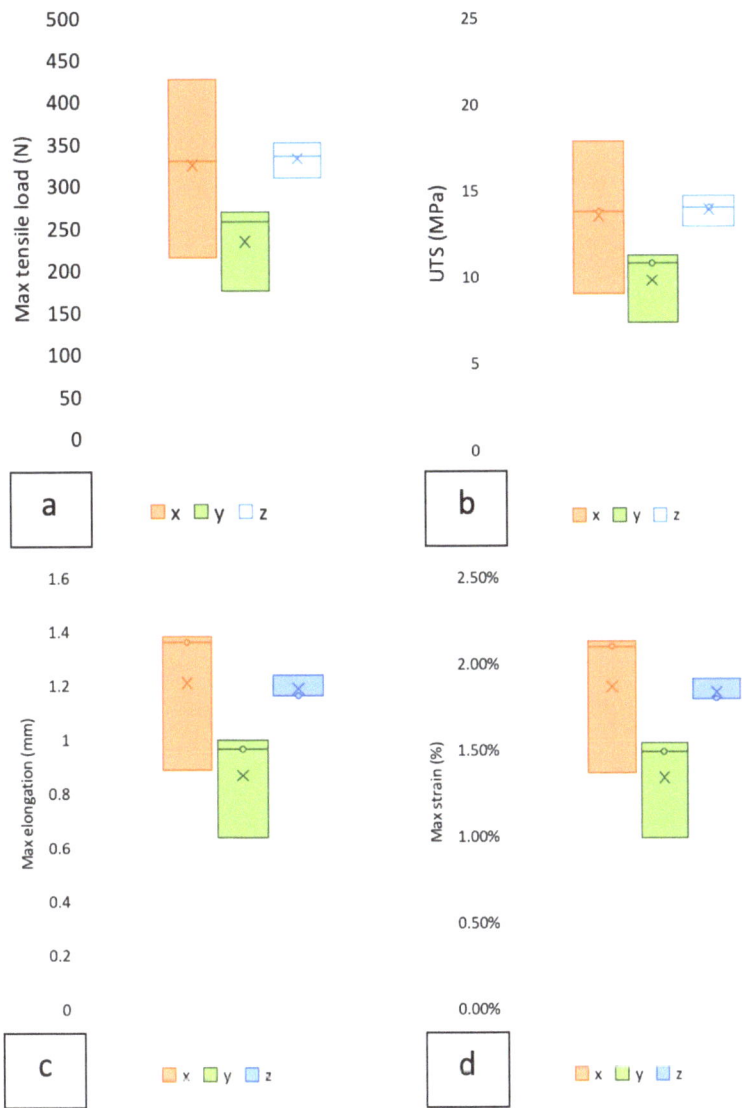

Figure 7. (**a**) Maximum tensile load, (**b**) ultimate tensile strength (UTS), (**c**) maximum elongation, and (**d**) maximum strain of tensile specimens with 0% weight nickel (Ni) across orientations X, Y, and Z.

From the findings above, it is evident that printing in the Z orientation offers superior results compared to printing in either the X or the Y orientation. As a result, further printing with a higher nickel wt.% exclusively occurred in the Z orientation. Tensile tests were then conducted on specimens with varying nickel mass ratios in the photoresin, including 2%, 4%, 6%, and 8%. The collected data include mechanical properties such as maximum tensile load, ultimate tensile stress, break elongation, and break strain. The results of these tensile tests are illustrated in Figure 9. The graph shows a gradual increase in maximum tensile load and ultimate tensile strength, with an increasing mass fraction of nickel within the composite material. As the nickel mass fraction increases to 1%, 2%, and 4%, there is a consistent increase in both the maximum load and tensile strength. However, the alterations in elongation and strain remain relatively marginal. At a 6%

nickel mass fraction, a conspicuous peak emerges in both the maximum load (568 N) and tensile strength (23.7 MPa), coinciding with a peak in elongation and strain, indicating optimal material performance (1.88 mm, 2.89%). Intriguingly, when the nickel mass fraction increased to 8%, a notable decrease was observed in the maximum load (457 N), tensile strength (18.8 MPa), elongation (1.74 mm), and strain (2.68%) compared to those of the 6% nickel fraction, signifying a less favorable performance.

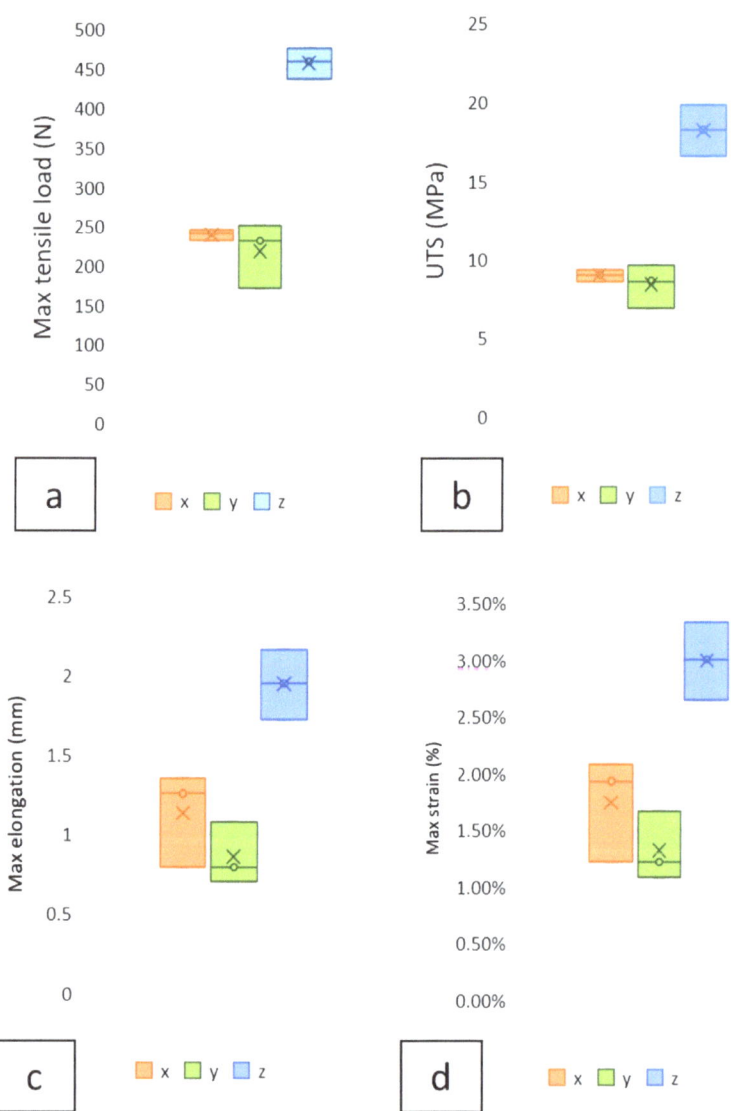

Figure 8. (**a**) Maximum tensile load, (**b**) ultimate tensile strength (UTS), (**c**) maximum elongation, and (**d**) maximum strain of tensile specimens with 1% weight nickel (Ni) across orientations X, Y, and Z.

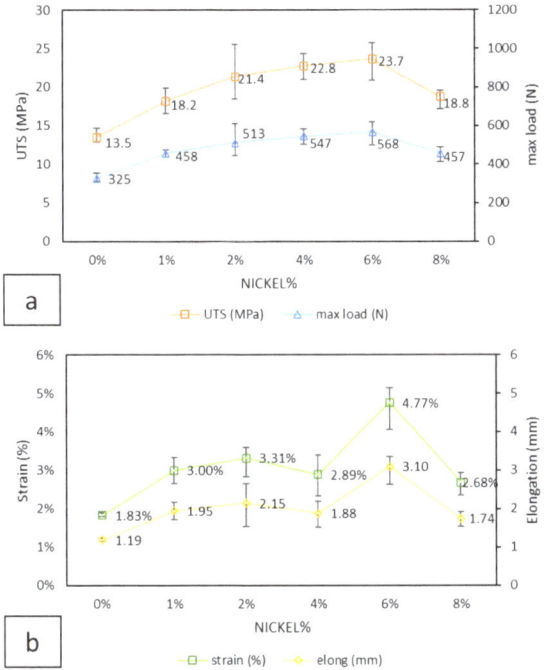

Figure 9. The progression of the mechanical characteristics of the resin–Ni micro composite compared to those of the Ni wt.% composite: (**a**) maximum tensile load and UTS; (**b**) maximum elongation and maximum strain.

3.4. Surface Roughness and Hardness

Surface roughness assessments revealed a notable increase in surface roughness following a 6% increase in the Ni content, as shown in Figure 10a. The hardness measurements exhibit marginal deviations in the Shore D values with increasing Ni wt.%, as depicted in Figure 10b. The marginal deviations in the hardness measurements with increasing Ni wt.% suggest that the DLP 3D printer can maintain a consistent printing precision despite the addition of metal particles. This is consistent with the findings of other studies, which have reported the high precision and accuracy of DLP 3D printers for composite material printing [28,29]. Similarly, a study reported that the surface roughness of a composite material increased with the addition of ceramic particles, which affected the printing quality and accuracy [30].

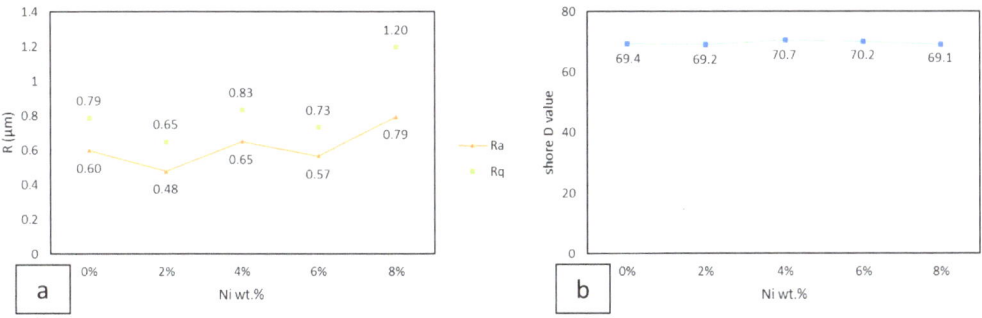

Figure 10. (**a**) Surface roughness and (**b**) shore hardness of samples with different Ni contents.

3.5. Microstructures and Fracture Mechanism

The observational findings from the printing orientations X and Y reveal a coarse texture in both the initial and final layers, closely resembling the printing surfaces, as outlined in Figure 11.

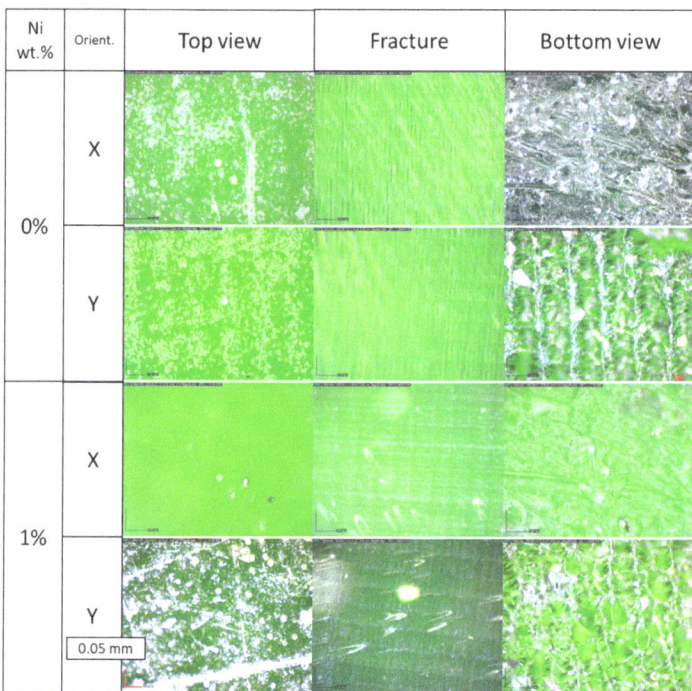

Figure 11. Microstructure of the resin–micro-Ni composite in the X- and Y-printed orientations.

Notably, nickel microparticles are suspended in the X orientation. This suspension of nickel microparticles in the X orientation suggests a more homogeneous dispersion of the filler within the resin matrix compared to the Y orientation. Further examination of the fracture region in both the X- and Y-orientations revealed a distinct layer-by-layer printing pattern indicative of brittle fracture.

The presence of nickel microparticles suspended in the resin is indicated by the formation of crazing in the resin. This finding is consistent with earlier research on the fracture behavior of polymer-based composites, in which the layered structure produced during the additive manufacturing process might contribute to preferential crack propagation along the printing interface [31]. Similarly, in both the X and Y orientations, the development of cavities on the top and bottom layers intensifies with nickel microparticles in the printed specimens in the Z orientation. Moreover, there is a gradual increase in the size and volume of the crazing observed in the fracture region, as outlined in Figure 12. The lack of nickel microparticles in the fracture area indicates a weakened bond between the particles and the matrix.

The micrographs in Figures 11 and 12 reveal a notable difference in the distribution of nickel (Ni) particles through the thickness of the additively manufactured composite samples. The bottom regions exhibit a higher concentration of dense Ni particles, compared to the top regions. This non-uniform particle distribution is likely attributable to particle settling effects during the printing process. Despite the relatively high viscosity of the photocurable resin used, the denser Ni particles may gradually sink and accumulate at the bottom of the resin–particle slurry prior to the curing of each layer. Alternatively, if a bottom-up printing sequence were to be employed, the initial layers would inherently have higher particle loadings from the as-mixed slurry before settling occurred. This particle segregation can have significant impacts on the local and overall mechanical response of the printed composites. The particle-rich bottom regions can be expected to exhibit higher stiffness and strength due to the increased reinforcement from the higher Ni content acting as rigid constraints on the polymer matrix. However, these regions may also suffer from reduced ductility and toughness compared to the polymer-rich top layers. Additionally, the highly loaded bottom layers face an increased propensity for particle agglomeration, defect formation, and compromised interfacial integrity—acting as precursors to premature failure.

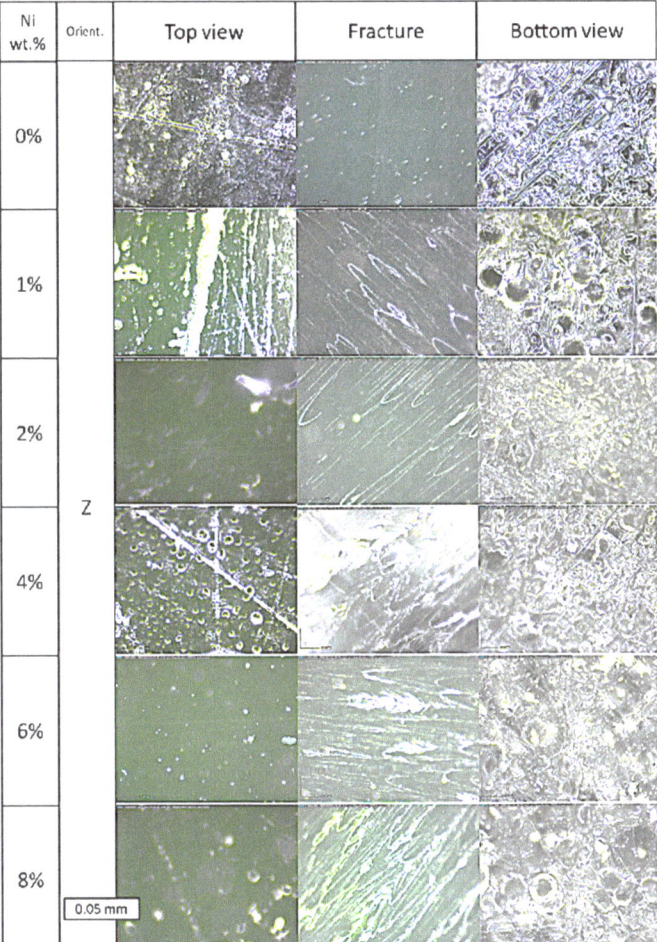

Figure 12. Microscopic structure of the resin–micro-Ni composite in the Z-printed orientation.

Figure 13 shows the SEM images of the plain resin composite and the resin–micro-Ni composite. The SEM images clearly show the successful incorporation of nickel (Ni) microparticles into the polymer resin matrix to form the resin–Ni composites. The Ni particles are well dispersed and embedded throughout the resin, indicating good interfacial bonding between the filler and matrix phases. It is well established that the content of the reinforcing filler phase plays a critical role in determining the mechanical performance of particulate-reinforced polymer composites. As the Ni particle loading increases, the composites can be expected to exhibit higher stiffness and strength due to increased constraint of the polymer chains by the rigid Ni particles. However, too high of a filler content can lead to particle agglomeration, increased defect concentration, and poor interfacial adhesion–resulting in premature failure and loss of toughness. The SEM images revealed typical particulate composite fracture features, suggesting reasonably good particle–matrix bonding aided by the photocuring process. However, some non-uniform Ni dispersion was noted, which could be a precursor to interfacial damage evolution.

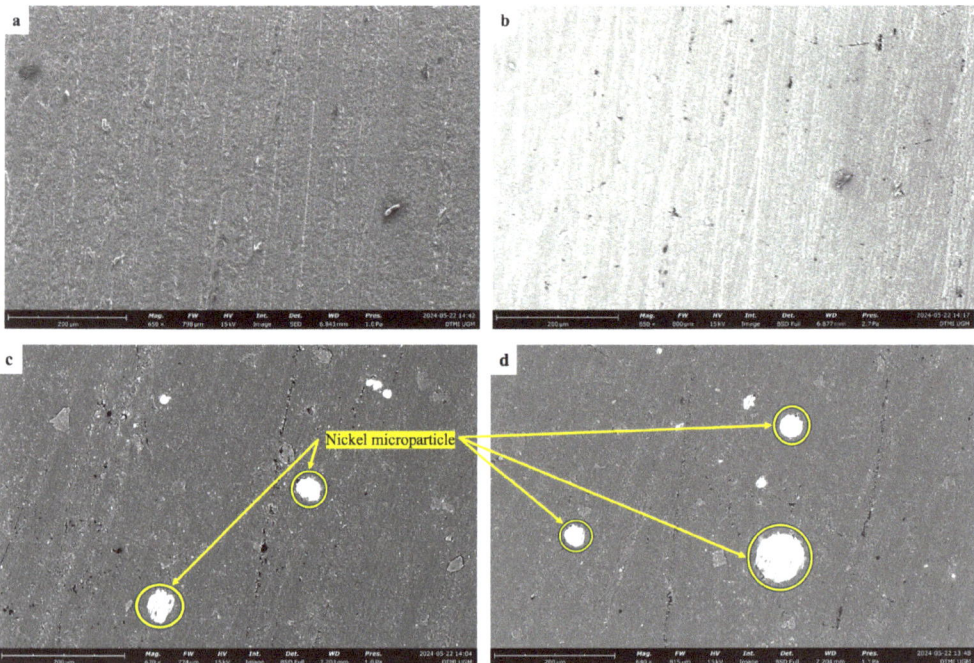

Figure 13. SEM images representing (**a**,**b**): resin composite, (**c**,**d**) resin–micro-Ni composite.

Figure 14a shows the plain resin with a uniform carbon and oxygen signal from the polymer matrix. For the resin–micro-Ni composite, Figure 14b reveals the interfacial regions where nickel particles are embedded within the carbon/oxygen-rich resin. The quality of this particle–matrix interface is crucial for mechanical integrity. The quality and characteristics of this particle–matrix interface play a pivotal role in dictating the efficiency of load transfer and determining the overall mechanical performance. Good interfacial bonding without defects or deleterious reactions is essential. Figure 14c depicts resin-rich areas containing isolated nickel particles, indicating some non-uniformity in the nickel dispersion. The presence of these individual nickel particles, rather than a continuous nickel phase, suggests that some level of non-uniformity or particle clustering occurred during the printing process. Such heterogeneities could serve as initiation sites for damage and premature failure. Finally, Figure 14d clearly distinguishes the dense nickel particulate reinforcement phase within the composite microstructure. The nickel particles appear dense and distinct, with well-defined morphologies and minimal interfacial phases or reactions evident. Preserving the intrinsic nature of the nickel particles is important for realizing their full reinforcing potential.

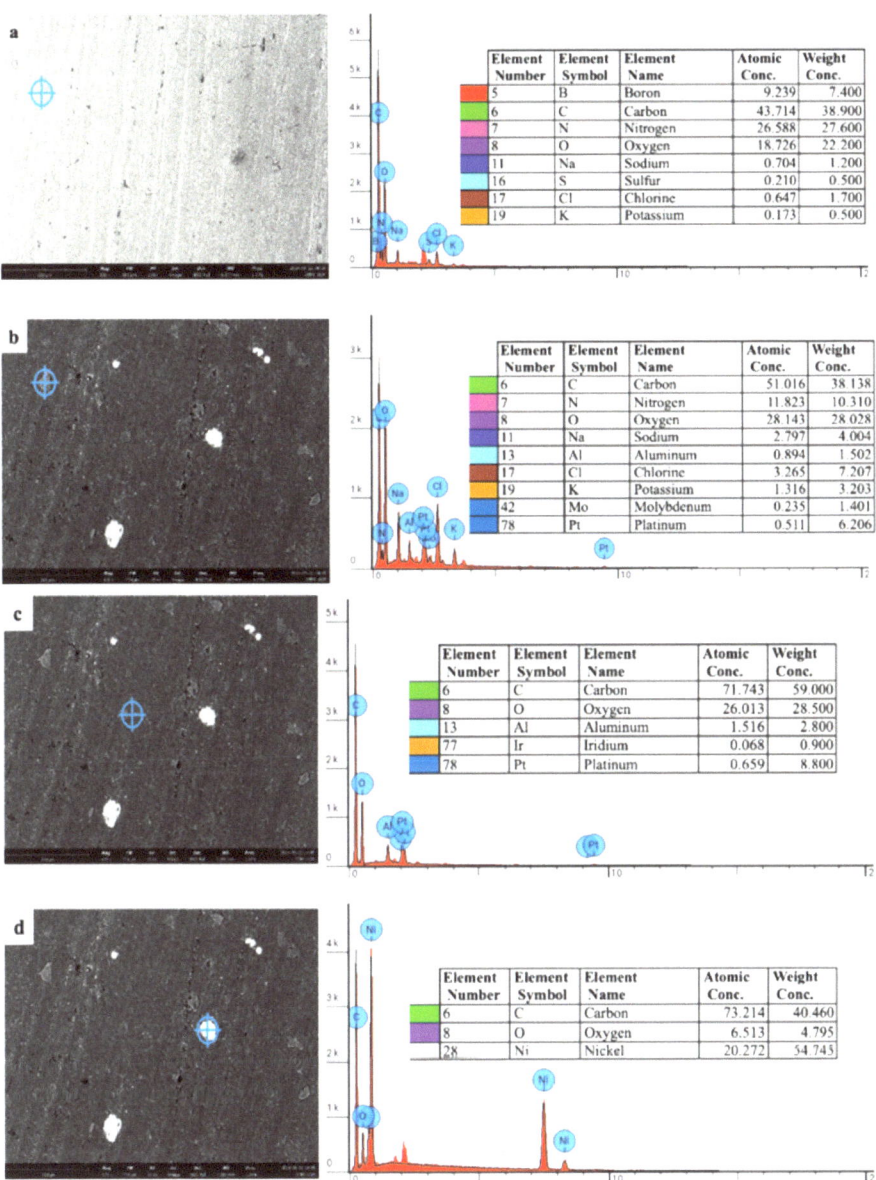

Figure 14. EDS images representing (**a**) resin composite, (**b**) interface region in resin–micro-Ni composite, (**c**) resin region in resin–micro-Ni composite, and (**d**) Ni in resin–micro-Ni composite.

The observed enhancements in tensile strength, elongation, and toughness with increasing Ni content up to 6 wt.% can be attributed to effective load transfer from the ductile resin matrix to the reinforcing Ni particles. The rigid Ni particles contribute to the constraint of the polymer chain mobility and plastic deformation processes. However, at higher Ni loadings above the optimum 6 wt.%, the formation of particle agglomerates and increased interfacial defects appears to degrade mechanical performance. In addition to filler content, the size of the Ni particles may also influence mechanical behavior. Smaller particle sizes provide more surface area for matrix–particle interactions and efficient load transfer. Composites with finer Ni particles could, therefore, offer higher stiffness and strength compared to those with larger particles at the same filler loading. However, very fine particle sizes can also raise concerns, like increased particle agglomeration tendency.

4. Conclusions

The research explores the effect of nickel microplastics on additively manufactured composite using the DLP technique. A mixture comprising 3D printing resin and nickel microparticles was prepared, with the concentration of nickel microparticles varying across different samples. The turbidity of the samples was analyzed, and the spectrophotometry test indicated a decrease in turbidity to approximately 50% within the first two minutes after the resin–nickel microparticle slurry was created. However, the slurry remains under stable conditions of 30% turbidity from 15 min to 60 min after mixing. The FT-IR spectrum of the resin without nickel exhibited distinct peaks at 1722, 1637, and 1110 cm^{-1}, corresponding to C=O, C=C, and C=O bonds, respectively. Upon adding 1, 2, and 4 wt.% nickel, the resin–Ni micro composites displayed similar peaks at 2360, 1670, and 1558 cm^{-1}, indicating consistent chemical compositions. The results from the tensile tests showed that the increase in the amount of nickel microparticles suspended in the resin translates to a peak in the mechanical characteristics at 6 wt.% nickel. The mechanical characteristics of the specimens increased slightly: the maximum tensile load was 568 N (74.7% increase), the UTS was 23.7 MPa (75.5% increase), the maximum elongation was 3.10 mm (160% increase), and the maximum strain was 4.77% (160% increase). Surface roughness assessments revealed a notable increase in surface roughness following a 6% increase in the Ni content. The hardness measurements exhibit marginal deviations in the Shore-D values with increasing Ni wt.%. Microscopic observation of the surfaces of the specimens revealed the fracture characteristics of a particulate composite but with uneven distributions of the nickel microparticles. The investigation demonstrated the considerable influence of changing nickel microparticle concentrations on the mechanical properties of the composite, with the best results reported at 6 wt.%. The study provides proper insights into the potential for incorporating nickel microparticles into the additive manufacturing of composite materials for various engineering applications.

Author Contributions: Conceptualization: M.A.M.; Methodology: B.S., V.V.K., L.S.; Investigation: B.S., L.S.; Validation: V.V.K., L.S.; Data curation: B.S., V.V.K., L.S.; Writing original draft—B.S., V.V.K., L.S., M.H., F.T., Y.D.R., H.A., M.A.M.; Writing- review and editing: B.S., V.V.K., M.A.M.; Supervision: M.A.M. All authors have read and agreed to the published version of the manuscript.

Funding: This research received funding from Hibah RIIM 2023–2024, Indonesia.

Data Availability Statement: The data supporting the findings of this study are available upon request.

Acknowledgments: All authors gratefully acknowledge the anonymous reviewers for their valuable comments, which have significantly contributed to improving the quality of the manuscript and shaping it into an excellent paper.

Conflicts of Interest: The authors declare no conflicts of interest.

Appendix A

Table A1. Properties of the photopolymers used in the experiment.

Material Properties	Values	Unit
Deformation temp	60 ± 5	°C
Vitrification temp	55 ± 5	°C
Maximum elongation	8–12	%
Activation wavelength	355–410	nm
Viscosity	150–350	MPa.s (25 °C)
Bending strength	40–50	MPa
Tensile strength	35–45	MPa

Table A2. Specification of DLP ANYCUBIC Photon Ultra Printer.

Machine Specifications	Values	Unit
Machine weight	4	kg
Printing dimensions	165 × 102.4 × 57.6	mm^3
Light source	DLP optical projector	-
Z-axis resolution	0.01	mm
Layer resolution	0.01~0.15	mm
Control panel	2.8	Inch
Data input	USB	-
Machine dimensions	383 × 222 × 227	mm^3
Printing technology	DLP (Digital Light Processing)	-
Max printing speed	60	mm/h
XY resolution	0.08	mm
Project resolution	1280 × 720	-
Power supply	12	W

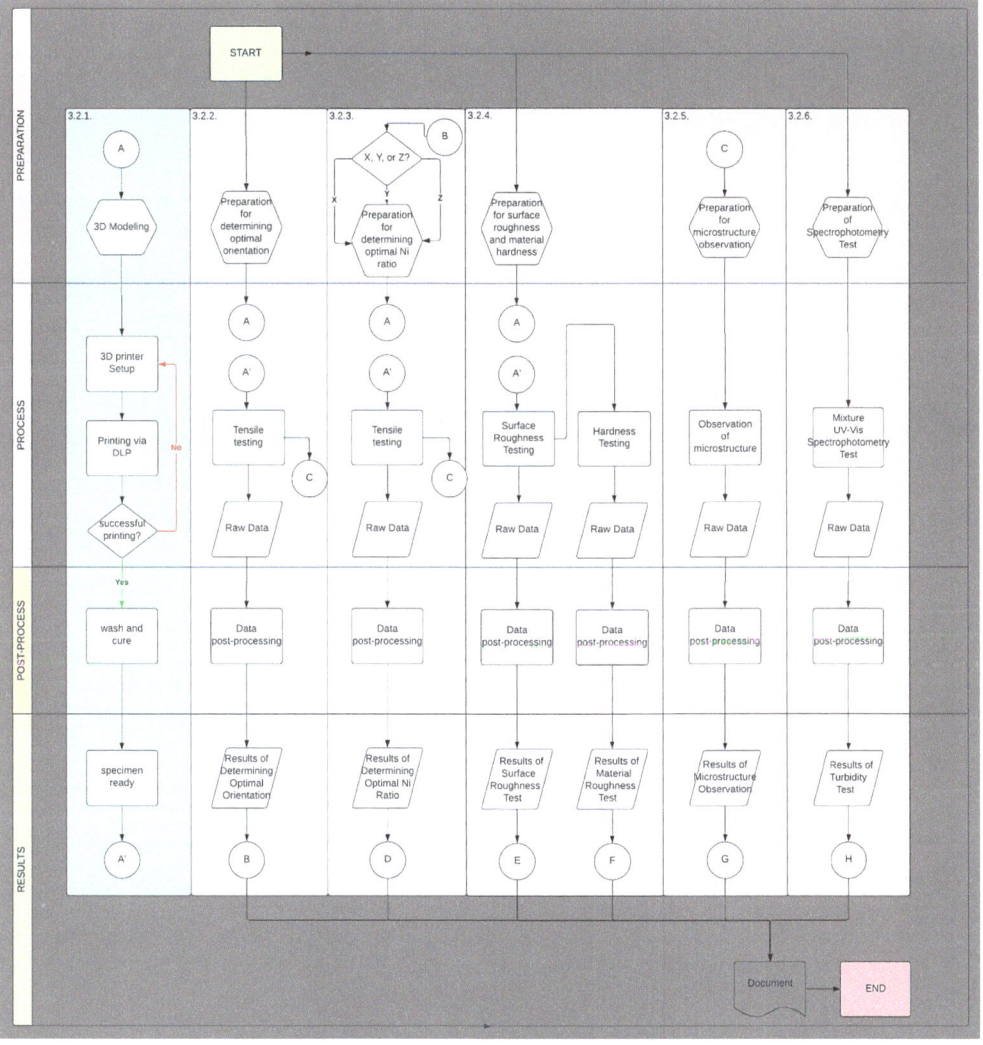

Figure A1. The flowchart diagram outlines the steps in manufacturing and testing.

References

1. Tran, T.Q.; Lee, J.K.Y.; Chinnappan, A.; Jayathilaka, W.; Ji, D.; Kumar, V.V.; Ramakrishna, S. Strong, lightweight, and highly conductive CNT/Au/Cu wires from sputtering and electroplating methods. *J. Mater. Sci. Technol.* **2020**, *40*, 99–106. [CrossRef]
2. Kumar, V.V.; Balaganesan, G.; Lee, J.K.Y.; Neisiany, R.E.; Surendran, S.; Ramakrishna, S. A review of recent advances in nanoengineered polymer composites. *Polymers* **2019**, *11*, 644. [CrossRef] [PubMed]
3. Parveez, B.; Kittur, M.; Badruddin, I.A.; Kamangar, S.; Hussien, M.; Umarfarooq, M. Scientific advancements in composite materials for aircraft applications: A review. *Polymers* **2022**, *14*, 5007. [CrossRef] [PubMed]
4. Krajangsawasdi, N.; Blok, L.G.; Hamerton, I.; Longana, M.L.; Woods, B.K.S.; Ivanov, D.S. Fused deposition modelling of fibre reinforced polymer composites: A parametric review. *J. Compos. Sci.* **2021**, *5*, 29. [CrossRef]
5. Cai, H.; Chen, Y. A Review of Print Heads for Fused Filament Fabrication of Continuous Carbon Fiber-Reinforced Composites. *Micromachines* **2024**, *15*, 432. [CrossRef] [PubMed]
6. Wu, C.; Yi, R.; Liu, Y.J.; He, Y.; Wang, C.C. Delta DLP 3D printing with large size. In Proceedings of the 2016 IEEE/RSJ International Conference on Intelligent Robots and Systems (IROS), Daejeon, Republic of Korea, 9–14 October 2016; IEEE: Piscataway, NJ, USA, 2016.
7. Petousis, M.; Michailidis, N.; Papadakis, V.; Mountakis, N.; Argyros, A.; Spiridaki, M.; Moutsopoulou, A.; Nasikas, N.K.; Vidakis, N. The impact of the glass microparticles features on the engineering response of isotactic polypropylene in material extrusion 3D printing. *Mater. Today Commun.* **2023**, *37*, 107204. [CrossRef]
8. Iyyadurai, J.; Arockiasamy, F.S.; Manickam, T.S.; Suyambulingam, I.; Siengchin, S.; Appadurai, M.; Raj, E.F.I. Revolutionizing polymer composites: Boosting mechanical strength, thermal stability, water resistance, and sound absorption of cissus quadrangularis stem fibers with nano silica. *Silicon* **2023**, *15*, 6407–6419. [CrossRef]
9. Yuan, S.; Shen, F.; Chua, C.K.; Zhou, K. Polymeric composites for powder-based additive manufacturing: Materials and applications. *Prog. Polym. Sci.* **2019**, *91*, 141–168. [CrossRef]
10. Aryaswara, L.G.; Kusni, M.; Wijanarko, D.; Muflikhun, M.A. Advanced properties and failure characteristics of hybrid GFRP-matrix thin laminates modified by micro glass powder filler for hard structure applications. *J. Eng. Res.* **2023**, *in press*. [CrossRef]
11. Vijayan, M.; Selladurai, V.; Vijay Kumar, V.; Balaganesan, G.; Marimuthu, K. Low-Velocity Impact Response of Nano-Silica Reinforced Aluminum/PU/GFRP Laminates. In *International Symposium on Plasticity and Impact Mechanics*; Springer: Berlin/Heidelberg, Germany, 2022.
12. Weng, Z.; Zhou, Y.; Lin, W.; Senthil, T.; Wu, L. Structure-property relationship of nano enhanced stereolithography resin for desktop SLA 3D printer. *Compos. Part A Appl. Sci. Manuf.* **2016**, *88*, 234–242. [CrossRef]
13. Nugraha, A.D.; Kumar, V.V.; Gautama, J.P.; Wiranata, A.; Mangunkusumo, K.G.H.; Rasyid, M.I.; Dzanzani, R.; Muflikhun, M.A. Investigating the Characteristics of Nano-Graphite Composites Additively Manufactured Using Stereolithography. *Polymers* **2024**, *16*, 1021. [CrossRef]
14. Ramesh, M.; Niranjana, K.; Selvan, M.T. Wear and Friction Behavior of Biocomposites Fabricated Through Additive Manufacturing. In *Tribological Properties, Performance and Applications of Biocomposites*; John Wiley & Sons: Hoboken, NJ, USA, 2024; pp. 219–246.
15. Ha, H.-Y.; Lee, T.-H.; Kim, S.-D.; Jang, J.H.; Moon, J. Improvement of the corrosion resistance by addition of Ni in lean duplex stainless steels. *Metals* **2020**, *10*, 891. [CrossRef]
16. Huang, Y.; Zeng, X.T.; Hu, X.F.; Liu, F.M. Corrosion resistance properties of electroless nickel composite coatings. *Electrochim. Acta* **2004**, *49*, 4313–4319. [CrossRef]
17. Yang, D.; Lei, Y.; Xie, J.; Shu, Z.; Zheng, X. The microbial corrosion behaviour of Ni-P plating by sulfate-reducing bacteria biofouling in seawater. *Mater. Technol.* **2019**, *34*, 444–454. [CrossRef]
18. Feng, Y.; Yang, H.; Yang, Z.; Hu, C.; Wu, C.; Wu, L. A review of the design, properties, applications, and prospects of Ni-based composite powders. *Mater. Des.* **2021**, *208*, 109945. [CrossRef]
19. Fantino, E.; Chiappone, A.; Roppolo, I.; Manfredi, D.; Bongiovanni, R.; Pirri, C.F.; Calignano, F. 3D Printing of Conductive Complex Structures with In Situ Generation of Silver Nanoparticles. *Adv. Mater.* **2016**, *28*, 3712–3717. [CrossRef]
20. Muflikhun, M.A.; Syahril, M.; Mamba'udin, A.; Santos, G.N.C. A novel of hybrid laminates additively manufactured via material extrusion–vat photopolymerization. *J. Eng. Res.* **2023**, *11*, 100146. [CrossRef]
21. Yunus, D.E.; Sohrabi, S.; He, R.; Shi, W.; Liu, Y. Acoustic patterning for 3D embedded electrically conductive wire in stereolithography. *J. Micromech. Microeng.* **2017**, *27*, 045016. [CrossRef]
22. *ISO 21920-2:2021*; Geometrical product specifications (GPS) — Surface texture: Profile, Edition 1. ISO: Geneva, Switzerland, 2021; pp. 1–78.
23. *ASTM_D2240-15*; Standard Test Method for Rubber Property—Durometer Hardness in ASTM D2240. ASTM: West Conshohocken, PA, USA, 2021; p. 13.
24. *ASTM_D638*; Standard Test Method for Tensile Properties of Plastic. ASTM: West Conshohocken, PA, USA, 2022; p. 17.
25. Ganguly, S.; Chakraborty, S. Sedimentation of nanoparticles in nanoscale colloidal suspensions. *Phys. Lett. A* **2011**, *375*, 2394–2399. [CrossRef]
26. Mazzoli, A.; Moriconi, G. Particle size, size distribution and morphological evaluation of glass fiber reinforced plastic (GRP) industrial by-product. *Micron* **2014**, *67*, 169–178. [CrossRef]
27. Chen, T.-H.; Wang, I.-H.; Lee, Y.-R.; Hsieh, T.-H. Mechanical property of polymer composites reinforced with nanomaterials. *Polym. Polym. Compos.* **2021**, *29*, 696–704. [CrossRef]

28. Deng, W.; Xie, D.; Liu, F.; Zhao, J.; Shen, L.; Tian, Z. DLP-based 3D printing for automated precision manufacturing. *Mob. Inf. Syst.* **2022**, *2022*, 2272699. [CrossRef]
29. Hanon, M.M.; Ghaly, A.; Zsidai, L.; Szakál, Z.; Szabó, I.; Kátai, L. Investigations of the mechanical properties of DLP 3D printed graphene/resin composites. *Acta Polytech. Hung.* **2021**, *18*, 143–161. [CrossRef]
30. Vrochari, A.D.; Petropoulou, A.; Chronopoulos, V.; Polydorou, O.; Massey, W.; Hellwig, E. Evaluation of surface roughness of ceramic and resin composite material used for conservative indirect restorations, after repolishing by intraoral means. *J. Prosthodont.* **2017**, *26*, 296–301. [CrossRef]
31. Brinckmann, S.A.; Young, J.C.; Fertig, R.S.; Frick, C.P. Effect of print direction on mechanical properties of 3D printed polymer-derived ceramics and their precursors. *Mater. Lett. X* **2023**, *17*, 100179. [CrossRef]

Disclaimer/Publisher's Note: The statements, opinions and data contained in all publications are solely those of the individual author(s) and contributor(s) and not of MDPI and/or the editor(s). MDPI and/or the editor(s) disclaim responsibility for any injury to people or property resulting from any ideas, methods, instructions or products referred to in the content.

Article

Effects of Infill Density and Pattern on the Tensile Mechanical Behavior of 3D-Printed Glycolyzed Polyethylene Terephthalate Reinforced with Carbon-Fiber Composites by the FDM Process

Mohamed Daly [1,2], Mostapha Tarfaoui [1,3,*], Mountasar Bouali [4] and Amine Bendarma [5,6]

1. National School of Advanced Techniques Brittany, IRDL, UMR CNRS 6027, F-29200 Brest, France; mohamed.daly@ensta-bretagne.org
2. Mechanics Laboratory of Sousse, National School of Engineering of Sousse, University of Sousse, Sousse 4023, Tunisia
3. Green Energy Park (IRESEN/UM6P), km2 R206, Benguerir 43150, Morocco
4. Aviation School of Borj el Amri, Bab Mnara, La Kasba, Tunis 1008, Tunisia; mos.bendahi@gmail.com
5. Laboratory for Sustainable Innovation and Applied Research, Universiapolis, Technical University of Agadir, Bab Al Madina, Qr Tilila, BP 8143, Agadir 80000, Morocco; b.amine@e-polytechnique.ma
6. Institute of Structural Analysis, Poznan University of Technology, Piotrowo 5, 60-965 Poznan, Poland
* Correspondence: mostapha.tarfaoui@ensta-bretagne.fr

Abstract: The impacts of infill patterns and densities on the mechanical characteristics of items created by material extrusion additive manufacturing systems were investigated in this study. It is crucial to comprehend how these variables impact a printed object's mechanical characteristics. This work examined two infill patterns and four densities of 3D-printed polyethylene terephthalate reinforced with carbon-fiber specimens for their tensile characteristics. Rectilinear and honeycomb infill designs were compared at 100%, while each had the following three infill densities: 20%, 50%, and 75%. As predicted, the findings revealed that as the infill densities increased, all analyzed infill patterns' tensile strengths and Young's moduli also increased. The design with a 75% honeycomb and 100% infill density has the highest Young's modulus and tensile strength. The honeycomb was the ideal infill pattern, with 75% and 100% densities, providing significant strength and stiffness.

Keywords: additive manufacturing; CF/PETG; infill pattern effect; infill density effect

1. Introduction

To improve the manufacturing process for all human endeavors, new technologies and solutions must be developed for civilization to advance. Traditionally, honeycombs or other composite materials are produced via extrusion, welding, or injection molding [1]. In this regard, additive manufacturing (AM) is a revolutionary technological advancement that is revolutionizing the production of goods. Mass customization, complex designs and geometries, waste reduction, supply chain simplification, quicker time to market, drastic assembly reduction, weight reduction (topology optimization), and low-volume manufacturing are just a few advantages of additive manufacturing (AM) over traditional manufacturing processes that are driving the revolution [2,3].

AM technology is now widely used in academic research as well as in several engineering applications, including the mechanical [4], biomedical [5–8], construction [9], aerospace, and food sectors [10]. Moreover, recycling and reusing thermoplastic composite materials at both high and low temperatures is known as additive manufacturing [11–13].

Binder jetting, direct energy deposition, material extrusion, material jetting, powder bed fusion, sheet lamination, and vat photopolymerization are the seven basic processing techniques used in AM based on printing technology [14].

Fused deposition modeling (FDM) is an extrusion-based technique for creating polymer-based models and structures [15]. The model that will be produced is transformed into a design model for the FDM printing technique and imported into the slicing program [16].

Several factors must be considered, including build orientation, nozzle diameter, printing speed, layer thickness, and extrusion temperature. The primary material used in FDM technology is a filament which is turned into a semi-liquid condition and injected into a nozzle that travels by the commands of the slicing program.

The material extruded from the nozzle is deposited layer-by-layer to print the entire model. Figure 1 depicts a schematic representation of the FDM process [17].

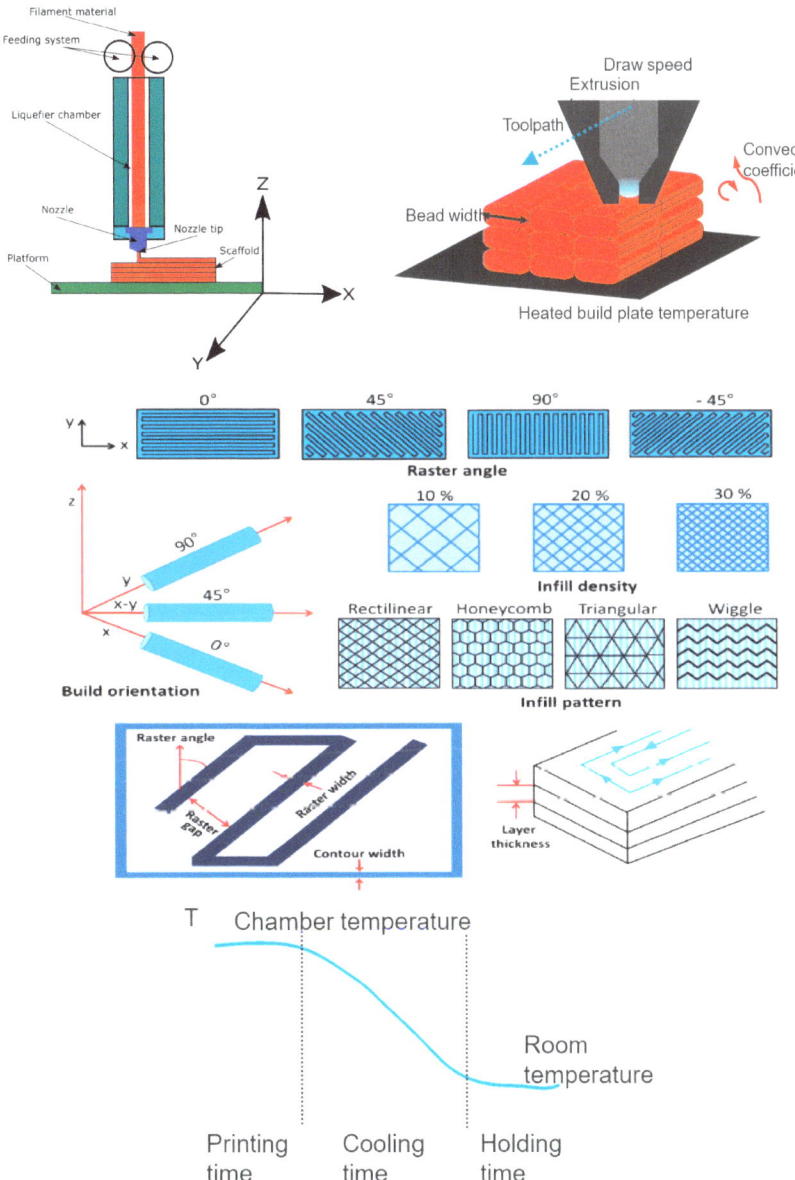

Figure 1. Schematic representation of the FDM process and its parameters.

When considering infill, one must consider the unique relationship between strength, cost, and print time. Every increase in an object's strength increases the printing cost and time. The key to successful infill use is finding a balance where sufficient strength is procured for an object's intended purpose while keeping the cost and time within reasonable bounds.

One of the most popular infill patterns is the 3D honeycomb infill. It is regarded as the most widely used and potent infill pattern. This standard infill pattern offers sufficient rigidity in all directions. It is also one of the most straightforward infill patterns to print, requiring only a tiny amount of bridging from a print head.

Several academic works have studied the static and dynamic behaviors of honeycombs made via additive manufacturing. The impacts of infill density and print orientation on the mechanical behavior of finished items have been the subject of several research works [18,19].

A filling density is the percentage by which a container's interior will be filled. Its primary influence is the overall resistance of the piece. However, the filling rate will be factored into the calculation of the print duration, as well as the final cost of the piece. The greater the density, the greater the amount of filament required. The filling density will also affect the weight of a piece.

A frequently asked question whether or not the importance of the filling density is related to the final resistance of a piece, and if so, why not use a 100% filling density for all impressions. In simple terms, the time it takes to print and the cost of consumables make using a 3D printer unprofitable. Furthermore, a 100% filling may cause a slight deformation in the printed piece, resulting in a change in the exterior linearity of the surface.

Furthermore, it is not only the filling density that influences the behavior of a 3D-printed sample. The evolution of the filling density is not always proportional to the evolution of the resistance, depending on the shape of the piece. Even if the filling density significantly impacts the final resistance, a filling density greater than 80% is not always recommended. According to 3DHubs, the resistance of a piece with a 50% filling rate increases by 25% compared to a piece with a 25% filling density. However, between a filling density of 50% and a filling density of 75%, the resistance increases by only 10%.

There are numerous geometries and shapes available for filling out samples. These themes each have their advantages and are intended for specific applications. The critical thing to remember when selecting a filling motif is to consider the use that will be made of the impression. Furthermore, the motif must be associated with a high fill rate to obtain satisfactory results.

Khan et al. [20] found that the honeycomb (also known as hexagonal) infill pattern for printed acrylonitrile butadiene styrene (ABS) specimens had the lowest tensile strength compared to rectilinear and concentric patterns. This could have been because more voids were printed with this infill pattern inside the specimen. Rismalia et al. [21] demonstrated that increasing infill densities improved the three infill patterns' tensile properties. Tensile properties are affected by infill patterns. Compared to the other two patterns, the concentrate infill pattern has the highest tensile properties, while the grid and tri-hexagonal patterns have similar levels.

Yang et al. [22] investigated the mechanical characteristics of additively manufactured 3D re-entrant honeycomb auxetic structures. By using additive manufacturing (AM), Fadida et al. [23] examined the mechanical aspects of samples of Ti6Al4V under static and dynamic compression. The findings demonstrated that when exposed to dynamic and static loads, the dense material created by a laser had superior resistance to the identical traditional material, but their ductility was equivalent.

The impact strength of PLA specimens created by additive manufacturing at various printing rates was investigated by Tsouknidas et al. [24]. The PLA sample was subjected to a compressive load. The investigation showed that printing at the slowest pace produced the highest compressive strength.

Using desktop 3D printing, Fernandez et al. [25] examined the impact of infill density on the tensile mechanical response. To determine how nonlinear scaling affected the stiffness of soft cellular structures, Wyatt et al. [26] conducted a study. The results showed a high correlation between the experimental findings and the finite element simulation.

A honeycomb structure constructed from polycaprolactone using additive manufacturing was shown by Zhang et al. [27] to recover up to 80% of its volume following a single compression to densification.

To evaluate the ability of the honeycomb structure Ti6Al4V produced by laser-engineered net shaping (LENSTM) to absorb impact energy, Dudka et al. [28] recently examined the static and dynamic behaviors of the structure. To completely comprehend the compressive behavior of 3D-printed thermoplastic polyurethane honeycombs with different densities, Simon and colleagues [29] conducted experimental research. The outcomes showed that the TPU constructions with different densities could offer adequate impact protection in challenging environmental circumstances.

A study by Yu et al. [30] showed that the analyzed additively manufactured gyroid samples had high dynamic compression-energy absorption capacities. In contrast, the analyzed gyroid structure's energy absorption was on par with that of a uniform gyroid structure.

To replace structures made with standard techniques with 3D-printed structures, it is fundamental to identify the optimal parameters for 3D printing; that is, printed structures must have properties that are equivalent to traditional structures. Many scientific and technical challenges must be met to achieve this goal, and they must go through the phases of characterization of these materials and quantify their performance. A detailed study of the bibliographical references showed no significant works on the mechanical behavior of honeycomb and rectilinear composites produced by 3D printing under quasi-static loads.

Using the INSTRON machine, a series of uniaxial tensile tests were carried out on polyethylene terephthalate reinforced with carbon-fiber (CF/PETG) composite specimens produced by the FDM technique in 3D printing. From the perspective of developing a new generation of lightweight materials with optimal mechanical performances, two parameters were studied in this paper: (1) the effect of the geometry (Nida vs. rectilinear) and (2) the filling density (20%, 50%, 75%, and 100%). This paper aimed to characterize the mechanical and fracture behaviors of the different types of specimens, particularly their Young's moduli, stiffness, maximum loads, and displacements at the breaks. A comparison between Nida and rectilinear fillers and an evaluation of the influence of the percentage of filler on the mechanical behavior of the 3D-printed parts was carried out.

This study intended to show how CF/PETG performs mechanically under tensile loading to strike a compromise between weight, resistance, and mechanical behavior. The material from which a structure is made, its cell shape, its relative density, and several other elements, such as the features of the manufacturing process, the structural boundary, and the loading circumstances, all affect how mechanically strong a structure will be. Similarly, the CF-PETEG product will have reduced tensile strength and yield strength if a designer decides to reduce the filling density from 100% to 75% to save on time and 3D-printed materials. In addition, the results demonstrate that the honeycomb filling pattern has the highest tensile and yield strengths when compared to straight filling. The tensile strengths of the honeycomb and 75% rectilinear fillings were marginally lower than that of the 100% filling.

2. Materials

PETG, also known as glycolic polyester, is a thermoplastic frequently used in additive manufacturing facilities because it combines the ease of PLA printing with the strength of ABS [31]. It is an amorphous plastic with an identical chemical make-up to polyethylene terephthalate, commonly known by its acronym PET, and it can be completely recycled. Glycol was added to lessen its fragility and brittle appearance.

As a result, PETG is a copolymer combining PET and glycol qualities. By including the latter, it is possible to lessen PET's overheating and, as a result, brittle appearance. PETG's hardness, chemical and impact resistance, transparency, and ductility make it ideal for 3D printing. It is a thermally stable material that is simple to extrude. In particular, it is valued for being compatible with food contact. Concerning its shortcomings, we note that the PETG requires a heating plate to prevent the warping issues seen with ABS [31].

It is better to utilize a BuildTak sheet to ensure the material hangs, even if the warping rate is modest. In comparison to PLA, it is also more prone to scratches. Also, it keeps well in a cold, dry atmosphere and can quickly absorb moisture. One must be aware that PETG is frequently reinforced with carbon fibers, increasing the part's rigidity while minimizing the weight of the 3D-printed components.

Our industrial partner provided this option in its selection of filaments. It has also created a carbon-fiber-reinforced PETG that increases rigidity while reducing the matrix's brittleness [31].

As shown in Table 1, the composite (CF/PETG) was chosen for this study as a material with a high mechanical performance. This substance is regarded as a high-tech engineering polymer.

Table 1. Mechanical properties of the CF/PETG.

Mechanical Properties of the CF/PETG		
Property	Value	Unit
Density	1.08	g/cm^3
Traction modulus	4700	MPa
Bending modulus	3800	MPa
Elongation at break	2	%
Stress at rupture	42	MPa
Poisson's ratio	0.4	-

3. Structural Design

The specimens displayed in this study were made using the FDM procedure. Figure 2 shows the FDM system operation diagram and the printing system used in our investigation of the printed specimens.

Table 2 summarizes the printing characteristics of all the samples. Two different CF/PETG composite configurations, Nida and rectilinear, were tested experimentally. The FDM process was used to create specimens with dimensions of 151.8 mm in length, 4 mm in thickness, and 21.8 mm in width. While fabricating the specimens, the infill densities were varied to better understand the mechanical properties. All specimens used to test the effect of filling density were made with the two filling patterns (NIDA and REC), with filling densities of 20%, 50%, 75%, and 100%, as shown in Figure 2.

Table 2. (A) The print properties of the CF/PETG and (B) the print statistics.

(A)	Print Properties of the CF/PETG	
Extrusion temperature	395	(°C)
Plate temperature	165	(°C)
Nozzle	0.5	(mm)
Print speed	40	(mm/s)
Layer thickness	0.3	(mm)

Table 2. *Cont.*

(B)		Number of Layers	Total Rows	Filament Required (mm)	Print Time
Rectilinear	20%	12	11,140	871	0 h, 40 mn, 36 s
	50%	12	12,369	1064	0 h, 47 mn, 01 s
	75%	12	12,369	1272	0 h, 47 mn, 01 s
Honeycomb	20%	12	12,832	882	0 h, 41 mn, 04 s
	50%	12	35,752	1226	0 h, 54 mn, 06 s
	75%	12	35,752	1515	0 h, 54 mn, 06 s
	100%	12	15,875	1438	1 h, 03 mn, 21 s

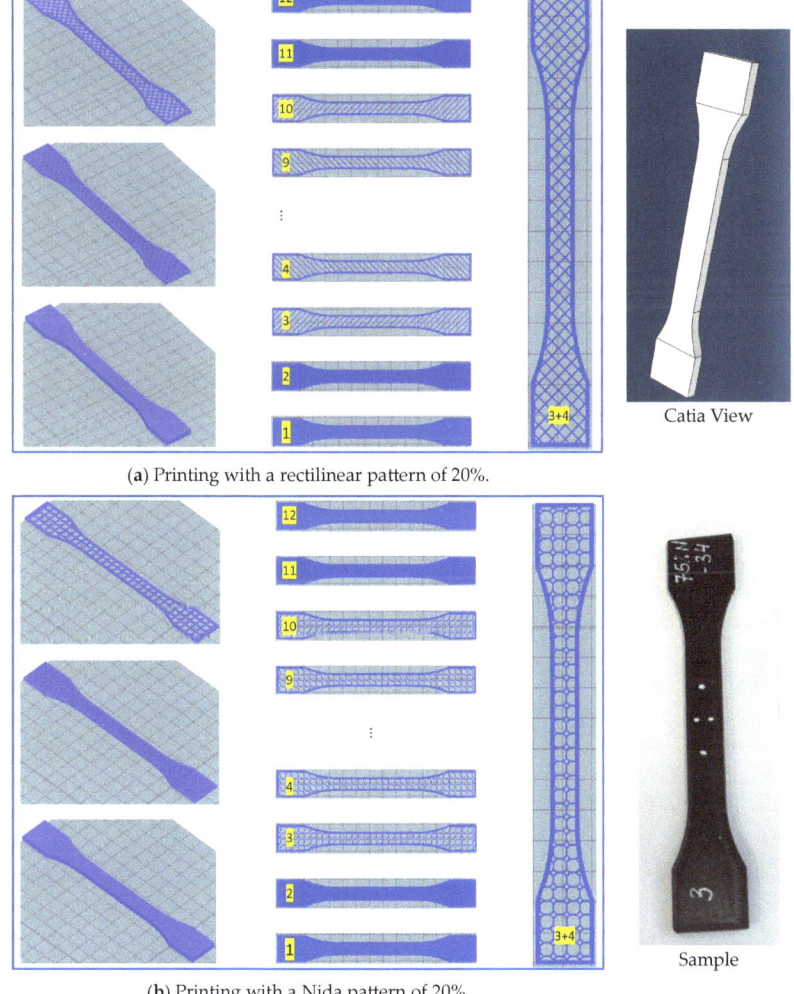

(**a**) Printing with a rectilinear pattern of 20%.

(**b**) Printing with a Nida pattern of 20%.

Figure 2. The 3D printing system and the differences between the two infill patterns.

As has already been said, the specifications of each print job should be considered when choosing a filler. Saving money and time without compromising the durability of a material is the fundamental justification for printing less than 100 percent of a filling piece. Table 2 shows the total number of printed pieces and a comparison of the various filling densities in terms of the amount of time and material required for printing.

4. Tensile Tests

The tensile test stands out as the most fundamental among the different mechanical tests used to fully comprehend and characterize the behaviors of materials. With its vital insights into the mechanical properties of materials, this specific testing technique is a cornerstone in materials research and engineering. The tensile test can identify critical properties that define a material's mechanical behavior. These essential characteristics include the ultimate tensile strength (UTS), indicating the most significant stress and strain a material can sustain without rupturing. The tensile test also offers information on the material's tensile strength, which measures the most incredible pressure it can withstand. Other crucial mechanical characteristics derived from the tensile test include yield strength, the stress point at which the material starts to demonstrate plastic deformation, and elongation at break, which reveals the material's ductility and capacity to deform before fracturing. The test also provides information on the material's stiffness (its Young's modulus) and lateral contraction in response to axial stretching (its Poisson's ratio). The tensile test procedure in this study entailed applying a continuous strain at a set rate to a specimen with a dumbbell shape.

The quasi-static tests were performed on an Instron electromechanical testing machine (a type 5585H universal traction machine) equipped with a 10 kN force cell and an INSTRON AVE 2663-821 model video extensometer (Figure 3 (1)). A computer ran the test through the Bluehill modular software (https://www.instron.com/en/products/materials-testing-software/bluehill-universal accessed on 1 March 2024) (Figure 3). The tensile force was applied to a sample until it broke at a constant speed according to the charging process. The two ends of the sample were clamped, and sliding was prevented by using sandpaper stuck onto the clamps. The samples were loaded with a displacement speed of 2 mm/min. All tests were performed at an ambient temperature and repeated three to five times on different specimens of the same shape to ensure repeatable results. Conducting tests to characterize the mechanical properties and measure the magnitude (stress and deformation) did not cause any particular problems.

Results

It was essential to establish and guarantee the reproducibility of the test procedure before carrying out the quasi-static tensile tests on the INSTRON machine. To this end, a thorough evaluation of each specimen was carried out to confirm the accuracy and reliability of the test results. Samples were subjected to the tensile test technique at least three times for this evaluation, with different filling patterns and densities at each iteration. In this way, we hoped to ensure that the results were not influenced by chance or external variables, guaranteeing subsequent test results' accuracy and consistency. This comprehensive testing technique underlined our commitment to the accuracy and reliability of our research results.

Figure 4 presents the force vs. displacement curves obtained for the different filling densities of the CF/PETG with honeycomb (NIDA) and rectilinear (REC) filling patterns. Figure 4 depicts a representative repeatable force vs. displacement variation for the different CF/PETG filling densities.

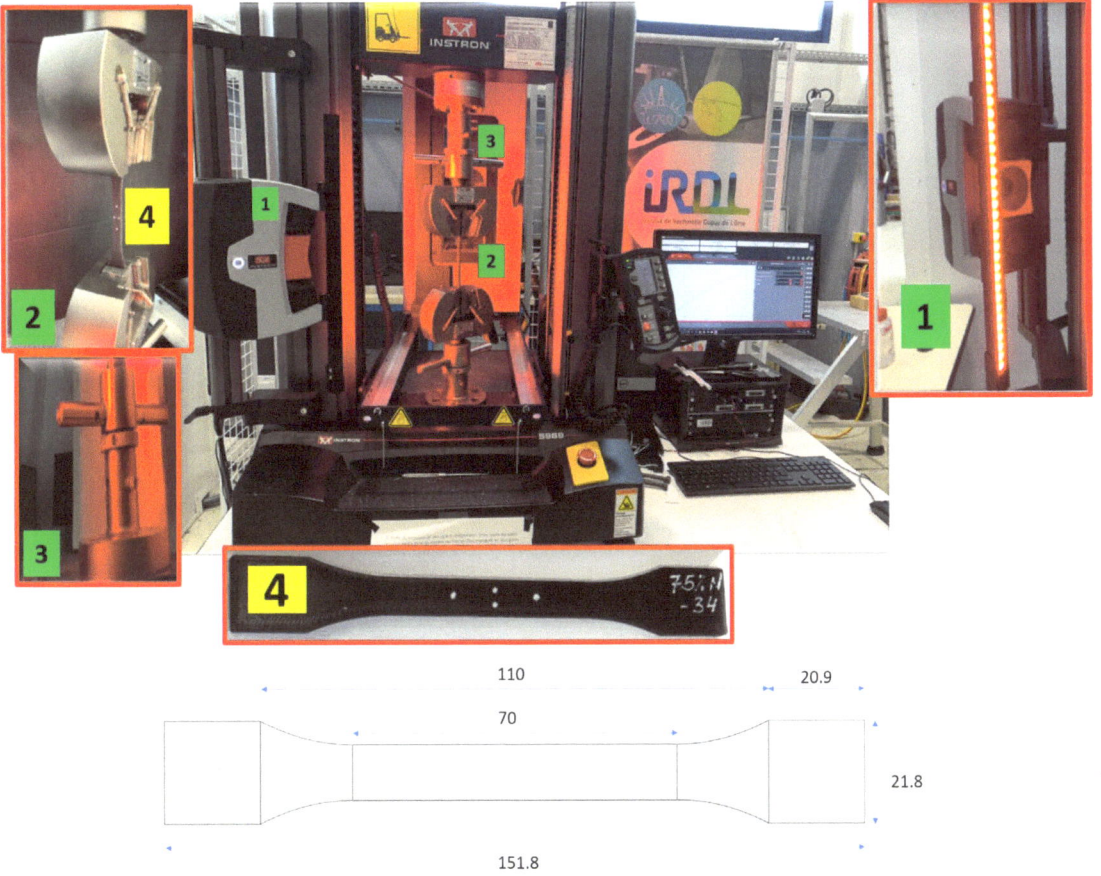

Figure 3. Components of the traction machine.

As seen in Figure 5, no particular infill pattern was formed by the infill densities in the 100% specimen created by the slicing program MakerBot Makerware. The data collected at the 100% infill density were used as the control variable for the comparative study.

Figure 5a,b depicts the behavior variations for the two filling patterns with different infill densities. It shows that specimens manufactured with 100% infill densities had the highest resistance in the tensile tests compared to the honeycomb and rectilinear filling specimens. The mechanical behavior of samples with densities of 20–50% infill decreased significantly. This relationship was nearly linear, and the peak stress was proportional to the infill densities in the honeycomb and rectilinear filling patterns. This finding indicated that the infill density significantly impacted the mechanical properties of the CF/PETG sample.

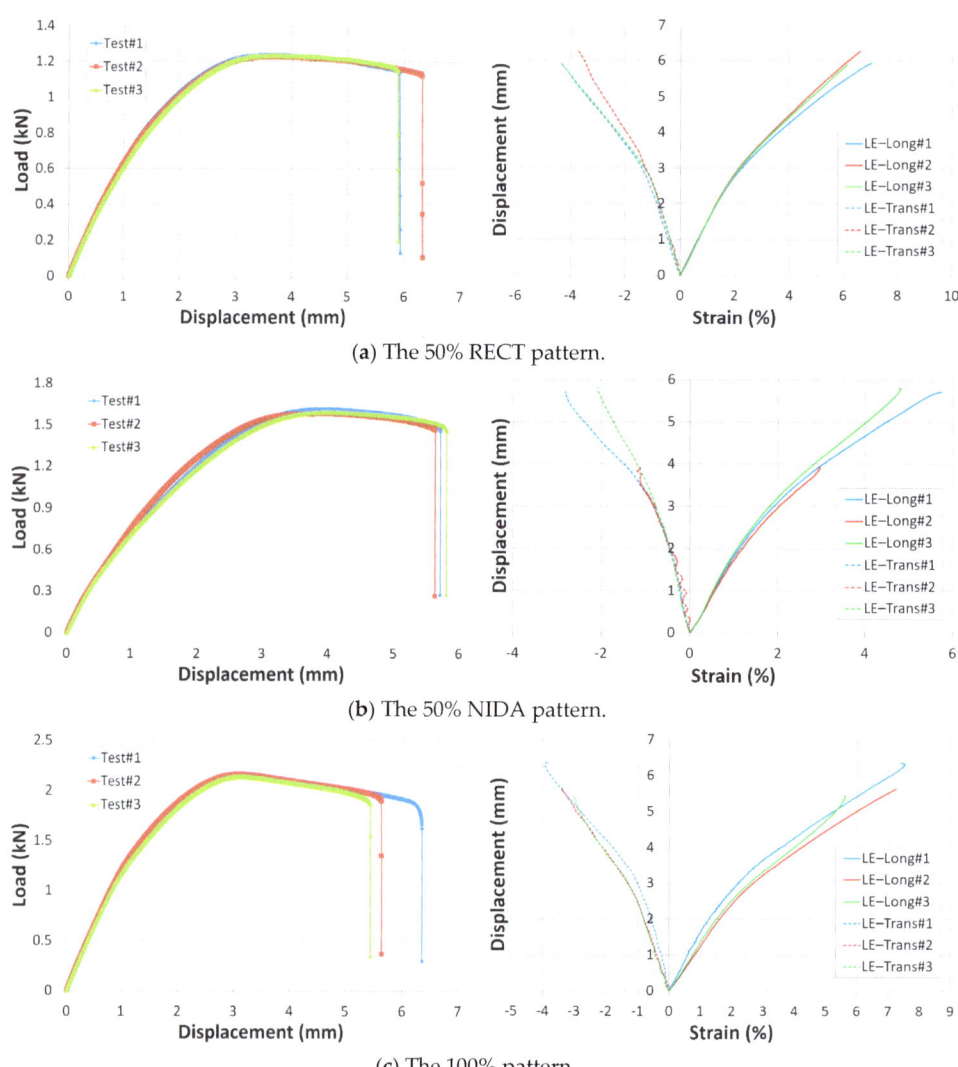

Figure 4. Examples of test reproducibility for the NIDA and RECT patterns.

The Young's moduli and tensile strengths rose with the increasing infill densities, according to the experimental results of the printed composite structures with various infill densities. These could be compared to the enhanced deformation resistance capabilities of the composite structures.

The ultimate tensile strength of the CF/PETG with the two infill patterns (Nida and REC) and the different density percentages (20%, 50%, 75%, and 100%) were tested using an Instron tensile machine with an additionally connected extensometer. The values with error bars are shown in Figure 5c and Tables 3 and 4.

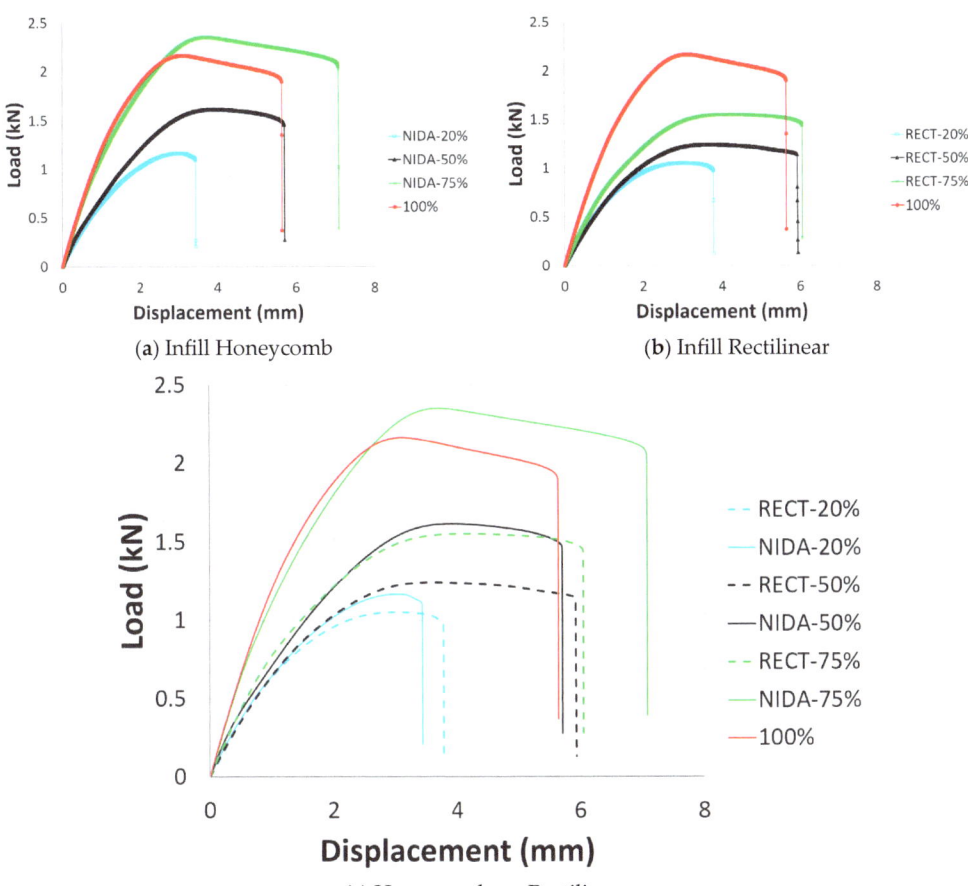

Figure 5. Force vs. displacement curve for the honeycomb and rectilinear infill patterns.

Table 3. Summary of the behavior results of the 3D-printed honeycomb samples.

		E (GPa)	Poisson's Ratio	U Max (mm)	F Max (kN)	LE_Longi Max (%)	Stiffness (N/mm)	UTS (Mpa)
20%	Average	2.276	0.46	3.36	1.16	3.02	830.267	28.957
	St-Dev	0.151	0.05	0.23	0.02	0.60	53.545	0.611
50%	Average	2.861	0.40	5.72	1.60	5.38	939.054	39.904
	St-Dev	0.071	0.04	0.09	0.02	0.51	13.679	0.409
75%	Average	3.730	0.38	6.24	2.34	6.00	1309.669	47.626
	St-Dev	0.174	0.04	0.79	0.02	1.23	29.067	0.594
100%	Average	3.545	0.49	5.81	2.14	6.80	1361.629	53.492
	St-Dev	0.601	0.03	0.48	0.02	1.03	6.203	0.611

Table 4. Summary of the behavior results of the 3D-printed rectilinear samples.

		E (Gpa)	Poisson's Ratio	U Max (mm)	F Max (kN)	LE_Longi Max (%)	Stiffness (N/mm)	UTS (MPa)
20%	Average	2.138	0.49	3.67	1.05	3.51	738.374	26.199
	St-Dev	0.046	0.01	0.23	0.00	0.73	15.345	0.040
50%	Average	2.105	0.59	6.35	1.23	7.73	737.611	30.843
	St-Dev	0.032	0.03	0.46	0.00	1.74	1.06115	0.118
75%	Average	2.607	0.47	5.90	1.54	5.56	900.791	38.564
	St-Dev	0.080	0.02	0.45	0.01	1.04	11.577	0.319
100%	Average	3.545	0.49	5.81	2.14	6.80	1361.629	53.492
	St-Dev	0.601	0.03	0.48	0.02	1.03	6.203	0.611

The ultimate tensile strengths of the CF/PETG samples at the different infill densities for the honeycomb filling method were 13.28, 18.31, and 21.85 MPa. The tensile strength increased progressively as the infill density continued to increase. The Young's modulus increased as the infill density increased, (2276, 2861, and 3730 MPa for 20%, 50%, and 75%, respectively). At a 100% infill density, the CF/PETG had the maximum Young's modulus of 4148.59 MPa.

The ultimate tensile strength of the CF/PETG at 50% and 75% infill densities for the rectilinear filling pattern were 30.84 and 38.56 MPa. At a 100% infill density, the CF/PETG had the maximum tensile strength of 53.49 MPa. At a 100% infill density, the annealed CF/PETG had the maximum Young's modulus of 4148.59 MPa. The Young's modulus increased as the infilling density increased, with 2690.31 and 3350.37 for 50% and 75%, respectively.

Improved deformation resistance capabilities may have caused the enhanced Young's moduli and tensile strengths observed with the increased infill densities. Notably, the higher infill densities reduced the gaps in the composite structures, increasing the cross-sectional areas that could effectively support the tensile loads [32].

Furthermore, these variabilities were related to the printed parts' orientations, preferred reinforcements, post-processing, and infill densities. The tensile modulus of each printed specimen was increased by transversal printing. The infill density significantly impacted the strength of the printed specimen. The space between each layer was too close at a 100% infill density, which may have increased the high bonding strength between each printed layer [33], as shown in Figure 6.

Figure 5a,b demonstrates that the elongations at the breaks of the composites increased with the decreasing filling densities for the two types of filling patterns, increasing the total relative energy absorption. The composite specimens with higher infill densities had higher stiffness and tensile strengths. Elongations, on the other hand, decreased as the infill densities increased.

A higher defect density with more connecting nodes may have caused a decrease in the elongation associated with the high infill densities in the composite constructions. All printed specimens contained manufacturing flaws in the internal nodes of the cells because of the numerous crossovers of fused filaments created by the printing process [34,35], as seen in Figure 5a,b.

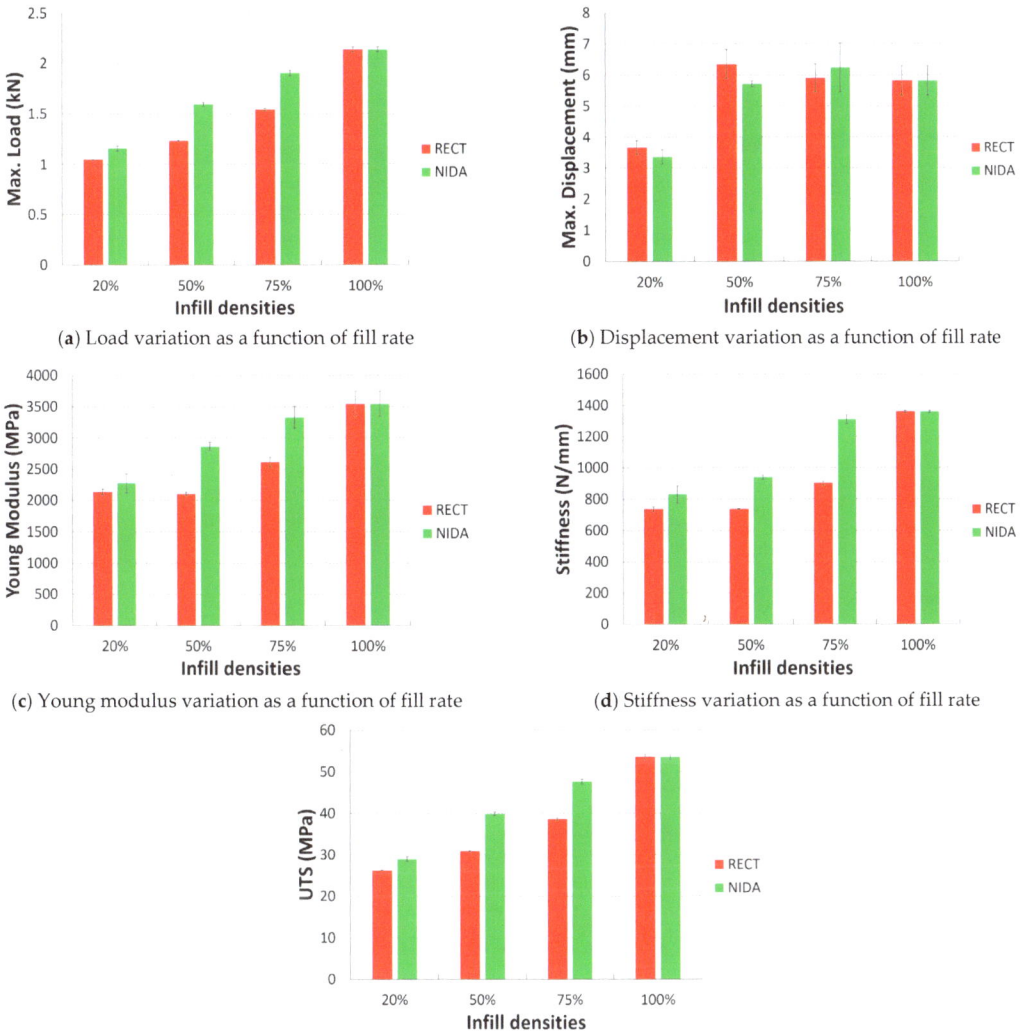

(a) Load variation as a function of fill rate

(b) Displacement variation as a function of fill rate

(c) Young modulus variation as a function of fill rate

(d) Stiffness variation as a function of fill rate

(e) Ultimate tensile strength variation (UTS) as a function of fill rate.

Figure 6. Comparative study between the honeycomb and rectilinear infill patterns.

The increased imperfection density and decreased elongation may have affected the higher infill density specimens [36]. Another reason for the increased elongation in the specimens with lower infill densities was the extended structure in each unit [37], as seen in Figure 5.

The maximal forces of the composite samples with the two distinct infill patterns resulted in similar stress distributions on the structures that were dependent on the infill densities during the static tensile deformation.

As a result, the infill density had little effect on the failure mode, except for the low-density percentage of 20%, which degraded in a short period of time.

As with the 50%, 75%, and 100% densities, a prominent unit structure would typically permit more rotation space and flexibility along the axial extension direction, resulting in a higher elongation for a composite construction.

As shown in Figure 5c, at varying infill densities, the composites with honeycomb infill structures were stiffer and less ductile than those filled with rectilinear units.

Figure 6 shows that the composite constructions with honeycomb infill patterns had better Young's moduli, and this was linked to the beam theory.

5. Conclusions

A filling must be selected for the unique requirements of a print job, as stated in the introduction. Therefore, filling density is decided based on the application needed.

One of the benefits of 3D printing FDM technology is that product lines can be produced with varying infill densities. This benefit reduces the time and material used and the cost of the finished product. After analyzing the tensile testing results, it was possible to conclude that:

- The infill types and densities affect the ultimate tensile strengths and yield strengths.
- The ultimate tensile strengths and yield strengths of all types of infill patterns increase as the density increases from 20% to 75%. Maximum strength is achieved with a 75% infill.
- The results showed that the honeycomb infill pattern had the highest ultimate tensile strength and yield strength when compared to the rectilinear infill.
- Compared to the 100% infill, the honeycomb and rectilinear infill patterns of 75% resulted in slightly lower ultimate tensile strengths.

The effects of infill patterns and densities on material properties were investigated in this paper only for tensile testing in one direction and for one printing orientation. Other factors to consider in future examinations include different printing orientations and testing in three directions. In addition, the effects of infill patterns and densities on other material properties (bending, pressure, hardness, and so on) should be investigated.

Furthermore, it will be necessary for companies to calculate the amount of filament to use and the printing time required due to filling in to determine if 3D printing remains profitable.

Author Contributions: The authors confirm that their contributions to the article were as follows: study design: M.D. and M.T.; experimental study: M.D.; data collection: M.D., M.T. and M.B.; analysis and interpretation of results: M.D., M.T., A.B. and M.B.; writing of the manuscript: M.D. and M.T. All authors have read and agreed to the published version of the manuscript.

Funding: This research received no external funding.

Data Availability Statement: Data are contained within the article.

Conflicts of Interest: The authors declare no conflict of interest. This manuscript is original, has not been published before, and is not currently being considered for publication elsewhere.

References

1. Grünewald, J.; Parlevliet, P.; Altstädt, V. Manufacturing of thermoplastic composite sandwich structures: A review of literature. *J. Thermoplast. Compos. Mater.* **2017**, *30*, 437–464. [CrossRef]
2. El Moumen, A.; Tarfaoui, M.; Lafdi, K. Additive manufacturing of polymer composites: Processing and modeling approaches. *Compos. Part B Eng.* **2019**, *171*, 166–182. [CrossRef]
3. Rouway, M.; Nachtane, M.; Tarfaoui, M.; Chakhchaoui, N.; Omari, L.E.H.; Fraija, F.; Cherkaoui, O. 3D printing: Rapid manufacturing of a new small-scale tidal turbine blade. *Int. J. Adv. Manuf. Technol.* **2021**, *115*, 61–76. [CrossRef]
4. Daly, M.; Tarfaoui, M.; Chihi, M.; Bouraoui, C. FDM technology and the effect of printing parameters on the tensile strength of ABS parts. *Int. J. Adv. Manuf. Technol.* **2023**, *126*, 5307–5323. [CrossRef]
5. Nachtane, M.; Tarfaoui, M.; Ledoux, Y.; Khammassi, S.; Leneveu, E.; Pelleter, J. Experimental investigation on the dynamic behavior of 3D printed CF-PEKK composite under cyclic uniaxial compression. *Compos. Struct.* **2020**, *247*, 112474. [CrossRef]
6. Tarfaoui, M.; Nachtane, M.; Goda, I.; Qureshi, Y.; Benyahia, H. 3D printing to support the shortage in personal protective equipment caused by COVID-19 pandemic. *Materials* **2020**, *13*, 3339. [CrossRef] [PubMed]
7. Javaid, M.; Haleem, A. Additive manufacturing applications in medical cases: A literature based review. *Alex. J. Med.* **2018**, *54*, 411–422. [CrossRef]

8. Goda, I.; Nachtane, M.; Qureshi, Y.; Benyahia, H.; Tarfaoui, M. COVID-19: Current challenges regarding medical healthcare supplies and their implications on the global additive manufacturing industry. *Proc. Inst. Mech. Eng. Part H J. Eng. Med.* **2020**, *236*, 613–627. [CrossRef] [PubMed]
9. Goda, I.; Reis, F.D.; Ganghoffer, J.-F. Limit analysis of lattices based on the asymptotic homogenization method and prediction of size effects in bone plastic collapse. In *Generalized Continua as Models for Classical and Advanced Materials*; Springer: Berlin/Heidelberg, Germany, 2016; pp. 179–211.
10. Goda, I.; Rahouadj, R.; Ganghoffer, J.-F. Size dependent static and dynamic behavior of trabecular bone based on micromechanical models of the trabecular architecture. *Int. J. Eng. Sci.* **2013**, *72*, 53–77. [CrossRef]
11. Picard, M.; Mohanty, A.K.; Misra, M. Recent advances in additive manufacturing of engineering thermoplastics: Challenges and opportunities. *RSC Adv.* **2020**, *10*, 36058–36089. [CrossRef]
12. Shanmugam, V.; Das, O.; Neisiany, R.E.; Babu, K.; Singh, S.; Hedenqvist, M.S.; Berto, F.; Ramakrishna, S. Polymer Recycling in Additive Manufacturing: An Opportunity for the Circular Economy. *Mater. Circ. Econ.* **2020**, *2*, 11. [CrossRef]
13. Yaragatti, N.; Patnaik, A. A review on additive manufacturing of polymers composites. *Mater. Today Proc.* **2020**, *44*, 4150–4157. [CrossRef]
14. Goda, I.; Ganghoffer, J.-F. 3D plastic collapse and brittle fracture surface models of trabecular bone from asymptotic homogenization method. *Int. J. Eng. Sci.* **2015**, *87*, 58–82. [CrossRef]
15. Rybachuk, M.; Mauger, C.A.; Fiedler, T.; Öchsner, A. Anisotropic mechanical properties of fused deposition modeled parts fabricated by using acrylonitrile butadiene styrene polymer. *J. Polym. Eng.* **2017**, *37*, 699–706. [CrossRef]
16. Mwema, F.M.; Akinlabi, E.T.; Mwema, F.M.; Akinlabi, E.T. Basics of fused deposition modelling (FDM). In *Fused Deposition Modeling: Strategies for Quality Enhancement*; Springer: Berlin/Heidelberg, Germany, 2020; pp. 1–15.
17. Shanmugam, V.; Das, O.; Babu, K.; Marimuthu, U.; Veerasimman, A.; Johnson, D.J.; Neisiany, R.E.; Hedenqvist, M.S.; Ramakrishna, S.; Berto, F. Fatigue behaviour of FDM-3D printed polymers, polymeric composites and architected cellular materials. *Int. J. Fatigue* **2021**, *143*, 106007. [CrossRef]
18. Tao, Y.; Wang, H.; Li, Z.; Li, P.; Shi, S.Q. Development and Application of Wood Flour-Filled Polylactic Acid Composite Filament for 3D Printing. *Materials* **2017**, *10*, 339. [CrossRef] [PubMed]
19. Abbas, T.; Othman, F.M.; Ali, H.B. Effect of infill Parameter on compression property in FDM Process. *Dimensions* **2017**, *12*, 24–25.
20. Alvarez, C.K.L.; Lagos, C.R.F.; Aizpun, M. Investigating the influence of infill percentage on the mechanical properties of fused deposition modelled ABS parts. *Ing. Investig.* **2016**, *36*, 110–116. [CrossRef]
21. Rismalia, M.; Hidajat, S.C.; Permana, I.G.R.; Hadisujoto, B.; Muslimin, M.; Triawan, F. Infill pattern and density effects on the tensile properties of 3D printed PLA material. *J. Phys. Conf. Ser.* **2019**, *1402*, 044041. [CrossRef]
22. Yang, L.; Harrysson, O.; West, H.; Cormier, D. Mechanical properties of 3D re-entrant honeycomb auxetic structures realized via additive manufacturing. *Int. J. Solids Struct.* **2015**, *69-70*, 475–490. [CrossRef]
23. Fadida, R.; Rittel, D.; Shirizly, A. Dynamic Mechanical Behavior of Additively Manufactured Ti6Al4V With Controlled Voids. *J. Appl. Mech.* **2015**, *82*, 041004. [CrossRef]
24. Tsouknidas, A.; Pantazopoulos, M.; Katsoulis, I.; Fasnakis, D.; Maropoulos, S.; Michailidis, N. Impact absorption capacity of 3D-printed components fabricated by fused deposition modelling. *Mater. Des.* **2016**, *102*, 41–44. [CrossRef]
25. Fernandez-Vicente, M.; Calle, W.; Ferrandiz, S.; Conejero, A. Effect of Infill Parameters on Tensile Mechanical Behavior in Desktop 3D Printing. *3D Print. Addit. Manuf.* **2016**, *3*, 183–192. [CrossRef]
26. Wyatt, H.; Safar, A.; Clarke, A.; Evans, S.L.; Mihai, L.A. Nonlinear scaling effects in the stiffness of soft cellular structures. *R. Soc. Open Sci.* **2019**, *6*, 181361. [CrossRef] [PubMed]
27. Zhang, P.; Arceneaux, D.J.; Khattab, A. Mechanical properties of 3D printed polycaprolactone honeycomb structure. *J. Appl. Polym. Sci.* **2017**, *135*, [CrossRef]
28. Antolak-Dudka, A.; Płatek, P.; Durejko, T.; Baranowski, P.; Małachowski, J.; Sarzyński, M.; Czujko, T. Static and Dynamic Loading Behavior of Ti6Al4V Honeycomb Structures Manufactured by Laser Engineered Net Shaping (LENS^{TM}) Technology. *Materials* **2019**, *12*, 1225. [CrossRef]
29. Bates, S.R.; Farrow, I.R.; Trask, R.S. Compressive behaviour of 3D printed thermoplastic polyurethane honeycombs with graded densities. *Mater. Des.* **2018**, *162*, 130–142. [CrossRef]
30. Yu, S.; Sun, J.; Bai, J. Investigation of functionally graded TPMS structures fabricated by additive manufacturing. *Mater. Des.* **2019**, *182*, 108021. [CrossRef]
31. Le Plastique PETG en Impression 3D. Available online: https://www.3dnatives.com/plastique-petg-18122019/#! (accessed on 1 August 2023).
32. Dawoud, M.; Taha, I.; Ebeid, S.J. Mechanical behaviour of ABS: An experimental study using FDM and injection moulding techniques. *J. Manuf. Process.* **2016**, *21*, 39–45. [CrossRef]
33. Kumar, K.S.; Soundararajan, R.; Shanthosh, G.; Saravanakumar, P.; Ratteesh, M. Augmenting effect of infill density and annealing on mechanical properties of PETG and CFPETG composites fabricated by FDM. *Mater. Today Proc.* **2020**, *45*, 2186–2191. [CrossRef]
34. Zhou, X.; Zhang, J.; Yang, S.; Berto, F. Compression-induced crack initiation and growth in flawed rocks: A review. *Fatigue Fract. Eng. Mater. Struct.* **2021**, *44*, 1681–1707. [CrossRef]
35. du Plessis, A.; Yadroitsev, I.; Yadroitsava, I.; Le Roux, S.G. X-Ray Microcomputed Tomography in Additive Manufacturing: A Review of the Current Technology and Applications. *3D Print. Addit. Manuf.* **2018**, *5*, 227–247. [CrossRef]

36. Wang, K.; Xie, X.; Wang, J.; Zhao, A.; Peng, Y.; Rao, Y. Effects of infill characteristics and strain rate on the deformation and failure properties of additively manufactured polyamide-based composite structures. *Results Phys.* **2020**, *18*, 103346. [CrossRef]
37. Lubombo, C.; Huneault, M.A. Effect of infill patterns on the mechanical performance of lightweight 3D-printed cellular PLA parts. *Mater. Today Commun.* **2018**, *17*, 214–228. [CrossRef]

Disclaimer/Publisher's Note: The statements, opinions and data contained in all publications are solely those of the individual author(s) and contributor(s) and not of MDPI and/or the editor(s). MDPI and/or the editor(s) disclaim responsibility for any injury to people or property resulting from any ideas, methods, instructions or products referred to in the content.

Article

Numerical and Experimental Characterisation of Polylactic Acid (PLA) Processed by Additive Manufacturing (AM): Bending and Tensile Tests

Mariana P. Salgueiro [1,2], Fábio A. M. Pereira [3], Carlos L. Faria [1,2], Eduardo B. Pereira [4], João A. P. P. Almeida [4], Teresa D. Campos [1,2], Chaari Fakher [5], Andrea Zille [6,*], Quyền Nguyễn [6] and Nuno Dourado [1,2]

1. CMEMS-UMinho, Campus de Azurém, University of Minho, 4800-058 Guimarães, Portugal; marianapereira19991@hotmail.pt (M.P.S.); carlosfaria@dem.uminho.pt (C.L.F.); teresa.ac.biome@gmail.com (T.D.C.); nunodourado@dem.uminho.pt (N.D.)
2. LABBELS, Associate Laboratory, 4710-057 Braga, Portugal
3. CITAB/UTAD, Departamento de Engenharias, Quinta de Prados, 5001-801 Vila Real, Portugal; famp@utad.pt
4. ISISE, IB-S, School of Engineering, Campus de Azurém, University of Minho, 4800-058 Guimarães, Portugal; eduardo.pereira@civil.uminho.pt (E.B.P.); japp.almeida@gmail.com (J.A.P.P.A.)
5. Mechanics, Modelling and Production Research Laboratory (LA2MP), National School of Engineers of Sfax, University of Sfax, Sfax 3047, Tunisia; fakher.chaari@gmail.com
6. 2C2T-Centro de Ciência e Tecnologia Têxtil, Universidade do Minho, 4800-058 Guimarães, Portugal; quyen@2c2t.uminho.pt
* Correspondence: azille@det.uminho.pt

Abstract: In additive manufacturing (AM), one of the most popular procedures is material extrusion (MEX). The materials and manufacturing parameters used in this process have a significant impact on a printed product's quality. The purpose of this work is to investigate the effects of infill percentage and filament orientation on the mechanical properties of printed structures. For this reason, the characterisation of polylactic acid (PLA) was done numerically using the finite element method and experimentally through mechanical tests. The experiments involved three-point bending and tensile tests. The results showed that mechanical performance is highly dependent on these processing parameters mainly when the infill percentage is less than 100%. The highest elastic modulus was exhibited for structures with filament align at 0° and 100% infill, while the lowest one was verified for specimen filament aligned at 0° and 30% infill. The results demonstrated that the process parameters have a significant impact on mechanical performance, particularly when the infill percentage is less than 100%. Structures with filament aligned at 0° and 100% infill showed the maximum elastic modulus, whereas specimens with filament oriented at 0° and 30% infill showed the lowest. The obtained numerical agreement indicated that an inverse method based only on the load–displacement curve can yield an accurate value for this material's elastic modulus.

Keywords: elastic modulus; experimental characterisation; finite element method; material extrusion

1. Introduction

Additive manufacturing (AM) first emerged in the 1980s for developing models and prototypes, expanding the digital chain beyond computer-aided design (CAD) or reverse engineering. The main advantages of additive manufacturing (AM) over other traditional manufacturing processes were rapidly identified, including the capacity to form complex geometries while reducing human interaction and product development time and cost [1,2]. Binder jetting (BJT), material extrusion (MEX), directed energy deposition (DED), material jetting (MJT), powder bed fusion (PBF), sheet lamination (SHL), and vat photopolymerization (VPP) are examples of additive manufacturing (AM) processes [3].

MEX is the second most widely used three-dimensional printing (3DP) technology because of its simplicity and accessibility of utilisation [4,5]. The advanced 3DP technique

for producing polymeric products is also referred to as fused filament fabrication or FFF. Through layer-by-layer melting and extrusion of a polymeric filament, this method produces three-dimensional structures. The fundamental idea behind the FFF production process is to melt the raw material to make it easier to form new shapes [6].

Low-melting-point polymers such as polylactic acid (PLA), polyamide (PA), acrylonitrile butadiene styrene (ABS), polyvinyl acetate (PVA), and polycarbonate (PC) are the principal components adopted in MEX [7]. In this technology, PLA is widely utilised. It is a biodegradable thermoplastic made from crops fermented like potatoes or maize. Its primary characteristics, when applied to FFF production techniques, are low glass transition temperature, high processability, low ductility, decreased resilience, and decreased volumetric contraction [8].

The material, structural elements, and manufacturing variables all affect the mechanical characteristics of MEX-printed parts. Among the manufacturing and structural characteristics are layer height, contour number, raster width, raster angle, infill percentage, fill pattern, air gap, build orientation, print speed, and extruder temperature. Component characteristics are influenced differently by various combinations and degrees of parameter variation (Figure 1) [9,10].

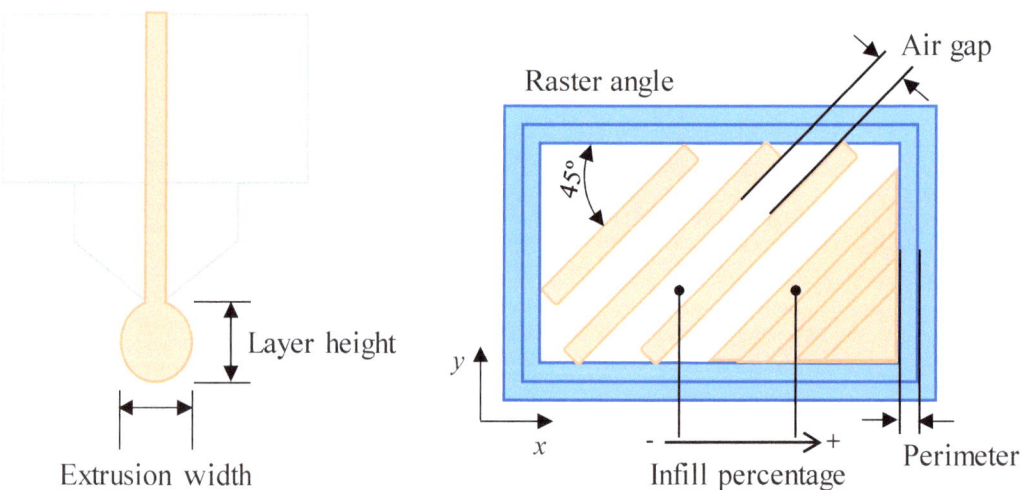

Figure 1. Schematic illustration of applicable analysed process parameters.

Numerous studies have been conducted over time to optimise and comprehend the influence of these parameters using a variety of materials, given the myriad combinations of parameters to be selected in this manufacturing technique.

Three responses, namely tensile, flexural, and impact strength, were examined by Sood et al. [11] in relation to five significant production parameters: layer thickness, orientation, raster angle, raster width, and air gap of printed ABS parts. The majority of these parameters can affect the printed parts' tensile characteristics, and porosity between layers can lead to weak structures, according to the authors' findings.

Baich et al. [12] considered the implications of infill concepts and printing time on the mechanical response (tensile, compressive, and flexural strength) of ABS printed parts. As expected, high infill density not only offers greater strength but also increases production costs due to the increased amount of material required.

Harpool and collaborators [13] studied the tensile strength of PLA components with four distinct infill patterns: rectilinear, diamond, hexagonal, and solid. According to the findings, the infill design has a considerable influence on the mechanical qualities of the printed objects.

The impact of process parameters, such as build direction, infill percentage, infill pattern, printing speed, extrusion temperature, and layer height, on the mechanical tensile properties of printed parts was experimentally investigated by Alafaghani et al. [14]. The authors discovered that the tensile properties are hardly affected by printing speed, infill density, or infill pattern.

Behzadnasab and Yousefi [15] investigated the effects of nozzle temperature settings on the mechanical behaviour of 3D-printed samples. The authors discovered that as the set nozzle temperature increased to 240 °C, the strength of PLA 3D-printed parts increased, although degradation was observed when nozzle temperatures were set higher than 240 °C polymer, which is the common upper bound of set temperature restraint of 3D printing of polymeric materials. Furthermore, the printing speed can have an impact on the mechanical properties of the printed parts. Setting a high printing speed can result in inadequate layer bonding, decreasing the mechanical strength of the parts [16].

Bardiya et al. [17] chose the three main process parameters to investigate: layer thickness, build orientation, and infill density. It was determined that the maximum impact strength of 3D-printed PLA parts can be achieved by employing a layer thickness of 0.2 mm, a layer orientation of 30°, and an infill density of 80%.

The study by Abeykoon et al. [18] was concentrated on the mechanical, thermal, and morphological properties of 3D-printed specimens under several processing conditions, including infill pattern, density, and speed, as well as with various printing materials. In the end, the best process parameters for the best performance of the various printing materials were a linear fill pattern, 100% infill density, 90 mm/s infill speed, and a set nozzle temperature of 215 °C.

In none of the above referred studies was the mechanical behaviour of 3D-printed components predicted using a combination of numerical modelling and experimental data. The cited authors used conventional methods to measure basic mechanical properties in the experimental characterisation. To date, there have been no studies examining the use of contactless full-field experimental techniques to assess the mechanical performance of 3D-printed materials and validate measurement protocols.

In this study, two experimental mechanical tests—the three-point bending (TPB) and tensile ones—were used to assess the impact of filament orientation and infill percentage on the overall stiffness. As a way to acquire tensile stress–strain curves and the fundamental mechanical properties of the material (PLA), two methods were combined to monitor discrete and full-field data. Numerical modelling was performed to simulate the influence of infill percentage on the PLA response for filament orientations both at 0° and 90°. Notably, this study goes beyond conventional approaches employed in previous studies by incorporating a combination of numerical modeling and experimental data, integrating non-contact full-field experimental techniques to evaluate the mechanical performance of 3D-printed materials. By doing so, this study aims to bridge the gap between theoretical predictions and practical results, thereby improving the understanding of the mechanical behaviour of 3D-printed components. Furthermore, this work introduces a unique focus on filament orientation and fill percentage, two crucial factors that influence the overall stiffness of 3D-printed structures.

2. Materials and Methods

2.1. Preparation of Specimens

Using a Creatbot® model F430 machine (3D printing technology, Zhengzhou, China), PLA samples were created using MEX technology. Flashforge Technology Co., Ltd. (Jinhua, China) supplied 1.75 mm diameter PLA filament (Primavalue®, 3DPrima, Malmö, Sweden). CAD models such as parallelepipedals and dogbones (Figure 2a,b) were developed in accordance with ISO 20753 and ISO 3167 type A, the corresponding testing standards for three-point bending (TPB) and tensile tests. The STL file format was adopted for slicing operations in the CreatWare V6.5.1 print control programme. G-code files were generated based on geometrical and printing settings (Table 1). Twenty-four specimens were printed to undergo TPB tests, three for each orientation with 30% and 100% infill. Thirty-two specimens were printed with four filament orientations (0°/90°, 45°/−45°, 0°, and 90°) and infill percentages of 30% and 100% for the tensile tests. Tensile tests were also conducted with 50% and 75% for filament orientations at 0° and 90° relative to the loading axis aiming to evaluate more deeply the impact of the elastic modulus with the infill percentage (Figure 3a,b). The elastic moduli E_1 and E_2 were identified through specimens with filaments oriented at 0° and 90°, respectively.

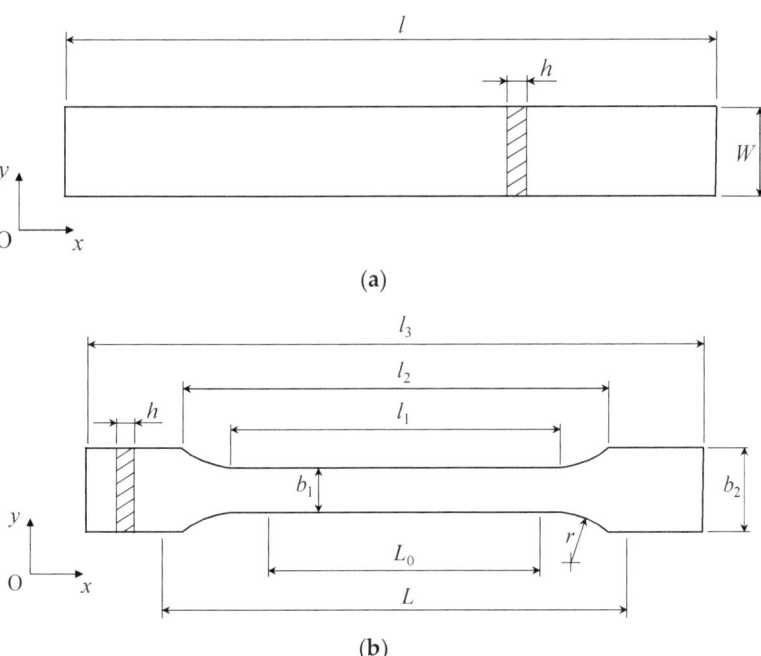

Figure 2. Specimen geometries: (**a**) three-point bending: $l = 80; h = 4; W = 10$ (mm) and (**b**) dogbone: $b_1 = 10; b_2 = 20; h = 4; l_1 = 80; l_2 = 110; l_3 = 170; r = 25; L_0 = 75; L = 115$ (mm).

Table 1. Printing parameters.

	Layer height (mm)	0.1
Quality	Extrusion width (mm)	0.4
	Flow (%)	100
	Perimeters	2
Fill	Top layers	4
	Print speed (mm/s)	40
Speed and Temperature	Bottom layers	4
	Print temperature (°C)	210
	Close bed after layer (°C)	100
	Bed temperature (°C)	45

Using varying orientations and infill percentages, specimens for TPB and tensile testing were printed (Figure 3), with the corresponding effective dimensions (width, length, and thickness) exhibited in Tables 2 and 3.

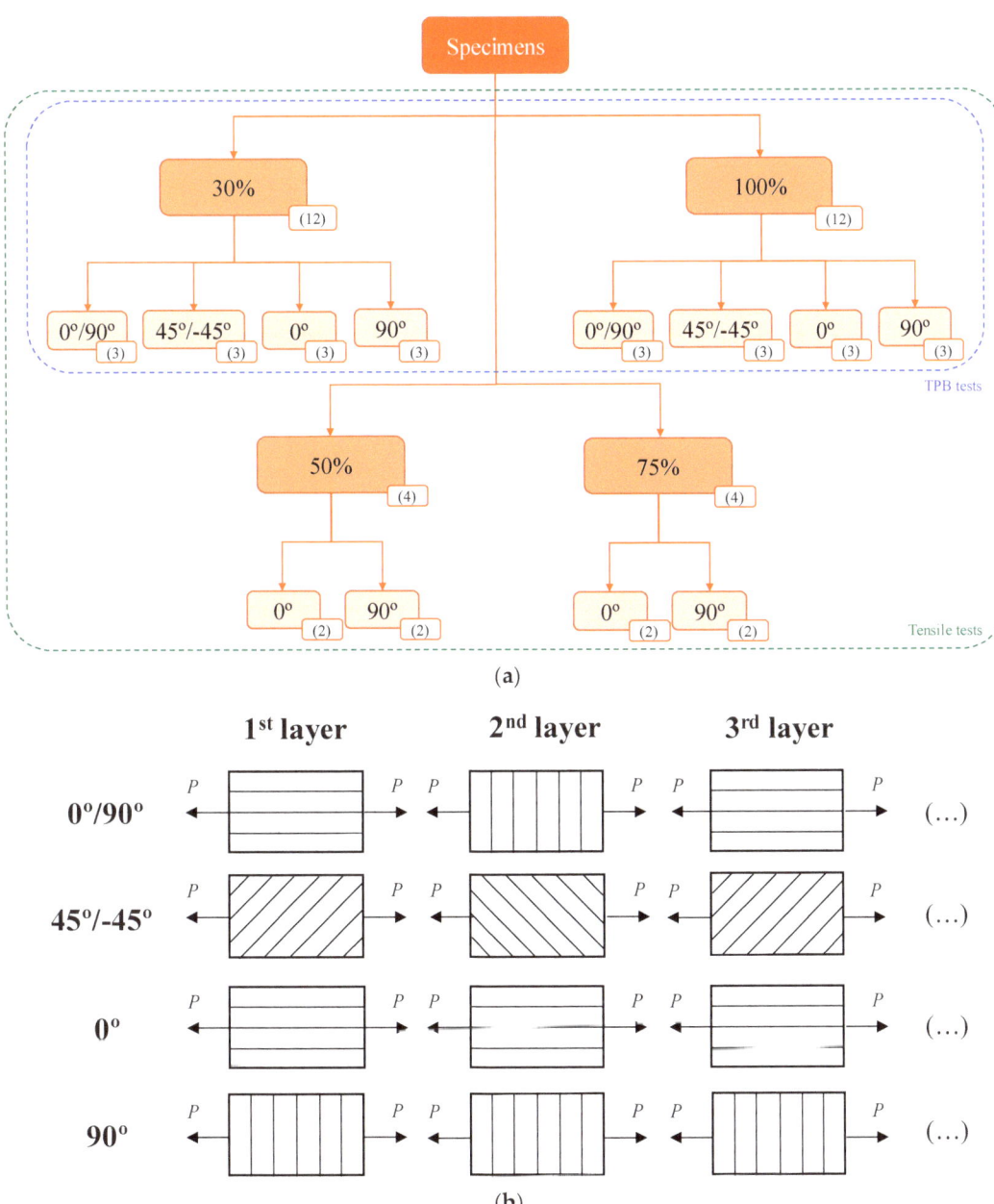

Figure 3. Specimens (**a**) infill percentage and filament orientations (data in parentheses represent specimen quantities); and (**b**) top-view of filament deposition (P represents the loading axis).

Table 2. Average values of specimen dimensions (height h and width W) printed with different orientations and infill percentages (values in mm).

			Orientations			
			0°/90°	45°/−45°	0°	90°
Infill	30%	h	4.28	4.28	4.22	4.28
		W	9.97	10.01	10.01	10.07
	100%	h	4.12	4.22	4.15	4.18
		W	9.97	9.96	9.99	9.99

Table 3. Average values of specimen dimensions (thickness h and width b_1) printed with different orientations and infill percentages (dimensions in mm).

			Orientations			
			0°/90°	45°/−45°	0°	90°
Infill	30%	h	4.25	4.39	4.19	4.22
		b_1	9.98	9.98	10.04	10.09
	100%	h	4.24	4.30	4.10	4.23
		b_1	10.04	10.10	10.08	10.15
	50%	h	-	-	4.20	4.24
		b_1	-	-	10.04	10.16
	75%	h	-	-	4.22	4.26
		b_1	-	-	10.08	10.10

2.2. Inverse Method

As previously stated, the goal of this work is to provide a simple strategy for identifying the elastic performance of PLA by combining experimental and numerical data through optimisation. Thus, the three-point bending (TPB) and tensile tests were selected as the two experimental tests. The chosen method entails identifying the printed PLA's elastic characteristics that mimic the experimentally determined mechanical response (P-d curve) (Figure 4).

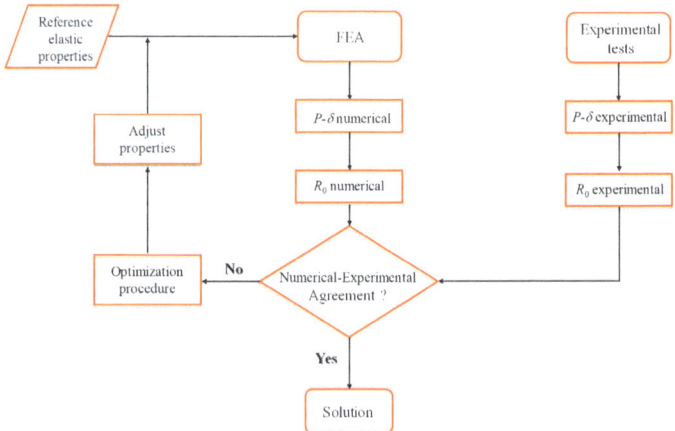

Figure 4. Flowchart of the adopted inverse method.

2.2.1. Three-Point Bending Tests

Bending tests were conducted according to the ISO 178 [19] standard using servo-electrical material testing equipment (MicroTester INSTRON 5848, Instron®, Norwood, MA, USA) under displacement control (5.0 mm/min). An acquisition frequency of 5 Hz was selected, and a 2 kN load cell was used. At room temperature (25 °C, 65°RH), the displacement (δ) and load (P) were recorded, and the loading span (L) was 60 mm. Twenty-four specimens were tested (for 30 and 100% infill percentages) in the linear elastic region (Figure 5).

Figure 5. Experimental setup of TPB test.

The modulus of elasticity along the longitudinal direction was obtained considering the Bernoulli–Euler beam theory,

$$E_x = \frac{R_0 L^3}{4 W h^3} \qquad (1)$$

where R_0 stands for the initial stiffness, W for the width, and h for the specimen height (Figure 2a).

2.2.2. Standard Tensile Tests

Tensile tests were accomplished at a displacement rate of 3 mm/min with a 25 kN load cell in an Instron® 8874 Universal Testing System. These tests were performed at room temperature (25 °C; 65°RH) following ISO 527-1 [20] & ISO 527-2 [21] standard directives. Using a gauge length L_0 of 25 mm, self-centering grips were implemented with a load-gripping distance L of 115 mm (Figure 2b). Tests were conducted on twenty-four specimens (Figure 6), with load (P) and displacement (d) being recorded at an acquisition frequency of 20 Hz.

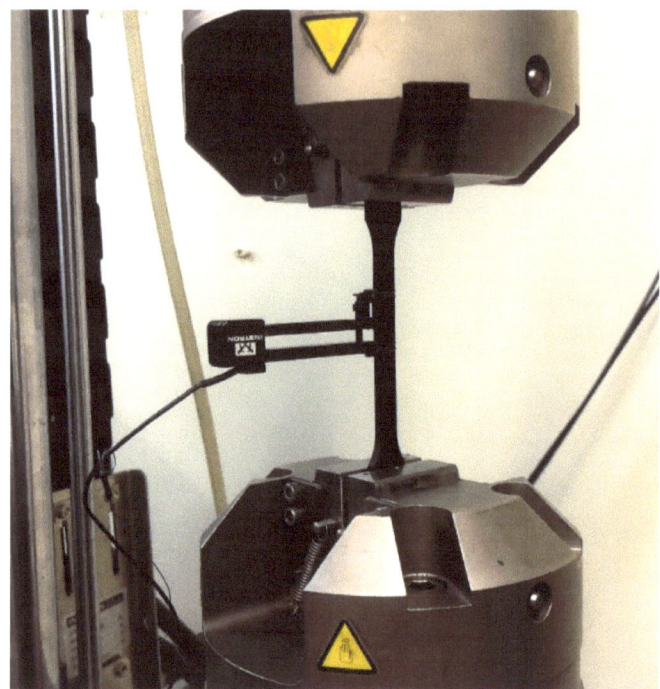

Figure 6. Experimental setup of standard tensile tests.

2.2.3. Tensile Tests with Digital Image Correlation

DIC measurements were employed at room temperature (25 °C; 65% RH) using an INSTRON® 5969 Universal Testing System equipped with a load cell of 5 kN and a displacement rate of 5 mm/min. The parameters that were established were load-gripping distance L (115 mm) and service length l_1 (80 mm) (Figure 2b). At the grips, the applied force P and displacement d were continuously measured at a 10 Hz acquisition rate. Since the strain field in the linear elastic domain must remain constant throughout the mechanical test, the central region of the specimen (located within l_1 in Figure 2b) is especially suited to measure elastic properties. This area was utilised specifically to create the speckle pattern due to the specimen surface being sufficiently flat and regular. To create a suitable carrier for DIC measurements, a fine matte coat was sprayed onto the specimen first, followed by an uneven spread of black marks (Figure 7). After this process, a distinct texture with the right isotropy and contrast could be defined due to the speckled pattern that was produced. In terms of the matching procedure, the reference image, or non-deformed configuration, is usually meshed into subsets designated as correlation domains, the size that determines the spatial resolution of the measurements. In this work, post-processing operations were carried out using the GOM Correlate 2018 software. The optical system was comprised of a TAMRON 24-70 MM F/2.8 lens (Saitama, Japan) and an 8-bit Nikon D800E camera (Tokyo, Japan) with a pixel resolution of 4912 × 7360 (Figure 8). The camera lens system was prudently placed in relation to the specimen surface, adopting a precision level, ensuring a working distance of 150 mm, which leads to a conversion factor of around 0.02 mm/pixel. By carefully opening the lens aperture (minimum depth of field), it was possible to focus and capture crisp photos of the speckled pattern. Still, to minimise diffraction effects (smallest apertures), the aperture was sealed throughout the loading procedure to produce a sufficient depth of field. By carefully controlling the shutter duration and illumination, motion blur and pixel saturation were prevented during movement. A specified 0.2 Hz was used for the picture capture rate, and 10–2 pixels (2 mm) was the displacement resolution achieved by setting the subset size and subset step to 15 × 15 pixels^2 and 11 × 11 pixels^2, respectively [22].

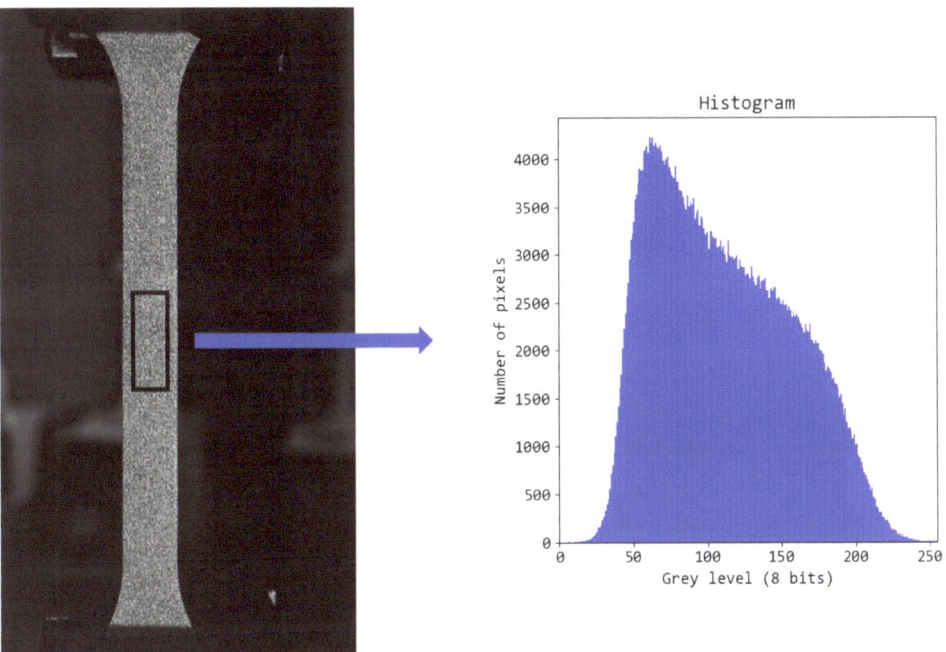

Figure 7. Typical speckle pattern for DIC measurements and corresponding grey level histogram.

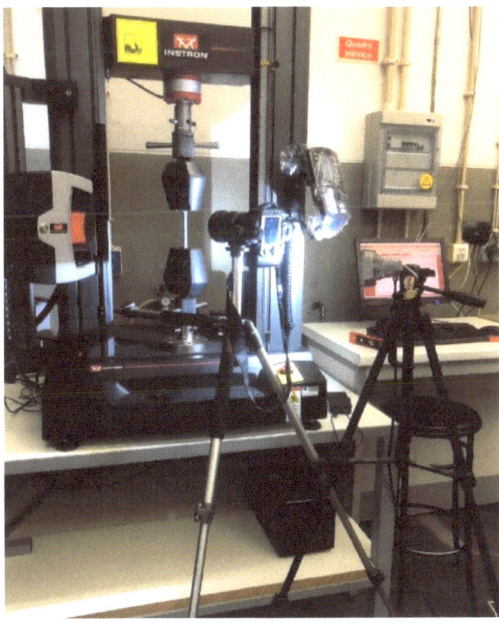

Figure 8. Experimental setup of tensile tests using DIC system.

3. Numerical Models

3.1. Three-Point Bending Model

A numerical model was developed to mimic the three-point bending test, taking advantage of geometrical and material symmetry, counting the actuator (Figure 9). Non-linear geometrical analyses were accomplished usinga model composed of 1070 20-node quadratic brick elements with 5896 nodes.

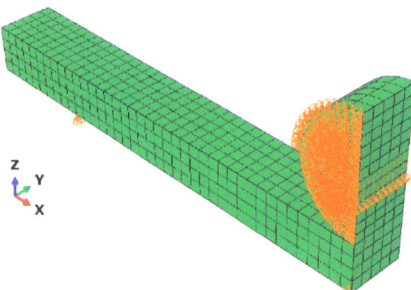

Figure 9. Three-point bending test model exhibiting the boundary conditions.

According to the mechanical test, boundary conditions (including symmetry) were imposed, and a vertical displacement of 5.0 mm was prescribed. The orthotropic behaviour of PLA-layered structures (Figure 3b) was modelled by taking each layer's thickness and filament orientation into account. Regarding 0°/90°, the first layer was oriented at 0°, the second at 90°, and the remaining ones resulting from the copies of these two. For a total of 40 layers in 4 mm (h in Figure 2a), the thickness and height of each layer were set to 0.1 mm. PLA specimens (structures) were modelled for orthotropy taking into account the elastic properties E_1, $E_2 = E_3$ (found experimentally in the section that follows), v_{12}, v_{13}, v_{23}, G_{12}, G_{13}, and G_{23} (Table 4). In order to replicate the mechanical behaviour of steel, elastic isotropy was taken into consideration for the actuator (E = 210 GPa, n = 0.3) [23].

Table 4. Elastic properties of the printed PLA [24].

v_{12}	v_{13}	v_{23}	G_{12} (MPa)	G_{13} (MPa)	G_{23} (MPa)
0.320	0.310	0.255	1019	1019	917

3.2. Tensile Test Model

Dogbone geometry was also simplified to a quarter model trying to decrease the computational cost. The 672 20-node quadratic brick elements were defined, with a total of 4197 nodes. Boundary conditions were defined to adequately mimic the mechanical test (Figure 8), with a vertical displacement of 5.0 mm (Figure 10). The orthotropic behaviour of PLA structures was mimicked using the same elastic properties as the TPB model.

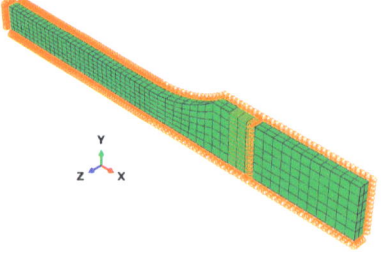

Figure 10. Numerical model of the tensile test showing the adopted boundary conditions.

4. Results and Discussion

4.1. Three-Point Bending Tests

Load–displacement curves (Figure 11) showed an impressive degree of consistency for each processing condition (infill and orientation), both in terms of initial stiffness (R_0) and ultimate load. This is the result of meticulous specimen preparation (implementing the same PLA coil), as well as extremely optimal conditions during the filament-making operation and the experimental work. Variances in the initial stiffness and ultimate load are perceived between processing conditions (0°/90°, 45°/−45°, 0°, and 90°, for 30% and 100% infill). Figures 11 and 12 show the impact of infill percentage and filament orientation on the initial stiffness along the specimen longitudinal direction. The stiffness resulting from 100% infill is greater than that obtained for 30% infill, regardless of the orientation of the filament. The stiffness is found to reach its maximum value for the 0° orientation at 30% infill, and then to reach its lowest value for 90°, 45°/−45°, and the stacking sequence 0°/90°. This indicates that, if the infill percentage is set to 30%, an increase in stiffness of 258% is achieved when the filament is aligned with the load (0°) compared to the orientation 90°. With respect to the 100% infill, the referred order hardly varies for 90° and 45°/−45° orientations. For the orientation 90° and 100% infill, this variation can be explained by the presence of a more extensive and pronounced filament fusion, which is not confirmed for the 30% infill.

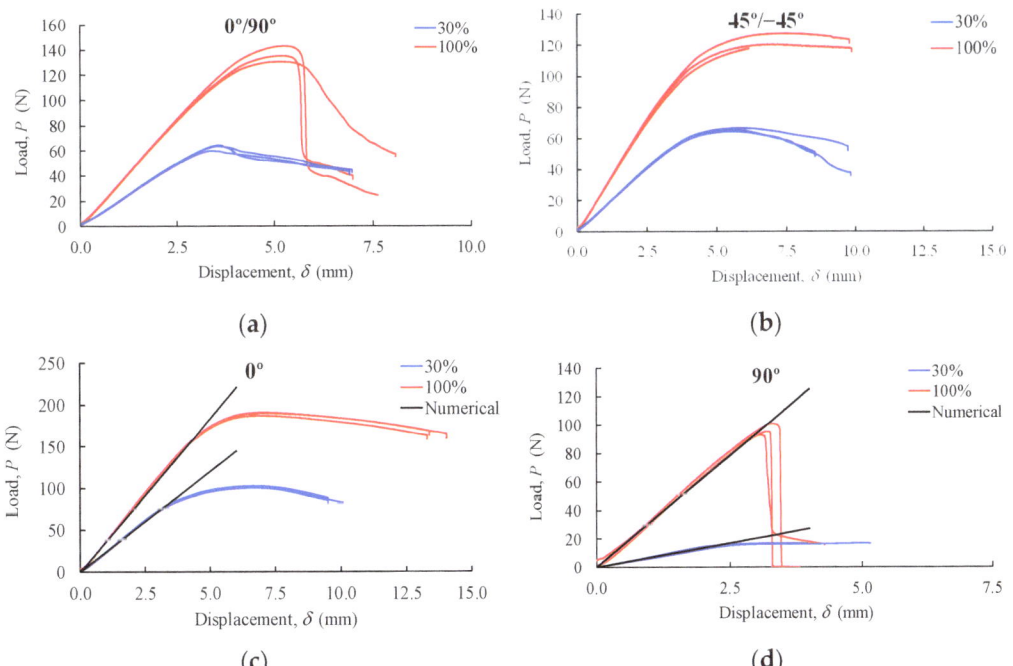

Figure 11. Experimental and numerical P-d curves for the orientations: (**a**) 0°/90°, (**b**) 45°/−45°, (**c**) 0°, and (**d**) 90° obtained in TPB tests.

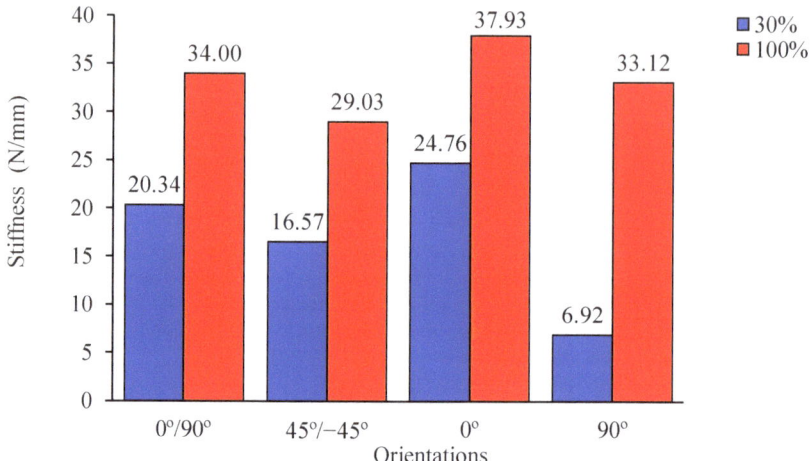

Figure 12. Initial stiffness for each filament orientation and infill percentage obtained in TPB tests.

4.2. Tensile Tests

Figure 13 shows the set of load–displacement curves attained in tensile tests using different orientations of filament (0°/90°, 45°/−45°, 0°, and 90°) and infill percentages (30%, 50%, 75%, and 100%). The results show excellent reproducibility for every printing condition once more, indicating steady conditions for material production and meticulous preparation of test specimens. The filament orientations clearly differ from one another, especially in terms of the percentage of infill. With reference to the load orientation, P, the stress–strain curves from specimens with filaments oriented at 0° and 90°, respectively, were used to determine the elastic moduli E_1 and E_2.

According to Table 5, the initial stiffness (R_0) and ultimate load (P_u) increase with infill percentage as well as filament alignment with the loading axis (from 90° to 0°). Despite the filament orientation, the stiffest structure is always the one with 100% infill (Figure 14). The smallest stiffness, on the other hand, is always the one corresponding to 30% infill, independent of the filament's deposition orientation. The stiffest specimens have 100% infill and are oriented at 0°, while the most flexible have 30% infill and are oriented at 45°/−45°.

Table 5. Resume of tensile tests.

Orientation (°)	Infill (%)	R_0 (N/mm)	P_u (N)
0/90	30	377	593
	100	1014	1676
45/−45	30	297	602
	100	990	1743
0	30	585	1256
	50	868	1774
	75	935	1838
	100	1088	2490
90	30	308	318
	50	346	308
	75	394	421
	100	1005	1549

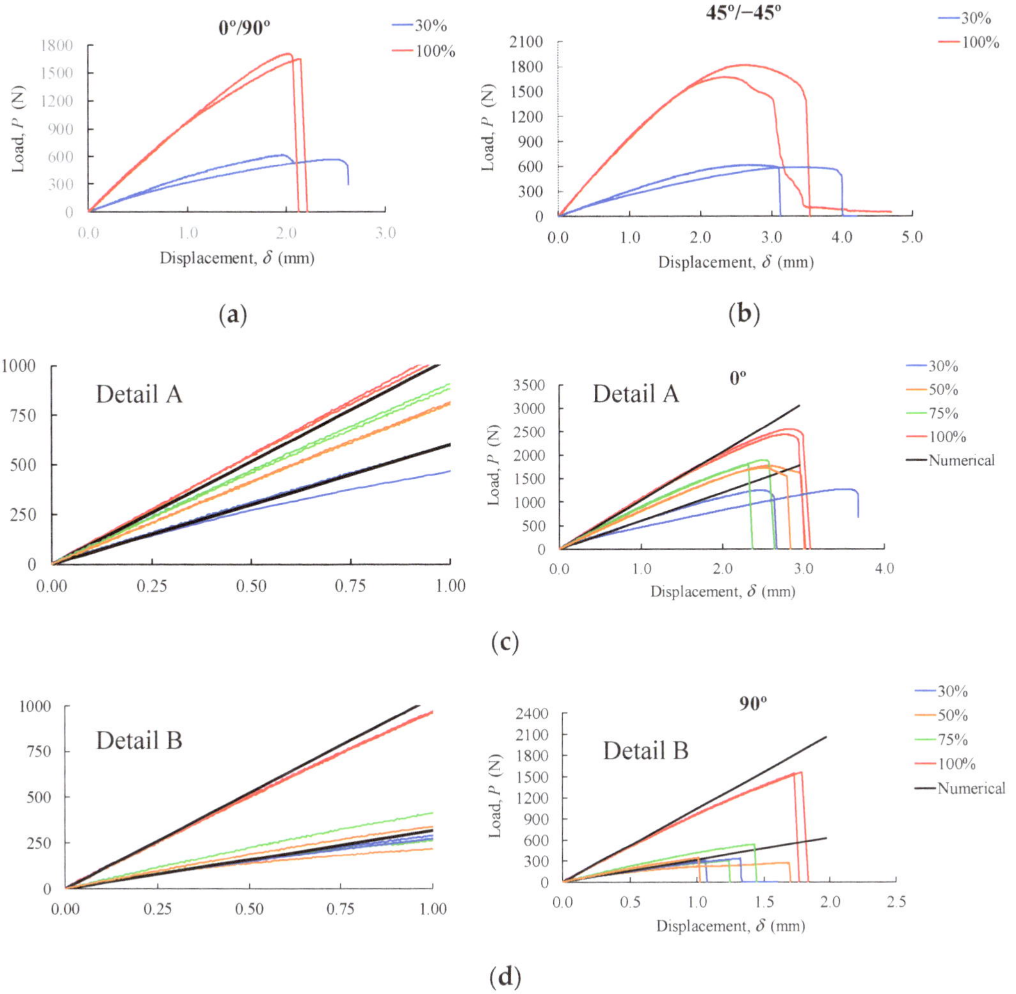

Figure 13. Experimental and numerical *P-d* curves for the orientations: (**a**) 0°/90°, (**b**) 45°/−45°, (**c**) 0°, and (**d**) 90° obtained in tensile tests.

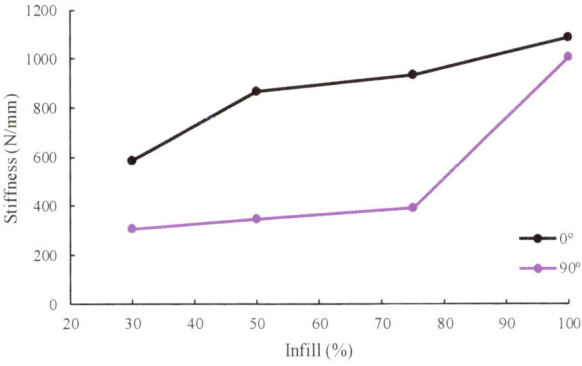

Figure 14. Evolution of initial stiffness with infill percentage.

Additionally, filaments oriented at 0 degrees show a gradual increase in stiffness, whereas orientations at 90° and 100% infill show a more pronounced increase in stiffness in comparison to the remaining infills. Similar behaviours may be noted for the ultimate load.

Another point worth mentioning is the ductile behaviour observed for structures oriented at 45°/−45° and 100% infill, which contrast with the remaining ones that exhibit a more brittle nature (Figure 13b). A filament that is equally oriented relative to the loading axis could explain this behaviour. In contrast to the ductile behaviour, specimens with 90° oriented filaments are clearly brittle (Figure 13d). A reasonable explanation for the referred brittleness could be due to the sudden initiation and propagation of damage at the filament interface, which in this case is perpendicular to the load direction.

For the combined orientations, i.e., 0°/90° and 45°/−45°, the stiffness should appear in the range obtained for the principal directions 0° and 90°. As Figure 15 shows, it is not possible to estimate the elastic response of 3DP PLA for combined orientations (e.g., 0°/90° and 45°/−45°) using test results obtained at 0° and 90°. It is confirmed that for 100% infill this behaviour is not verified.

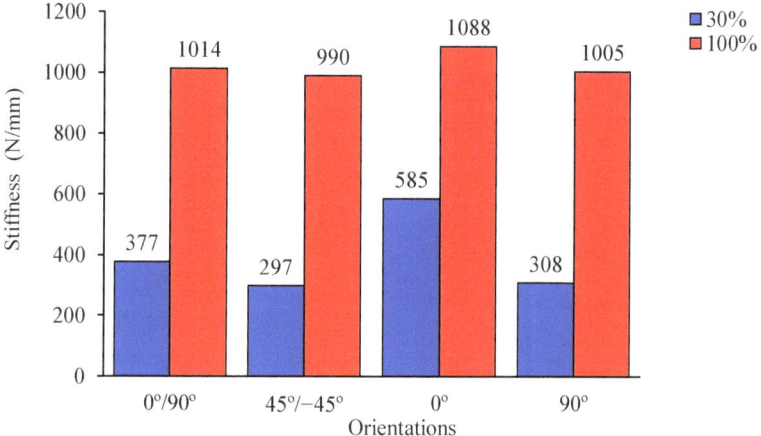

Figure 15. Stiffness for each filament orientation and infill percentage obtained in tensile tests.

Using the DIC technique, full-field measurements of strain were carried out in both the x and y directions in the specimen's central region (Figure 2b), displaying uniform measurements in the linear–elastic domain. The strain field in the x-direction obtained in the linear–elastic domain at three loading points on the stress–strain curve's ascending branch is displayed in Figure 16.

Furthermore, in specimens for which DIC measurements were not performed, a strain gauge was employed within the service length l_1 for 30% and 100% infill, for all filament orientations (Figure 17). Although DIC is a more time-consuming technique compared to the employment of a strain gauge, it allows verifying whether the strain-field in the specimen central region is homogeneous. This technique was applied to a limited number of specimens, namely those with extreme values of infill (30% and 100%), since in those conditions it is possible to reveal more notorious mechanical differences in the material mechanical response. For the remaining specimens, with infill percentages of 30%, 50%, 75%, and 100%, the less time-consuming technique (strain gauge) was employed. The plotted curve for the strain gauge data was created by taking mean values into consideration. Once more, it is evident that the outcomes are highly consistent with one another. There is a discernible variation between the infill percentages, as would be predicted based on earlier findings. Effectively, more compact structures (100% infill) have higher tensile strength and elastic stiffness than less compact ones (30% infill).

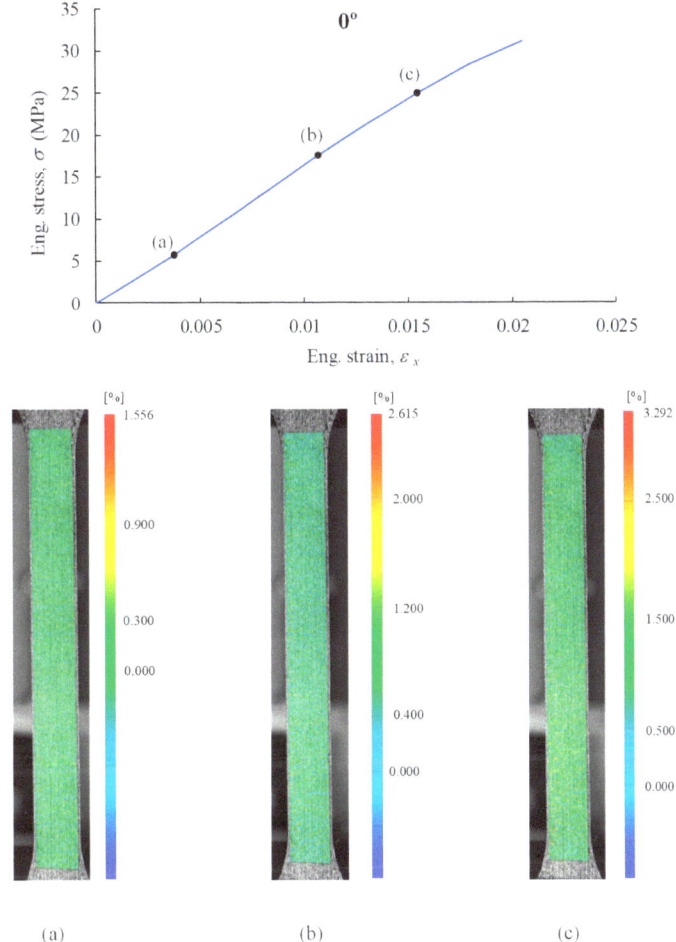

Figure 16. Strain field along the loading direction in tensile test relatively to 0° orientation and 30% infill obtained with DIC technique. (**a**), (**b**) and (**c**) correspond to indentified points in the above stress-strain curve.

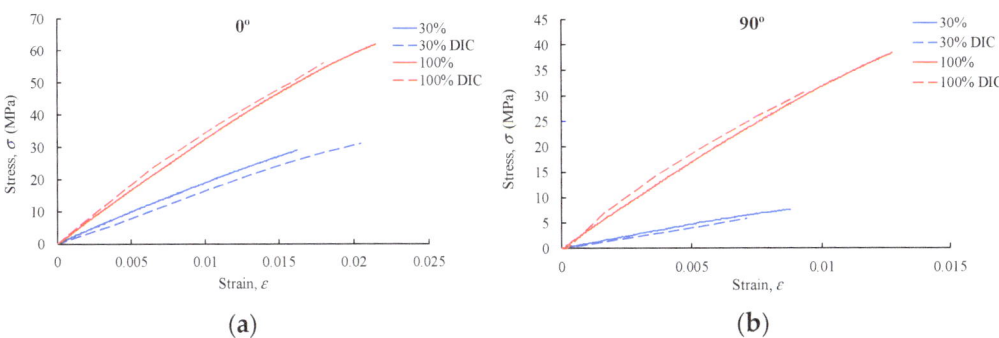

Figure 17. Stress–strain curves for the orientations: (**a**) 0° and (**b**) 90° obtained in standard tensile tests and with DIC.

The elastic modulus (E) was then obtained for filament orientated at 0° and 90° (Figure 17). One can notice that the obtained value of E is slightly higher in tests using a strain gauge (standard tests) compared to those carried out with DIC (13% on average). However, the latter technique yields more information (full-field measurements) than the former, based on measurements conducted in two specimen sections (gauge fittings).

Figure 18 reveals that the modulus of elasticity for 100% infill remains higher than that for 30%, regardless of filament orientation. Specimens with filaments oriented at 0° to the loading axis exhibit a higher elastic modulus than those oriented at 90°. Furthermore, differences among 30% infill are greater than those observed for 100% infill. A plausible reason for the behaviour observed for 30% infill in the orientation 0° may be due to a more effective contribution of the less constrained aligned filaments than the fully filled (fused filament) structures, i.e., 100%.

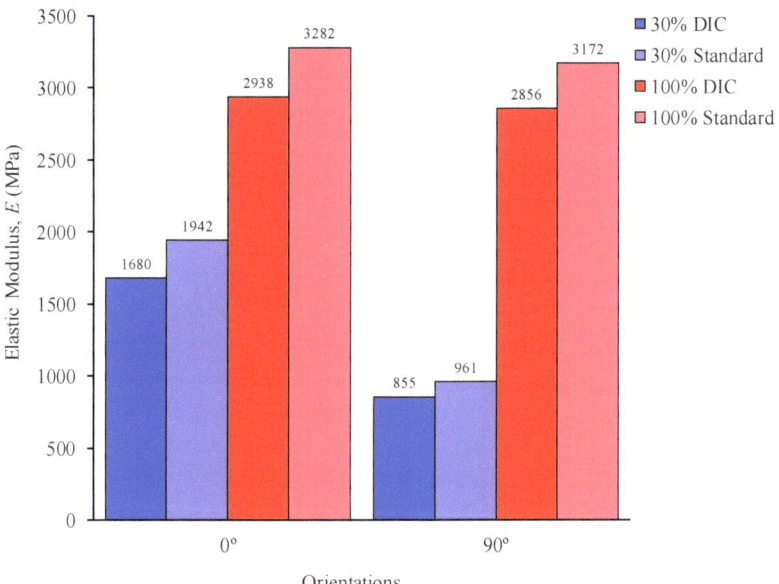

Figure 18. Elastic modulus corresponding to filament orientation considering 30% and 100% infill relatively standard tensile tests and with DIC (MPa).

The elastic modulus was determined for filament aligned at 0° and 90° (Figure 19) thus enabling more precise analysis of the impact of the infill percentages (i.e., 50% and 75% infill, in addition to 30% and 100%), leading to E_1 and E_2, respectively.

As a result, it was possible to determine that, for both orientations, the elastic modulus consistently increases with the infill percentage. Regarding the 90° filament orientation, only the first three infill percentages (30% to 75%) show a consistent trend, with the 100% infill showing a noticeably higher percentage. Again, the highest elastic modulus is obtained for the filament oriented at 0° (i.e., E_1) and 100% infill, while the lowest is obtained for 90° and 30% infill.

For the orientation at 90° (i.e., E_2), a less pronounced increase of the elastic modulus is observed in the infill interval 30–75%, compared to the one observed at 0° (i.e., E_1). In contrast to the remaining structures, the orientation at 90° exhibits a much more pronounced growth rate of E in the 75% to 100% transition than the orientation at 0°. This difference may be attributed to a greater amount of filament contact and a corresponding decrease in the percentage of air gap. It is possible to conclude that 100% infill structures are less affected by filament orientation than the others.

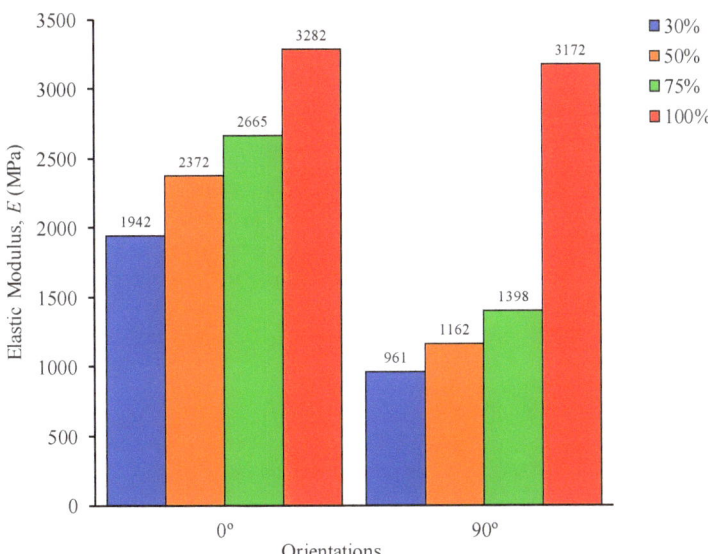

Figure 19. Elastic modulus for 0° and 90° filament orientation considering 30%, 50%, 75% and 100% infill relatively to standard tensile tests (values in MPa).

Poisson's ratio values ($n_{yx} = -e_y/e_x$) were measured for filaments oriented at 0° and 90° (n_{21} and n_{12}, respectively), with 30% and 100% infill (Figure 20), i.e., structures evaluated with DIC. The result, that the Poisson's ratio is lower in 30% infill than in 100% infill, is consistent with a more compliant structure when the number of filaments per unit volume is lower along the transverse direction (y-direction in Figure 2b). Figure 20 indicates that the orientation at 90° and 100% infill yielded the highest Poisson's ratio value. This may be justified by the structure's greater sensitivity to longitudinal loads resulting in transversal deformation for a considerable condensed material, i.e., 100% infill. Since less-compact structures (30% infill) exhibit more air gaps, such behaviour is not observed.

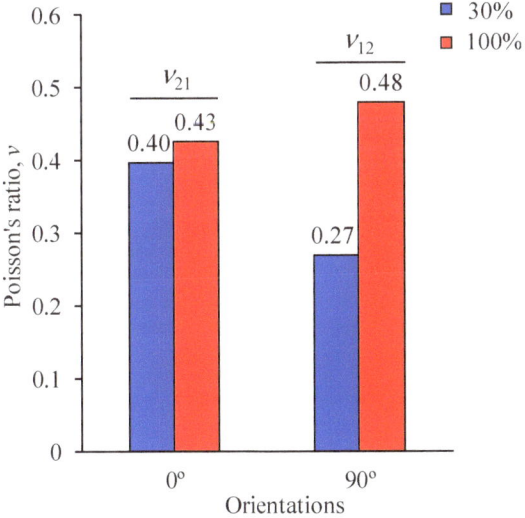

Figure 20. Poisson's ratio for different filaments orientation and infill corresponding to tensile tests using DIC analysis.

4.3. Numerical Modelling

For both the filament orientations of $0°$ and $90°$, the elastic properties selected for the FE analysis to simulate the TPB test were meant to reproduce the initial stiffness R_0 that was measured experimentally (Figure 11). For the orientations $0°$ and $90°$ (Figure 11c,d), the values of the elastic moduli E_1 and E_2 were set to the ones obtained experimentally using Equation (1), taking into account the remaining parameters as in Table 4. The numerical approximation obtained was found to be reasonably close to the experimental data.

Analogously, DIC's experimental results were used to set the elastic properties that were used to simulate the tensile test. Replicating the experimental elastic stiffness with accuracy, a distinct linear response (*P-d* curve) was obtained (Figure 13c,d). Notably, the obtained numerical agreement suggests that the elastic modulus of this material can be accurately determined using an inverse method that solely relies on the load–displacement curve.

5. Conclusions

Three-point bending tests were adopted to evaluate the initial stiffness of polylactic acid (PLA)-made parts obtained using material extrusion (MEX). The orientations of the tests were $0°/90°$, $45°/-45°$, $0°$, and $90°$, with 30% and 100% infill. Both the infill percentage and the measurements related to the filament orientation showed consistent variations. These results led to the conclusion that stiffness measurements in PLA parts are dependent upon the percentage of infill, specifically within the range of 30% to 100%. Apart from the previously mentioned experimental protocol, tensile tests were carried out using standard procedures (strain-gauge) for the studied orientations, with additional infill percentages in the range of 30–100% for filament oriented at $0°$ and $90°$. To assess the elastic properties (Young moduli), digital image correlation technique (DIC) was used due to the accuracy of those methods depending on the homogeneity of the strain field. The subsequent findings demonstrated that using standard operating procedures made it possible to measure elastic properties accurately. Standard methods were then used to assess the influence of the infill on the elastic moduli for the investigated filament orientations in greater detail. This study also demonstrated that the failure mode (ductile or brittle) in 3D-printed PLA parts under tensile loading is determined by the filament arrangement (combined orientations) rather than the infill percentage. Remarkably, the effect of material orientation on the elastic modulus with regard to the loading axis can be determined more precisely using the uniaxial tensile test than is possible with the TPB test. Recognizing that TPB test results are highly sensitive to loading span provides a very plausible explanation for this. In addition, the elastic modulus was shown to increase with the infill percentage regardless of filament orientation. A clear linear response (*P-δ* curve) was obtained by properly reproducing the experimental elastic stiffness. By achieving a strict numerical agreement on the *P-δ* curves that emerged from the experimental testing, the inverse approach provided for the numerical determination of the material's modulus of elasticity.

Author Contributions: M.P.S.: Methodology, Investigation, Writing—original draft, Writing—review & editing. F.A.M.P.: Methodology, Software, Supervision. C.L.F.: Methodology, Conceptualization. E.B.P. and J.A.P.P.Λ.: Conceptualisation, Methodology, Software. T.D.C.: Data curation, Investigation. Q.N.: Formal Analysis. C.F.: Software. A.Z.: Funding acquisition, Investigation. N.D.: Conceptualisation, Methodology, Writing—review & editing, Supervision. All authors have read and agreed to the published version of the manuscript.

Funding: National Innovation Agency (ANI) for MSc grant of Mariana Salgueiro n° POCI-01-0247-FEDER-039733 and Portuguese Foundations for Science and Technology. This project was co-financed by European Regional Development Fund (ERDF) through SI&IDT Projects in the framework of co-hosting—Competitiveness and Internationalisation Operational Programme (CIOP)—COMPETE 2020, Portugal 2020, with the National Innovation Agency (ANI) as the Intermediate Partner. Fabio Pereira acknowledges the Portuguese Foundation for Science and Technology, under the project UIDB/04033/2020. Mariana Salgueiro and Andrea Zille acknowledge the European Commission and the National Innovation Agency (ANI) for the financial support through the project "ARCHKNIT: Innovative smart textile interfaces for architectural applications", Ref.: POCI-01-0247-FEDER-039733. This project was co-financed by European Regional Development Fund (ERDF) through SI&IDT Projects in the framework of co-hosting—Competitiveness and Internationalisation Operational Programme (CIOP)—COMPETE 2020, Portugal 2020, with the National Innovation Agency (ANI) as the Intermediate Partner. Nuno Dourado acknowledges FCT for the conceded financial support through the reference project UID/EEA/04436/2019 and "Programa bilateral de Portugal com a Tunísia". Charii Fakher acknowledges the « Fondation pour la Recherche Scientifique" for the conceded financial support through "Programa bilateral de Portugal com a Tunísia".

Data Availability Statement: The data supporting the findings of this study are available within the article.

Conflicts of Interest: The authors declare no conflicts of interest.

References

1. Gardan, J. *Additive Manufacturing Technologies for Polymer Composites: State-of-the-Art and Future Trends*; Elsevier: Amsterdam, The Netherlands, 2020. [CrossRef]
2. Ashley, S. Rapid prototyping systems. *Mech. Eng.* **1991**, *113*, 34–36.
3. Vaezi, M.; Seitz, H.; Yang, S. A review on 3D micro-additive manufacturing technologies. *Int. J. Adv. Manuf. Technol.* **2012**, *67*, 1721–1754. [CrossRef]
4. Jain, P.; Kuthe, A.M. Feasibility Study of Manufacturing Using Rapid Prototyping: FDM Approach. *Procedia Eng.* **2013**, *63*, 4–11. [CrossRef]
5. Awasthi, P.; Banerjee, S.S. Fused deposition modeling of thermoplastic elastomeric materials: Challenges and opportunities. *Addit. Manuf.* **2021**, *46*, 102177. [CrossRef]
6. Cano-Vicent, A.; Tambuwala, M.M.; Hassan, S.S.; Barh, D.; Aljabali, A.A.A.; Birkett, M.; Arjunan, A.; Serrano-Aroca, Á. Fused deposition modelling: Current status, methodology, applications and future prospects. *Addit. Manuf.* **2021**, *47*, 102378. [CrossRef]
7. Liu, Z.; Wang, Y.; Wu, B.; Cui, C.; Guo, Y.; Yan, C. A critical review of fused deposition modeling 3D printing technology in manufacturing polylactic acid parts. *Int. J. Adv. Manuf. Technol.* **2019**, *102*, 2877–2889. [CrossRef]
8. Relvas, C. O Mundo Da Impressão 3d E Do Fabrico Digital, Publindúst. 2017. Available online: www.engebook.com (accessed on 7 February 2022).
9. Afonso, J.A.; Alves, J.L.; Caldas, G.; Gouveia, B.P.; Santana, L.; Belinha, J. Influence of 3D printing process parameters on the mechanical properties and mass of PLA parts and predictive models. *Rapid Prototyp. J.* **2021**, *27*, 487–495. [CrossRef]
10. Onwubolu, G.C.; Rayegani, F. Characterization and Optimization of Mechanical Properties of ABS Parts Manufactured by the Fused Deposition Modelling Process. *Int. J. Manuf. Eng.* **2014**, *2014*, 598531. [CrossRef]
11. Sood, A.K.; Ohdar, R.K.; Mahapatra, S.S. Parametric appraisal of mechanical property of fused deposition modelling processed parts. *Mater. Des.* **2010**, *31*, 287–295. [CrossRef]
12. Baich, L.; Manogharan, G.; Marie, H. Study of infill print design on production cost-time of 3D printed ABS parts. *Int. J. Rapid Manuf.* **2015**, *5*, 308. [CrossRef]
13. Harpool, T.D. Observing the Effect of Infill Shapes on the Tensile Characteristics of 3D Printed Plastic Parts. 2016. Available online: https://soar.wichita.edu/handle/10057/13504 (accessed on 20 January 2023).
14. Alafaghani, A.; Qattawi, A.; Alrawi, B.; Guzman, A. Experimental Optimization of Fused Deposition Modelling Processing Parameters: A Design-for-Manufacturing Approach. *Procedia Manuf.* **2017**, *10*, 791–803. [CrossRef]
15. Behzadnasab, M.; Yousefi, A.A. Effects of 3D printer nozzle head temperature on the physical and mechanical properties of PLA based product. In Proceedings of the 12th International Seminar on Polymer Science and Technology, Tehran, Iran, 2–5 November 2016.
16. Johansson, F. *Optimizing Fused Filament Fabrication 3D Printing for Durability Tensile Properties & Layer Bonding*; Blekinge Institute of Technology: Karlskrona, Sweden, 2016.
17. Bardiya, S.; Jerald, J.; Satheeshkumar, V. Effect of process parameters on the impact strength of fused filament fabricated (FFF) polylactic acid (PLA) parts. *Mater. Today Proc.* **2021**, *41*, 1103–1106. [CrossRef]
18. Abeykoon, C.; Sri-Amphorn, P.; Fernando, A. Optimization of fused deposition modeling parameters for improved PLA and ABS 3D printed structures. *Int. J. Light. Mater. Manuf.* **2020**, *3*, 284–297. [CrossRef]

19. *ISO 178, 2019. DIN EN ISO ISO 178*; Plastics-Determination of Flexural Properties. International Standard Organization Std.: Geneva, Switzerland, 2019. Available online: https://www.iso.org/standard/70513.html (accessed on 7 February 2022).
20. *ISO 527-1, 2012. DIN EN ISO 527-1*; Plastics-Determination of Tensile Properties, Part 1: General Principles. International Standard Organization Std.: Geneva, Switzerland, 2012. Available online: https://www.iso.org/fr/standard/56045.html (accessed on 7 February 2022).
21. *ISO 527-2, 2012. DIN EN ISO 527-2*; Plastics-Determination of Tensile Properties, Part 2: Test Conditions for Moulding and Extrusion Plastics. International Standard Organization Std.: Geneva, Switzerland, 2012.
22. Xavier, J.; De Jesus, A.M.P.; Morais, J.J.L.; Pinto, J.M.T. Stereovision measurements on evaluating the modulus of elasticity of wood by compression tests parallel to the grain. *Constr. Build. Mater.* **2012**, *26*, 207–215. [CrossRef]
23. Gere, J.M.; Goodno, B.J. *Mechanics of Materials*; Cengage Learning: Boston, MA, USA, 2009.
24. Gonabadi, H.; Chen, Y.; Yadav, A.; Bull, S. Investigation of the effect of raster angle, build orientation, and infill density on the elastic response of 3D printed parts using finite element microstructural modeling and homogenization techniques. *Int. J. Adv. Manuf. Technol.* **2022**, *118*, 1485–1510. [CrossRef]

Disclaimer/Publisher's Note: The statements, opinions and data contained in all publications are solely those of the individual author(s) and contributor(s) and not of MDPI and/or the editor(s). MDPI and/or the editor(s) disclaim responsibility for any injury to people or property resulting from any ideas, methods, instructions or products referred to in the content.

Article

Additive Manufacturing and Characterization of Sustainable Wood Fiber-Reinforced Green Composites

Christopher Billings, Ridwan Siddique, Benjamin Sherwood, Joshua Hall and Yingtao Liu *

School of Aerospace and Mechanical Engineering, University of Oklahoma, 865 Asp Ave., Norman, OK 73019, USA; christopherbillings@ou.edu (C.B.); ridwan.y.siddique-1@ou.edu (R.S.); benjamin.d.sherwood-1@ou.edu (B.S.); joshua.d.hall@ou.edu (J.H.)
* Correspondence: yingtao@ou.edu; Tel.: +1-(405)-325-3663

Abstract: Enhancing mechanical properties of environmentally friendly and renewable polymers by the introduction of natural fibers not only paves the way for developing sustainable composites but also enables new opportunities in advanced additive manufacturing (AM). In this paper, wood fibers, as a versatile renewable resource of cellulose, are integrated within bio-based polylactic acid (PLA) polymer for the development and 3D printing of sustainable and recycle green composites using fused deposition modeling (FDM) technology. The 3D-printed composites are comprehensively characterized to understand critical materials properties, including density, porosity, microstructures, tensile modulus, and ultimate strength. Non-contact digital image correlation (DIC) technology is employed to understand local stress and strain concentration during mechanical testing. The validated FDB-based AM process is employed to print honeycombs, woven bowls, and frame bins to demonstrate the manufacturing capability. The performance of 3D-printed honeycombs is tested under compressive loads with DIC to fully evaluate the mechanical performance and failure mechanism of ultra-light honeycomb structures. The research outcomes can be used to guide the design and optimization of AM-processed composite structures in a broad range of engineering applications.

Keywords: wood fiber; polylactic acid; composites; additive manufacturing; 3D printing; sustainability; recyclable; fused deposition modeling

---

Citation: Billings, C.; Siddique, R.; Sherwood, B.; Hall, J.; Liu, Y. Additive Manufacturing and Characterization of Sustainable Wood Fiber-Reinforced Green Composites. *J. Compos. Sci.* **2023**, *7*, 489. https://doi.org/10.3390/jcs7120489

Academic Editor: Yuan Chen

Received: 11 October 2023
Revised: 8 November 2023
Accepted: 23 November 2023
Published: 26 November 2023

Copyright: © 2023 by the authors. Licensee MDPI, Basel, Switzerland. This article is an open access article distributed under the terms and conditions of the Creative Commons Attribution (CC BY) license (https://creativecommons.org/licenses/by/4.0/).

1. Introduction

In recent years, there has been a rapid surge in the advancement of eco-friendly green composites, signifying a groundbreaking transformation in materials science and engineering [1–3]. These advancements have been driven by the increasing global imperative to mitigate environmental challenges and transition towards more sustainable manufacturing practices. Sustainable composites, often derived from renewable, bio-based resources, offer multifaceted advantages, including reduced carbon footprint, reduced reliance on non-renewable resources, and properties comparable to traditional composites [4]. Additionally, they open new possibilities for closed-loop recycling and circular economies, reducing waste and promoting efficient resource utilization. Now that global challenges intensify, such as climate change, resource scarcity, and environmental pollution, sustainable composites emerge not only as an alternative but also as an essential evolution in material design, aligning technological progress with environmental stewardship [5].

Natural fibers, such as wood fibers, cotton fibers, bamboo fibers, and silk, play a pivotal role in the development of sustainable green composites [6–8]. These plant-based and animal-based fibers, derived from renewable sources, such as wood, hemp, flax, and jute, offer significant environmentally friendly advantages due to their biodegradability, recyclability, low carbon footprint, cost-effectiveness, and minimum generation of non-recyclable waste. The dominant chemical composition of natural fibers usually contains lignin, cellulose, and hemicellulose, which can be used as the filler materials within a polymer matrix as the renewable resource to produce green composites [9–11]. Natural fibers can enhance the mechanical properties of green composites, such as strength and stiffness, while maintaining lightweight and sustainable characteristics. This combination of eco-friendliness, renewability, low cost, and outstanding performance makes natural fibers the essential components in the development of sustainable green composites, significantly contributing to a greener and more environmentally responsible future for materials science and engineering. However, the integration of natural fibers and polymers for the development of green composites requires processing optimization considering the natural characteristics of both materials. For example, most natural fibers are hydrophilic, whereas many polymers are hydrophobic in nature [12]. This mismatch can lead to poor interfacial adhesion between the fibers and the matrix, potentially reducing the mechanical properties of green composites. Additionally, natural fibers tend to absorb moisture from the environment, leading to a negative impact on the mechanical and dimensional stability of composites [13]. To improve compatibility between fibers and polymers, surface treatments, such as alkali treatment, silane treatment, or the use of coupling agents, are usually needed. Therefore, the development of novel processing and manufacturing technologies are urgently needed for the broad applications of sustainable green composites.

Additive manufacturing (AM) offers promising solutions to prepare, process, and fabricate sustainable composites that combine renewability with beneficial functionalities for broad engineering applications. Natural fibers, recycled materials, and biodegradable polymers have been integrated and applied to the 3D printing processes, yielding composites with reduced environmental impacts [14–16]. Multiple AM processes, including fused deposition modeling (FDM), selective laser sintering (SLS), stereolithography (SLA), and direct ink writing (DIW), have been studied for the development of innovative printing techniques that can optimize sustainable composite fabrication [17–21]. As one of the most common methods for 3D printing of green composites, FDM-based AM uses filaments comprising natural fibers or recycled materials blended with biodegradable polymers and extruded materials layer by layer to create intricate 3D products and components. FDM's versatility allows for the precise placement of reinforcing fibers within the printed object, enhancing its mechanical strength while maintaining an eco-friendly profile. For example, Cali et al. employed FDM-based 3D printing technology and processed five organic biocomposite filaments, including polylactic acid (PLA) polymer with hemp, weed, tomato, carob, and pruned fillers. Each natural agricultural additive generated different mechanical/physical properties, such as tensile strength, elasticity, density, porosity, and a strong visual and tactile identity [18]. Although FDM-based 3D printing excels in processing thermoplastic biopolymers, polymers in other forms, such as powders, require additional AM approaches for 3D printing of sustainable composites.

SLS-based 3D printing technology provides an alternative solution for the AM of sustainable composites. In SLS, a high-powered laser selectively sinters powdered sustainable composite materials layer by layer, allowing for the creation of intricate and robust structures [22]. SLS technology not only enhances the mechanical properties of these sustainable composites but also minimizes material wastage as unused powders that can be recycled for future prints. This approach aligns perfectly with the growing emphasis on sustainable practices in various industries, such as aerospace, healthcare, and architecture, where the production of lightweight, strong, and eco-friendly components is of paramount importance [23,24]. Additionally, there is a growing emphasis on the development of sustainable composite filaments and powders for commercial 3D printers, broadening the accessibility of these materials to a wider audience [25]. Recently, Idrises et al. reported an investigation to develop the AM of prosopis chilensis and polyethersulfone composite using SLS-based 3D printing. A comprehensive mechanical characterization was carried out to fully understand key parameters, such as bending and tensile strengths. Additionally, post-processing infiltration was employed to further enhance the performance of 3D-printed green composites [26].

Bio-based polymer resins derived from renewable resources have been developed for the 3D printing of green composites using the SLA and DIW AM methods [27–29].

Natural products, such as soybean oil, linseed oil, and even starch, have been investigated for resin development. Micro and nano natural fibers, such as cellulose nanocrystals, can be employed to further enhance the mechanical properties of bio-based resin [30]. However, SLA- and DIW-based 3D printing of green composites are limited by a few technical challenges, including effective bonding between the fiber and polymer matrix and consistent rheological properties for high printability throughout the entire printing process.

Although significant efforts have been made to develop novel materials and manufacturing processes for sustainable composites, challenges related to material compatibility, mechanical properties, and post-processing techniques still limit the broad engineering applications of certain green composites. Additionally, it is urgently needed to identify the material–process–structure–property relationship of green composites so that certain knowledge can be applied to the design and development of novel products using eco-friendly and recyclable green composites. Moreover, substantial progress is urgently needed to further improve environmental friendliness and customizable manufacturing processes with potential applications in industries, including aerospace, automotive, and consumer goods.

In this paper, we reported an investigation of 3D printing of sustainable composites composed of PLA and wood fibers. The novelty of this paper focuses on the identification of material–process–structure–property relationships for AM-processed green composites and their extended applications on honeycomb structures. FDM-based 3D printing technology was employed for the AM process of wood fiber-enhanced composites. Critical properties of the 3D-printed green composites, including density, porosity, and tensile strength and modulus, were systematically characterized. Microstructures were characterized using both optical microscope and scanning electron microscope (SEM). Non-contact and full-field digital image correlation (DIC) technology was employed to obtain accurate local strain concentration. Additionally, the wood fiber-enhanced composites were used to 3D print honeycomb structures, and DIC technology was employed to identify the failure mechanism of FDM 3D-printed green composites. The research outcomes of this paper can be useful for further investigation of sustainable composites as the core material for potential ultra-light sandwich composite applications in aerospace, automotive, and mechanical applications.

2. Materials and Methods

2.1. Materials

Unless otherwise stated, all materials were used as received. Walnut wood fiber-reinforced PLA filament and pristine PLA filament were purchased from Amolen (Shenzhen, China). According to the supplier's data, this filament included 30 wt.% wood fibers in the PLA matrix.

2.2. FDM-Based 3D Printing of Wood Fiber-Reinforced Composites

In this paper, a FlashForge Creator Max 2 Independent Dual Extruder 3D Printer (Rowland Heights, CA, USA) was employed as the AM platform to fabricate green composite samples using the wood fiber-reinforced PLA matrix composite filament. The geometry of the dogbone composite samples followed ASTM D638 type V standard [31]. To ensure consistency and accuracy, all printed dogbones were configured to have 100% infill density. To identify the material–process–property relationship, three printing orientations, including 0°, 45°, and 90° orientations, were studied in this paper, aiming to understand the impact of potential fiber alignment along printing direction on composite properties. Additionally, the successful FDM printing parameters were further employed to print honeycombs, woven bowls, and wireframe bins to demonstrate the AM capability of the wood fiber-reinforced composite materials.

2.3. Property Characterization and Mechanical Testing

The microstructures of the 3D-printed wood fiber reinforced composites were characterized using a Keyence VHX-7000 optical microscope (Itasca, IL, USA) and a Thermo Fisher

Scientific Quattro SEM (Waltham, MA, USA). The microscopic images were used to identify wood fiber dispersion and potential void size embedded within the 3D-printed composites.

The density cup method was first employed as a robust and effective technique to quantify the density of 3D-printed composite samples. The measured densities were instrumental in our analysis, as they served as key parameters for characterizing the porosity of the composite materials. Five samples from each type of the printed composites with 0°, 45°, and 90° orientations were tested. By comparing the measured densities of 3D-printed composites to the densities of the filament, we were able to calculate the porosity of the composites using Equation (1) below:

$$Po = \left(1 - \frac{D_c}{D_f}\right) \times 100\% \quad (1)$$

where Po is the calculated composite porosity, D_c is the measured density of 3D-printed sustainable composites, and D_f is the measured density of original composite filament.

The printed dogbone samples were characterized to understand the mechanical properties of these 3D-printed sustainable composites under tensile loads and to evaluate the effects of printing orientations. Wood fiber composites with the printing orientation of 0° and 90° orientations were tested using an Instron 5969 Dual Column Mechanical Testing system (Norwood, MA, USA) under quasi-static loads with the load rate of 1 mm/min. To ensure the repeatability of the testing results, three experiments were conducted for each type of sample, following the same testing procedures. Additionally, the full-field 3D strain fields were measured using an ARAMIS DIC system during all the mechanical tests. All the composite samples were carefully prepared by creating the stochastic speckle pattern using an air brush, ensuring high contrast and randomness for optimal DIC tracking during the tensile tests. The mechanical experimental setup consisted of the Instron tensile testing machine equipped with a pair of high-resolution camera system positioned perpendicularly to the specimen's gauge length, ensuring a clear field of view throughout the entire testing process. Ambient lighting conditions were controlled to minimize reflections and ensure consistent illumination. Post-testing, the captured images were processed using the specialized DIC software provided by the vendor. The software algorithm tracked the deformation of the speckle patterns, allowing the extraction of strain distributions across the specimen's gauge length. The derived data provided a comprehensive understanding of the material's mechanical response, showcasing the potential of DIC in capturing local strain heterogeneities during tensile loading. The mechanical property of the 3D-printed composites was compared to that of the pristine PLA samples printed using the same FDM 3D printer and printing parameters. Critical tensile properties, including the Young's modulus, maximum elongation, and ultimate tensile strength, were all analyzed in this study.

2.4. Evaluation of Mechanical Performance of 3D-Printed Honeycomb Structures

Hexagonal honeycombs were 3D printed and experimentally characterized in this study. Each cell of the hexagonal honeycomb had 10 mm inner diameter, 1 mm wall thickness, and 10 mm wall heights. The mechanical properties of honeycombs could depend on their cell geometry, material properties, and the relative density of the honeycomb. The 3D CAD model and the 3D-printed composite honeycombs are shown in Figure 1. To fully understand the effects of wood fiber composites on the performance of honeycombs, the cell geometry and relative density of the honeycomb were maintained consistent in this study.

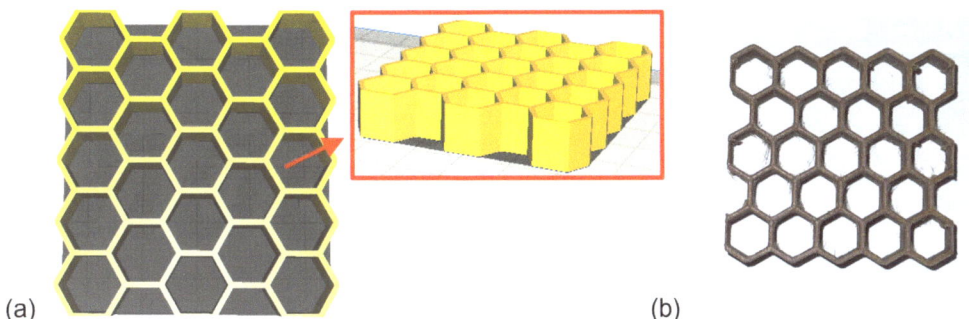

Figure 1. Hexagonal composite honeycombs, which were 3D printed and tested in this study; (**a**) 3D CAD model of the honeycomb (top view and isometric view); (**b**) 3D-printed composite sample (top view).

The mechanical performance of 3D-printed honeycombs was tested under uniaxial compressive loads. The 3D strain fields were measured using the DIC system to fully understand the failure mechanism of honeycombs. The same experimental DIC setup and sample preparation procedures were employed for all testing using honeycomb samples. The stochastic speckle patterns were painted on front, side, and top surfaces of each honeycomb samples, enabling the 3D strain field measurement and local stress/strain concentration measurement. Quasi-static load with the load rate of 1 mm/min was employed to avoid any impact of dynamic loads. The 3D-printed honeycombs were tested in three different orientations to fully understand their failure mechanism in three different loading directions.

3. Results and Discussion

3.1. Microstructural Characterization of Wood Fiber Reinforced Composites

As shown in Figure 2, wood fiber reinforcements were uniformly distributed within the 3D-printed sustainable composites without any evident agglomerations, indicating the improvement of mechanical properties provided by the wood fiber fillers within the PLA matrix. Additionally, embedded pores were observed in the optical and SEM images. To better understand the characteristics of the pores, ImageJ software was used to analyze the geometries of the pores. The measured diameters of the embedded pores were in the range of 35 μm to 80 μm. Although 100% infiltration density was used during the design and G-code generation for FDM-based 3D printing, the pores were potentially generated due to the under-extrusion or over-extrusion, high print speed, and layer height. Under-extrusion could cause incomplete layers, whereas over-extrusion could lead to bulging and poor adhesion between adjacent lines. Fast print speed could reduce adhesion between layers and create pores in printed composites. A reduction in layer height could potentially produce smoother surface and denser composites but would increase print time and reduce print efficiency.

3.2. Density and Porosity of 3D-Printed Wood Fiber Composites

The printing orientations of the three types of composite samples are shown in Figure 3a. In the case of the 0° print orientation, the printing direction was perfectly aligned with the gauge length of the dogbone samples, thus running parallel to it. Conversely, the 90° print orientation indicated that the printing direction was perpendicular to the gauge length of the dogbone samples. For the 45° print orientation, the angle between the print direction and the gauge length direction was precisely 45°, demonstrating an oblique arrangement and potential design flexibility for FDM-processed AM products. Importantly, it is worth noting that the print direction remained consistent for all layers within each sample, ensuring uniformity throughout the fabrication process.

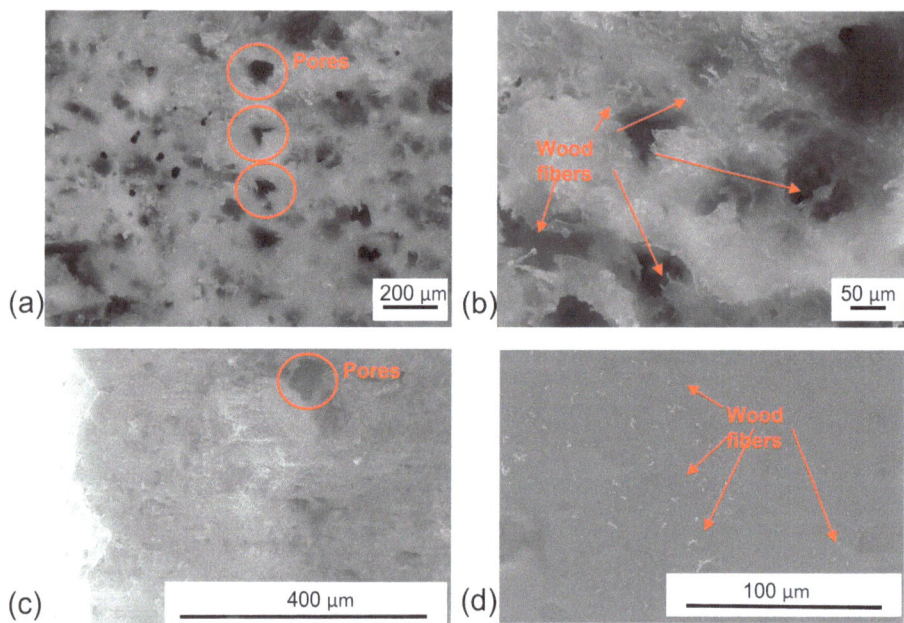

Figure 2. Microscopy images of wood fiber-reinforced composites; (**a,b**) optical microscopy images of cross-section of fractured composites; (**c**) SEM image of the cross-section of fractured composites; (**d**) SEM image of 3D-printed composite surface.

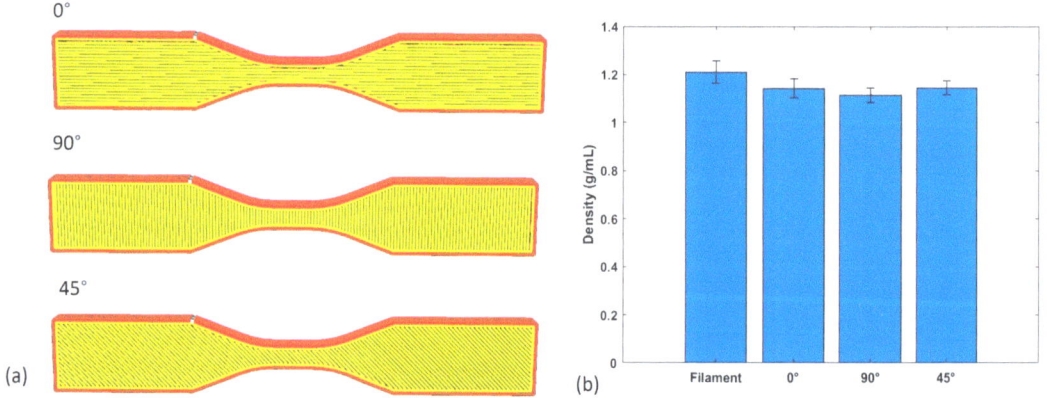

Figure 3. (**a**) 3D print orientations of FDM-processed composites; (**b**) Measured densities of composite filament and 3D-printed samples.

The measured density served as a direct indicator of the average porosity within the 3D-printed composites in this study. Considering the fabrication of the wood fiber composite filament involved an extrusion process conducted under high-temperature and high-pressure conditions, it is a valid assumption that the filament achieved a state of full density. Therefore, the density of this filament was adopted as the reference for comparative analysis of the densities of the 3D-printed composites. The average densities and related standard deviations of the three types of composites is shown in Figure 3b. It is noted that the average densities of composites with 0°, 90°, and 45° print orientations were 1.14 g/mL, 1.12 g/mL, and 1.15 g/mL, respectively, and the density of the composite filament was 1.21 g/mL. According to Equation (1), the calculated porosities of these three types of 3D-printed composites were 5.79%, 7.44%, and 4.96%. The low porosity of the FDM-processed composites indicated that the 3D-printed materials and parts should have outstanding mechanical properties and performance.

3.3. Mechanical Properties of 3D-Printed Composites

The tensile properties of the 3D-printed composites and pristine PLA polymers with print orientations of 0° and 90° were characterized to evaluate the material's tensile properties, particularly its strength and elasticity, using dogbone shape samples following the ASTM D638 standard. Quasi-static tensile loads were applied and DIC images were taken simultaneously during all the tensile tests. Three experiments were conducted for each type of sample to ensure the repeatability of the experimental results. The DIC technology not only allowed the measurement of 3D strain fields but also identified the local stress and strain concentrations using tensile testing, indicating the damage initiation and growth of the sample for detailed fracture analysis. Compared to pristine PLA, wood fiber-reinforced composites showed improved ultimate tensile strengths due to the reinforcement of uniformly dispersed wood fiber in the PLA matrix, as shown in Figure 4a. Notably, when the printing direction was set to 0°, the ultimate tensile strength of wood fiber composite was 38.91 MPa, which resulted in a 10.79% increase compared to the ultimate tensile strength of pristine PLA of 35.12 MPa. Similarly, when the print orientation was set at 90°, the ultimate tensile strength of wood fiber composite was 36.27 MPa, which was also 5.84% increase compared to the ultimate tensile strength of pristine PLA of 34.27 MPa. Notable improvements were observed in the Young's modulus of the 3D-printed specimens. For the wood fiber-reinforced composites aligned at 0°, the Young's modulus was measured at 2.79 GPa, indicating a 21.83% increase from the PLA sample with 2.29 GPa Young's modulus. Additionally, the wood fiber-reinforced composites oriented at 90° exhibited a Young's modulus of 2.67 GPa, which was a 30.88% increase compared to the PLA dogbone samples printed in the same orientation, with a Young's modulus of 2.04 GPa.

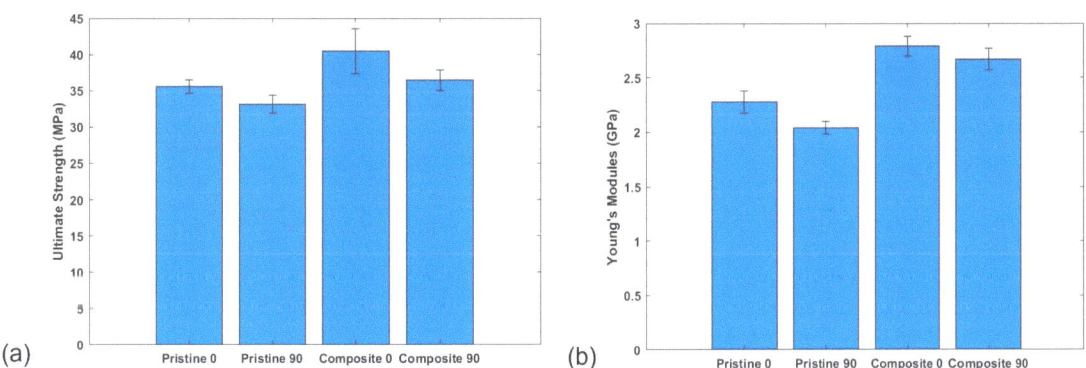

Figure 4. Tensile test results of FDM-processed dogbone samples using pristine PLA and wood fiber-reinforced composites; (**a**) comparison of ultimate tensile strengths; (**b**) comparison of the Young's modulus.

These enhancements of mechanical properties in 3D-printed composites could be due to multiple factors, such as the alignment of wood fibers along the printing direction as well as the strong adhesion between wood fiber and PLA polymers. Aligning wood fibers along the 3D printing direction is critical to improve mechanical performance of 3D-printed sustainable composites. During FDM-based 3D printing, the nozzle's movement creates shear stress within the extrude composite as it is pushed through the nozzle and deposited onto the build platform. This shear stress improves the alignment of wood fibers within the polymer matrix material. Additionally, there is a shearing action between the newly deposited layer and the previous layer, which also encourages alignment of wood fibers parallel to the layer interfaces. Other printing parameters, such as extrusion speed, nozzle temperature, and layer height, should be adjusted to control the shear forces applied to the composites during printing. Optimizing these parameters can help achieve the desired wood fiber alignment. These findings emphasized the critical role that printing orientation and beneficial fillers played in determining the mechanical performance of 3D-printed sustainable composite structures, providing valuable insights for optimizing design and AM fabrication processes for the development of future sustainable and recyclable composites.

Non-contact DIC strain field measurement technology was employed during all the tensile tests to ensure the precise strain measurement in the gauge section of each dogbone sample. The employed DIC system utilized a pair of synchronized cameras to capture images of the specimen surface cameras. Before loading or deformation, a random speckle pattern was painted on the sample surface, serving as a reference image. During deformation, the cameras simultaneously captured images, which were then compared to the reference images using sophisticated image correlation algorithms. By analyzing the displacement of the speckle patterns between the reference and deformed images, full-field 3D surface deformations and, subsequently, the strain fields were derived. The precision and accuracy of this method depended upon various factors, including the quality of the speckle pattern, camera resolution, and the calibration process.

In this study, as the local strains in 3D space were all measured simultaneously using the DIC, local stress and strain concentrations were clearly identified for the prediction of potential fracture locations. As shown in Figure 5a, the local stress and strain concentration was clearly identified near the top of the gauge section of pristine PLA dogbone sample after the local stress passed the composite yield stress, generating plastic deformation in the local section of the dogbone gauge area. The final fracture also happened at the stress concentration location, as shown in Figure 5b. Similar performance was observed in PLA printed in 90° print orientation as well as the wood fiber composite samples, as shown in Figure 5c,d and Figure 6. Additionally, the average tensile strain ϵ_y in the vertical direction was used for the generation of stress–strain curves. The average tensile strain measured from the DIC system was more accurate than the average strain recorded from the Instron mechanical testing system due to the potential experimental errors introduced by the tab section of the dogbone samples.

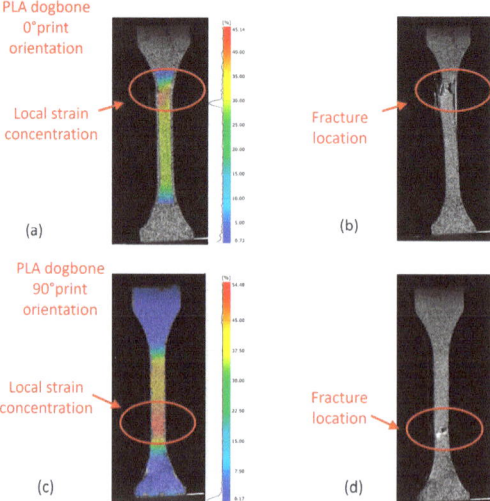

Figure 5. (**a**) DIC image of tensile strain ϵ_y in PLA sample with 0° print orientation; (**b**) fracture location matching with the local strain concentration in PLA sample with 0° print orientation; (**c**) DIC image of tensile strain ϵ_y in PLA sample with 90° print orientation; (**d**) fracture location matching with the local strain concentration in PLA sample with 90° print orientation.

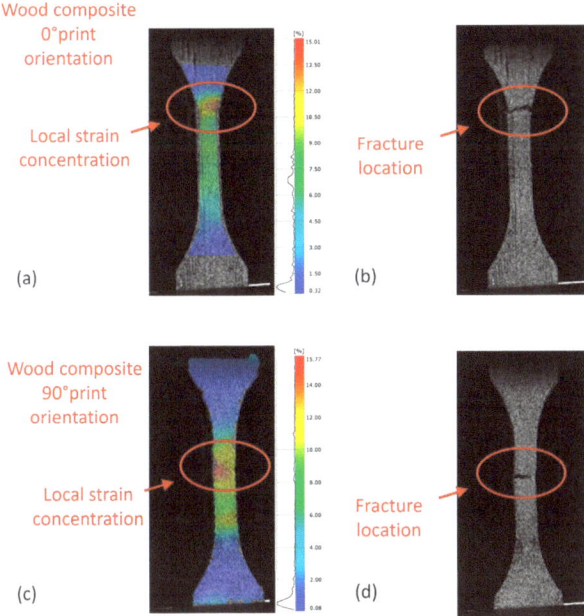

Figure 6. (**a**) DIC image of tensile strain ϵ_y in wood fiber composite sample with 0° print orientation; (**b**) fracture location matching with the local strain concentration in wood fiber composite sample with 0° print orientation; (**c**) DIC image of tensile strain ϵ_y in wood fiber composite sample with 90° print orientation; (**d**) fracture location matching with the local strain concentration in wood fiber composite sample with 90° print orientation.

3.4. Mechanical Performance and Fracture Mechanism of Honeycomb Structures

Honeycombs have been well recognized as ultra-lightweight structures with outstanding mechanical properties, resulting in numerous promising applications in the fields of architecture, automotive, aerospace, marine, and space applications. Utilizing natural fiber composites in honeycomb design and development can significantly enhance the sustainability and recyclability of honeycombs. However, novel fabrication and in-depth understanding of the novel composite honeycombs' mechanical properties and performance is still urgently needed.

The 3D-printed hexagonal honeycombs were first tested by placing the flat surface of the sidewall on the Instron machine, as shown in Figure 7a. The top surface of the honeycomb was measured by the DIC system to obtain the 3D strain fields during the mechanical testing. Local normal and shear strains in 3D space were measured, and the equivalent strain, such as the Von Mises strain, could be calculated for further analysis. All the tested samples showed the 45° failure mode during their post-yielding behavior, as shown in Figure 7b. The local load condition of the joint section of three adjacent cells is shown in Figure 7c. The applied compressive load caused compressive load applied on the joint area and further compressed two adjacent cell walls, resulting in the local torsional load on the joint. Due to the local torsional load applied at the joint of adjacent honeycomb cells, the joint rotated as the overall compressive load was applied on the top and bottom surfaces, resulting in local stress concentrations and fracture by the end of each test, as shown in Figure 7e. The entire 3D strain field was recorded, as shown in Figure 7d.

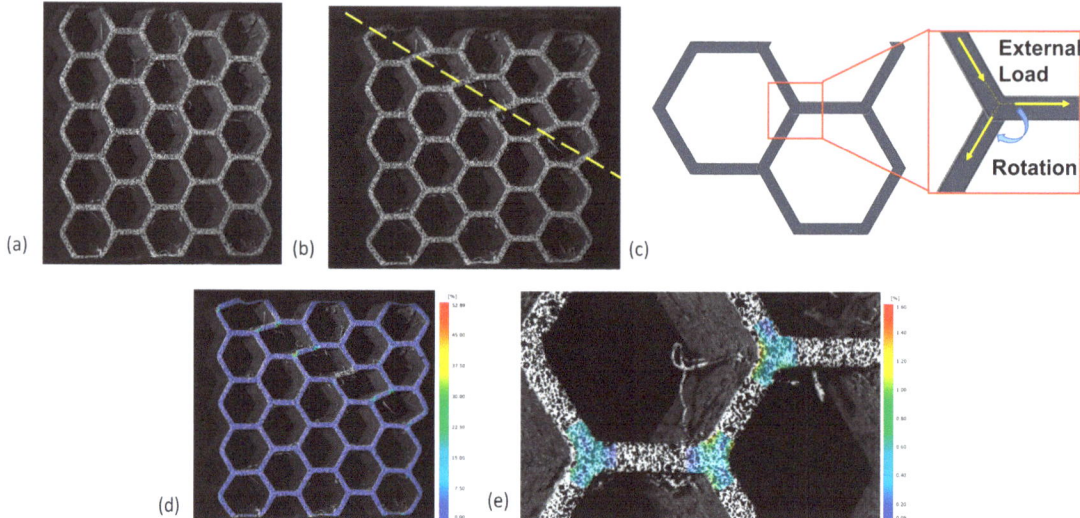

Figure 7. (**a**) Composite honeycomb with painted random speckle pattern for DIC and compressive test, flat surface on test plate; (**b**) compressed composite honeycomb showing the failure mode; (**c**) load free body diagram showing the applied load at the corner of three adjacent cell units; (**d**) full Von Mises strain field of the compressed honeycomb; (**e**) stress concentration at the joint section of three adjacent hexagonal cells.

To fully evaluate the performance of the 3D-printed hexagonal honeycombs under compressive loads, new samples with the same dimensions were tested by placing the corners of sidewall on the Instron test plate, resulting in point contact and concentrated compressive loads during the test, as shown in Figure 8a. The failure mode of the new compressive tests is shown in Figure 8b. As the external load was vertically applied down to the joint of adjacent cells, the symmetrical cell structures equally distributed the load to each side of the cell walls, leading to the horizontal collapse of the honeycomb. The local loads applied at each joint section of three cells is shown in Figure 8c. The deformation in the vertical direction of the tested honeycomb sample is shown in Figure 8d. The top three rows of hexagonal cells were all collapsed after the compressive test; however, the third row showed the most deformation. The local stress concentration of the joint of three adjacent cell walls is shown in Figure 8e.

The comparison of the honeycomb performance in the two load directions revealed that the 3D-printed hexagonal honeycombs could better distribute to the entire structure when the applied load was parallel to the cell walls. Therefore, increased compressive load could be carried by the honeycombs before collapse and structural failures. The recorded force and displacement data of the two types of experiments proved this conclusion, as shown in Figure 8f. Although initial stress concentrations were generated at each corner contacting the test plates, the applied compressive force was distributed throughout the honeycomb, leading to the 4.3% increase in the maximum load capacity.

When the compressive load was applied along the height of the honeycomb, the 3D-printed samples showed the highest load capacity and structural stability, as shown in Figure 9a. Due to the large contact area, all the applied loads could be uniformly distributed throughout the honeycomb structures. Additionally, the cell walls could serve as the stiffeners and further stabilize the honeycomb structure, leading to improved overall structural strengths. The collapsed honeycomb and the equivalent Von Mises strain field after compressive tests are shown in Figure 9b. The applied force and displacement data are shown in Figure 9c. Experimental results indicated that the 3D-printed honeycombs carried more than 83 times of the compressive load when the load was applied along the height direction. It is noted that different failure modes, such as buckling, could potentially happen if the height to cell diameter increases when the designs of honeycomb structures change.

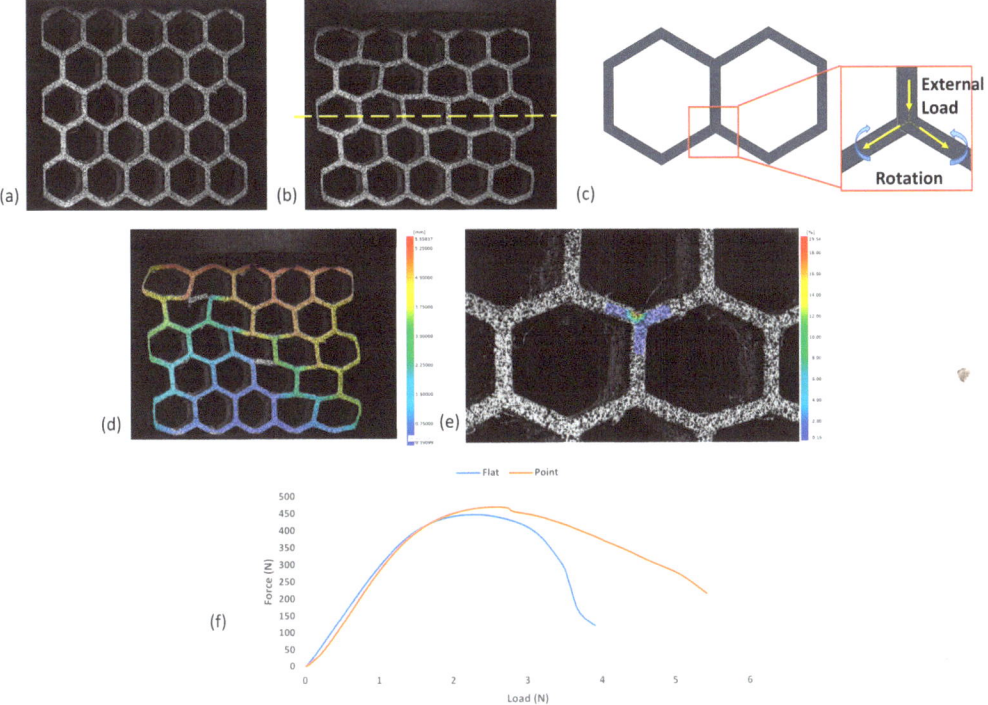

Figure 8. (**a**) Composite honeycomb with corners on test plate; (**b**) image of compressive honeycomb showing the failure mode of horizontal collapse; (**c**) local free-body diagram showing the applied loads at the joint section of three adjacent honeycomb cells; (**d**) full deformation field in the vertical direction; (**e**) local stress concentration that resulted in local cell collapse; (**f**) force and displacement data of the composite honeycombs under two different compressive load directions.

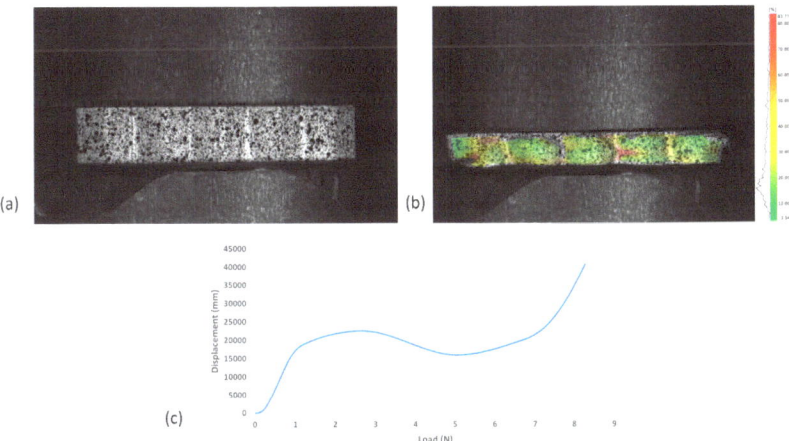

Figure 9. (**a**) 3D-printed honeycomb with compressive load applied along the height direction; (**b**) DIC strain field showing the stress concentration before collapse; (**c**) force and displacement data of a typical compressive test.

3.5. Demonstration of FDM-Based 3D Printing Capability Using Wood Fiber Composites

It is critical to demonstrate the FDM-based 3D printing technology using the wood fiber-reinforced PLA matrix composite with complex geometries. This type of test print can not only be used as the proof-of-concept for sustainable composite AM fabrication but also is part of the process to evaluate the manufacturing readiness levels before real engineering practice. Compared to traditional petroleum-based polymer and composites, wood fiber-reinforced PLA composites can be fully manufactured using recyclable and renewable bio-based materials, presenting an eco-friendly solution for current challenges in sustainability, recyclability, and carbon footprint reduction. Additionally, the demonstration of FDM using sustainable composites allows for intricate geometrical designs that were previously impossible with conventional manufacturing methods. Such AM capabilities pave new ways for innovative product design and multifunctional composite development.

Two CAD models were employed for the test print and demonstration of FDM-based AM technology using the wood fiber-reinforced PLA matrix composites. As shown in Figure 10, both CAD models have complex geometries and a significant number of hollow structures. The woven bowl CAD model has multiple geometrical features, such as hanging walls and thin contact points of the woven structures. Similarly, the frame bin CAD model also requires precise manipulation of hollow structures and joins. The successful prints of the two models showed that commercial FDM printers can potentially achieve the required structural accuracy, integrity, and quality.

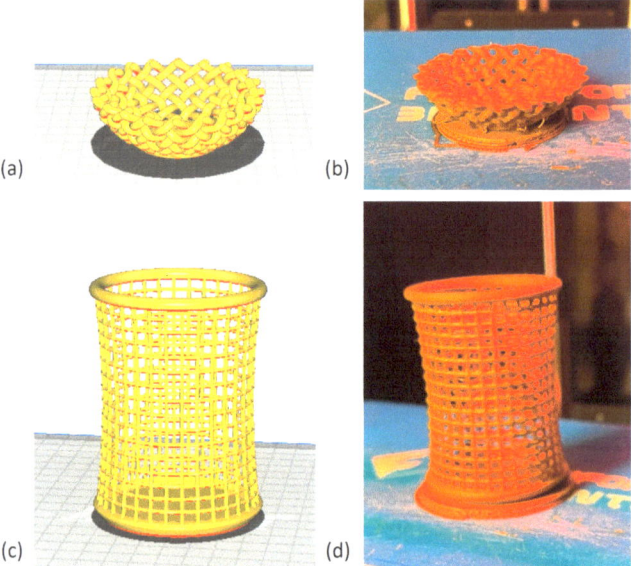

Figure 10. (**a**) CAD model of woven bowl; (**b**) image of FDM 3D-printed woven bowl using wood fiber composites; (**c**) CAD model of frame bin; (**d**) image of FDM 3D-printed frame bin.

4. Conclusions

FDM-based AM technology was employed to 3D print sustainable composites using wood fiber and PLA in this study. The quality of the 3D-printed eco-friendly composites was experimentally characterized by testing multiple critical properties, such as density, porosity, pore size, wood fiber dispersion quality, and tensile properties for the validation of FDM parameters for 3D printing. The experimental results indicated that the alignment of wood fibers within PLA polymer resulted in enhanced mechanical performance. Additionally, the validated FDM process was employed to 3D print honeycombs, woven bowl, and frame bins to further demonstrate the AM capability of products with complex geometries. The 3D-printed honeycombs were further tested under compressive loads in three different directions to fully understand the performance of the developed wood fiber composites in complex geometries and loading conditions. The experimental results, particularly the 3D full strain fields, demonstrated the failure mode of honeycombs under different load directions. The outcomes of this paper can be further applied to guide the design and 3D printing of wood fiber-reinforced sustainable composites and structures, providing a novel solution to fabricate environmentally friendly products with complex geometries and functionalities for broad engineering applications.

Author Contributions: Conceptualization, C.B., R.S. and Y.L.; validation, R.S., C.B., B.S. and J.H.; formal analysis, R.S. and Y.L.; writing—original draft preparation, C.B. and R.S.; writing—review and editing, Y.L.; supervision, Y.L. All authors have read and agreed to the published version of the manuscript.

Funding: This research was partially funded by the Oklahoma Aerospace and Defense Innovation Institute (OADII) and the Data Institute for Societal Challenges (DISC), Office of the Vice President for Research and Partnerships, University of Oklahoma.

Data Availability Statement: The data that support the findings of this study are available on request from the corresponding author Yingtao Liu.

Conflicts of Interest: The authors declare no conflict of interest. The funders had no role in the design of the study; in the collection, analyses, or interpretation of data; in the writing of the manuscript; or in the decision to publish the results.

References

1. La Mantia, F.; Morreale, M. Green composites: A brief review. *Compos. Part A Appl. Sci. Manuf.* **2011**, *42*, 579–588. [CrossRef]
2. Rangappa, S.M.; Siengchin, S.; Dhakal, H.N. Green-composites: Ecofriendly and sustainability. *Appl. Sci. Eng. Prog.* **2020**, *13*, 183–184. [CrossRef]
3. Kopparthy, S.D.S.; Netravali, A.N. Green composites for structural applications. *Compos. Part C Open Access* **2021**, *6*, 100169. [CrossRef]
4. Khalil, H.A.; Bhat, A.; Yusra, A.I. Green composites from sustainable cellulose nanofibrils: A review. *Carbohydr. Polym.* **2012**, *87*, 963–979. [CrossRef]
5. Vazquez-Nunez, E.; Avecilla-Ramírez, A.M.; Vergara-Porras, B.; López-Cuellar, M.d.R. Green composites and their contribution toward sustainability: A review. *Polym. Polym. Compos.* **2021**, *29* (Suppl. S9), S1588–S1608. [CrossRef]
6. Faruk, O.; Bledzki, A.K.; Fink, H.-P.; Sain, M. Progress report on natural fiber reinforced composites. *Macromol. Mater. Eng.* **2014**, *299*, 9–26. [CrossRef]
7. Girijappa, Y.G.T.; Rangappa, S.M.; Parameswaranpillai, J.; Siengchin, S. Natural fibers as sustainable and renewable resource for development of eco-friendly composites: A comprehensive review. *Front. Mater.* **2019**, *6*, 226. [CrossRef]
8. Kim, S.-J.; Moon, J.-B.; Kim, G.-H.; Ha, C.-S. Mechanical properties of polypropylene/natural fiber composites: Comparison of wood fiber and cotton fiber. *Polym. Test.* **2008**, *27*, 801–806. [CrossRef]
9. Karimah, A.; Ridho, M.R.; Munawar, S.S.; Adi, D.S.; Ismadi; Damayanti, R.; Subiyanto, B.; Fatriasari, W.; Fudholi, A. A review on natural fibers for development of eco friendly bio composite: Characteristics, and utilizations. *J. Mater. Res. Technol.* **2021**, *13*, 2442–2458. [CrossRef]
10. Thomas, S.; Paul, S.A.; Pothan, L.A.; Deepa, B. Natural fibres: Structure, properties and applications. *Cellul. Fibers Bio-Nano-Polym. Compos. Green Chem. Technol.* **2011**, 3–42.
11. Duan, J.; Reddy, K.O.; Ashok, B.; Cai, J.; Zhang, L.; Rajulu, A.V. Effects of spent tea leaf powder on the properties and functions of cellulose green composite films. *J. Environ. Chem. Eng.* **2016**, *4*, 440–448. [CrossRef]
12. Kalia, S.; Kaith, B.; Kaur, I. Pretreatments of natural fibers and their application as reinforcing material in polymer composites—A review. *Polym. Eng. Sci.* **2009**, *49*, 1253–1272. [CrossRef]
13. Sethi, S.; Ray, B.C. Environmental effects on fibre reinforced polymeric composites: Evolving reasons and remarks on interfacial strength and stability. *Adv. Colloid Interface Sci.* **2015**, *217*, 43–67. [CrossRef] [PubMed]
14. Nguyen, N.A.; Barnes, S.H.; Bowland, C.C.; Meek, K.M.; Littrell, K.C.; Keum, J.K.; Naskar, A.K. A path for lignin valorization via additive manufacturing of high-performance sustainable composites with enhanced 3D printability. *Sci. Adv.* **2018**, *4*, eaat4967. [CrossRef] [PubMed]
15. Tonk, R. Natural fibers for sustainable additive manufacturing: A state of the art review. *Mater. Today: Proc.* **2021**, *37*, 3087–3090. [CrossRef]
16. Shanmugam, V.; Das, O.; Neisiany, R.E.; Babu, K.; Singh, S.; Hedenqvist, M.S.; Berto, F.; Ramakrishna, S. Polymer recycling in additive manufacturing: An opportunity for the circular economy. *Mater. Circ. Econ.* **2020**, *2*, 1–11. [CrossRef]

17. Scaffaro, R.; Gulino, E.F.; Citarrella, M.C.; Maio, A. Green composites based on hedysarum coronarium with outstanding FDM printability and mechanical performance. *Polymers* **2022**, *14*, 1198. [CrossRef] [PubMed]
18. Calì, M.; Pascoletti, G.; Gaeta, M.; Milazzo, G.; Ambu, R. A new generation of bio-composite thermoplastic filaments for a more sustainable design of parts manufactured by FDM. *Appl. Sci.* **2020**, *10*, 5852. [CrossRef]
19. Michaud, P.; Pateloup, V.; Tarabeux, J.; Alzina, A.; André, D.; Chartier, T. Numerical prediction of elastic properties for alumina green parts printed by stereolithography process. *J. Eur. Ceram. Soc.* **2021**, *41*, 2002–2015. [CrossRef]
20. Billings, C.; Siddique, R.; Liu, Y. Photocurable Polymer-Based 3D Printing: Advanced Flexible Strain Sensors for Human Kinematics Monitoring. *Polymers* **2023**, *15*, 4170. [CrossRef]
21. Agustiany, E.A.; Rasyidur Ridho, M.; Rahmi, D.N.M.; Madyaratri, E.W.; Falah, F.; Lubis, M.A.R.; Solihat, N.N.; Syamani, F.A.; Karungamye, P.; Sohail, A.; et al. Recent developments in lignin modification and its application in lignin-based green composites: A review. *Polym. Compos.* **2022**, *43*, 4848–4865. [CrossRef]
22. Zhang, H.; Guo, Y.; Jiang, K.; Bourell, D.L.; Zhao, D.; Yu, Y.; Wang, P.; Li, Z. A review of selective laser sintering of wood-plastic composites. In Proceedings of the 2016 International Solid Freeform Fabrication Symposium, Austin, TX, USA, 8–10 August 2016.
23. Arif, Z.U.; Khalid, M.Y.; Noroozi, R.; Hossain, M.; Shi, H.H.; Tariq, A.; Ramakrishna, S.; Umer, R. Additive manufacturing of sustainable biomaterials for biomedical applications. *Asian J. Pharm. Sci.* **2023**, 100812. [CrossRef]
24. Li, Y.; Ren, X.; Zhu, L.; Li, C. Biomass 3D Printing: Principles, Materials, Post-Processing and Applications. *Polymers* **2023**, *15*, 2692. [CrossRef] [PubMed]
25. Das, O.; Babu, K.; Shanmugam, V.; Sykam, K.; Tebyetekerwa, M.; Neisiany, R.E.; Försth, M.; Sas, G.; Gonzalez-Libreros, J.; Capezza, A.J.; et al. Natural and industrial wastes for sustainable and renewable polymer composites. *Renew. Sustain. Energy Rev.* **2022**, *158*, 112054. [CrossRef]
26. Idriss, A.I.; Li, J.; Wang, Y.; Guo, Y.; Elfaki, E.A.; Adam, S.A. Selective laser sintering (SLS) and post-processing of prosopis chilensis/polyethersulfone composite (PCPC). *Materials* **2020**, *13*, 3034. [CrossRef] [PubMed]
27. Fombuena, V.; Bernardi, L.; Fenollar, O.; Boronat, T.; Balart, R. Characterization of green composites from biobased epoxy matrices and bio-fillers derived from seashell wastes. *Mater. Des.* **2014**, *57*, 168–174. [CrossRef]
28. Lascano, D.; Garcia-Garcia, D.; Rojas-Lema, S.; Quiles-Carrillo, L.; Balart, R.; Boronat, T. Manufacturing and characterization of green composites with partially biobased epoxy resin and flaxseed flour wastes. *Appl. Sci.* **2020**, *10*, 3688. [CrossRef]
29. Jagadeesh, D.; Kanny, K.; Prashantha, K. A review on research and development of green composites from plant protein-based polymers. *Polym. Compos.* **2017**, *38*, 1504–1518. [CrossRef]
30. Calvino, C.; Macke, N.; Kato, R.; Rowan, S.J. Development, processing and applications of bio-sourced cellulose nanocrystal composites. *Prog. Polym. Sci.* **2020**, *103*, 101221. [CrossRef]
31. Miller, A.; Brown, C.; Warner, G. Guidance on the use of existing ASTM polymer testing standards for ABS parts fabricated using FFF. *Smart Sustain. Manuf. Syst.* **2019**, *3*, 122–138. [CrossRef]

Disclaimer/Publisher's Note: The statements, opinions and data contained in all publications are solely those of the individual author(s) and contributor(s) and not of MDPI and/or the editor(s). MDPI and/or the editor(s) disclaim responsibility for any injury to people or property resulting from any ideas, methods, instructions or products referred to in the content.

Article

Additively Manufactured Multifunctional Composite Parts with the Help of Coextrusion Continuous Carbon Fiber: Study of Feasibility to Print Self-Sensing without Doped Raw Material

Anthonin Demarbaix [1,*], Imi Ochana [1], Julien Levrie [1], Isaque Coutinho [2], Sebastião Simões Cunha, Jr. [2] and Marc Moonens [1]

[1] Science and Technology Research Unit, Haute Ecole Provinciale de Hainaut Condorcet, Square Hierneaux 2, 6000 Charleroi, Belgium; marc.moonens@condorcet.be (M.M.)

[2] Mechanical Engineering Institute, Federal University of Itajubá, Avenida BPS, 1303, Bairro Pinheirinho, Itajubá 37500-903, Brazil

* Correspondence: anthonin.demarbaix@condorcet.be

Abstract: Nowadays, the additive manufacturing of multifunctional materials is booming. The fused deposition modeling (FDM) process is widely used thanks to the ease with which multimaterial parts can be printed. The main limitation of this process is the mechanical properties of the parts obtained. New continuous-fiber FDM printers significantly improve mechanical properties. Another limitation is the repeatability of the process. This paper proposes to explore the feasibility of printing parts in continuous carbon fiber and using this fiber as an indicator thanks to the electrical properties of the carbon fiber. The placement of the fiber in the part is based on the paths of a strain gauge. The results show that the resistivity evolves linearly during the elastic period. The gauge factor (GF) increases when the number of passes in the manufacturing plane is low, but repeatability is impacted. However, no correlation is possible during the plastic deformation of the sample. For an equivalent length of carbon fiber, it is preferable to have a strategy of superimposing layers of carbon fiber rather than a single-plane strategy. The mechanical properties remain equivalent but the variation in the electrical signal is greater when the layers are superimposed.

Keywords: additive manufacturing composite; smart material; structural health monitoring

1. Introduction

Additive manufacturing (AM) enables the production of mechanical parts with complex geometries in various sectors, such as automotive, robotics, aeronautics, and aerospace. This technology enables the use of just the right material to fulfill the required functions. This saves weight and raw materials while minimizing the assembly of multiple parts. These advantages make it a very competitive alternative to other manufacturing processes when it comes to small/medium production runs. This is one of the reasons why additive technology is gaining ground in the industrial world, particularly in aerospace and aeronautics, with the aim of lightening aircraft and thus reducing the carbon footprint of flights. AM is currently developing exponentially, with a growth rate of around 20%/year [1].

Fused deposition modeling (FDM) or fused filament fabrication (FFF) is a material extrusion (ME) 3D printing method for polymers and fiber-reinforced composites. This technology has been significantly growing especially in the aerospace, automobile, and medical industries. The main advantages of the FDM method are its reliability, low maintenance required, low investment cost, wide low-cost filament material availability, and cost-effectiveness, and it is highly customizable. However, it is limited to low-melting-point materials, and it is also a slow printing process. Single-screw extruders are usually used in mass-production applications where pure polymers are used as raw materials. There are

also a few commercially available filaments with continuous fiber already impregnated to an extrudable thermoplastic. For more complex applications, where precision and better properties are needed, twin-screw extruders can be used for blending two or more materials. A high degree of dispersion between the polymer matrix and the filler materials can be obtained. However, printers have also been modified to be able to coextrude in a single nozzle the fibers and the thermoplastic filaments that are fed separately. Thermoplastics are the most used materials in FDM systems. Their main advantages are their low cost and melting point. From the literature, these materials' tensile strength can range from 1.5 to 150 MPa [2]. However, pure polymers do not present enough mechanical properties for structural applications. Due to the low strength of pure polymers, they can be either filled or reinforced to improve their mechanical properties. Acrylonitrile butadiene styrene (ABS), polylactide (PLA), and polyamide (PA) nylon are widely used as matrix materials. Their low melting point is the main reason for their vast applicability.

Other functions can also be sought in printed materials, such as electrical conductivity. Ryan et al. [3] are particularly interested in the conductive functionality that can be achieved using the FDM process. The use of conductive filaments has a negative impact on mechanical properties. Conductive polymers are therefore also a limiting factor in development because of the poor mechanical properties obtained. The authors recommend the use of a second polymer to obtain good mechanical properties as well as decent conductivity [3].

The composite can also be filled with particles or short fiberswhere the fibers are already impregnated on the polymer matrix. Short fibers can be twice as strong as pure plastic, and there are some commercially available filaments with impregnation that, normally, are filled with short carbon fibers. On the other hand, continuous-fiber-reinforced composites can be 30 times as strong as pure plastic [4].

Short-fiber reinforcements provide better tensile modulus than unreinforced, but the tensile strength is not improved. Continuous filament reinforcement composite polymers (CFCRPs) offer superior properties and are normally fabricated by expensive methods [4]. The FDM method allows printing CFCRPs in complex geometries [5,6]. On the other hand, the specimens fabricated by this method present lower tensile strength than other methods, which is mostly because of the poor polymer–fiber adhesion [6]. While the aerospace industry tolerates porosity rates under 1 vol%, the porosity of continuous-fiber-reinforced thermoplastic composites (CFRTPCs) is 5–10 vol% [5].

The introduction of fiber into a matrix makes it possible to improve mechanical properties. The 3D-printed part can be given electrical properties to make it a multifunctional composite [6].

In the case of multifunctional composite structures, the active material does not only fulfill its function as a load-bearing structure, but also performs additional functions. The fiber used in the polymer adds functions to the composite part [7].

Introducing carbon fiber into 3D-printed parts also provides electrical properties. These electrical properties will also have an impact, enabling the part to obtain other functional properties, such as a sensor, for example. One of the main functions that can be conferred on the composite structure is self-sensing, which makes it possible to become aware of the state of the structure [7] without the aid of an integrated or mounted device. The main principles of self-sensing are direct piezoelectricity, thermoelectricity, and piezoresistivity [8].

Bekas et al. [9] have established a literature review with a summary of the research efforts for the development and characterization of 3D-printed multifunction composites. In 3D printing, functionality mainly consists of loading the raw material with carbon nanotubes or black carbon to provide conductive or sensing properties [9].

Kim et al. [10] propose to print a sensor using the dual-nozzle fused deposition method (FDM): one with a commercial filament and another with a functional nanocomposite filament (carbon nanotubes (CNTs) and thermoplastic polyurethane (TPU)). The second nozzle enables the filled filament to be deposited directly on the workpiece, creating a functional sensor without assembly. These filaments can then be used to target the sensor zones of the printed part, enabling them to be connected to these parts. The sensor thus offers signal information in three directions, and the difference in resistivity can thus give an indication of the force exerted. Kim et al. [10] mention that additive manufacturing anisotropy leads to a difference in recovered signals. It is possible to correlate the strain exerted by a difference in resistivity to thus obtain a sensor. A similar experiment was carried out with printed strain gauges, again using the electrical properties of a thermoplastic doped with carbon particles, taking into account the influence of temperature variations on the material's resistance value [11].

Gackowski et al. [12] demonstrate multifunctional 3D printing for nylon with the addition of another piezoresistive material for structural health monitoring. The concept of 3D printing nylon structures with embedded piezoresistive sensors of carbon nanotubes and short carbon fibers is investigated. Modifications in electrical resistance can be detected in tensile, flexural, and indentation tests, up to and including material failure. The order of magnitude of the resistivity of the printed sensor corresponds to the order of semiconductor magnitude [12].

Another alternative is using carbon fiber. Georgopoulou et al. [13] propose to embed manually the filament in an elastomer to obtain a piezoresistive sensor. The elastomer is printed directly onto the carbon fiber to create a sensor that is assembled on the robot arm. This is then controlled using an Arduino to obtain the position of the robot arm. The electrical signal can then be analyzed to verify the position of the robot arm and match the signal of a human finger to that of a robot finger. This method shows that when the fiber is displaced longitudinally, it is possible to obtain an electrical signal that is repeatable and directly related to the imposed angle. Nevertheless, the operations required to obtain the sensor are complex and require an assembly step on the robotic arm [13].

The additive process makes it easier to introduce carbon fibers into the part during the process. This has led to renewed interest in the correlation between electrical resistivity and deformation. Yao et al. [14] evaluate the embedding of manually inserted carbon fiber during the manufacturing process. They show that a link exists between mechanical and electrical behavior. The elastic zone can be detected by the resistivity measured on the carbon fibers [14]. When a defect appears in the composite, the electrical signal is disturbed, causing a change in the direction of the slope. In fact, a change in the direction of the slope makes it possible to detect the breakage of the first fiber in the part and is therefore an indicator of damage to the part [15].

Yao et al. [14] define the gauge factor (GF) as the ratio of the relative change in electrical resistance to the relative elongation:

$$\text{GF} = \frac{\frac{\Delta R}{R_0}}{\frac{\Delta L}{L_0}}, \qquad (1)$$

where ΔR the change in resistance, R_0 the initial resistance (in Ω), ΔL the change in length, and L_0 (in m) is the initial length. The calculation of the GF in the elastic period shows a certain stability. In the case of the study of a carbon fiber (3K) with a 20% PLA filling, the GF over the elastic period is 0.59 ± 0.13 [14].

This shows the interest in using continuous fiber to obtain mechanical properties while at the same time providing interesting electrical properties. FDM printing of continuous fibers is relatively recent. Kabir et al. [16] mention that the first continuous-fiber composite FDM appeared in 2014 by Markforged. There are two nozzles in the printer: one for preimpregnated fiber filaments and the other for pure plastic filaments. The matrix used is nylon (PA), and the reinforcing fibers can be carbon, glass, and aramid. The two separate nozzles are assembled on the same printhead. A cutting mechanism is used on top of the printhead, and a feeding mechanism is used to push the fiber filament into the nozzle. In this technology, paths for continuous fiber are limited to predefined trajectories [17].

Galos et al. [18] study the electrical conductivity of a composite obtained by Markforged technology. They show that there is an impact on conductivity that can be measured, mainly due to the process, which causes breakage during passage through the nozzle for printers not using the coextrusion principle. Electrical conductivity is measured using silver paint applied to the ends of the sample. This paint is then brought into contact with a copper electrode. It is not possible to access the carbon fiber to connect directly to it. This highlights the fact that additive technology has a significant impact on this aspect. This technology is a hindrance to the manufacture of an intelligent part. In the context of the additive forming of a thermoplastic continuous-fiber composite, Galos et al. [18] also show that thermal post-treatment is essential to improve mechanical properties.

Other technologies, such as Anisoprint, are entering the continuous-fiber FDM market. Luxembourg-based Anisoprint has developed a 3D printer that is based on the patented technology of the coextrusion of a continuous composite-reinforcing fiber with a thermoplastic polymer. The high-performance physicomechanical properties of the material are ensured by the high-volume fraction of reinforcing fibers in the material, good adhesion between binder and fibers, fiber straightness and continuity, and reliable impregnation [19].

The Anisoprint print head consists of a coextrusion head. The composite extruder introduces a thermoplastic filament which is heated to its extrusion temperature to enable the fiber to be deposited along the desired path. A cutter is placed at the entrance to the heating element, enabling the fiber to be cut when the frame or another layer is changed.

In addition, Aura's slicing software allows great flexibility by giving you control over each print head. Paths within the printed part can be made freely. Indeed, masks can be used with Aura Premium to carry out an additional model. This model is added to the work area to intersect with the base model and to change the internal structure with or without carbon fiber [4].

The aim of this paper is to use a continuous-fiber FDM printer to manufacture multifunctional 3D parts. To this end, FDM coextrusion technology is used to impart electrical properties through continuous carbon fiber, enabling a 3D sensor part to be manufactured without human intervention. This study shows the feasibility of embedding continuous carbon fiber along a defined path within a 3D part. The defined path is based on the characteristic paths observed in snaking strain gauges. This study also shows the impact of fiber introduction on both mechanical and resistive properties, and whether the resistive signal can be a reliable indicator to predict the failure of a 3D-printed part under load. This approach can be used for soft robotics to detect the gripping of a clamp or in structural health monitoring for autonomous drones to check the health of the landing gear before carrying out a new mission.

2. Materials and Methods

The tests carried out were tensile tests monitored using carbon fiber samples obtained by 3D composite coextrusion printing.

2.1. Methods

Figure 1 shows the experimental setup. Tensile tests were carried out on a Zwick/Roëll Z2.5 testing machine equipped with a 2.5 kN cell, using a simple tensile program based on ASTM D638 TYP IV [20].

A RIGOL DM3058 multimeter connected to the terminals of the resistive element for each test was used to monitor the resistance evolution at a sampling rate of two measurements per second. The 4-wire measurement technique was used to take these readings: a constant current of 1 mA was injected via 2 wires, and only the voltage drop across the resistive element was recorded by the voltmeter. The resistance value was then calculated by the multimeter and recorded.

Figure 1. Experimental setup.

2.2. Specimens

Manufacturing was carried out on a Composer A4 from Anisoprint using their Smooth PA nylon. The nozzle dedicated to the composite receives the strand of the continuous carbon fiber (CCF-1.5k) on the one hand and the filament of the thermoplastic resin (CFC PA) in which it is embedded on the other. The filaments are Anisoprint (Esch-sur-Alzette, Luxembourg) commercial filaments coproduced with Polymaker (Shanghai, China). The diameter of the pure filament extrusion head is 0.4 mm, whereas the diameter of the composite extrusion head is 0.8 mm. The nozzle temperature is 265 °C with a feed speed of 45 mm/s. The temperature of the build plate is 60 °C.

Figure 2 shows the dimensions of the specimen printed in conformity with ASTM D638. Inside this specimen, the carbon fiber is deposited using Aura software V2.4.7 to create a zig-zag path by pulling the end of the fiber out of the specimen based on the strain gauges.

Figure 2. ASTM D638 test specimen (unit in mm)

Inside the specimen, two distinct variants were created to increase the length of the fiber and visualize the influence and potentially improve the resolution of the measurement. Figure 3 shows the carbon fiber resistive element in a U-shape or W-shape. The number of paths in the manufacturing planc is therefore 2 for the U-shape and 4 for the W-shape. Specimens without resistive elements were also produced to visualize the influence of the introduction of resistive elements on mechanical properties. The choice of paths in U and W is mainly a choice of spatial dimensions within the standard specimen.

These trajectories were added to the base model of the specimen using the mask function. Continuous carbon fiber filling was imposed in this internal structure. The filling at the jaws was denser to ensure that the break occurred in the working area. Figure 4 shows the printing strategy used with Aura Premium's mask function.

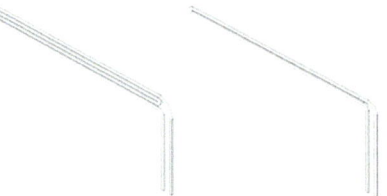

Figure 3. Resistive element type W (**left**) and type U (**right**).

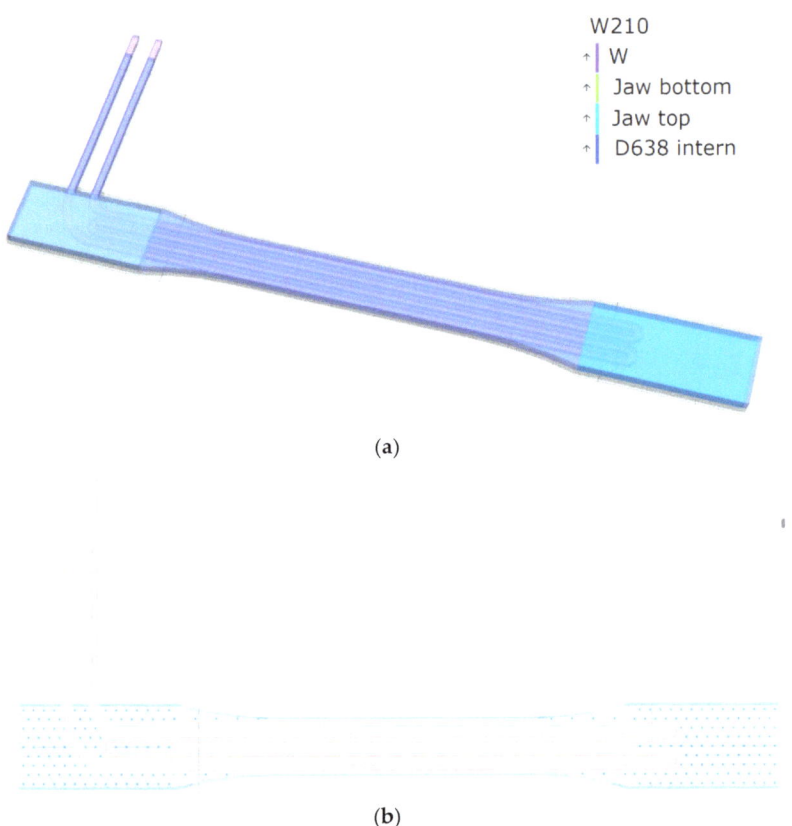

Figure 4. Printing strategy for the W210 specimen using Aura Premium: (**a**) use of the mask option in CAM software; (**b**) result of the internal structure.

In addition to these two designs, the internal fill ratio varies from 10 to 30%, and the thickness of a resistive element is either 2 layers or 4 layers, with each layer containing 2 filaments of 1500 carbon fibers. In addition to these specimens, single fiber-free samples were printed with a fill rate of 10% and 30%; the number of samples for each configuration is 3. Table 1 summarizes the parameters used for the feasibility study.

The coextruded fibers on the outside of the specimen were embedded in the thermoplastic. A chemical treatment (methanoic acid: CH_2O_2) was applied to the volumes added to the standard specimen to expose the fibers. The ends of the fibers were then coated with conductive silver paint to optimize the quality of electrical contact.

Finally, the strands were held together by clamping in two brass screw connectors forming the terminals of the resistive element. This method ensures a stable resistance value on the multimeter.

Table 1. Summary of the different printing parameters used for the tests.

	Z10	Z30	W210	U210	U230	U430
Number of fiber paths (X-Y Plan)	0	0	4	2	2	2
Number of fiber layers (Z Plan)	0	0	2	2	2	4
Infill (%)	10	30	10	10	30	30

3. Results and Discussion

This section shows the results obtained during the tests, together with a discussion of the results. This section is divided into three parts:

- Without carbon fiber to see if the tests are repeatable
- With W-shaped carbon fiber, which is the longest path in the printing plane and relates to at most one strain gauge
- With U- and W-shaped carbon fiber, and the addition of several layers of carbon to visualize trends.

3.1. Without Resistive Elements

Figure 5 shows the tensile curves of the non-fiber series tests. It can be seen that each series offers a specific trajectory, and that the curves of the tests in the same series are similar to each other until failure.

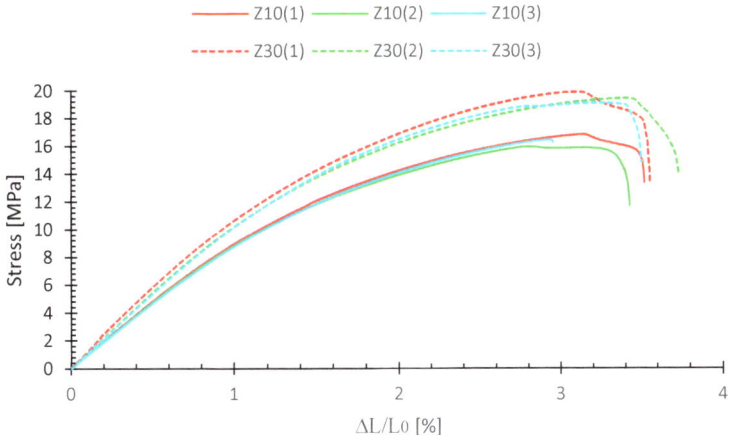

Figure 5. Tensile curves for specimens without carbon fiber.

It can also be seen that the mechanical properties are slightly improved by moving from a filling density of 10% to 30%.

At 10% (Z10), the average Rm is 16.42 MPa with a standard deviation of 0.35 Mpa and a Young's modulus of 0.96 Gpa.

At 30% (Z30), the average Rm is 19.42 MPa with a standard deviation of 0.32 MPa and a Young's modulus of 1.15 GPa.

Mechanical properties follow the trends expected for 3D-printed parts. Basic samples without carbon fiber continue to show a repeatable evolution with a Young's modulus of 0.96 GPa and an Rm of 16.42 MPa for a 10% internal fill. The repeatable tests show that the chosen printing parameters are acceptable. Ali et al. [21] show that increasing the filling of the PA structure increases the fracture stress in a rectilinear curve. As expected, when the material is added internally, mechanical properties are improved. Young's modulus is similar for a series of specimens, but tensile strength varies slightly. This is mainly due to adhesion between the printed layers.

3.2. With W-Shaped Carbon Fiber

Figure 6 shows the evolution of stress (curve with the annotation "m" for mechanical) and relative resistivity (curve with the annotation "r" for resistivity) as a function of relative elongation for the W-shape sample with 10% internal density and 2 layers of resistive elements. There are three distinct cases of evolution in resistivity.

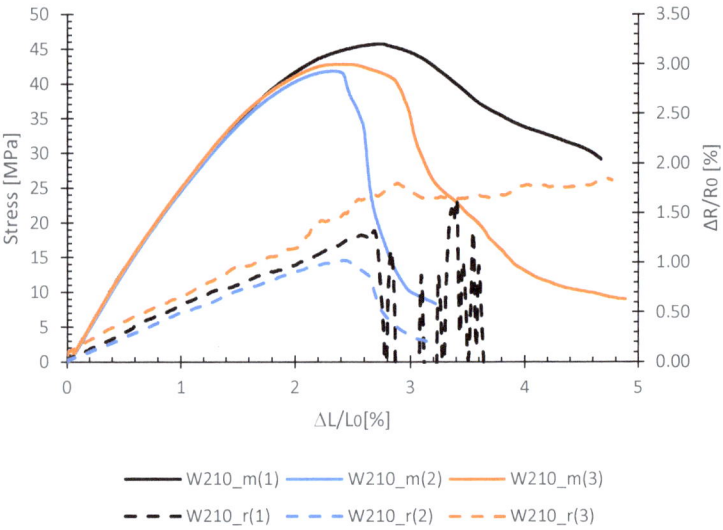

Figure 6. Tensile curves for specimens with carbon fiber in W-shape and 10% intern infill.

The mean Young's modulus is 2727.30 MPa with a standard deviation of 53.59 MPa. While the average Rm is 43.49 MPa with a standard deviation of 2 MPa. In the elastic period, the trend is similar for all 3 curves, but significant variability is visible in the plastic period up to failure. The introduction of continuous carbon fiber leads to a significant improvement in mechanical properties as demonstrated by Kabir et al. [16].

In terms of relative resistivity, the average initial resistivity is 46.46 Ω with a standard deviation of 0.53 Ω. The three curves in the elastic zone also show a similar trend. The average GF at 1.6% elongation is 0.53 with a standard deviation of 0.07. The value of the average gauge factor is very similar to those obtained in a similar experiment by Yao et al. [14], but with specimens printed in another material, PLA, and fitted with only 3000 fibers inserted manually during manufacture. A single pass of the fiber through the sample. The average GF in the elastic zone is 0.59 with a standard deviation of 0.13.

However, after 2% relative elongation, the electrical signal shows different signals. Curve W210-r(2) has a maximum of 1% before decreasing to 0%. Curve W210-r(3) continues to follow the trend of the plastic zone and continues to grow. Finally, Curve W210-r(1) shows significant oscillations in the signal. Two distinct zones are visible: behavior in the elastic zone and behavior in the plastic zone. Behavior in the elastic zone is similar to that observed by Galos et al. [18], i.e., a linear resistivity zone in relation to relative elongation.

In this zone, the adhesion of the carbon fiber to the matrix produces a repeatable signal response directly related to the chosen shape.

Several cases are visible in the plastic deformation zone:

- Perfect adhesion between fiber and matrix (W210_r(1) curve) with progressive breakage: a strong variation in the resistive signal is present due to progressive fiber breakage inside the filament adhered to the matrix. A signal is still visible because the connection is ensured by the remaining fibers.

- Perfect adhesion between fiber and matrix (W210_r(2) curve) with sharp breakage: the resistive signal is the image of mechanical behavior, with a maximum detected in the same zone as for stress. Sample rupture is marked, with perfect breakage of the carbon fiber leading to loss of signal.
- Poor adhesion between fiber and matrix (W210_r(3) curve): the resistive signal continues to increase as the fiber no longer adheres to the matrix. This means that the fiber slides during tensile stress, and that the signal obtained is only the behavior of the fiber tension. Drop-out mainly takes place in the plastic zone, as the behavior of the matrix changes in this area.

3.3. With U- and W-Shaped Carbon Fiber

Figure 7 shows the evolution of stress and resistivity relatives in relation to the elongation relative for different cases (U-shape/W-shape).

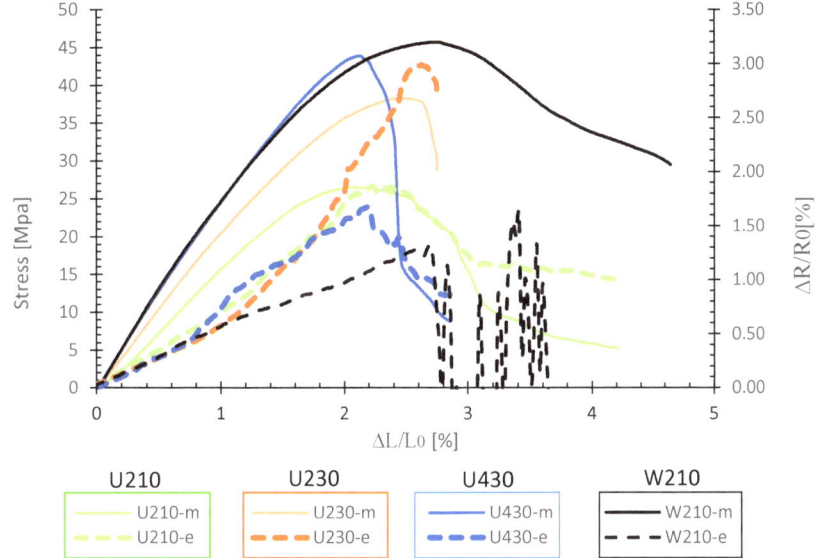

Figure 7. Tensile curves for specimens with carbon fiber in W and U-shape.

Table 2 gives a summary of the results obtained during the test campaign, including mechanical properties, resistivity and gauge factor.

Table 2. Summary of the various indicators means collected during the experimental campaign.

	W210	U210	U230	U430
R0 (Ω)	46.46	29.18	30.87	19.41
GF at 1.6%	0.503	0.717	0.745	0.752
E (GPa)	2.27	1.69	2.04	2.75
Rm (MPa)	43.49	24.57	31.75	47.65

An increase in internal density from 10% to 30% leads to improved mechanical properties, as shown by the U210-m and U230-m curves. Young's modulus increases from 1.69 GPa to 2.04 GPa while Rm increases from 24.57 MPa to 31.75 MPa. Increasing the resistive element layer also leads to improved mechanical properties, with a Young's modulus of 2.75 GPa and an Rm of 47.65 MPa.

The double passage of two layers of the resistive element (U-shape) results in an increase of over 700 MPa compared with no continuous fiber. By doubling the passage through the sample (W-shape), the increase is over 1.3 GPa. This shows that carbon fiber makes a significant contribution to mechanical performance. The number of layers of carbon fiber improves the mechanical behavior of the sample. At constant carbon fiber lengths, i.e., W210 and U430, the superposition of 2 layers results in improved mechanical properties than a double passage in the same plane of the sample. Stresses are shared between the four layers, resulting in greater strength than with two layers.

For an equivalent fiber length of U430 and W210, the elastic behavior is similar. The fibers are positioned on the different layers in the same direction as the stress. This proves that the mechanical stress is distributed on the carbon fibers and not on the matrix. The mechanical behavior is therefore improved following the introduction of the fibers, but the doubling of the passage in a plane or the superposition has no influence on this improvement.

The change in shape of the resistive element from W to U leads to a deterioration in mechanical properties, with a reduction in Young's modulus of 1.64 GPa and an Rm of 18.5 MPa. In order to obtain a similar Young's modulus and Rm with the U-shape, the number of layers of resistive elements must be doubled and the internal filling increased by 20%. This leads to an increase in the length of the resistive element, influencing the initial resistance. This is justified by Pouillet's law, which says:

$$R_0 = \frac{\rho l}{S}, \qquad (2)$$

where ρ is the resistivity (in Ωm), l (in m) is the length of the resistive element, and S (in m^2) is the cross-section of the resistive element. In the case of the W-shape, the length is twice as long as the U-shape, thus increasing resistance.

This has an impact on relative resistivity, as evidenced by the slope of the curve for the W-shape. At constant current, increasing the length of the resistive element leads to a decrease in relative resistivity. This demonstrates the sensitivity of the resistive signal, which is directly related to the length of the resistive element. In the case of the U-shape, detection of the fiber's disengagement from the matrix is more easily visible due to the sensitivity of the signal.

The average GF at 1.6% elongation is 0.53 for W210-r, while the other GFs are close to 0.717. The GF of sample U210-r is 0.789, that of sample U230-r is 0.745, and that of sample U430-r is 0.752. This is represented by curves with the same inclination for the U-shape over the elastic period, whereas the inclination of the W-shape curve is less significant.

The results shown in Table 2 were obtained with Smooth PA specimens through which 6000 or 12,000 fibers passed twice (U-shape) or four times (W-shape). The average GF in the elastic zone for U-shape is 0.717 with a standard deviation of 0.10. For the W-shape, the average gauge factor is 0.503 with a standard deviation of 0.07.

Although the matrix is not identical to PLA/Nylon proposed by Yao et al. [14], the results obtained during the test campaign are comparable. GF is not directly influenced by the number of fibers in the sample. However, the placement of the resistive element in the material is important. Young's modulus is higher when the number of passes through the cross-section increases, which results in improved mechanical properties. This leads directly to a reduction in GF, as the relative elongation will be greater during the same load.

On the other hand, the standard deviation suggests that repeatability is greater when the number of passes through the section is increased. This can be explained by the automation of fiber placement in the sample compared to Yao's approach [14]. Nevertheless, it also has a significant impact on the adhesion of the fiber to the matrix. Having several passes increases the probability of the fiber being adhered to the matrix, resulting in a lower standard deviation.

The main limitation is the serpentine strategy that can be inserted into the sample. The number of passes is limited by the size of the part. In the case of this study, it was impossible to explore a serpentine longer than W because of the space required.

4. Conclusions

Nowadays, the additive manufacturing of smart materials is a growing trend. Fused deposition modeling (FDM) is a manufacturing process that enables composites to be produced easily using two extrusion heads. The main limitation is the mechanical properties obtained, which are relatively weak but also difficult to repeat because of the defects obtained during printing. This paper proposes the use of a continuous carbon fiber coextrusion FDM printer to overcome this limitation. This study focused on the feasibility of using continuous carbon fiber not only as a reinforcement, but also as an indicator of the state of health of the part.

The fiber is positioned along a path similar to that of a strain gauge so that the resistive signal can be exploited as a response to constant deformation to detect a defect before breakage. Two paths were analyzed: a U-shaped path and a W-shaped path.

This feasibility study highlighted the following trends:

- The printing of a continuous carbon fiber path using coextrusion technology not only strengthens the printed part, but also allows the fiber to be used as an indicator of the state of health thanks to the electrical properties of the carbon fiber;
- The adhesion of the fiber to the matrix does not affect the electrical signal during the elastic period. After this period, adhesion between fiber and matrix is not ensured;
- The length of carbon fiber introduced into the part greatly improves the mechanical properties of the printed part;
- For the same length of carbon fiber, the fiber placement strategy has no significant influence on the mechanical properties;
- For the same length of carbon fiber, the fiber placement strategy has a considerable influence on the electrical properties;

In perspective, an optimization of the parameters of the number of fibers placed in the manufacturing plane and the height can be achieved with the maintained mechanical properties. This is to achieve the objective of having a resistive signal that is as sensitive as possible to elongation while retaining similar mechanical properties.

Author Contributions: Conceptualization, A.D. and I.O.; methodology, A.D. and J.L.; software, I.O.; validation, A.D., I.O., and S.S.C.J.; formal analysis, A.D.; investigation, I.O. and J.L; resources, A.D. and M.M.; writing—original draft preparation, A.D. and I.C.; writing—review and editing, A.D., I.O., S.S.C.J. and M.M; visualization, M.M.; supervision, A.D. All authors have read and agreed to the published version of the manuscript.

Funding: This research was funded by the Government of the French Community (named Fédération Wallonie-Bruxelles) in Belgium related to a FRHE project called THERMPOCOMP.

Institutional Review Board Statement: Not applicable.

Informed Consent Statement: Not applicable.

Data Availability Statement: No new data were created.

Acknowledgments: The authors would like to thank the Laboratory of Polymeric and Composite Materials of UMONS for access to the Zwick/Roëll Z2.5 testing machine.

Conflicts of Interest: The authors declare no conflict of interest.

References

1. Demoly, F.; André, J.-C. Impression 4D: Promesses ou Futur Opérationnel? In *Mécanique | Fabrication Additive—Impression 3D*; Technique de L'ingénieur: Paris, France, 2021. [CrossRef]
2. Dizon, J.R.C.; Espera, A.H., Jr.; Chen, Q.; Advincula, R.C. Mechanical characterization of 3D-printed polymers. *Addit. Manuf.* **2018**, *20*, 44–67. [CrossRef]

3. Ryan, K.R.; Down, M.P.; Hurst, N.J.; Keefe, E.M.; Banks, C.E. Additive manufacturing (3D printing) of electrically conductive polymers and polymer nanocomposites and their applications. *eScience* **2022**, *2*, 365–381. [CrossRef]
4. Anisoprint. Available online: https://support.anisoprint.com/design/introduction-in-composites/ (accessed on 28 June 2023).
5. Wang, Y.; Zhang, G.; Ren, H.; Liu, G.; Xiong, Y. Fabrication strategy for joints in 3D printed continuous fibre reinforced composite lattice structures. *Compos. Commun.* **2022**, *30*, 101080. [CrossRef]
6. Dickson, A.N.; Barry, J.N.; McDonnell, K.A.; Dowling, D.P. Fabrication of continuous carbon, glass and Kevlar fibre reinforced polymer composites using additive manufacturing. *Addit. Manuf.* **2017**, *16*, 146–152. [CrossRef]
7. Ahmed, O.; Wang, X.; Tran, M.-V.; Ismadi, M.-Z. Advancements in fibre-reinforced polymer composite materials damage detection methods: Towards achieving energy-efficient SHM systems. *Compos. Part B Eng.* **2021**, *223*, 109136. [CrossRef]
8. Nauman, S. Piezoresistive Sensing Approaches for Structural Health Monitoring of Polymer Composites—A Review. *Eng* **2021**, *2*, 197–226. [CrossRef]
9. Bekas, D.G.; Hou, Y.; Liu, Y.; Panesar, A. 3D printing to enable multifunctionality in polymer-based composites: A review. *Compos. Part B Eng.* **2019**, *179*, 107540. [CrossRef]
10. Kim, K.; Park, J.; Suh, J.-H.; Kim, M.; Jeong, Y.; Park, I. 3D printing of multiaxial force sensors using carbon nanotube (CNT)/thermoplastic polyurethane (TPU) filaments. *Sens. Actuators A Phys.* **2017**, *263*, 493–500. [CrossRef]
11. Lanzolla, A.M.L.; Attivissimo, F.; Percoco, G.; Ragolia, M.A.; Stano, G.; Nisio, A.D. Additive Manufacturing for Sensors: Piezoresistive Strain Gauge with Temperature Compensation. *Appl. Sci.* **2022**, *12*, 8607. [CrossRef]
12. Gackowski, B.M.; Goh, G.D.; Sharma, M.; Idapalapati, S. Additive manufacturing of nylon composites with embedded multi-material piezoresistive strain sensors for structural health monitoring. *Compos. Part B Eng.* **2023**, *261*, 110796. [CrossRef]
13. Georgopoulou, A.; Michel, S.; Vanderborght, B.; Clemens, F. Piezoresistive sensor fibre composites based on silicone elastomers for the monitoring of the position of a robot arm. *Sens. Actuators A Phys.* **2021**, *317*, 112433. [CrossRef]
14. Yao, X.; Luan, C.; Zhang, D.; Lan, L.; Fu, J. Evaluation of carbon fibre-embedded 3D printed structures for strengthening and structural-health monitoring. *Mater. Des.* **2016**, *114*, 424–432. [CrossRef]
15. Güemes, A.; Fernandez-Lopez, A.; Pozo, A.R.; Sierra-Pérez, J. Structural Health Monitoring for Advanced Composite Structures: A Review. *J. Compos. Sci.* **2020**, *4*, 13. [CrossRef]
16. Kabir, S.M.F.; Mathur, K.; Seyam, A.-F.M. A critical review on 3D printed continuous fiber-reinforced composites: History, mechanism, materials and properties. *Compos. Struct.* **2020**, *232*, 111476. [CrossRef]
17. Zhuo, P.; Li, S.; Ashcroft, I.A.; Jones, A.I. Material extrusion additive manufacturing of continuous fibre reinforced polymer matrix composites: A review and outlook. *Compos. Part B Eng.* **2021**, *224*, 109143. [CrossRef]
18. Galos, J.; Hu, Y.; Ravindran, A.R.; Ladani, R.B.; Mouritz, A.P. Electrical properties of 3D printed continuous carbon fibre composites made using the FDM process. *Compos. Part A Appl. Sci. Manuf.* **2021**, *151*, 106661. [CrossRef]
19. Anisoprint introduces a new way of composite materials manufacturing. *Reinf. Plast.* **2021**, *63*. [CrossRef]
20. *ASTM D638-22*; Standard Test Method for Tensile Properties of Plastics. ASTM International: West Conshohocken, PA, USA, 2022.
21. Ali, L.F.; Raghul, R.; Ram, M.Y.M.; Reddy, V.H.; Kanna, N.S. Evaluation of the polyamide's mechanical properties for varying infill percentage in FDM process. *Mater. Today Proc.* **2022**, *68*, 2509–2514. [CrossRef]

Disclaimer/Publisher's Note: The statements, opinions and data contained in all publications are solely those of the individual author(s) and contributor(s) and not of MDPI and/or the editor(s). MDPI and/or the editor(s) disclaim responsibility for any injury to people or property resulting from any ideas, methods, instructions or products referred to in the content.

Article

Enhancing Strength and Toughness of Aluminum Laminated Composites through Hybrid Reinforcement Using Dispersion Engineering

Behzad Sadeghi [1,*], Pasquale Cavaliere [1] and Behzad Sadeghian [2]

[1] Department of Innovation Engineering, University of Salento, Via per Arnesano, 73100 Lecce, Italy; pasquale.cavaliere@unisalento.it

[2] Department of Materials Engineering, Isfahan University of Technology, Isfahan 84156-83111, Iran; behzadsadeghian91@gmail.com

* Correspondence: b.sadeghi2020@gmail.com

Abstract: In this work, we propose a hybrid approach to solve the challenge of balancing strength and ductility in aluminum (Al) matrix composites. While some elements of our approach have been used in previous studies, such as in situ synthesis and ex situ augmentation, our work is innovative as it combines these techniques with specialized equipment to achieve success. We synthesized nanoscale Al_3BC particles in situ using ultra-fine particles by incorporating carbon nanotubes (CNTs) into elemental powder mixtures, followed by mechanical activation and annealing, to obtain granular (UFG) Al. The resulting in situ nanoscale Al_3BC particles are uniformly dispersed within the UFG Al particles, resulting in improved strength and strain hardening. By innovating the unique combination of nanoscale Al_3BC particles synthesized in situ in UFG Al, we enabled better integration with the matrix and a strong interface. This combination provides a balance of strength and flexibility, which represents a major breakthrough in the study of composites. (Al_3BC, CNT)/UFG Al composites exhibit simultaneous increases in strength (394 MPa) and total elongation (19.7%), indicating increased strength and suggesting that there are promising strengthening effects of in situ/ex situ reinforcement that benefit from the uniform dispersion and the strong interface with the matrix. Potential applications include lightweight and high-strength components for use in aerospace and automotive industries, as well as structural materials for use in advanced mechanical systems that require both high strength and toughness.

Keywords: laminated composites; aluminum matrix composite; nanoscale Al_3BC; CNT; in situ; dispersion engineering; strength; toughness

Citation: Sadeghi, B.; Cavaliere, P.; Sadeghian, B. Enhancing Strength and Toughness of Aluminum Laminated Composites through Hybrid Reinforcement Using Dispersion Engineering. *J. Compos. Sci.* **2023**, *7*, 332. https://doi.org/10.3390/jcs7080332

Academic Editor: Yuan Chen

Received: 31 July 2023
Revised: 11 August 2023
Accepted: 14 August 2023
Published: 16 August 2023

Copyright: © 2023 by the authors. Licensee MDPI, Basel, Switzerland. This article is an open access article distributed under the terms and conditions of the Creative Commons Attribution (CC BY) license (https://creativecommons.org/licenses/by/4.0/).

1. Introduction

Hybrid reinforcement is a novel and significant strategy, the core idea of which is the use of hybrid reinforcements to exert respective advantages and fabricate advanced MMCs that are currently motivated by the purpose of overcoming the strength–ductility trade-off [1–5]. Due to the higher efficiency of nano-sized reinforcements compared to the counterpart micron-sized reinforcements [6], a step forward has been taken in the pursuit of simultaneous improvement of strength and ductility. Although the use of ex situ non-reinforcement always brings the dilemma of uniform dispersion in MMCs [7,8]. Along with all of the advantages of using ex situ reinforcements, deleterious aspects, such as the formation of incoherent interfaces between the reinforcement-matrix, problems such as stress concentration and deformation discontinuity at the microscale scale during the deformation process emerge [9,10]. Therefore, it seems that the simultaneous use of the benefits of in situ and ex situ reinforcements as part of a hybrid reinforcement strategy would be one of the optimal ways to realize the simultaneous improvement of strength and ductility in MMCs.

Recently, the laminated structures that imitate the structure of nacre were noticeably succussed to provide materials that were both strong and damage tolerant [11,12]. Furthermore, a reinforcement strategy based on the core idea of using hybrid reinforcements to exert their respective advantages to achieve synergistic effects, leading to excellent overall performance, has been demonstrated [4,5]. In addition, the grain refinement of the matrix grains, which is apparently one of the best ways of simultaneously improving the strength–ductility relationship, could not overcome the dislocation annihilation at grain boundaries (GBs), which leads to fast exhaustion of the strain-hardening capacity [13]. To cope with this issue, by employing a graphene (GNS)–copper (Cu) hybrid material, exceptional tensile strength and ductility has been achieved using a heterogenous structure [14]. Creating a laminated structure accompanied by hybrid reinforcement significantly contributes to hetero-deformation-induced (HDI) stress strengthening and sustained strain hardening, generating the key mechanical properties of GNS-Cu/Al. Moreover, high-performance reinforcement and a tailored architecture has been gained using a reduced graphene oxide (RGO)–CNT hybrid/Al composite prepared through a composite flake assembly process due to the formation of a planar network of RGO and CNT, which improves the load transfer efficiency between the matrix and the reinforcement in composites [5]. Recently, a strong and tough AMC reinforcement with graphene oxide (GO)–CNT hybrid was prepared via powder metallurgy [4]. The strength of the composites was improved via the synergistic effect of carbon nanotubes using in situ Al_4C_3 and GO. More recently, superior tensile properties of AMC reinforced using both in situ Al_2O_3 nanoparticles and ex situ GNSs, which were developed by manipulating the PM route [15]. The composite possesses a high tensile strength of 464 MPa and appreciable amount of ductility (8.9%), which result from the combination effect of grain refinement, in situ Al_2O_3 nanoparticles, and ex situ GNSs. It was inferred that the presence of a synergetic strategy between in situ and ex situ reinforcements, accompanied by a tailored architecture, could provide a significant contribution to both strengthening and toughening mechanisms to achieve the best combination of strength and ductility. Promisingly, the configuration of a microstructure in mesoscale and uniform spatial distribution of hybrid reinforcements offers great potential regarding the tuning of the mechanical properties of advanced AMCs [16]. Motivated by the above considerations, an easy and innovative route on the basis of elemental powder strategy via in situ synthesis was employed to introduce nano-Al_3BC into Al matrix composite reinforced with CNTs to attain a significant synergy of high strength and toughness. This method differs from those used by other researchers in several ways: 1—High-speed cutting and coating: High-speed cutting is used to break the CNT group, and B is used as a new step in manufacturing. This step allows carbon nanotubes to be dispersed within the Al matrix, which can improve the mechanical properties. 2—Micro–micro rolling (MMR) process: The MMR process involves micro-scale rolling of composite materials and appears to be a unique and innovative process used in the production of laminated composites. This process will introduce microstructures and textures that can help improve the properties of (Al_3BC, CNT)/Al composites. 3—In Situ Formation of Al_3BC: The in situ formation of Al_3BC particles in composites through the reaction of Al particles with B and carbon elements and subsequent annealing distinguishes this method from another method based on the addition of preformed particles. This in situ synthesis provides better control of the size, distribution, and coupling between the material and the matrix. 4—Low growth of Al_3BC: This method relies on using a solid-state reaction to produce low-growth Al_3BC particles, resulting in nanoscale Al_3BC particles with unique properties being produced. The presence of Al_4C_3 can also assist in the formation of submicron-sized heterogeneous core layers, such as Al_3BC particles. 5—Minimal damage to CNTs: Experiments show that CNTs suffer little or no major damage during pre-dispersion, extrusion, and MMR, which suggests that a careful and well-optimized approach is used to preserve the integrity of CNTs in composites. We demonstrated that a synergetic strengthening of intragranular and intergranular occurs in (Al_3BC, CNT)/Al laminated composites via in situ nano-Al_3BC, CNTs, Al_4C_3, and Al_2O_3, respectively. Specifically, strong bonding between Al_3BC and Al inside of the Al grain's interior with its uniform dispersion, as well as a significant decrease in the dislocation annihilation at GBs due to the CNTs, leads to the significant storage capability of dislocations with increased strain hardening and, consequently, uniform tensile ductility at high flow stresses.

The fabrication of (Al_3BC,CNT)/Al composites includes the high-speed shearing process for breaking CNTs clusters and B agglomerations using a molar ratio of Al/B/CNT = 3/1/1, which uniformly coats the surface of the Al flakes (Figure 1a) and the micro–micro rolling (MMR) process (Figure 1b) [17,18] produces the laminated composites (details of the experimental process are given in the Supplementary Materials). Diffraction peaks of Al_3BC and Al were detected using (Al_3BC,CNT)/Al composites (Figure 2b), while only Al peaks were found in (B,CNT)/Al composites (Figure 2a). The formation of Al_3BC was mostly caused by the reaction Al particles with the B element and the carbon formed due to the decomposition of stearic acid during the fabrication process, as well as subsequent annealing [19]. Indeed, due to the appearance of some areas in which boron and carbon atoms are supersaturated, Al_3BC particles are in situ homogeneously nucleated in a solid state with a limited growth rate. In fact, due to the very low solubility of boron and carbon atoms in Al, especially during solid-state reactions [20,21], the growth processes of Al_3BC particles are soon compromised by the deficiency in carbon and boron atoms; consequently, nanoscale Al_3BC particle are formed. However, the presence of Al_4C_3 effectively contributes to heterogeneously nucleated platelet Al_3BC particles of submicron size via the liquid–solid reaction method [22,23]. For this concept, it should be noted that the characteristic peaks of the Al_4C_3 phase have not been observed in (CNT,B)/Al composite powder due to either very low relative content or a lack of formation.

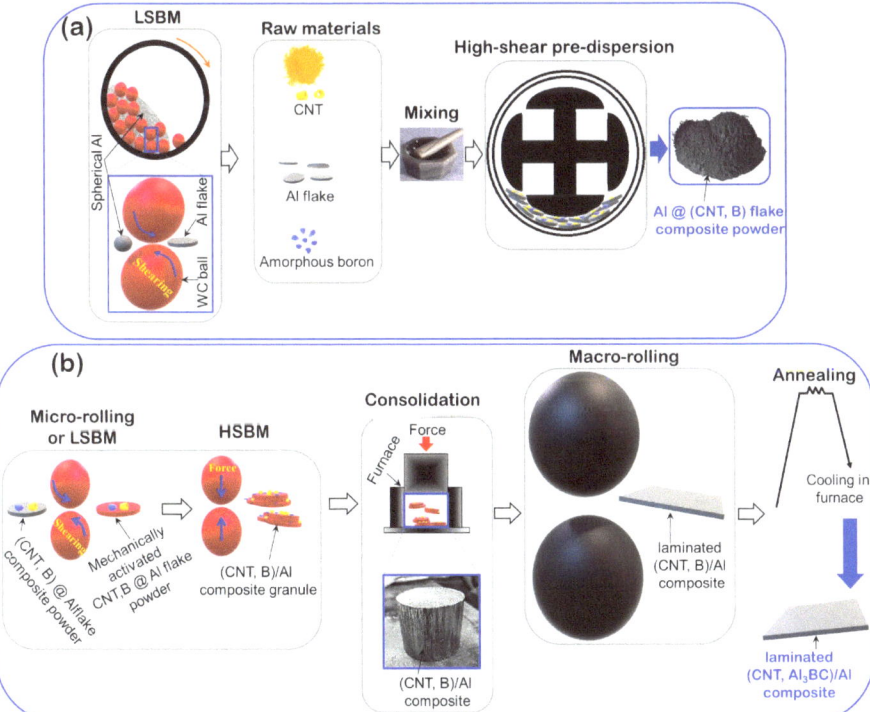

Figure 1. Schematic diagram of (**a**) the preparation of flaky-shape Al/(CNT,B) composite powder and (**b**) the fabrication of laminated (CNT,Al_3BC)/Al composites via the micro–micro rolling (MMR) process [17].

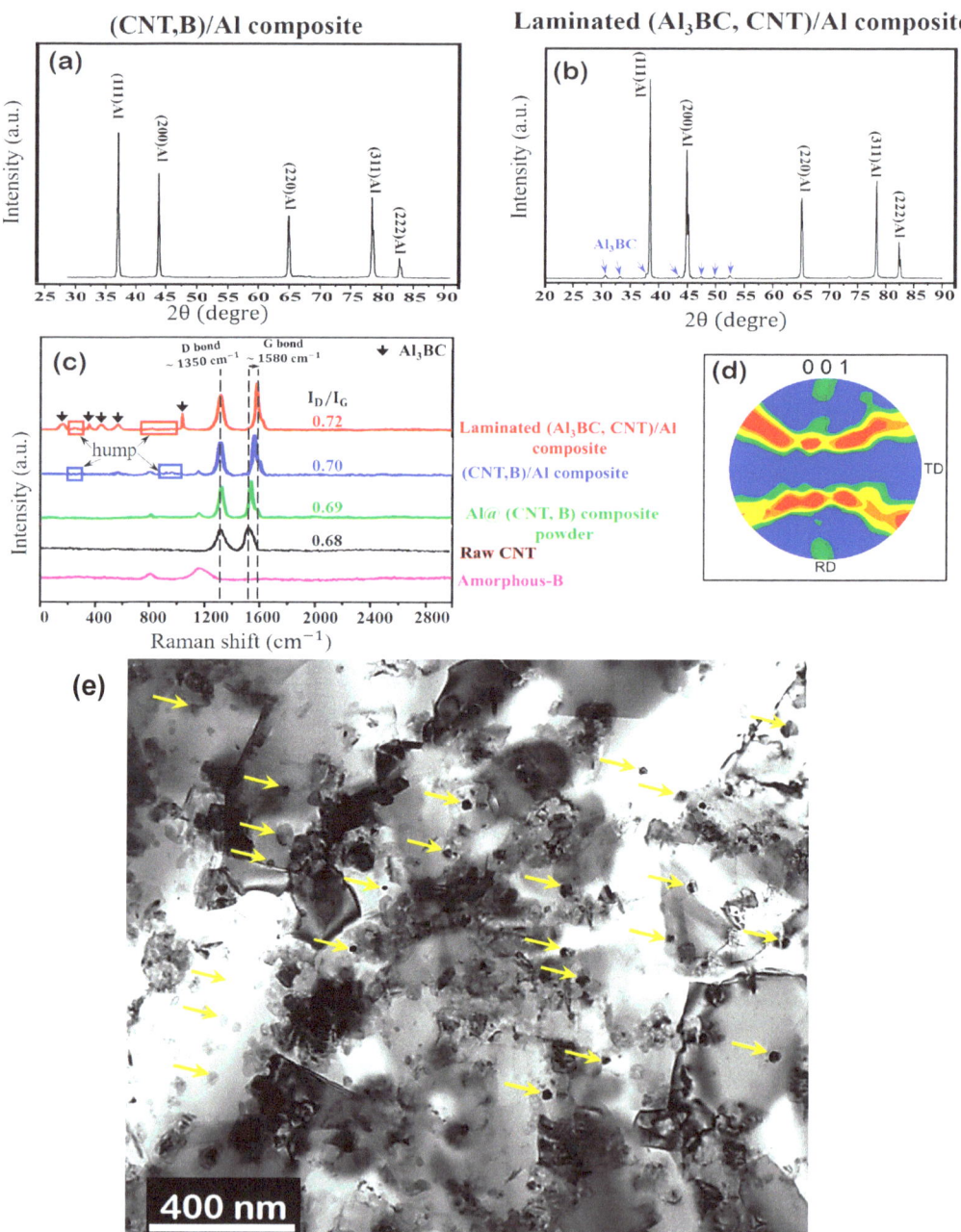

Figure 2. (**a**,**b**) XRD spectra and (**c**) Raman spectra of the studied materials; (**d**) (001) pole figure of a typical (Al$_3$BC,CNT)/Al laminated composite, (**e**) TEM image of a cross section showing the uniformity of the Al$_3$BC particle (yellow arrows) dispersion within the UFG Al matrix. (All of the tested sample planes were perpendicular to the RD).

In addition, no specific peak of Al_4C_3 was detected in either the (CNT,B)/Al composite or the (Al_3BC, CNT)/Al laminated composite. This observation is likely due to there being either no reaction between broken CNTs or a low relative content. Comparing the intensities of the peaks, it can be observed that the ratio of {111} intensity to {200} intensity increased after the MMR process, indicating the development of a strong <111> fiber texture. This belief was verified using a typical (001) pole figure, as shown in Figure 2d. As shown in Figure 2c, the relative intensity ratio of the I_D/I_G of the (Al_3BC,CNT)/Al composite increased to 0.72, while the ratio was 0.7, 0.69, and 0.68 for (CNT,B)/Al composite, (CNT,B)/Al composite powder, and raw CNTs, respectively. These results imply that there was almost no serious damage to the CNTs during the pre-dispersion, extrusion, and MMR processes. Additionally, only a peak shift of G-band from $\sim$1580 cm^{-1} to $\sim$1594 cm^{-1} was semi-quantitatively detected in the (Al_3BC, CNT)/Al composite, indicating that it might have originated from the infiltration of Al atoms in CNTs, causing a slight distortion of the sp^2 bonding structures of CNTs. The characteristic peaks of Al_4C_3 [24,25] were not detected in either the (CNT,B)/Al composite or the (Al_3BC,CNT)/Al composite. However, a few sluggish humps appeared in the composites, indicating that slight interface reaction may still exist. Indeed, the formation of Al_4C_3 can result from CNTs that have undergone partial damage, as well as atomic carbon originating from stearic acid when subjected to elevated sintering temperatures. It seems that thanks to the flaky-shaped Al building blocks and the increased available effective surface, the deeper the CNTs are embedded into the Al particles, the smaller the growth value of the I_D/I_G ratio, indicating better protection of the CNTs in both the (CNT,B)/Al composite and the (Al_3BC,CNT)/Al composite [8,26]. Additionally, the TEM image in Figure 2e is provided to demonstrate the distribution of Al_3BC particles within the UFG Al. The criteria used to evaluate uniformity encompassed the absence of localized clustering or agglomeration of Al_3BC particles, ensuring that their distribution was not sporadic but homogeneously spread across the UFG Al. Uniformity was assessed through various characterization techniques, including the TEM image shown in Figure 2e, which, despite its limitations in representing the entire sample, provides a visual indication of the general distribution trend. Furthermore, the uniform distribution of Al_3BC particles contributed to consistent mechanical reinforcement across the composite material. This result occurred because the interactions between dislocations and Al_3BC particles were optimized when particles were evenly distributed, leading to enhanced strengthening effects.

Figure 3a demonstrates an obvious laminated structure with measured grain sizes of approximately 190 and 545 nm. Such an observation is more likely to be associated with the reduced mobility of GB and the subsequent restriction of grain growth caused by the presence of CNTs at the GBs. Furthermore, a significant number of elongated grains exhibited a strong <111> texture, with only a few grains being oriented in the <001> direction. This finding indicates a pronounced restriction of slip, both within the Al grains and at the GBs. The microstructure reveals the presence of a laminated structure with approximately 40% LAGB content. Among these LAGBs, approximately 12% had an angle of $\theta < 3°$, while 28% had an angle of $3° < \theta < 15°$. These LAGBs played a crucial role in strengthening the material through dislocations and the specific strengthening mechanism associated with LAGBs. This outcome is illustrated in Figure 3b. Additionally, approximately 60% of HAGBs were present, most of which were decorated by CNTs, along with some γ-Al_2O_3 nanoparticles. These HAGBs were highly effective in impeding the movement of dislocations, forcing them to tangle and accumulate within the grain interior and near the boundaries. This process resulted in a very high dislocation density of about 1.5×10^{16} m^{-2}, as indicated by the dashed ovals shown in Figure 3c.

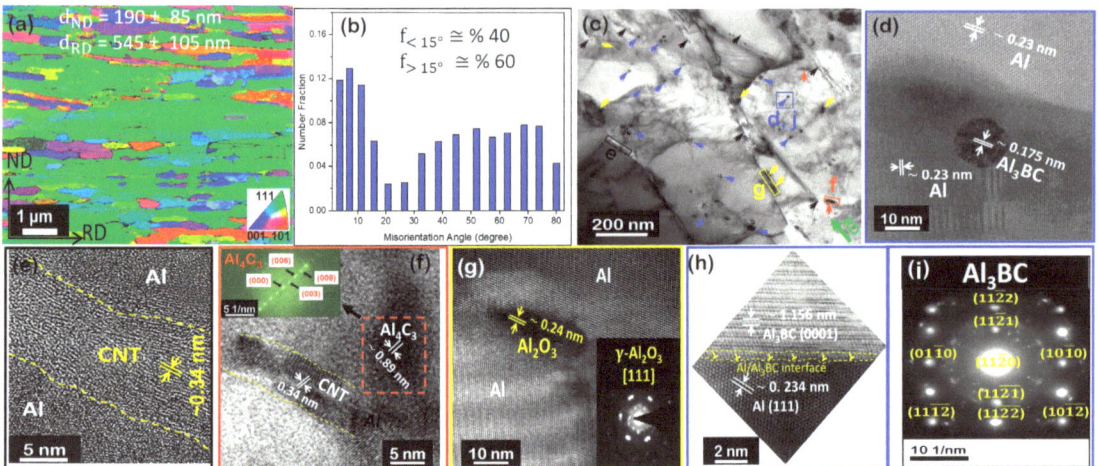

Figure 3. (a) EBSD image of (Al$_3$BC, CNT)/Al laminated composite, (b) the distribution of the grain boundary misorientation angle (θ), (c) a TEM image showing the elongated Al grains containing different reinforcement elements, (d) a STEM image showing a nanoscale Al$_3$BC in Al matrix, (e–h), HRTEM images of (e) the CNT-Al interface, (f) a typical structure of CNTs with small interfacial Al$_4$C$_3$, (g) the structure of γ-Al$_2$O$_3$, (h) the Al$_3$BC–Al interface formed in the grain interior of Al, and (i) the relative fast Fourier transforms of Al$_3$BC. The dashed ovals show dislocation entanglements. The red, black, blue, and yellow arrows show Al$_4$C$_3$, CNTs, Al$_3$BC, and γ-Al$_2$O$_3$, respectively.

Upon closer examination of Figure 3c, numerous spherical gray particles in the size range of 15–25 nm can be observed. These particles have been identified as Al$_3$BC based on the STEM image provided in Figure 3d,j. In thermodynamic terms, it is important to note that Al$_3$BC has a negative formation enthalpy of approximately −56.34 KJ/mol [27], which is lower than that of Al$_4$C$_3$ (~−36 KJ/mol [27]). This observation, coupled with the availability of boron atoms that can form bonds with carbon atoms at damaged CNT tips and the presence of amorphous carbon coating on CNTs, suggests that annealing heat treatment at 800 °C for 3 h in a flowing argon environment would be ideal for the in situ synthesis of Al$_3$BC nanoparticles. It is expected that the formation of Al$_3$BC is primarily influenced by the diffusion rate of carbon and boron in the Al matrix. This diffusion rate typically increases in line with temperature in the solid state. Therefore, it is more likely that no Al$_3$BC would form during hot extrusion, as the high temperature would enhance the diffusion and potentially deplete the available boron atoms. In such a scenario, it is possible that a small number of nanorod-shaped Al$_4$C$_3$ particles may form due to exposure to high temperatures when boron atoms become scarce. In other words, the content of Al$_4$C$_3$ in this study is not only influenced by temperature, as previously reported by [21], but also strongly depends on the availability of free boron and carbon atoms. The lattice fringes observed in individual Al$_3$BC, CNT, Al$_4$C$_3$, and Al$_2$O$_3$ particles in the composites had approximate spacing values of 0.175 nm, 0.34 nm, 0.89 nm, and 0.24 nm, respectively. These spacing values roughly correspond to the (110) plane of Al$_3$BC, the (1120) plane of graphite, the (003) plane of Al$_4$C$_3$, and the (311) plane of γ-Al$_2$O$_3$, as shown in Figure 3d–g. In conclusion, nanoscale Al$_3$BC particles with sizes ranging from 6 to 25 nm have been successfully synthesized in the composites. The size range of the Al$_3$BC nanoparticles is comparable to those previously synthesized via liquid–solid reaction [23] and self-propagating high-temperature synthesis [28]. In contrast to the complexity and costliness of other proposed routes, the synthesis route proposed in this study is feasible and simple. It involves the direct formation of nanoscale Al$_3$BC through the reaction between dissolved boron and carbon atoms and the Al matrix, which is achieved through the ball milling, rolling, and annealing processes. The resulting nanorods of in situ Al$_4$C$_3$

exhibit strong interfacial bonding, good dispersion within the aluminum grain interior, and a complete single-crystal structure. In the (Al_3BC, CNT)/Al composite, the Al_3BC particles are in situ synthesized and embedded within the Al matrix. The orientation relationship between the Al_3BC particles and the Al matrix is represented by (0111) Al_3BC/(111) Al, which means that the crystallographic planes of Al_3BC align with those of the Al matrix. This alignment allows a better fit between the two materials, resulting in fewer mismatch dislocations at the interface. Thus, the in situ-Al_3BC/Al interface demonstrates a strong bonding effect, which effectively minimizes the presence of mismatch dislocations at the interface (Figure 3h), corroborating the findings reported in previous studies [29,30]. The strong bonding strength, along with the preferred orientation of the elongated Al grains, significantly contributes to the intragranular strengthening effects of Al_3BC on the Al matrix. The majority of CNTs are embedded in the Al matrix and aligned in the RD, indicating a strong and firm bond between the CNTs and the Al matrix. The presence of a clean interface between the CNTs and Al, which lacks any voids or interfacial products (Figure 3e), suggests that both the preservation of the CNT structure and a strong bonding between Al and CNTs occur, which contribute to load transfer and strengthening. However, a small amount of nanoscale rod-like Al_4C_3 is formed in situ through the reaction of partially damaged CNTs with the Al matrix (indicated via the red arrows shown in Figure 3c). The preferential nucleation of interfacial Al_4C_3 at the open edges of the CNTs suggests that these open edges are more reactive than other parts of the CNTs, and these results are consistent with previous studies' findings [31,32]. It appears that the formation of Al_4C_3 could be related to the shortening and breaking of CNTs, which, in turn, increases the number of carbon atoms available to react with Al and B, leading to the production of in situ Al_3BC nanoparticles. The presence of Al_3C_4 in the composite has an effect on the strength and ductility of the product. Al_3C_4 is a special phase with its own mechanical properties that contribute to all composite materials. The presence of Al_3C_4 nanorods can act as an additional reinforcement in the composite. These nanorods may impede dislocation motion and contribute to forest hardening, thus enhancing the strength of the composite. In addition, strong interfacial bonding can be achieved via the formation of an appropriate amount of Al_4C_3, which can promote load transfer and enhance the fracture elongation of the CNT/Al composite. The size of Al_4C_3 is the most important factor affecting the mechanical properties of the composite. In this study, the interfacial bond was improved, while the mean diameters of the short Al_4C_3 rods in composite were as small as ~30–35 nm; therefore, it causes more outstanding elongation (19.7%). As it is a strong combination, it is not easy for microcracks to form and propagate at the interface during plastic deformation. The effect of Al_3C_4 on curing is more complex and needs further analysis and research. However, since Al_4C_3 is a brittle phase, excess Al_4C_3 negatively affects the mechanical properties of the composite.

Additionally, spherical nanoparticles of γ-Al_2O_3 can be observed within both the interior of the Al grains and at the grain boundaries (indicated via yellow arrows, as shown in Figure 3c). These nanoparticles are the result of the transformation of the native amorphous (am)–Al_2O_3 skin that is broken during processing at temperatures above 450 °C [33]. Therefore, the oxygen introduced during the fabrication process predominantly exists in the form of oxides, specifically γ-Al_2O_3. Considering that the native Al_2O_3 skin on the surface of the nanoflakes is approximately 4 nm thick, the volume fraction of Al_2O_3 is estimated to be about 2.6% (8/300 × 100%) of the total volume of the 300-nanometer nanoflake Al powder, which aligns with the value reported in [33]. By assuming that all oxygen present in am–Al_2O_3 is completely transformed into γ-Al_2O_3 after annealing at 800 °C, it is found that the oxygen content also remains virtually unchanged after annealing. The nano-Al_3BC and Al_4C_3 play important roles in generating high effective stress via forest dislocation cutting and Orowan strengthening, while the CNTs and nano-Al_2O_3 effectively provide high back stress via the accumulation and hindering of dislocation annihilation at the interfaces and GBs. Enhanced strength in composites can be achieved through promoted movable dislocation–reinforcement interactions, leading to improved performance based on the Orowan strengthening phenomenon [34]. The Orowan strengthening occurs a result of the force exerted on a dislocation to bypass particles within the matrix [35]. In this context, Al_3BC nanoparticles are uniformly dispersed throughout the laminated UFG Al grain, with the interparticle spacing estimated to be approximately 170–190 nm, which is close to the particle diameter of about 190 nm. The intragranular Al_3BC nanoparticles and Al_4C_3 in nanorod form could store mobile dislocations in the form of Orowan loops in the internal regions of Al grains during straining, which more effectively provides HDI stress hardening and forest dislocation hardening. This result is consistent with those reported by Liu et al. [35]. This outcome can reduce the ratio of Orowan strengthening to total strengthening. Considering the overall ultimate tensile strength, other strengthening mechanisms come into play. These mechanisms include forest hardening, which arises from dislocation–dislocation interactions that occur due to the relatively higher dislocation density present, and GB strengthening, which results from the presence of UFG Al grains. Both of these factors significantly contribute to the overall enhancement of the strength of material. The presence of nanoparticles (Al_3BC, Al_2O_3) or even Al_4C_3 nanorods causes mechanical incompatibility and promotes dislocation formation (mostly misorientation at less than 3°, which is taken into account as dislocations [36,37]), resulting in stronger dislocation–dislocation interactions and forest dislocation strengthening. Also, during deformation, GBs prevent dislocation movement due to the misorientation of adjacent grains. Higher misorientation causes resistance to movement [38,39]. In addition, UFG Al has a large GB relative to the grain size, resulting in greater resistance to movement and higher stress. Dislocations from LAGBs (approximately 28%) result in a combination of easy sliding and depositing in the Al lamella, resulting in a combination of high elongation while maintaining high tensile strength. This finding is consistent with the literature mentioned above. Indeed, the presence of a high dislocation density, which is estimated to be approximately 8.13×10^{15} m^{-2}, is primarily attributed to the inhomogeneous deformation between the matrix and the reinforcements. This high dislocation density plays a significant role in generating non-directional short-range local stresses, which facilitate dislocation movement. Moreover, it leads to long-range interactions with mobile dislocations, thereby exerting a directional long-range back stress that opposes the applied stress, especially during forward loading (as illustrated in Figure 4).

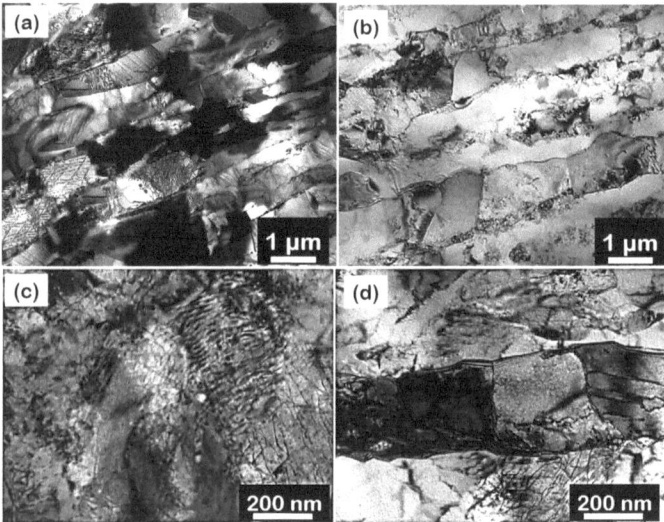

Figure 4. TEM images of (**a**,**c**) (Al_3BC,CNT)/Al and (**b**,**d**) CNT/Al laminated composites showing high density of dislocations.

Figure 5a presents the tensile stress–strain curves of the (Al$_3$BC,CNT)/Al laminated composites and CNT/Al composites. It is evident that the (Al$_3$BC,CNT)/Al laminated composites exhibit significantly higher strength than the CNT/Al composites. This result can be attributed to the combined intra- and inter-granular effects of the reinforcements, as well as the laminated structure of the composites. Furthermore, Figure 5b illustrates the variation in the strain-hardening rate as a function of true strain, demonstrating the superior maintenance of a high strain-hardening rate in the (Al$_3$BC,CNT)/Al laminated composites. The improved dislocation storage capability of ultra-fine-grained (UFG) Al in the (Al$_3$BC,CNT)/Al laminated composites is attributed to the laminated structure and the in situ formation of intragranular Al$_3$BC nanoparticles. Specifically, the (Al$_3$BC,CNT)/Al laminated composites exhibit a strength of 394 MPa, a uniform elongation of 13.2%, and a total elongation of 19.7%, which is approximately 29% higher in strength and shows an increase in the total elongation from 17.7 to 19.5% compared to the CNT/Al composites. Moreover, the uniform elongation of the (Al$_3$BC,CNT)/Al laminated composites is 13.2%, whereas for PM (powder metallurgy) 16 vol.% B$_4$C/UFG Al with an average grain size of 500 nm, it is only 3.5% [40]. Additionally, for PM (0.5 wt.% CNT, 2 vol.% γ-Al$_2$O$_3$)/UFG Al with a unique intragranular dispersion of ultra-short CNTs, the uniform elongation is approximately 5.9% [41]. Considering the similarity of the in situ reinforcements found in the (Al$_3$BC,CNT)/UFG Al and the 15 vol.% Al$_3$BC/UFG Al composites [29], the superior strength and ductility of the (Al$_3$BC,CNT)/UFG Al laminated composite can be attributed to the synergistic contribution of intra- (e.g., in situ formed Al$_3$BC, partially formed Al$_4$C$_3$) and inter-granular strengthening (e.g., CNT, γ-Al$_2$O$_3$). The presence of in situ formed nano-dispersoids, particularly the spherical nanoparticles of Al$_3$BC (Figure 5c), contributes to the synergetic intra- and inter-granular reinforcements within the elongated Al grains. This finding highlights the promising strengthening effects of Al$_3$BC, which benefits from the strong interface with the matrix. The GBs decorated by CNTs and some γ-Al$_2$O$_3$ (~2 wt.%) effectively lose their efficiency through dislocation annihilation, as dislocations nucleate in the neighboring matrix grain near the GBs. This process reduces the magnitude of stress concentration [42] and results in an increase in the critical resolved shear stress due to harder dislocation motion. This finding supports the notion that dislocation multiplication and accumulation, rather than annihilation through recovery, are the governing mechanisms. This result can be attributed to the profound role of in situ reinforcements, which mostly formed a strong interface in the Al grain interior. Recently, a significant high combination of ultimate tensile strength and ductility as a consequence of the presence of CNTs and Al$_4$C$_3$, as well as a high dislocation density (~8.13×10^{15} m^{-2}), was demonstrated [43]. Most importantly, these dislocations were mostly caused by inhomogeneous deformation between the matrix and the reinforcements; significantly, both produce the non-directional short-range local stress required for a dislocation to move and dramatically provide long-range interactions via mobile dislocations, exerting a directional long-range back stress opposed to the stress applied in the case of forward loading.

Incorporating both Al$_3$BC and Al$_2$O$_3$ improves the mechanical properties of the composite material. This improvement is mostly achieved through Zener drag, which occurs when dispersed particles are randomly distributed within the material matrix. The effectiveness of Zener drag depends on various factors, such as the nature, geometry, size, spacing, and volume fraction of the particles [44]. In our composite, the presence of these nanoparticles effectively prevents the migration of both HAGBs and LAGBs. As a result, dynamic and static recrystallization is hindered, and grain growth is restrained. This characteristic allows the composite to retain its mechanical properties in advanced Al-based MMCs. The introduction of CNTs and some γ-Al$_2$O$_3$ to GBs significantly reduces the dislocation annihilation efficiency, thereby contributing to the occurrence of back stress hardening. On the other hand, reinforcements embedded within the grain interior of Al enhance dislocation blocking and forest hardening, leading to the creation of an advanced approach to achieving ultra-high strengthening and toughening efficiency through intragranlar/intergranular dispersion engineering.

Additionally, the integration of CNTs within the GBs significantly hinders the movement of dislocations across the interface between Al and CNTs [45]. This obstruction of GB motion is more effective when using CNTs due to their elongated nanofiber-shaped structure compared to nearly spherical nature of nanoparticles [46]. The unique morphology of CNTs provides a strong impediment to dislocation movement within the composite. By reinforcing the material, CNTs synergistically enhance its strength through related mechanisms, such as load distribution and dislocation inhibition, while preserving the inherent ductility of the Al matrix.

Therefore, the combined effects of Al_3BC, Al_2O_3, and CNTs contribute significantly to the overall improvement in the strength and ductility of the composite. These materials work together through various mechanisms, including Zener drag, dislocation inhibition, and load distribution, to enhance the mechanical properties of the composite while maintaining the malleability of the Al matrix. This multifaceted enhancement strategy offers possibilities to create advanced materials with superior mechanical performances that make them suitable for a wide range of applications. This advancement represents a progression beyond the work of Jiang et al. [10]. The presence of a well-arranged array of in situ-formed hybrid nanoparticle-rich zones, surrounded by CNT-decorated GBs, synergistically enhances both the strength and ductility of the composite material.

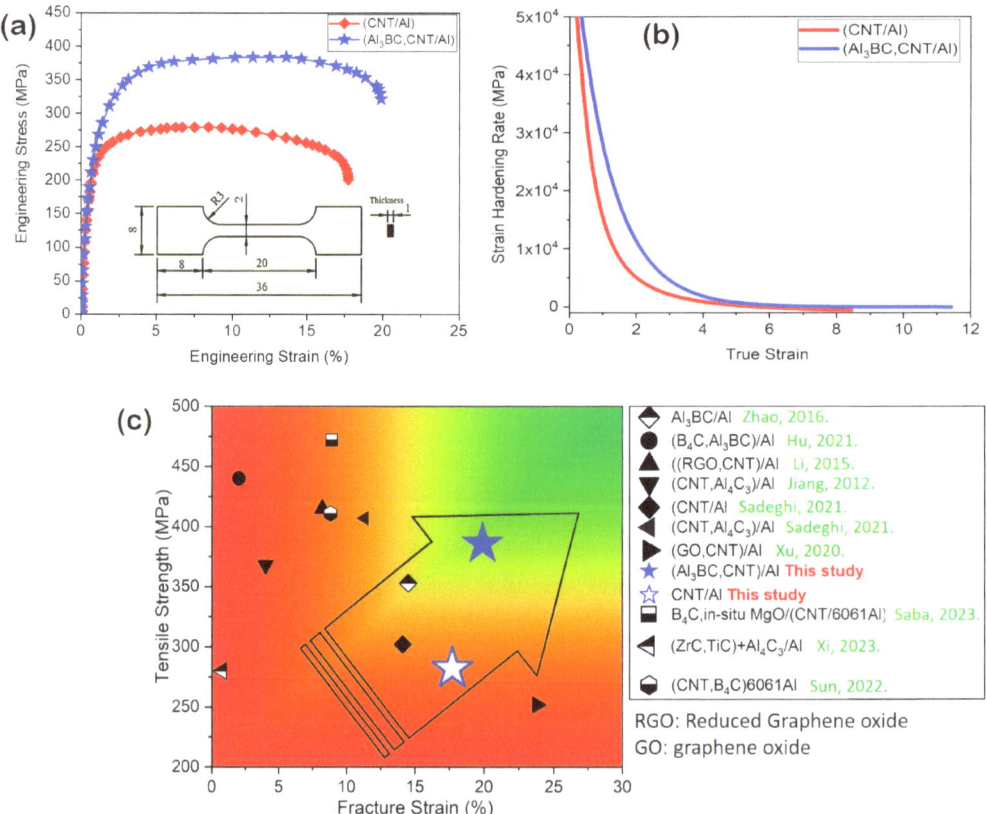

Figure 5. (**a**) Engineering tensile stress–strain curves, (**b**) strain-hardening rate curves of (Al_3BC,CNT/Al) and CNT/Al composites and (**c**) tensile properties of Al composites fabricated via various solid state-based approaches [2,5,27,29,43,47–50]. The arrow shows the direction of the development of the stronger and tougher composites.

Several future directions and potential applications can be explored. Composite materials have a good combination of high strength and ductility, as well as applications in many industries. They are used in lightweight, aerospace, and automotive performance applications to improve fuel efficiency and reduce emissions. In addition, the mechanical properties of composite materials in the medical field facilitate the production of implants and medical devices. To further advance the practical applications of the (Al_3BC, CNT)/Al composite, more research can focus on improving the manufacturing process to develop products using better technology. Exploring the differences and combinations between Al_3BC and CNTs can provide composite materials with unique properties for different applications. Additionally, studying the behavior of composites in different environments, such as high-temperature or corrosive environments, will provide insights into specific applications.

2. Conclusions

In summary, our work presented a new (Al_3BC, CNT)/UFG Al composite sheet that exhibits good support–length properties, having a strength of about 394 MPa and an elongation of about 19.7%. The composite achieved a uniform interparticle/interparticle distribution using a dispersion engineering approach that distinguishes it from most AMCs. The success of this composite could be attributed to the good relationship between the in situ nano-Al_3BC particles in the UFG Al particles and the Al_4C_3 formed at the end of the GB. These materials provided more significant support than conventional CNT/UFG Al composites. Through a careful combination of in situ and ex situ reinforcement techniques, we provided support distribution, leading to the development of dislocation blocks, stress recovery, and forest hardening. This process leads to an increase in the strength and ductility of the composite. The success of our combination and the surprising properties of this unique combination opened up new possibilities for research into and further development of advanced materials with improved equipment and general applications. Overall, the exceptional mechanical properties of this composite make it a promising material for use in various real-world applications. For instance, its high strength and strain-hardening capacity can be advantageous for the development of lightweight and resilient structures for use in aerospace, defense, medical device, and engineering system applications.

Supplementary Materials: The following supporting information can be downloaded at: https://www.mdpi.com/article/10.3390/jcs7080332/s1. References [51–54] are cited in the supplementary materials.

Author Contributions: Methodology: B.S. (Behzad Sadeghi), B.S. (Behzad Sadeghian); Validation: B.S. (Behzad Sadeghi); Investigation: B.S. (Behzad Sadeghi), P.C.; Writing—original draft: B.S. (Behzad Sadeghi), B.S. (Behzad Sadeghian); Writing—review & editing: B.S. (Behzad Sadeghi), P.C.; Formal analysis: P.C.; Supervision: P.C.; Funding acquisition: P.C.; Software: B.S. (Behzad Sadeghian); Data curation: B.S. (Behzad Sadeghian). All authors have read and agreed to the published version of the manuscript.

Funding: This research received no external funding.

Institutional Review Board Statement: Not applicable.

Informed Consent Statement: Not applicable.

Acknowledgments: The authors would like to express their sincere gratitude to Niloofar Ebrahimzadeh Esfahani for her invaluable contributions and support during scientific discussions throughout the course of this study. Her insightful insights and thoughtful discussions greatly enriched the quality of our work. Additionally, the authors extend their heartfelt thanks to Nikzad Sadeghi for his invaluable support in providing innovative ideas and engaging discussions that significantly contributed to the development of this research. Their collaborative input and dedication were instrumental in shaping the direction and outcomes of this study.

Conflicts of Interest: The authors declare that they have no conflict of interest.

Impact Statement: A unique approach involving the incorporation of elemental powder mixture was employed for the first time to create a hybrid nanoreinforcement strategy, resulting in the successful attainment of a remarkable combination of strength and ductility in CNT/Al composites.

References

1. Sadeghi, B.; Cavaliere, P.; Pruncu, C.I.; Balog, M.; Marques de Castro, M.; Chahal, R. Architectural design of advanced Al matrix composites: A review of recent developments. *Crit. Rev. Solid State Mater. Sci.* **2022**, 1–71. [CrossRef]
2. Sun, H.; Saba, F.; Fan, G.; Tan, Z.; Li, Z. Micro/nano-reinforcements in bimodal-grained matrix: A heterostructure strategy for toughening particulate reinforced metal matrix composites. *Scr. Mater.* **2022**, *217*, 114774. [CrossRef]
3. Luo, F.; Jiang, X.; Sun, H.; Mo, D.; Zhang, Y.; Shu, R.; Xue, L. Microstructures, mechanical and thermal properties of diamonds and graphene hybrid reinforced laminated Cu matrix composites by vacuum hot pressing. *Vacuum* **2023**, *207*, 111610. [CrossRef]
4. Xu, Z.Y.; Li, C.J.; Wang, Z.; Fang, D.; Gao, P.; Tao, J.M.; Yi, J.H.; Eckert, J. Balancing the strength and ductility of graphene oxide-carbon nanotube hybrid reinforced Al matrix composites with bimodal grain distribution. *Mater. Sci. Eng. A* **2020**, *796*, 140067. [CrossRef]
5. Li, Z.; Fan, G.; Guo, Q.; Li, Z.; Su, Y.; Zhang, D. Synergistic strengthening effect of graphene-carbon nanotube hybrid structure in Al matrix composites. *Carbon* **2015**, *95*, 419–427. [CrossRef]
6. Chen, L.-Y.; Xu, J.-Q.; Choi, H.; Pozuelo, M.; Ma, X.; Bhowmick, S.; Yang, J.-M.; Mathaudhu, S.; Li, X.-C. Processing and properties of magnesium containing a dense uniform dispersion of nanoparticles. *Nature* **2015**, *528*, 539–543. [CrossRef]
7. Fan, G.; Liu, Q.; Kondo, A.; Naito, M.; Kushimoto, K.; Kano, J.; Tan, Z.; Li, Z. Self-assembly of nanoparticles and flake powders by flake design strategy via dry particle coating. *Powder Technol.* **2023**, *418*, 118294. [CrossRef]
8. Sadeghi, B.; Fan, G.; Tan, Z.; Li, Z.; Kondo, A.; Naito, M. Smart Mechanical Powder Processing for Producing Carbon Nanotube Reinforced Al Matrix Composites. *KONA Powder Part. J.* **2022**, *39*, 2022004. [CrossRef]
9. Ma, L.; Zhang, X.; Pu, B.; Zhao, D.; He, C.; Zhao, N. Achieving the strength-ductility balance of boron nitride nanosheets/Al composite by designing the synergistic transition interface and intragranular reinforcement distribution. *Compos. Part B Eng.* **2022**, *246*, 110243. [CrossRef]
10. Jiang, L.; Yang, H.; Yee, J.K.; Mo, X.; Topping, T.; Lavernia, E.J.; Schoenung, J.M. Toughening of Al matrix nanocomposites via spatial arrays of boron carbide spherical nanoparticles. *Acta Mater.* **2016**, *103*, 128–140. [CrossRef]
11. Zhu, Y.; Wu, X. Heterostructured materials. *Prog. Mater. Sci.* **2023**, *131*, 101019. [CrossRef]
12. Lu, L.; Zhao, H. Progress in Strengthening and Toughening Mechanisms of Heterogeneous Nanostructured Metals. *Acta Metall. Sin.* **2022**, *58*, 1360–1370. [CrossRef]
13. Ma, X.L.; Huang, C.X.; Xu, W.Z.; Zhou, H.; Wu, X.L.; Zhu, Y.T. Strain hardening and ductility in a coarse-grain/nanostructure laminate material. *Scr. Mater.* **2015**, *103*, 57–60. [CrossRef]
14. Pu, B.; Zhang, X.; Chen, X.; Lin, X.; Zhao, D.; Shi, C.; Liu, E.; Sha, J.; He, C.; Zhao, N. Exceptional mechanical properties of Al matrix composites with heterogeneous structure induced by in-situ graphene nanosheet-Cu hybrids. *Compos. Part B Eng.* **2022**, *234*, 109731. [CrossRef]
15. Wan, J.; Yang, J.; Zhou, X.; Chen, B.; Shen, J.; Kondoh, K.; Li, J. Superior tensile properties of graphene/Al composites assisted by in-situ alumina nanoparticles. *Carbon* **2023**, *204*, 447–455. [CrossRef]
16. Zhang, Y.; Li, X. Bioinspired, graphene/Al$_2$O$_3$ doubly reinforced Al composites with high strength and toughness. *Nano Lett.* **2017**, *17*, 6907–6915. [CrossRef]
17. Sadeghi, B.; Cavaliere, P.; Balog, M.; Pruncu, C.I.; Shabani, A. Microstructure dependent dislocation density evolution in micro-macro rolled Al$_2$O$_3$/Al laminated composite. *Mater. Sci. Eng. A* **2022**, *830*, 142317. [CrossRef]
18. Sadeghi, B.; Cavaliere, P.; Pruncu, C.I. Architecture dependent strengthening mechanisms in graphene/Al heterogeneous lamellar composites. *Mater. Charact.* **2022**, *188*, 111913. [CrossRef]
19. Kleiner, S.; Bertocco, F.; Khalid, F.; Beffort, O. Decomposition of process control agent during mechanical milling and its influence on displacement reactions in the Al–TiO$_2$ system. *Mater. Chem. Phys.* **2005**, *89*, 362–366. [CrossRef]
20. Dabouz, R.; Bendoumia, M.; Belaid, L.; Azzaz, M. Dissolution of Al 6%wt c Mixture Using Mechanical Alloying. *Defect Diffus. Forum* **2019**, *391*, 82–87. [CrossRef]
21. Kubota, M.; Cizek, P. Synthesis of Al$_3$BC from mechanically milled and spark plasma sintered Al–MgB$_2$ composite materials. *J. Alloys Compd.* **2008**, *457*, 209–215. [CrossRef]
22. Zhao, Y.; Ma, X.; Zhao, X.; Chen, H.; Liu, X. Enhanced aging kinetic of Al$_3$BC/6061 Al composites and its micro-mechanism. *J. Alloys Compd.* **2017**, *726*, 1053–1061. [CrossRef]
23. Zhao, Y.; Qian, Z.; Liu, X. Identification of novel dual-scale Al$_3$BC particles in Al based composites. *Mater. Des.* **2016**, *93*, 283–290. [CrossRef]
24. Jiang, Y.; Tan, Z.; Fan, G.; Zhang, Z.; Xiong, D.-B.; Guo, Q.; Li, Z.; Zhang, D. Nucleation and growth mechanisms of interfacial carbide in graphene nanosheet/Al composites. *Carbon* **2020**, *161*, 17–24. [CrossRef]
25. Sadeghi, B.; Cavaliere, P. CNTs reinforced Al-based composites produced via modified flake powder metallurgy. *J. Mater. Sci.* **2022**, *57*, 2550–2566. [CrossRef]

26. Wei, H.; Li, Z.Q.; Xiong, D.B.; Tan, Z.Q.; Fan, G.L.; Qin, Z.; Zhang, D. Towards strong and stiff carbon nanotube-reinforced high-strength Al alloy composites through a microlaminated architecture design. *Scr. Mater.* **2014**, *75*, 30–33. [CrossRef]
27. Hu, Q.; Guo, W.; Xiao, P.; Yao, J. First-principles investigation of mechanical, electronic, dynamical, and thermodynamic properties of Al$_3$BC. *Phys. B Condens. Matter* **2021**, *616*, 413127. [CrossRef]
28. Tsuchida, T.; Kan, T. Synthesis of Al$_3$BC in air from mechanically activated Al/B/C powder mixtures. *J. Eur. Ceram. Soc.* **1999**, *19*, 1795–1799. [CrossRef]
29. Zhao, Y.; Qian, Z.; Ma, X.; Chen, H.; Gao, T.; Wu, Y.; Liu, X. Unveiling the Semicoherent Interface with Definite Orientation Relationships between Reinforcements and Matrix in Novel Al$_3$BC/Al Composites. *ACS Appl. Mater. Interfaces* **2016**, *8*, 28194–28201. [CrossRef]
30. Zhao, Y.; Ma, X.; Chen, H.; Zhao, X.; Liu, X. Preferred orientation and interfacial structure in extruded nano-Al$_3$BC/6061 Al. *Mater. Des.* **2017**, *131*, 23–31. [CrossRef]
31. Chen, B.; Jia, L.; Li, S.; Imai, H.; Takahashi, M.; Kondoh, K. In Situ Synthesized Al$_4$C$_3$ Nanorods with Excellent Strengthening Effect in Al Matrix Composites. *Adv. Eng. Mater.* **2014**, *16*, 972–975. [CrossRef]
32. Zhou, W.; Yamaguchi, T.; Kikuchi, K.; Nomura, N.; Kawasaki, A. Effectively enhanced load transfer by interfacial reactions in multi-walled carbon nanotube reinforced Al matrix composites. *Acta Mater.* **2017**, *125*, 369–376. [CrossRef]
33. Balog, M.; Krizik, P.; Nosko, M.; Hajovska, Z.; Victoria Castro Riglos, M.; Rajner, W.; Liu, D.-S.; Simancik, F. Forged HITEMAL: Al-based MMCs strengthened with nanometric thick Al$_2$O$_3$ skeleton. *Mater. Sci. Eng. A* **2014**, *613*, 82–90. [CrossRef]
34. Li, C.; Mei, Q.; Li, J.; Chen, F.; Ma, Y.; Mei, X. Hall-Petch relations and strengthening of Al-ZnO composites in view of grain size relative to interparticle spacing. *Scr. Mater.* **2018**, *153*, 27–30. [CrossRef]
35. Liu, K.; Su, Y.; Wang, X.; Cai, Y.; Cao, H.; Ouyang, Q.; Zhang, D. Achieving simultaneous enhancement of strength and ductility in Al matrix composites by employing the synergetic strengthening effect of micro- and nano-SiCps. *Compos. Part B Eng.* **2022**, *248*, 110350. [CrossRef]
36. Jiang, J.; Britton, T.B.; Wilkinson, A.J. Measurement of geometrically necessary dislocation density with high resolution electron backscatter diffraction: Effects of detector binning and step size. *Ultramicroscopy* **2013**, *125*, 1–9. [CrossRef]
37. Jiang, J.; Britton, T.B.; Wilkinson, A.J. Evolution of dislocation density distributions in copper during tensile deformation. *Acta Mater.* **2013**, *61*, 7227–7239. [CrossRef]
38. Ma, K.; Liu, Z.Y.; Liu, K.; Chen, X.G.; Xiao, B.L.; Ma, Z.Y. Structure optimization for improving the strength and ductility of heterogeneous carbon nanotube/Al–Cu–Mg composites. *Carbon* **2021**, *178*, 190–201. [CrossRef]
39. Lu, F.; Nie, J.; Ma, X.; Li, Y.; Jiang, Z.; Zhang, Y.; Zhao, Y.; Liu, X. Simultaneously improving the tensile strength and ductility of the AlNp/Al composites by the particle's hierarchical structure with bimodal distribution and nano-network. *Mater. Sci. Eng. A* **2020**, *770*, 138519. [CrossRef]
40. Kai, X.; Li, Z.; Fan, G.; Guo, Q.; Tan, Z.; Zhang, W.; Su, Y.; Lu, W.; Moon, W.-J.; Zhang, D. Strong and ductile particulate reinforced ultrafine-grained metallic composites fabricated by flake powder metallurgy. *Scr. Mater.* **2013**, *68*, 555–558. [CrossRef]
41. Liu, Q.; Fan, G.; Tan, Z.; Guo, Q.; Xiong, D.; Su, Y.; Li, Z.; Zhang, D. Reinforcement with intragranular dispersion of carbon nanotubes in Al matrix composites. *Compos. Part B Eng.* **2021**, *217*, 108915. [CrossRef]
42. Li, Z.; Zhao, L.; Guo, Q.; Li, Z.; Fan, G.; Guo, C.; Zhang, D. Enhanced dislocation obstruction in nanolaminated graphene/Cu composite as revealed by stress relaxation experiments. *Scr. Mater.* **2017**, *131*, 67–71. [CrossRef]
43. Sadeghi, B.; Tan, Z.; Qi, J.; Li, Z.; Min, X.; Yue, Z.; Fan, G. Enhanced mechanical properties of CNT/Al composite through tailoring grain interior/grain boundary affected zones. *Compos. Part B Eng.* **2021**, *223*, 109133. [CrossRef]
44. Huang, K.; Logé, R. *Zener Pinning*; Elsevier: Amsterdam, The Netherlands, 2016.
45. Fu, Y.; Xu, R.; Yuan, C.; Tan, Z.; Fan, G.; Ji, G.; Xiong, D.-B.; Guo, Q.; Li, Z.; Zhang, D. Strain Rate Sensitivity and Deformation Mechanism of Carbon Nanotubes Reinforced Al Composites. *Metall. Mater. Trans. A* **2019**, *50*, 3544–3554. [CrossRef]
46. Ma, P.-C.; Siddiqui, N.A.; Marom, G.; Kim, J.-K. Dispersion and functionalization of carbon nanotubes for polymer-based nanocomposites: A review. *Compos. Part A Appl. Sci. Manuf.* **2010**, *41*, 1345–1367. [CrossRef]
47. Jiang, L.; Li, Z.Q.; Fan, G.L.; Cao, L.L.; Zhang, D. Strong and ductile carbon nanotube/Al bulk nanolaminated composites with two-dimensional alignment of carbon nanotubes. *Scr. Mater.* **2012**, *66*, 331–334. [CrossRef]
48. Sadeghi, B.; Qi, J.; Min, X.; Cavaliere, P. Modelling of strain rate dependent dislocation behavior of CNT/Al composites based on grain interior/grain boundary affected zone (GI/GBAZ). *Mater. Sci. Eng. A* **2021**, *820*, 141547. [CrossRef]
49. Saba, F.; Sun, H.; Fan, G.; Tan, Z.; Xiong, D.-B.; Li, Z.; Li, Z. Strength-ductility synergy induced by high-density stacking faults in Al alloy composites with micro/nano hybrid reinforcements. *Compos. Part A Appl. Sci. Manuf.* **2023**, *173*, 107700. [CrossRef]
50. Xi, L.; Feng, L.; Gu, D.; Prashanth, K.G.; Kaban, I.; Wang, R.; Xiong, K.; Sarac, B.; Eckert, J. Microstructure formation and mechanical performance of micro-nanoscale ceramic reinforced Al matrix composites manufactured by laser powder bed fusion. *J. Alloys Compd.* **2023**, *939*, 168803. [CrossRef]
51. He, C.; Zhao, N.; Shi, C.; Song, S. Mechanical properties and microstructures of carbon nanotube-reinforced Al matrix composite fabricated by in situ chemical vapor deposition. *J. Alloys Compd.* **2009**, *487*, 258–262. [CrossRef]
52. Yan, L.P.; Tan, Z.Q.; Ji, G.; Li, Z.Q.; Fan, G.L.; Schryvers, D.; Shan, A.D.; Zhang, D. A quantitative method to characterize the Al4C3-formed interfacial reaction: The case study of MWCNT/Al composites. *Mater. Charact.* **2016**, *112*, 213–218. [CrossRef]

53. Williamson, G.; Smallman, R., III. Dislocation densities in some annealed and cold-worked metals from measurements on the X-ray debye-scherrer spectrum. *Philos. Mag.* **1956**, *1*, 34–46. [CrossRef]
54. Lahiri, D.; Bakshi, S.R.; Keshri, A.K.; Liu, Y.; Agarwal, A. Dual strengthening mechanisms induced by carbon nanotubes in roll bonded aluminum composites. *Mater. Sci. Eng. A* **2009**, *523*, 263–270. [CrossRef]

Disclaimer/Publisher's Note: The statements, opinions and data contained in all publications are solely those of the individual author(s) and contributor(s) and not of MDPI and/or the editor(s). MDPI and/or the editor(s) disclaim responsibility for any injury to people or property resulting from any ideas, methods, instructions or products referred to in the content.

Article

Extrusion-Based Additively Manufactured PAEK and PAEK/CF Polymer Composites Performance: Role of Process Parameters on Strength, Toughness and Deflection at Failure

S. Sharafi [1], M. H. Santare [1], J. Gerdes [2] and S. G. Advani [1,*]

[1] Department of Mechanical Engineering, Center for Composite Materials, University of Delaware, Newark, DE 19711, USA
[2] Weapons Sciences Division, DEVCOM ARL, Aberdeen Proving Ground, Harford County, MD 21005, USA
* Correspondence: advani@udel.edu

Abstract: Poly aryl-ether-ketone (PAEK) belongs to a family of high-performance semicrystalline polymers exhibiting outstanding material properties at high temperatures, making them suitable candidates for metallic part replacement in different industries such as aviation, oil and gas, chemical, and biomedical. Fused filament fabrication is an additive manufacturing (AM) method that can be used to produce intricate PAEK and PAEK composite parts and to tailor their mechanical properties such as stiffness, strength and deflection at failure. In this work, we present a methodology to identify the layer design and process parameters that will have the highest potential to affect the mechanical properties of additively manufactured parts, using our previously developed multiscale modeling framework. Five samples for each of the ten identified process conditions were fabricated using a Roboze-Argo 500 version 2 with heated chamber and dual extruder nozzle. The manufactured PAEK and PAEK/carbon fiber samples were tested until failure in an Instron, using a video extensometer system. Each sample was prepared with a speckle pattern for post analysis using digital image correlation (DIC) to measure the strain and displacement over its entire surface. The raster angle and the presence of fibers had the largest influence on the mechanical properties of the AM manufactured parts, and the resulting properties were comparable to the mechanical properties of injection molded parts.

Keywords: high temperature polymer composites performance; additive manufacturing (FFF); design of experiment using multiscale modeling; tensile testing using digital image correlation (DIC); material characterization (DSC,TGA); PAEK and PAEK/CF fracture toughness

Citation: Sharafi, S.; Santare, M.H.; Gerdes, J.; Advani, S.G. Extrusion-Based Additively Manufactured PAEK and PAEK/CF Polymer Composites Performance: Role of Process Parameters on Strength, Toughness and Deflection at Failure. *J. Compos. Sci.* **2023**, *7*, 157. https://doi.org/10.3390/jcs7040157

Academic Editor: Yuan Chen

Received: 1 March 2023
Revised: 29 March 2023
Accepted: 4 April 2023
Published: 11 April 2023

Copyright: © 2023 by the authors. Licensee MDPI, Basel, Switzerland. This article is an open access article distributed under the terms and conditions of the Creative Commons Attribution (CC BY) license (https://creativecommons.org/licenses/by/4.0/).

1. Introduction

Since the advent of 3D printing in different industries, the technology has seen increasing use in the past decade for either prototyping or part replacement. Due to the layer-by-layer manufacturing technique, one can tune the mechanical, biocompatibility or even surface properties at the microscale to tailor macroscale properties at a level not possible with traditional manufacturing methods such as extrusion, injection or compression molding and pultrusion. The material, chemical and physical properties as well as flow behavior and solidification characteristics affect the overall layer properties. 3D printers with environmental temperature control at the layer scale offer the unique capability to ensure thermal stability and layer adhesion while maintaining minimal warpage and shrinkage. Low-temperature materials such as ABS, PLA and PC can be easily used for the purpose of prototyping and proof of concept without the need to optimize print parameters or use sophisticated printers with environment control. Process optimization methods for low-temperature carbon fiber polymer composites such as CF reinforced ABS or PLA with different fiber contents have been the subject of many studies for part replacements [1–4]. However, there is still a need to optimize mechanical properties for high-temperature materials such as PAEK polymer which can be additively manufactured in temperature-controlled FDM printers [5].

The majority of research conducted in characterizing the behavior of high-temperature polymers is focused on understanding their material behavior, including the rheology, wettability and solidification of the extrudate layers which results in the formation of bead-to-bead bonding and layer-to-layer adhesion [6–12]. This process is complex due to existence of various challenges affecting material behavior at elevated temperatures, when molten polymer is deposited through a small nozzle orifice with a diameter ranging from 0.1 mm to 1.2 mm. The optimal print speed for a selected print temperature is defined through the relationship between viscosity and temperature [13]. The failure to consider melt flow behavior may result in a wide range of responses, including under-extrusion, nozzle cloggage, a rough surface finish, dimensional inaccuracy and low mechanical performance [10]. Another challenge is to consider the nature of polymer melt due to the orientation of the polymeric chains as soon as it extrudes out of the nozzle. Additionally, it is important to consider the wettability of the polymer chains and their interdiffusion behavior to ensure minimal residual stress and warpage [6]. Other limiting factors are solidification and bond adhesion, which are affected by the polymer chain's interdiffusion or interlocking phenomena, coupled with ability to form crystallites in the case of semicrystalline polymers in which the cooling rate can greatly affect the degree of crystallinity [14]. In general, higher deposition temperatures are usually favorable due to the lowering of the viscosity, while faster cooling rates such as rapid quenching limit the formation of perfectly ordered crystalline regions [11]. However, the addition of carbon fiber assists mechanical properties but hinders processing at lower temperatures. Thus, the processing conditions that influence the rheological behavior of the material are critical, and should be taken into account to ensure fewer defects and better bond formation [13].

Most widespread applications of PAEK and PAEK with carbon fibers (PAEK/CF) are in the aviation industry, where replacement of metallic parts is favored because AM of composites can save weight, time and cost by avoiding traditional tooling requirements such as injection molds. The challenge is to maintain the equivalent mechanical performance of the PAEK/CF additively manufactured parts. With the emergence of new machines capable of 3D printing PAEK polymer and polymer composites, application of this class of polymers have extended beyond aviation industries to other sectors such as chemical plants, oil and gas and electronic sectors which benefit from the fact that this group of polymers are chemically, biologically and electronically inert [15]. The most recent applications are focused in the biomedical industry, where biocompatible PAEK polymers are FDA approved, with proven high-performance properties in surgical tool manufacturing applications alongside orthopedic implants with tunable properties [3,4,16].

Trial and error AM experimentation for optimizing the process parameters of PAEK polymers during fabrication is an inefficient and time-consuming approach [17,18]. There is a lack of well-defined and comprehensive methods to predict macroscopic response based on microscopic parameters and optimize them to improve mechanical performance [5]. A recent review article [19] indicated a broad experimental approach to improving PAEK mechanical performance by modifying the surface or addition of different reinforcements such as carbon fiber, ceramic or even metallic-based particles. However, layer design and process parameters during AM are not thoroughly considered and investigated. The fused filament additive manufacturing process parameters that will play a role are either environmental or 3D printer parameters such as (1) nozzle diameter, (2) nozzle shape, (3) oven temperature (i.e., bed temperature), (4) deposition speed and (5) fan speed, or slicing parameters such as (1) raster angle (the angle between two consecutive beads), (2) infill density (how much material is deposited in each layer), (3) infill shape (layer architecture and porosity), (4) layer height (thickness of each layer), (5) infill overlap (the overlap between exterior seam for each layer and infill structure), (6) extrusion width (the width of extrudate leaving the nozzle), (7) shells (the number of continuous lines forming perimeter of each layer), and (8) support (the number of homogenous +/−45 degree 100%

infill layers at the top and the bottom of each part). Small changes in nozzle shape [20] and diameter have previously been shown to have a negligible effect [11]; therefore, we have not considered them in our analysis. However, we have considered the change in the layer height, which indirectly has a similar effect as the nozzle size. The deposition speed affects the temperature change during processing, and is considered to have a similar effect on mechanical properties as temperature. Therefore, considering the effect of deposition speed and layer height indirectly for oven temperature and nozzle size, only nine out of the thirteen aforementioned parameters are needed to investigate the direct effect on the mechanical performance of fused filament fabricated parts.

Conducting experiments for all nine of these parameters using full fractional factorial with three intensities (low, medium, high) would require 3^9 (19,683) experiments. Considering the redundancy present in the full factorial design, we decided to employ a screening method to identify critical parameters that have the most influence on the properties. Screening methods based on Plackett–Burman or on Resolution III fractional factorial design focus only on the main parameters or any confounding variables without considering their intensities [21]. For instance, the Plackett–Burman approach suggests the choice of any number of experiments divisible by four, while the Resolution III fractional factorial design only focuses on main factors with correlation factors [22]. Neither of these screening methods are ideal for the present study, as optimizing parameters and their intensities is not feasible considering the required number of experimental studies. Therefore, there is a need for an offline optimization tool to help down-select parameters and their intensities prior to any attempt to design an experiment [23].

Our modeling framework, unlike other methods [24–26] which are based on geometry simplification and multiple test trials to define representative volume elements (RVEs) for a specific design, uses the actual geometry of the part and loading conditions to predict the macroscopic response [27]. The other drawbacks of previous modeling methods lie in their applications, which are limited due to the unknown nature of load transfer and interface strength and lack direct correlation between experimental results and the proposed modeling framework. Our formulation is based on a constitutive, phenomenological, continuum-based model which only requires two sets of experiments for calibration; one to define the RVE and one to assign the material model at the RVE level. Then, the calibrated model can be used to optimize the target macro-property. In the current study, the target property chosen is deflection at failure, and it is optimized by varying the processing parameters.

In this paper, by using the proposed modeling framework, we conducted an offline numerical study to design the part and down-select the key processing parameters that have the greatest effect on the target property (deflection at failure). Then, we fabricated ten sets of dog bone samples for experimental characterization. In each set, only one of the ten parameters was varied. We conducted tension tests along with digital image correlation (DIC) for all the samples. This allowed us to determine the differences in elastic properties over the assigned gauge length, and identify the process parameters that have the largest effect on these properties. Finally, the results are presented, followed by discussion and conclusion.

2. Offline Study Using Multiscale Modeling Framework

Using our previously developed multiscale modeling approach [27], we have considered three intensities for each parameter in a sensitivity analysis of a total of six layer design parameters out of nine target parameters, as shown in Table 1. In fact, the effect of environmental parameters such as bed and nozzle temperature as well as fan speed are excluded, because creep studies and crystallization kinetics would be needed to develop a comprehensive FE model to translate the effect of these parameters at multiscale. Therefore, multiscale modeling that uses material properties assigned to the RVE at the microscale, combined with previously set boundary condition (B.C.) in ANSYS software, is used to solve for the deflection at failure for each corresponding process design. The G-code, which defines the processing procedure, is developed using the selected processing parameters and their intensities, with the aim of predicting the resulting deflection at failure.

Table 1. Optimized layer design parameters using multiscale modeling for three intensities, in which the one in bold is chosen as the optimum intensity.

Parameters	Intensity	Defection at Failure (mm)
Extrusion width %	60	6.25
	105	7.5
	120	8.4
Infill overlap %	10	12.7
	90	7.5
	60	10.1
Layer height (nozzle size) mm	0.1	7.98
	0.25	7.58
	0.3	7.44
Infill shape	Wiggle	6.95
	Rectilinear	6.63
	Full honeycomb	7.83
Solid layers	3	7.2
	0	8.67
	2	7.82
No. of shells	2	11.31
	3	8.43
	0	13.12

3. Design of Experiment with Down Selected Parameters

Our multiscale model predictions, as discussed in the previous section, allowed us to identify parameters that most influence the deflection at failure and find their optimal values. Another important objective in our offline analysis was to identify the best combination of parameters to achieve homogenous layer architecture, and therefore properties to produce parts that are more comparable with traditional manufacturing processes such as injection modeling. Using this approach, we conducted an experimental study with the down-selected parameters listed in Table 2.

Table 2. The highlighted parameter is the only one changed for ten experimental process conditions. The first process condition is conducted with optimal intensity.

Condition Process	Temp Nozzle °C	Speed (mm/s)	Raster Angle (Z)	Density Infill (%)	Infill Shape	Layer Height (mm)	Infill Over-lap	Extrusion Width (%)	Bed Temp °C	Number of Shells
1	430 °C	2200	0/90	100%	Rectilinear	0.21	90%	105	160 °C	3
2	430 °C	2200	+−45	100%	Rectilinear	0.21	90%	105	160 °C	3
3	430 °C	2200	0/90	70%	Rectilinear	0.21	90%	105	160 °C	3
4	430 °C	2200	0/90	100%	Full Honeycomb	0.21	90%	105	160 °C	3
5	430 °C	2200	0/90	100%	Rectilinear	0.12	90%	105	160 °C	3
6	430 °C	2200	0/90	100%	Rectilinear	0.21	10%	105	160 °C	3
7	430 °C	2200	0/90	100%	Rectilinear	0.21	90%	120	160 °C	3
8	430 °C	2200	0/90	100%	Rectilinear	0.21	90%	105	200 °C	3
9	450 °C	2200	0/90	100%	Rectilinear	0.21	90%	105	160 °C	3
10	450 °C	2200	0/90	100%	Rectilinear	0.21	90%	105	200 °C	3

4. Experimental Method

4.1. Preparation of PAEK and PAEK/CF (10% wt.) Samples

A high-temperature Roboze-Argo 500 version 2, AM machine with heated chamber and dual extruder nozzle with 0.4 mm diameter was used to prepare dog bone samples for testing to determine the mechanical properties and deflection at failure. Solidworks software was used to design and prepare the STL digital files to 3D print the samples. Then, the generated STL files were sliced using Simplify3D software to populate the G-code files, which were loaded into the Roboze high temperature 3D printer to additively manufacture the dog bone samples. The nominal dimensions of the samples manufactured for the tensile test are based on ASTM D638-Type V [28], with a thickness of 3.18 mm.

4.2. Material Characterization

The thermal transition and enthalpies of pure PAEK and PAEK/CF filaments as well as thermal history of selected processes are listed in Table 3, alongside the thermogravimetric analysis.

Table 3. Average thermal transitions and enthalpies for PAEK and PAEK/CF and their thermal history under different processing conditions. All samples were manufactured with oven temperature of 160 °C, except process 4, for which the oven temperature was 180 °C.

#	Material Condition	Glass Transition	Melting Point	ΔHf (J/g)	χc (%)
1	PAEK Filament (1.75 mm diameter)	146.0 °C	338.0 °C	42.3	32.0
2	PAEK/CF Filament (10 wt%) (1.75 mm diameter)	138.0 °C	345.0 °C	31.1	23.5
3	Process 1-PAEK-sample	146.2 °C	339.5 °C	35.3	26.7
4	Process 1-PAEK-sample	146.5 °C	341.6 °C	32.7	24.7
5	Process 7-PAEK-sample	149.3 °C	341.7 °C	40.5	30.7
6	Process 9-PAEK-sample	147.5 °C	340.0 °C	34.0	25.8
7	Process 10-PAEK-sample	148.2 °C	340.8 °C	36.9	28.0
8	Process 9-PAEK/CF-sample	143.5 °C	340.1 °C	36.0	27.2
9	Process 10-PAEK/CF-sample	143.9 °C	340.8 °C	38.2	29.0

4.2.1. PAEK and PAEK/CF Rheology

Rheological properties are benchmarked from manufacturer thermal analysis results compared to the recently published literature [29] to identify the category of PAEK polymer, which is PAEK-A5. We conducted DSC analysis to compare thermal transitions and confirmed the type of PAEK polymer used, as shown in Figure 1. Published viscoelastic data [29] at 370 °C–390 °C indicate its stability at melt state; however, at temperatures beyond 390 °C, more branching and improvement in processability is expected. TGA analysis was conducted to confirm that processing of 3D printed PAEK or PAEK/CF in the range of 420 °C–450 °C did not show evidence of degradation. To additively manufacture at elevated temperatures, the knowledge of thermal history and the role of cooling rate on degree of crystallinity would be essential.

4.2.2. Thermal Analysis Using DSC

DSC analysis is a major characterization technique to study effect of processing parameters as well as cooling rate on degree of crystallinity. Measuring the area under the melting curve is a standard method to evaluate thermal history of printed samples made under different processing conditions shown in Table 2. The effect of cooling rate on degree of crystallinity of PAEK and PAEK/CF samples, shown in Figure 2, can form the basis to develop a crystallization kinetics model but is beyond the scope of this study.

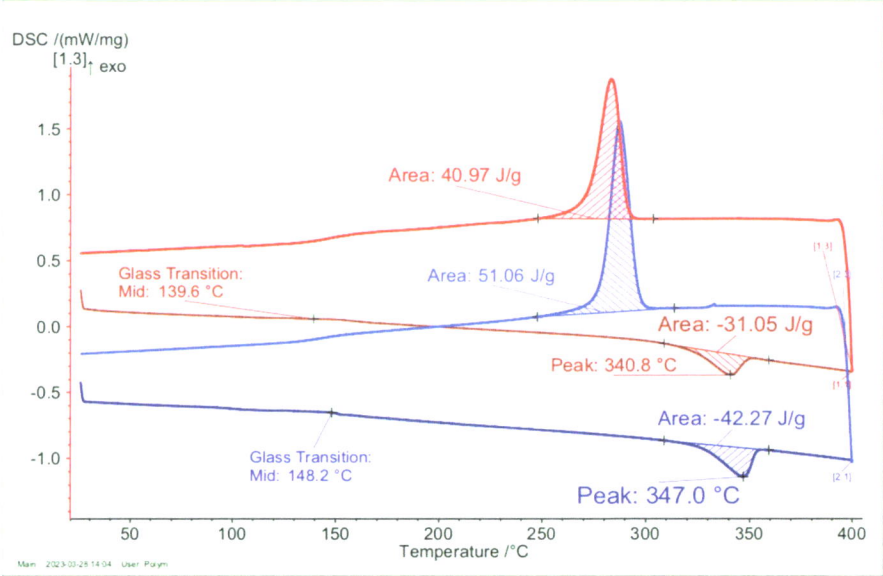

Figure 1. DSC curves for PAEK and PAEK/CF filaments at a heating rate of 5 K/min are shown in red and blue, respectively, followed by cooling curves at 20 K/min. The area under the melting curve can be used to calculate the degree of crystallinity.

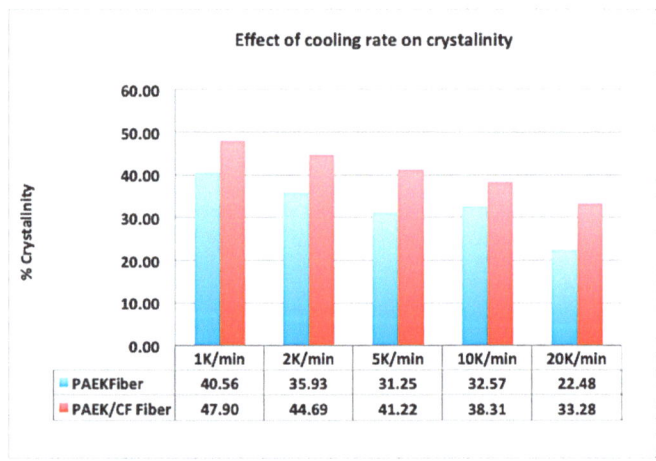

Figure 2. The percent crystallinity for different cooling rates is calculated by measuring the area under the curve as shown in Figure 1 (J/g).

Details of sample preparation: PAEK and PAEK/CF filaments are tested using a DSC Proteous 214 machine using encapsulated samples 9–11 mg in pierced aluminum pans. After heating the samples at 5 K/min from room temperature to 400 °C, followed by a 5 min isothermal step to ensure that samples are completely melted, they are then cooled at 20 K/min under a nitrogen flow rate of 40 mL/min. This procedure was used to obtain the thermal properties of various samples with different thermal histories, as shown in Table 3.

Figure 1 shows the DSC curves for PAEK and PAEK/CF filaments. To calculate the degree of crystallinity, the measured area under the melting curves is divided by heat of fusion for the complete crystalline sample, which is 131.9 J/g, from the Roboze data sheet [30].

Table 3 shows the thermal transitions of filaments from the manufacturer data sheet, which are consistent with the DSC results from Figure 1. The average thermal history of three samples from each material condition is reported in rows 3–9. The results in row 4 compared to row 3 show that increasing the oven temperature from 160 °C to 180 °C, which will promote annealing, has a negligible effect on the degree of crystallinity. On the other hand, the highest degree of crystallinity is obtained for row 5, in which a higher compaction pressure between beads as a result of an increase in extrusion width promoted crystallinity. A similar trend is expected in the case of PAEK/CF, as higher crystallinity merely elevates performance and promotes inhomogeneity in 3D-printed parts, and facilitates flaw-driven phenomena such as fractures, which result in lower deflection at failure and toughness.

Based on our DSC analysis at different cooling rates, as shown in Figure 2, the highest degree of crystallinity is related to the cooling rate of 1 K/min both for PAEK and PAEK/CF filaments. Degree of crystallinity for this set of samples that have undergone repetitive heating and cooling cycles as shown in Figure 2 is lower than the one studied in Table 3 with only one heating and cooling cycle (rows 1 and 2). The lower degree of crystallinity may be due to disrupted polymer chain arrangement specially for PAEK/CF polymer composite which requires further investigation.

Cooling rates slower than 20 K/min is not achievable with the current print speed of 40 mm/s, since it takes less than 40 s for the polymer melt to leave the nozzle. This in fact shows that in the best case scenario at our suggested print speed, as shown in Table 3 for one cycle of heating and cooling of PAEK and PAEK/CF, the degree of crystallinity is 32.0 and 23.5 percent respectively.

To summarize, the cooling rate as well as presence of any additional component such as carbon fiber in the polymer melt can change crystallinity in traditional manufacturing [13] with lower porosity. However, finding a direct correlation between increments in crystallinity with mechanical properties in 3D-printed parts is merely plausible. There is a need for comprehensive study in this field to understand how mechanical properties are affected by crystallinity and interface strength in 3D-printed parts. We believe the presence of an excessive number of voids and weak interfaces cancels the positive impact that improved crystallinity can have. Some other factors such as higher extrusion width can change the pressure inside each layer and elevate crystallinity as well, as seen in the case of process 7 for the PAEK samples.

4.2.3. Thermogravimetric Analysis

The evolution of dried PAEK vs. PAEK/CF samples under high-temperature conditions is evaluated through a heat ramp from 250 °C to 1000 °C, at a rate of 10 K/min, using TG209F1 Libra. In both cases, the tests are performed under nitrogen atmospheres at 10 mL.min^{-1}, for 1 h and 40 min. Encapsulated PAEK and PAEK fiber samples ranged from 20 to 30 mg, respectively.

As shown in Figure 3, the initiation of degradation for PAEK and PAEK/CF occurred at 504.2 °C and 499.2 °C, respectively; however, at temperatures beyond 499.2 °C, samples showed 5% loss in weight, which is considered the initiation of degradation. The addition of CF favors degradation by shifting the degradation temperature ~5 degrees, as improved thermal conduction promotes heat absorption. This temperature gap increases to almost 13.6 degrees at the final stage of decomposition, in which more CF ash remains, thereby aiding heat transfer at higher rates. The results show evidence of decomposition at temperatures above 504.2 °C when samples are kept at these temperatures for more than 50 min. In the majority of 3D printing cases, the time it takes for polymeric filaments to stay at deposition temperatures in the range 420 °C–450 °C is less than few minutes. In the case of PAEK/CF composites, it is not advisable to print samples at temperatures above 450 °C due to their complex rheological behavior [29] and the proximity to the degradation temperature. Therefore, decomposition is not of concern within a defined range for nozzle temperature; however, it is essential to know decomposition temperatures and the duration of stability of PAEK polymer composites, due to their application at elevated temperatures.

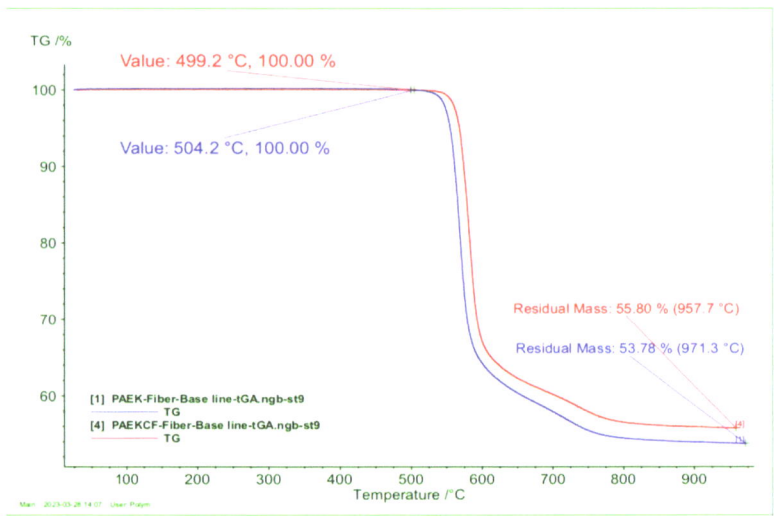

Figure 3. TGA curves of PAEK vs. PAEK/CF from 25 to 1000 C at 10 K/min in an inert atmosphere.

4.3. Tensile Testing Using DIC-Equipped Instron

A total of five tensile test samples of PAEK polymer from each print condition listed in Table 2 were fabricated. Following addition of carbon fiber to the polymeric matrix, prints for the process 9 and process 10 were replicated, and parts at 470 °C with different raster angles and oven temperatures were printed in order to justify our TGA analysis (Figure 3) and evaluate the performance of parts made beyond our suggested optimum deposition temperature.

The tensile tests were performed using an Instron 4448 machine with a 10 kN load cell capacity at a rate of 1 mm/min. The ASTM standard [31] for V-type samples with which failure happens at less than 5 min suggests using slightly lower strain rates to allow for 3D-printed parts with a high degree of inherent porosity compared to traditional manufacturing to achieve better polymer chain stretching and real application. The lower strain rates help to better understand material behavior at the high temperatures at which polymeric relaxation mechanisms activate.

The local strain was measured with DIC using a video extensometer and following the steps shown in Figure 4; the final values of the deflection at failure were calculated from the Instron displacement data. The Instron's recorded force and the nominal cross-sectional area of the samples were used to calculate engineering stress at each time step. The video extensometer and digital image correlation (DIC) software was used to measure the true strain of samples, which was then used to calculate the modulus. The average result of the five samples with standard deviations are reported for each process condition. We manufactured and tested only V-type samples. However, we confirmed through our DIC measurements that the strain was uniform across the sample during mechanical testing of the samples until failure. In addition, from Table 3, we can conclude that the effect of thermal history is relatively small.

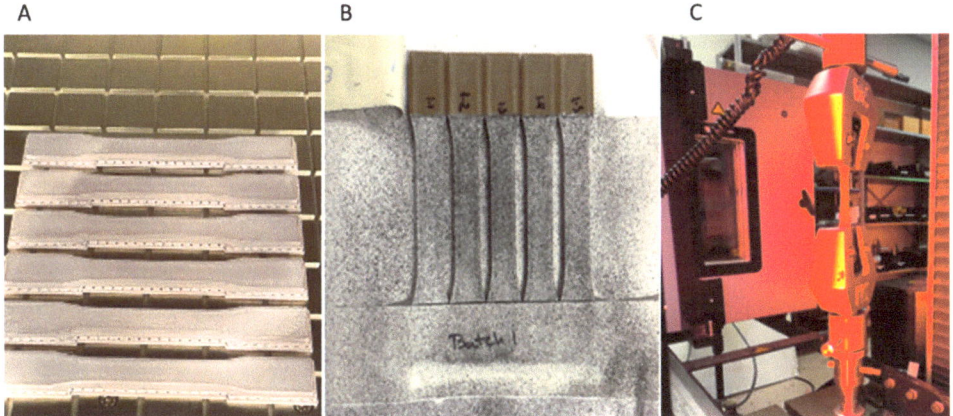

Figure 4. (**A**) printed dog bone coupons; (**B**) coupons were painted with a dot pattern so DIC could measure strain variations; (**C**) DIC-equipped Instron to measure strain using a video extensometer.

5. Results and Discussion

5.1. Effect of Temperature and Deposition Speed on PAEK Polymer Composites

Most of the reported process optimization efforts in the literature are limited to evaluating the effect of temperature and print speed, and do not consider optimizing the layer parameters; this motivates us to further broaden our study in this field. In those studies, the elastic modulus for PAEK is reported to be between 2.3 and 3.0 GPa, with 4.3 to 4.9 +/− 0.24 mm elongation at break [5] and a flexural modulus of 2.43 GPa [3], based on the processing parameters such as raster angle and print speed that were used. Additionally, as suggested by the authors, the best 3D-printed parts are processed at 400 °C, with a maximum modulus of 3.0 GPa and a strength of 77 MPa, compared to the injection molded parts with a maximum modulus and a strength of 3.6 GPa and 98 MPa, respectively [5]. Another important factor that affects the part performance discussed in the literature is annealing, in which the annealed parts showed comparable results to injection molded parts, with a maximum modulus of 3.55 GPa and 97 MPa strength [5] for the samples with a lower degree of crystallinity; however, we have not studied the effect of annealing in manufactured parts. Although the effect of temperature on crystallinity and degradation in PAEK and PAEK-CF polymer composites and their performance has been broadly studied [32], the role of deposition speed is widely neglected because it can be correlated to the temperature change. At higher deposition speeds, parts remain in a molten state for an extended time, considering the constant cooling rate. Although the addition of the next layer at elevated temperatures favors better interlayer adhesion and bond strength [33], there is concern for the possibility of degradation at temperatures beyond 450 °C, as shown in TGA; degradation is seen after 50 min at 499.2 °C. Therefore, to achieve better dimensional accuracy and to avoid degradation, for parts that are made at higher temperatures beyond 400 °C, a lower deposition speed is advisable to improve bond adhesion and aide in decreasing the entrapped air during the printing of PAEK polymer composites.

5.2. Multiscale Modeling for Layer Design Optimization

The proposed modeling framework [27] can be applied to any printer once it is characterized, and our main contribution is the development of a methodology to down-select and relate the important processing conditions to mechanical properties. The model prediction results for each process condition, shown in Table 2 and used as an optimum method for down-selecting parameters, are shown in Figure 5 and compared with the outcomes of our experimental studies. Two particular parameters, i.e., the number of outline shells and support layers which we had optimized using our proposed multiscale modeling, are not varied in this study to ensure greater part homogeneity. All optimized values for parameters are used in process condition 1 as a baseline, and other process conditions are defined by varying one parameter at a time, as listed in Table 2. For instance, for process condition 2, only the raster angle is changed from the 0/90th degree in process 1 to the 45/−45th degree, which results in higher toughness and deflection at failure.

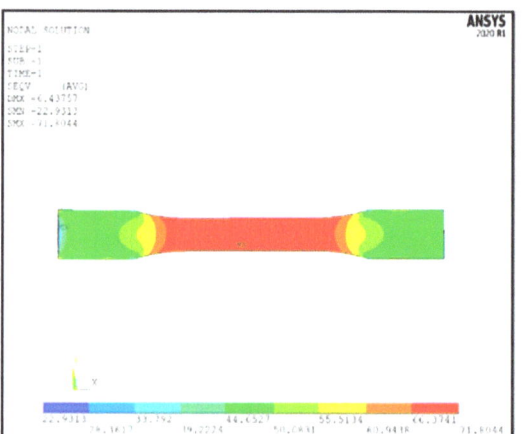 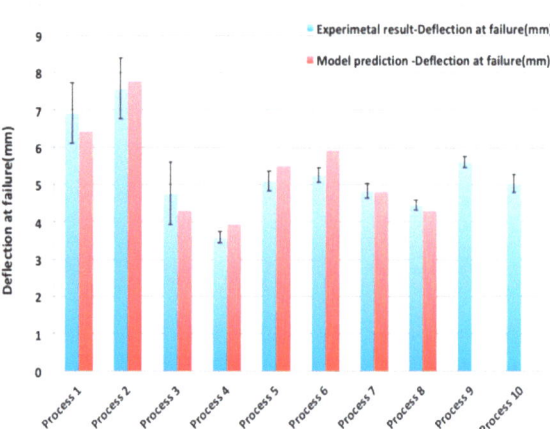

Figure 5. Simulation results for predicted deflection at failure for process is shown on the left (6.44 mm) and other processes are similarly modeled and compared with experimental results, as shown on the right; however, processes 9 and 10 are left out, as the effect of temperature is not considered due to lack of creep studies. Simulation results are based on a material model which may have a 1 to 7.5% error [27].

Infill density and shape, introduced in process condition 3 and 4, affect the amount of material incorporated into each layer, which impacts mechanical properties. In general, some infill patterns such as full hexagonal (HC) are chosen to introduce a higher degree of porosity compared to the compact rectilinear shape (RC). The higher infill density in RC suggests higher strength and stiffness due to lower porosity.

The role of the interfaces formed between layers in 3D-printed parts, which act similarly to weld lines in traditional manufacturing, is not well studied. By changing the layer height or nozzle size, the number of interfaces formed is determined for a particular thickness part. Considering the relationship between layer height and nozzle diameter, the optimum value for layer height is in the range of [50–60%] of the nozzle size. For instance, the optimum layer height for the 0.4 mm nozzle to achieve high strength and adhesion is between (0.2–0.24) mm. In general, the greater the number of weld lines, the lower the part's strength and stiffness. Therefore, larger layer height implies higher strength for the same thickness of the part. Raster angle and layer height can change the shape of voids and number of weld lines formed at each cross-section, as represented in Figure 6. The top section of this figure is related to a sample made with a finer layer height at 0/90 alternating degrees (process 5; Table 2), while the bottom part shows the sample with 45/−45 alternating angles as well as a larger layer height. A higher strength and stiffness as well as deflection at failure were achieved for the bottom image (process 2; Table 2) with a lower number of weld lines.

The parameter studied in process condition 7, referred to as extrusion width in the slicing software, Simplify3D, controls the amount of extrudate deposited at each location; it is controlled by nozzle diameter and extrudate rheology. A higher percentage of extrudate width improves the die-swelling effect in polymers and bead-to-bead compaction; DSC results show that samples made under this condition have higher crystallinity. Another parameter introduced in process condition 8, which is inversely correlated with extrusion width, is infill overlap, a higher value of which provides higher local deposition; therefore, a generally better part performance is implied. A poor choice of values for this parameter can cause inhomogeneity through the thickness, resulting in premature failure and lower overall mechanical properties, as shown in Figure 7.

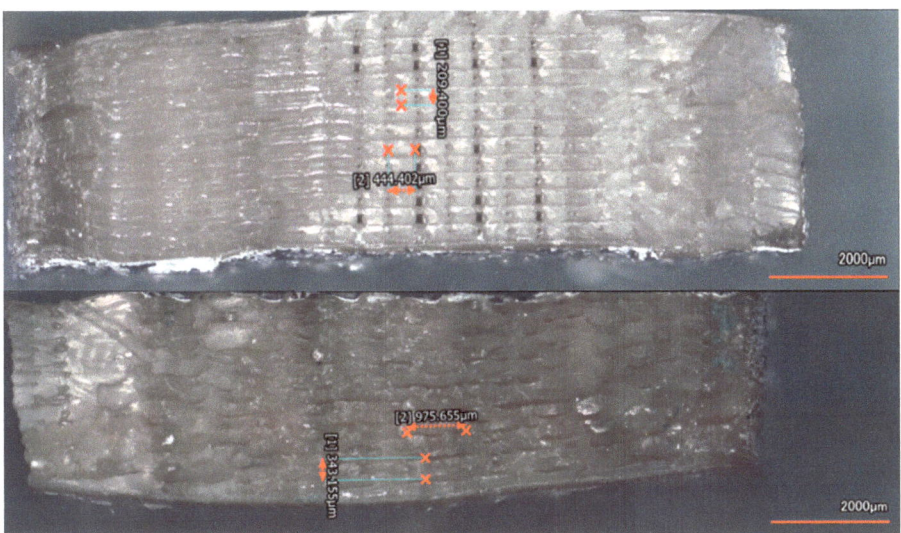

Figure 6. Cross-section of a sample made with 0.12 mm layer height and a 0/90 raster angle, contrasted with a part made with 0.21 mm layer height and a 45/45 raster angle.

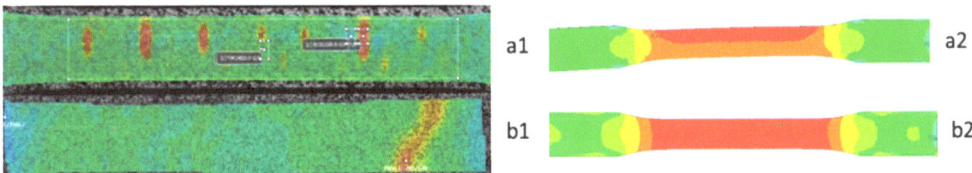

Figure 7. The simulation result for homogenous layer design (**b2**) compared with DIC from experimental results (**b1**). In contrast, DIC for experimental results of the inhomogeneous layer design (**a1**) is compared with simulation results (**a2**).

5.3. Change of Mechanical Properties with Layer Design Parameters

Changes in processing parameters such as infill percentage (process condition 3) and pattern shape (process condition 4) can affect void content by introducing more gaps between the beads, and this can result in a dramatic drop in all aspects of mechanical properties, as shown in Figure 8. Another important aspect of part performance is interface strength, the effect of which is studied by changing the layer height or infill overlap. The former parameter affects the number of interfaces between consecutive layers, and the latter parameter affects the bond formation and adhesion. A 57% decrease in layer height is shown to cause a 10% reduction in overall strength, a 30% decrease in deflection at failure and 20% reduction in tensile toughness. Another contributing factor in void content is the possibility of entrapping air during layer-by-layer deposition. The wider extrusion width used in process condition 7 may contribute to this phenomenon by delaying the cooling process, which results in improved crystallinity, as shown in Table 3. In contrast, increasing deposition temperature, as in process conditions in 9 and 10, helps to reduce viscosity with a similar degree of crystallinity compared to process 7, allowing the release of entrapped air, thereby lowering porosity. This in turn improves strength, toughness, and deflection at failure, and has been shown to have little or no effect on the stiffness, as shown in Figure 8. PAEK polymer is a semicrystalline polymer, and similar to other high-temperature polymers, its toughness reduces with increasing crystallinity [22]. The PAEK polymer processed at 430 °C shows higher ductility and deflection at failure compared to the sample processed at 450 °C. Raster angle, unlike temperature, has a negligible effect on strength of PAEK, and the 0/90 angle has shown consistency in improving strength (~0.5%) compared to the 45/−45th degree.

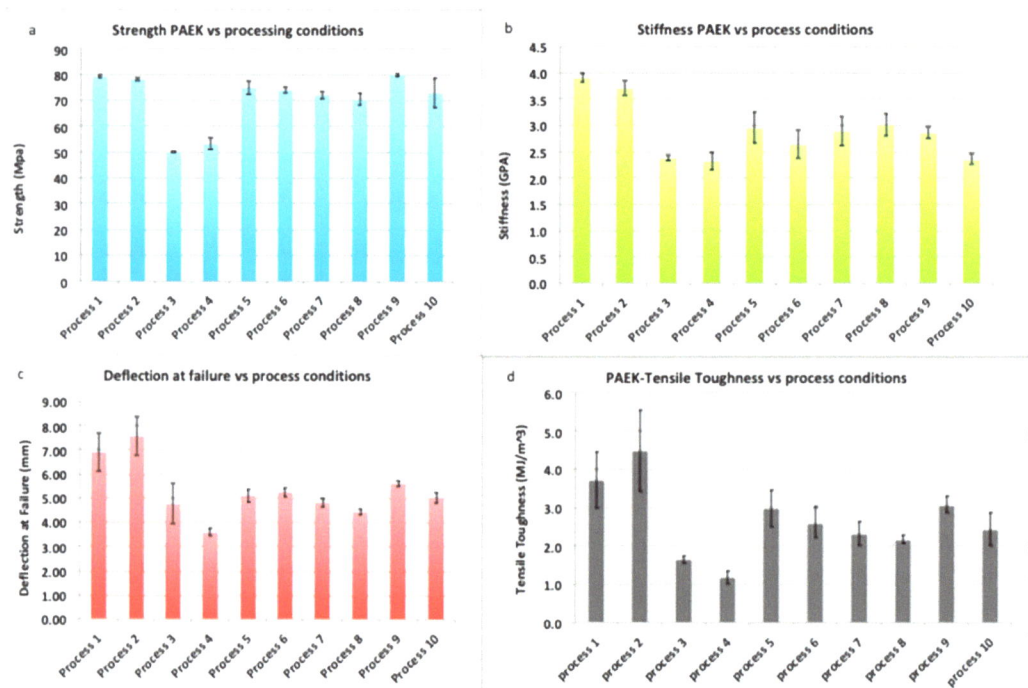

Figure 8. Measured (**a**) strength, (**b**) stiffness, (**c**) deflection at failure, and (**d**) toughness, with standard deviations. The X-axis label refers to the different process conditions, as listed in Table 2.

The PEAK coupons fabricated at 450 °C show little effect due to change in raster angle (80 ± 1.2) MPa. The strength of PAEK polymer parts changes significantly with change in layer density, and less so with a change in raster angle and temperature, as can be seen from Figure 8. Samples with raster angles of 0/90 that are made at 450 °C show the highest strength of (120 ± 2.5) MPa. The deflection at failure is reduced by 50% compared to the PAEK sample made in optimal conditions, as shown in Figure 8, to (3.52 ± 0.012)mm, because of brittleness caused by a higher degree of crystallinity [8].

5.4. Change of Mechanical Properties with Addition of Carbon Fiber and Role of Temperature

The addition of carbon fiber acts as reinforcement for the PAEK matrix, and as expected, increases the stiffness and strength when processing temperature is optimum. As shown in Figure 9, the addition of 10 wt% carbon fiber reinforcement (with an aspect ratio of 26.3) to the PAEK polymer improved mechanical stiffness and strength by 50% and 60%, respectively, over the neat PAEK polymer at 430 °C for raster angles of 0/90 and +−45. This results in a quasi-brittle behavior, with the addition of PAEK/CF showing a lower deflection at failure and toughness compared to neat PAEK samples.

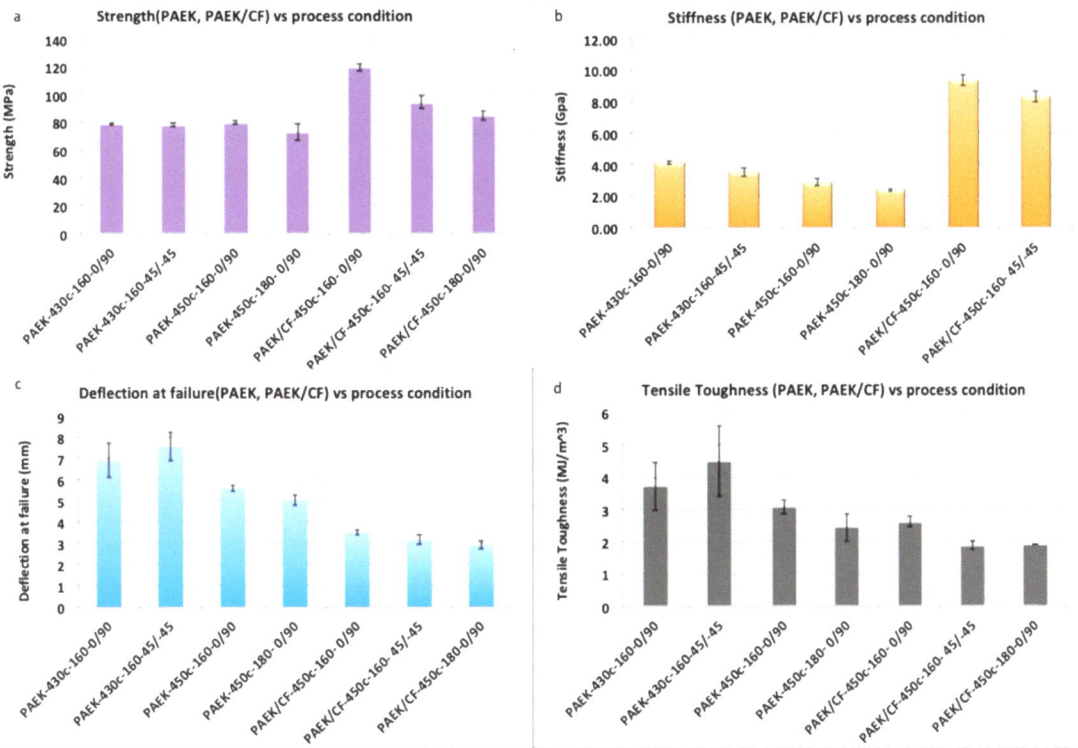

Figure 9. PAEK and PAEK/CF composites' mechanical properties: (**a**) strength, (**b**) stiffness, (**c**) deflection at failure, and (**d**) toughness. The X-axis label shows the performance of PAEK or PAEK/CF at nozzle temeparatures (430 °C and 450 °C) and oven temperatures (160 °C & 180 °C) for different raster angles (45/−45 & 0/90).

On the other hand, the effect of an optimum choice of range of processing temperatures for PAEK/CF polymer composites is shown in Figure 10 to be an essential element impacting mechanical properties. At 450 °C, PAEK/CF samples showed the highest mechanical properties compared to samples made at 470 °C, with increases in strength of 15% and 40% for raster angles +−45 and 0/90, respectively. However, samples made at 470 °C with a 45/−45 raster angle at a lower oven temperature 160 °C showed a 20% improvement in strength compared to PAEK CF made at 450 °C with the same raster angle and oven temperature. However, samples made at 450 °C and with a 45/−45 raster angle at 160 °C oven temperature have the highest strength of all the different conditions.

Considering the mechanical performance of PAEK/CF samples made at 470 °C, we believe PAEK/CF filaments undergo complex rheological behavior during deposition at this temperature; entrapment of more air with a fast speed of deposition results in initiation of a higher degree of porosity and poorer mechanical performance. However, further analysis using characterization methods such as Micro CT and SEM is essential to verify our claim.

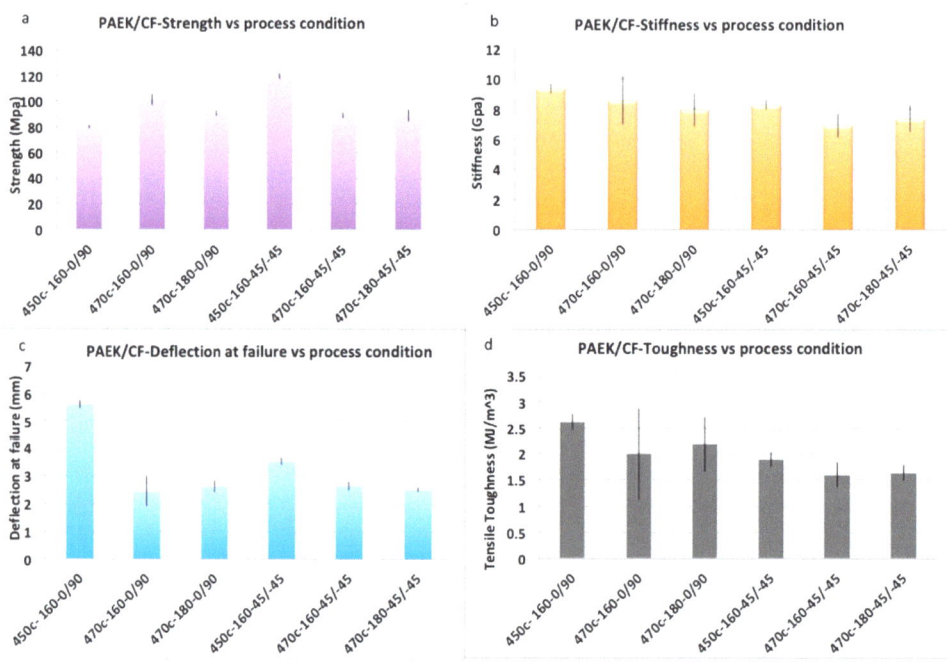

Figure 10. PAEK/CF samples made at 470 °C show a drop in mechanical properties, (**a**) strength, (**b**) stiffness, (**c**) deflection at failure, and (**d**) toughness, compared to samples made at 450 °C. The X-axis label shows the performance of PAEK or PAEK/CF at different nozzle temeparatures (430 °C and 450 °C) and oven temperatures (160 °C & 180 °C), for different raster angles (45/−45 & 0/90).

6. Conclusions

PAEK polymers and their composites have comparable properties to metallic parts, which make them good candidates for weight reduction in many applications. Additive manufacturing of PAEK composites has opened up the possibility of producing complex shapes, with control over their internal features and architecture achievable through the use of multiple processing parameters through slicing software such as Simplify3D to define layer architecture and assembly. To obtain the greatest benefit from AM manufacturing of PAEK polymers, there is a need to control parameters based on the target application to achieve optimum performance. Our modeling framework can relate processing details to local material properties, which can in turn be used to fabricate parts with preferred macro-properties and high repeatability. We have conducted an offline numerical design study using our multiscale modeling approach to identify the effect of each parameter on deflection at failure, and to optimize processing parameters and their intensities prior to the designing of the experiments. The main goal for this step was to achieve maximum homogeneity in the combination of parameters, while obtaining comprehensive knowledge on how to tailor mechanical properties at microscale to achieve high stiffness and strength at the macroscale. Using the optimized process, condition 1, as the baseline process, we down-selected ten experimental process conditions to check their effects on stiffness, strength, toughness and failure. Parameters such as infill density and infill shape as well as higher bed temperature can have an adverse effect on toughness by reducing deflection at failure and stiffness; however, strength has been shown to be relatively independent of porosity and adhesion. The higher content of porosity induced by infill shape or lower infill percentage can also negatively impact the strength of material. In general, failure is driven by both strength along the loading direction and interface adhesion, which are related to the density of material [5], and the number of interfaces and resulting porosity through the thickness. All of these are affected by the control of the manufacturing process.

Author Contributions: Conceptualization, S.S.; Methodology, S.S.; Software, S.S.; Validation, S.S., S.G.A. and M.H.S.; Formal Analysis, S.S., M.H.S. and S.G.A.; Data Curation, S.S.; Writing—Original Draft Preparation, S.S.; Writing—Review & Editing, S.G.A., M.H.S. and J.G.; Visualization, S.S.; Supervision, S.G.A. and M.H.S.; Funding Acquisition, J.G. and S.G.A. All authors have read and agreed to the published version of the manuscript.

Funding: This research was sponsored by DEVCOM ARL and was accomplished under Cooperative Agreement Number W911NF-18-2-0299. The views and conclusions contained in this document are those of the authors and should not be interpreted as representing the official policies, either expressed or implied, of DEVCOM ARL or the U.S. Government. The U.S. Government is authorized to reproduce and distribute re-prints for Government purposes notwithstanding any copyright notation herein.

Data Availability Statement: Data is unavailable to the public due to privacy restrictions.

Acknowledgments: The authors would like to express gratitude to Aristedes Yiournas for his invaluable assistance in extracting the DIC result data. Additionally, thanks to Nicola Caringella, Nic Cantwell, and Josh Elmer from the Roboze Company for their help in troubleshooting the Roboze 3D printer and printing some of the samples.

Conflicts of Interest: The authors have no conflict of interest to declare. All co-authors have seen and agree with the contents of the manuscript. We certify that the submission is original work and is not under review at any other publication.

References

1. Honigmann, P.; Sharma, N.; Okolo, B.; Popp, U.; Msallem, B.; Thieringer, F.M. Patient-specific surgical implants made of 3D printed PEEK: Material, technology, and scope of surgical application. *BioMed Res. Int.* **2018**, *2018*, 4520636. [CrossRef]
2. Harding, M.J.; Brady, S.; O'Connor, H.; Lopez-Rodriguez, R.; Edwards, M.D.; Tracy, S.; Dowling, D.; Gibson, G.; Girard, K.P.; Ferguson, S. 3D printing of PEEK reactors for flow chemistry and continuous chemical processing. *React. Chem. Eng.* **2020**, *5*, 728–735. [CrossRef]
3. Vaezi, M.; Yang, S. Extrusion-based additive manufacturing of PEEK for biomedical applications. *Virtual Phys. Prototyp.* **2015**, *10*, 123–135. [CrossRef]
4. Kurtz, S.M. Chapter 15—Development and Clinical Performance of PEEK Intervertebral Cages. In *Plastics Design Library*, 2nd ed.; Kurtz, S.M., Ed.; William Andrew Publishing: Norwich, NY, USA, 2019; pp. 263–280. ISBN 978-0-12-812524-3.
5. El Magri, A.; El Mabrouk, K.; Vaudreuil, S.; Chibane, H.; Touhami, M.E. Optimization of printing parameters for improvement of mechanical and thermal performances of 3D printed poly (ether ether ketone) parts. *J. Appl. Polym. Sci.* **2020**, *137*, 49087. [CrossRef]
6. Liu, P.; Kunc, V. Effect of 3D printing conditions on the micro- and macrostructure and properties of high-performance thermoplastic composites. In *Woodhead Publishing Series in Composites Science and Engineering*; Friedrich, K., Walter, R., Soutis, C., Advani, S.G., Fiedler, I.H.B., Eds.; Woodhead Publishing: Sawston, CA, USA, 2020; pp. 65–86. ISBN 978-0-12-819535-2.
7. Lepoivre, A.; Boyard, N.; Levy, A.; Sobotka, V. Heat Transfer and Adhesion Study for the FFF Additive Manufacturing Process. *Procedia Manuf.* **2020**, *47*, 948–955. [CrossRef]
8. Yang, D.; Cao, Y.; Zhang, Z.; Yin, Y.; Li, D. Effects of crystallinity control on mechanical properties of 3D-printed short-carbon-fiber-reinforced polyether ether ketone composites. *Polym. Test.* **2021**, *97*, 107149. [CrossRef]
9. Mackay, M.E. The importance of rheological behavior in the additive manufacturing technique material extrusion. *J. Rheol.* **2018**, *62*, 1549–1561. [CrossRef]
10. Das, A.; Gilmer, E.L.; Biria, S.; Bortner, M.J. Importance of polymer rheology on material extrusion additive manufacturing: Correlating process physics to print properties. *ACS Appl. Polym. Mater.* **2021**, *3*, 1218–1249. [CrossRef]
11. de Paula Santos, L.F.; Alderliesten, R.; Kok, W.; Ribeiro, B.; de Oliveira, J.B.; Costa, M.L.; Botelho, E.C. The influence of carbon nanotube buckypaper/poly (ether imide) mats on the thermal properties of poly (ether imide) and poly (aryl ether ketone)/carbon fiber laminates. *Diam. Relat. Mater.* **2021**, *116*, 108421. [CrossRef]
12. Das, A.; Chatham, C.A.; Fallon, J.J.; Zawaski, C.E.; Gilmer, E.L.; Williams, C.B.; Bortner, M.J. Current understanding and challenges in high temperature additive manufacturing of engineering thermoplastic polymers. *Addit. Manuf.* **2020**, *34*, 101218. [CrossRef]
13. Bonmatin, M.; Chabert, F.; Bernhart, G.; Djilali, T. Rheological and crystallization behaviors of low processing temperature poly(aryl ether ketone). *J. Appl. Polym. Sci.* **2021**, *138*, 51402. [CrossRef]

14. Yi, N.; Davies, R.; Chaplin, A.; McCutchion, P.; Ghita, O. Slow and fast crystallising poly aryl ether ketones (PAEKs) in 3D printing: Crystallisation kinetics, morphology, and mechanical properties. *Addit. Manuf.* **2021**, *39*, 101843. [CrossRef]
15. Francis, J.N.; Banerjee, I.; Chugh, A.; Singh, J. Additive manufacturing of polyetheretherketone and its composites: A review. *Polym. Compos.* **2022**, *43*, 5802–5819. [CrossRef]
16. Berg-Johansen, B.; Lovald, S.; Altiok, E.; Kurtz, S.M. Chapter 17—Applications of Polyetheretherketone in Arthroscopy. In *Plastics Design Library*, 2nd ed.; Kurtz, S.M., Ed.; William Andrew Publishing: Norwich, NY, USA, 2019; pp. 291–300. ISBN 978-0-12-812524-3.
17. Wang, P.; Zou, B.; Ding, S.; Li, L.; Huang, C. Effects of FDM-3D printing parameters on mechanical properties and microstructure of CF/PEEK and GF/PEEK. *Chin. J. Aeronaut.* **2021**, *34*, 236–246. [CrossRef]
18. Miri, S.; Kalman, J.; Canart, J.-P.; Spangler, J.; Fayazbakhsh, K. Tensile and thermal properties of low-melt poly aryl ether ketone reinforced with continuous carbon fiber manufactured by robotic 3D printing. *Int. J. Adv. Manuf. Technol.* **2022**, *122*, 1041–1053. [CrossRef]
19. Pazhamannil, R.V.; Edacherian, A. Property enhancement approaches of fused filament fabrication technology: A review. *Polym. Eng. Sci.* **2022**, *62*, 1356–1376. [CrossRef]
20. Papon, M.E.A.; Haque, A.; Sharif, M.A.R. Effect of nozzle geometry on melt flow simulation and structural property of thermoplastic nanocomposites in fused deposition modeling. In Proceedings of the American Society for Composites, Thirty-Second Technical Conference, Tucson, AZ, USA, 22–25 October 2017.
21. Samuels, M.L.; Witmer, J.A.; Schaffner, A.A. *Statistics for the Life Sciences*, 7th ed.; Pearson: London, UK, 2016.
22. Mehmet-Alkan, A.A.; Hay, J.N. The crystallinity of poly (ether ether ketone). *Polymer* **1992**, *33*, 3527–3530. [CrossRef]
23. Oehlert, G. *A First Course in Design and Analysis of Experiments*; W.H Freeman&Co: New York, NY, USA, 2000.
24. Hasanov, S.; Gupta, A.; Alifui-Segbaya, F.; Fidan, I. Hierarchical homogenization and experimental evaluation of functionally graded materials manufactured by the fused filament fabrication process. *Compos. Struct.* **2021**, *275*, 114488. [CrossRef]
25. Nasirov, A.; Hasanov, S.; Fidan, I. Prediction of mechanical properties of fused deposition modeling made parts using multiscale modeling and classical laminate theory. In Proceedings of the 30th Annual International Solid Freeform Fabrication Symposium-An Additive Manufacturing Conference, Austin, TX, USA, 12–14 August 2019; Volume 1376.
26. Nasirov, A.; Gupta, A.; Hasanov, S.; Fidan, I. Three-scale asymptotic homogenization of short fiber reinforced additively manufactured polymer composites. *Compos. Part B Eng.* **2020**, *202*, 108269. [CrossRef]
27. Sharafi, S.; Santare, M.H.; Gerdes, J.; Advani, S.G. A multiscale modeling approach of the Fused Filament Fabrication process to predict the mechanical response of 3D printed parts. *Addit. Manuf.* **2022**, *51*, 102597. [CrossRef]
28. *ASTM D638-14*; Standard Test Method for Tensile Properties of Plastics. ASTM International: West Conshohocken, PA, USA, 2014.
29. White, K.; Sue, H.-J.; Bremner, T. Rheological characterization and differentiation in PAEK materials. In Proceedings of the High Performance Thermoplastics and Composites for Oil and Gas Applications, Houston, TX, USA, 11–12 October 2011.
30. Roboze. The Highest Performing Super Polymers and Composites in the World. Available online: https://www.roboze.com/en/3d-printing-materials/ (accessed on 10 October 2022).
31. *ASTM D6671/D6671M-19*; Standard Test Method for Mixed Mode I-Mode II Interlaminar Fracture Toughness of Unidirectional Fiber Reinforced Polymer Matrix Composites. ASTM International: West Conshohocken, PA, USA, 2019. Available online: www.astm.org (accessed on 1 May 2021).
32. Wang, P.; Zou, B.; Ding, S.; Huang, C.; Shi, Z.; Ma, Y.; Yao, P. Preparation of short CF/GF reinforced PEEK composite filaments and their comprehensive properties evaluation for FDM-3D printing. *Compos. Part B Eng.* **2020**, *198*, 108175. [CrossRef]
33. Sharafi, S.; Santare, M.H.; Gerdes, J.; Advani, S.G. A review of factors that influence the fracture toughness of extrusion-based additively manufactured polymer and polymer composites. *Addit. Manuf.* **2021**, *38*, 101830. [CrossRef]

Disclaimer/Publisher's Note: The statements, opinions and data contained in all publications are solely those of the individual author(s) and contributor(s) and not of MDPI and/or the editor(s). MDPI and/or the editor(s) disclaim responsibility for any injury to people or property resulting from any ideas, methods, instructions or products referred to in the content.

Article

A Systematic Approach to Determine the Cutting Parameters of AM Green Zirconia in Finish Milling

Laurent Spitaels, Hugo Dantinne, Julien Bossu, Edouard Rivière-Lorphèvre and François Ducobu *

Machine Design and Production Engineering Lab, Research Institute for Science and Material Engineering, UMONS, 7000 Mons, Belgium
* Correspondence: Francois.Ducobu@umons.ac.be

Abstract: Additive manufacturing (AM) opens new possibilities of obtaining ceramic green parts with a tailored complex design at low cost. Meeting the requirements of highly demanding industries (aeronautical and biomedical, for example) is still challenging, even for machining. Hybrid machines can solve this problem by combining the advantages of both additive and subtractive processes. However, little information is currently available to determine the milling parameters of additively fabricated ceramic green parts. This article proposes a systematic approach to experimentally determine the cutting parameters of green AM zirconia parts. Three tools, one dedicated to thermoplastics, one to composites, and a universal tool, were tested. The tool–material couple standard (NF E 66-520-5) was followed. The lower cost and repeatable generation of smooth surfaces (Ra < 1.6 µm) without material pull-out were the main goals of the study. The universal tool showed few repeatable working points without material pull-out, while the two other tools gave satisfying results. The thermoplastic tool ensured repeatable results of Ra < 0.8 µm at a four times lower cost than the composite tool. Moreover, it exhibited a larger chip thickness range (from 0.003 mm to 0.036 mm). Nevertheless, it generated an uncut zone that must be considered when planning the milling operations.

Keywords: hybrid machine; green ceramic; milling; additive manufacturing; material extrusion; finishing operations

1. Introduction

1.1. Context

Ceramics and advanced ceramics, such as zirconia, are key materials in a wide range of sectors, such as the electrical, mechanical, chemical, and biomedical industries [1,2]. From the large choice offered by the ceramic materials, 3Y-TZP (Yttrium Tetragonal Zirconia Polycrystal) exhibits very interesting mechanical properties. Indeed, its fracture toughness (4–12 MPa$\sqrt{m}$) and flexural strength (500–1800 MPa) are among the highest available in commercial ceramics [3]. These properties, added to its high chemical resistance, make zirconia very attractive for the oil and gas, biomedical and tooling sectors [1,3,4].

The conventional processing route to manufacture ceramics is usually made of four main steps: the powder synthesis, the shaping of the part (obtention of a green body), and its debinding (obtention of a brown body) and sintering (obtention of a white body) operations. However, for demanding applications (such as contact, for example), a very good surface topography (Ra < 1.6 µm) can be required, and this leads to additional finishing operations performed after sintering (usually consisting of polishing, grinding, or lapping) [3]. However, these operations can represent up to 80% of the total production costs of the part [3]. Indeed, when the finishing operation is performed on the sintered part, it has already acquired the final material properties and requires dedicated tools and careful operators. Moreover, machining operations can lead to micro-cracks and surface defects that decrease the final part's properties [4]. Machining the part at the green stage (just after its shaping) is, consequently, an interesting solution. Indeed, at the green state, the part mostly exhibits the mechanical properties of its binder, which is usually a thermoplastic polymer [5]. Consequently, green machining operations are less expensive and easier to perform while reducing the risk of generating macro defects like cracks, which can lead to the premature failure of the part during its service [4,5].

One of the major drawbacks of the conventional ceramic production route is the impossibility of generating complex designs with, for example, dedicated pore size and interconnectivity or a very fine structure [2,6]. Moreover, the costs to produce a single part with a non-standard design are too high [7], discounting the possibility of serial and personalized production as in Industry 4.0 [8]. Conventional processes are then only economically viable for intermediate, and large series production [3]. Such characteristics (complex tailored design) are required for specific applications, such as in the biomedical sector [7,9]. Non-conventional processes, such as additive manufacturing (AM), are bringing an answer to these problems and offer great hopes and promises [2,10]. Additive processes, in opposition to subtractive processes, enable the production of geometrically complex parts in near-net shapes [11]. Moreover, these processes allow the generation of parts with less material than conventional processes (than in machining, for example) [11] and without the need to invest in costly pieces of equipment specific to a single part design, such as moulds for Ceramic Injection Moulding (CIM) and Metal Injection Moulding (MIM) [2,6].

In the seven families of AM processes classified by the ISO/ASTM 52900 standard, material extrusion is one of the most promising in a ten-year horizon, according to a SmarTech Analysis of 2018 [12]. Indeed, this technology allows the relatively low-cost fabrication of ceramics and metal parts through the use of highly-filled polymer filaments containing ceramic or metal particles [2,8]. The material extrusion method using a fused filament is named Fused Deposition Modelling (FDM) [7]. One of the major limitations of the AM processes is their feedstock availability, and cost [8]. The pellets of the Powder Injection Moulding (PIM) processes can solve this issue since they have been used since the 1970s in the industry [13]. The processes using the FDM method with pellets are named Pellet Additive Manufacturing (PAM). This kind of process does not require high capital expenditure investment (CAPEX), is compatible with every material used in the conventional injection processes (even metals and ceramics), and can be inserted inside an existing workshop using already available MIM or CIM machines [14]. The PAM process produces parts in a green state. Indeed, the ceramic powder in the pellets is mixed with an organic or inorganic binder [9]. The PAM process is referred to as indirect since the shaping and sintering operations are decoupled [9]. The part shaped by the PAM process will then require debinding and sintering steps to achieve its final density and mechanical properties.

Even if the PAM process allows the generation of complex designs impossible to obtain using the conventional ceramic forming processes, these exhibit rough surfaces compared to the other AM processes due to the inherent staircase effect [8,15]. The high arithmetic roughness (from 9 µm to 40 µm [16]) reduces the fatigue resistance of the parts, and their tribological properties [17,18]. Some conventional machining operations (such as turning and milling) can solve the problem but require simple designs [5,19]. Indeed, complex parts exhibiting an internal lattice structure cannot be machined even with 5-axis machining since the inner surfaces of the part are unattainable by the cutting tool. Moreover, when applied to fully dense parts obtained after sintering, the machining operations are difficult and very expensive, as explained earlier [19].

All these reasons motivate the development of methods combining the advantages of the AM processes and machining. Such machines are called hybrid [5,20] and are currently subject to a strong industrial interest [21]. Commercial hybrid machines combining additive and subtractive processes already exist. They allow the generation of metal parts with Direct Energy Deposition (DED) or Selective Laser Sintering (SLS) [5,20]. However, their price is very high since the machining operations are made in a dense part, then requiring a robust and complex construction while dealing with low material removal rates and challenging tool wear. A solution to this problem is to machine the part when it is in a green state [5,22]. Indeed, the part in the green state requires less energy to be machined than dense parts since it behaves as a pseudo-plastic material [5,22]. This effect, added to lower thermal stresses, allow the reduction of tool wear and consequently avoids the need for expensive tools [5].

In the case of ceramic parts, hybrid machines are not yet commercialised. Moreover, as announced in a recent review [19], there is a lack of data for the machining of materials shaped by additive manufacturing, except for highly used alloys, such as Ti6Al4V. The determination of the most suitable cutting parameters in green ceramics obtained using the PAM process will then allow the fabrication of a hybrid machine able to produce ceramic parts with tight tolerances and smooth surface topography to be foreseen.

1.2. NF E 66-520-5 Standard

The cutting parameters of ductile materials can be experimentally determined using the tool–material couple standard NF E 66-520-5 [23]. This standard provides a method using mainly the specific cutting energy to quantify the ability of a tool (generation of an appropriate surface topography while avoiding catastrophic damage mode of the tool) to realise machining operations on a specific material. The standard also advices to take into account the surface topography when the tool will be used for finishing operations, as in this work.

Green ceramics do not behave as ductile materials. Consequently, the standard cannot be directly applied. However, recent papers [4,22] have already used the standard to obtain ranges of cutting parameters and to compare the behaviour of a dedicated material to ductile materials. This paper uses the tool–material couple standard as a general guide to determine the suitable cutting parameters for a determined tool. As described in the standard dedicated to milling operations, four cutting parameters were used: the cutting speed (v_c), the feed per tooth (f_z), the depth of cut (a_p), and the radial depth of cut (a_e).

The standard proposes to perform six different experiments [23]:

- Selection of a stable operating point (qualification test);
- Determination of $v_{c,min}$;
- Determination of minimal and maximal chip thickness ($h_{min} - h_{max}$);
- Determination of limiting data;
- Wear tests;
- Tests to determine the auxiliary parameters.

The experiments of the standard allow the determination of the working range for $v_{c,min}$, f_z, a_p, and a_e. When machining is performed with a tool within its working range, it is repeatable, i.e., two different machining operations conducted with the same parameters lead to the same results in terms of specific cutting energy (the power needed to remove a given volume of material) and surface topography. Achieving this repeatability is required to foresee reliable finishing operations.

1.3. Motivation and Objective of the Paper

Even though green ceramic machining has been performed for decades [3], no standardised methods exist to determine the appropriate cutting conditions for such a material. The tool–material couple standard can be a relevant guide, even though it was designed for ductile materials as demonstrated in the literature [4,22]. Moreover, the PAM process opens new possibilities in terms of geometrical complexity for ceramic parts thanks to its flexibility of feedstock [7,14]. However, the surface topographies, as well as dimensional and geometrical tolerances generated by this process, still exhibit high arithmetic roughness (Ra > 1.6 µm), which makes them not suitable for contact applications. Finishing the parts using conventional processes, such as milling, is then mandatory.

In this context, the objective of this paper is the determination of suitable cutting parameters for the finishing of green ceramic parts obtained with the PAM process. To guarantee low costs of processing, three standard milling tools were selected. One is dedicated to thermoplastics, another to composites, and the last is suitable for a large set of materials (universal tool). The tool–material couple standard [23] was used as a guide to establishing the different cutting conditions. This paper is the first to apply a systematic and objective method to determine the cutting parameters of AM green ceramics. The main objectives in terms of surface topography are to ensure Ra < 1.6 µm and a surface without material pull-out.

2. Materials and Methods

2.1. Manufacturing of Parts

The geometry of the parts is a cube on top of a cylinder. The cylinder has a diameter of 15 mm and a height of 15 mm while the cube exhibits dimensions of 20 mm × 20 mm × 20 mm. Machining is performed on the cube while the part's fixture is ensured by the cylinder. A fillet of 3 mm links both geometrical entities and decreases the risk of deformation after printing. Figure 1 shows the part's geometry as well as its reference axes. The Z axis was set along the cylinder's main axis direction, and the X and Y axes were aligned with two edges of the cube.

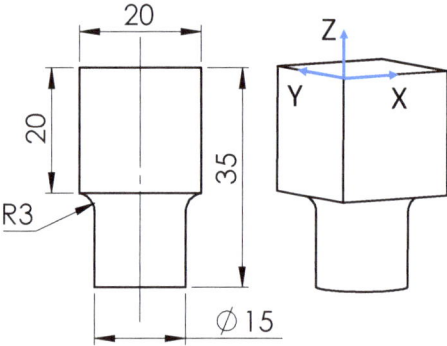

Figure 1. Design of the parts (in mm).

The printing of the parts was performed on a Pollen AM Series MC (Ivry-sur-Seine, France). This PAM printer was fed with K2015 pellets from Inmatec (Rheinbach, Germany). This feedstock is composed of a mix of polyamide binder (15% in mass) and zirconium oxide (85% in mass) partially stabilised with 3 mol% of yttrium oxide (Y_2O_3). The resulting material exhibits a density of 6000 kg/m^3. The zirconium oxide used in the feedstock has a d_{50} of 0.5 µm and comes from the Tosoh Corporation (Tokyo, Japan). This feedstock is developed for the CIM industry and requires a temperature of injection between 110 °C and 150 °C while ensuring a back pressure above 50 bar. After its shaping, the part will require a two-stage debinding before its sintering operation. The first stage is chemical with acetone, while the second is thermal (temperature up to 325 °C). The final sintering operation should be performed at 1400 °C in air. The resulting material has a black colour after its sintering operation.

The printer was used with a 1 mm nozzle diameter. The layer thickness was set at 0.35 mm while the first layer thickness was set at 0.17 mm. The extrusion temperature stood at 165 °C while the build platform temperature was set at 35 °C. An amount of 100% of the infill was selected with a concentric strategy deposition. The build direction was set along the part's Z axis, and the cubic base was the first to be printed. This avoided the use of support. The printing of a part took about 25 min. SolidWorks version 2021 was used to create the CAD file, while Cura version 4.1 generated the sliced file. The STL format was selected to transfer the CAD from SolidWorks to Cura. Figure 2 shows a part after its fabrication, while Figure 3 gives the surface topography obtained. As can be seen in the pictures, the part exhibits a very rough surface topography. Its arithmetic roughness was estimated to be higher than 12.5 µm using a viso-tactile roughness comparator, demonstrating the need for finishing to achieve the required arithmetic roughness.

Figure 2. As built part before milling.

Figure 3. Surface topography of the as built part before milling.

2.2. Milling and Data Acquisition

The milling operations were performed on a Stäubli TX200 (Pfäffikon, Switzerland) fitted with a Teknomotor ATC71 electrospindle (Quero, Italy). The spindle exhibits a maximal power of 7.8 kW and a maximal rotational speed (N) of 24,000 rpm. Figure 4 gives an overview of the cutting configuration.

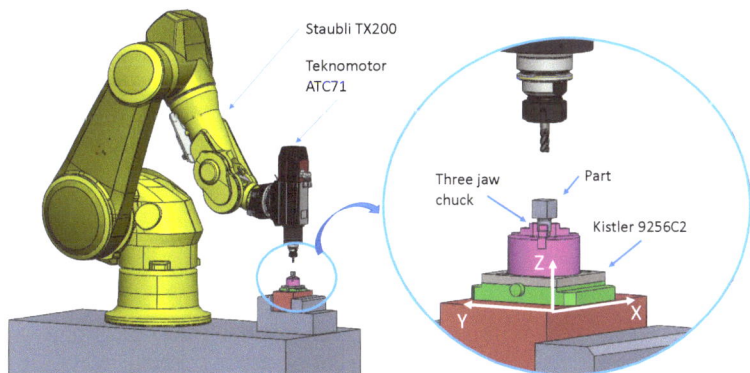

Figure 4. Cutting configuration.

Three different tools with a 6-mm diameter were selected. Their diameter was chosen to ensure a high enough cutting speed since the maximal spindle speed was 24,000 rpm. Table 1 gives the main characteristics of the selected tools, such as their diameter (D), number of teeth (z), maximal axial depth of cut ($a_{p,max}$), supplier reference and relative price compared to the tool dedicated to the thermoplastics (reference price). In the rest of the paper, and for the sake of simplicity, each tool is called by the material for which it was designed (e.g., thermoplastic tool). The geometry of each tool is shown in Figure 5.

Table 1. Main characteristics of the selected tools.

Type of Tool	D (mm)	z	$a_{p,max}$ (mm)	Price	Supplier/Reference
Thermoplastics	6	3	19	1.00	Hoffmann/209425 6
Composites	6	10	18	3.64	SECO/871060.0-DURA
Universal	6	3	14	1.63	SECO/553060SZ3.0-SIRON

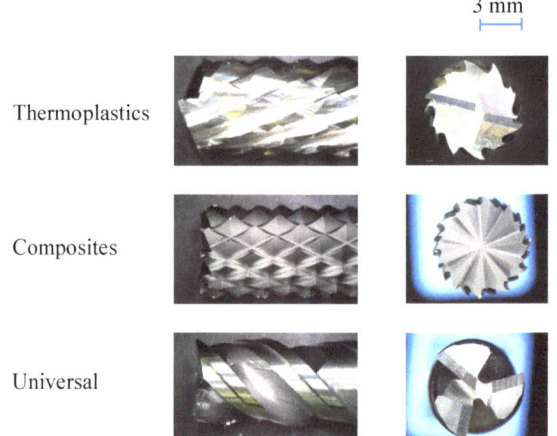

Figure 5. Geometry of the thermoplastic, composite, and universal tools.

The desired cutting parameters were related to the finishing of AM parts. As a consequence, 3 mm axial depth (a_p) and 0.5 mm radial depth of cut (a_e) were selected. Indeed, in finishing operations, only the external layer had to be removed. Straight shoulders were milled on the part along its X axis.

The part was clamped into a three-jaw chuck, itself rigidly attached to a Kistler 9256C2 force sensor (see Figure 4, Winterthur, Switzerland). The cutting forced signals in the X (feed direction), Y, and Z directions were recorded using a Kistler charge amplifier 5070A (Winterthur, Switzerland). A data acquisition system Kistler 5697A2 (Winterthur, Switzerland) and a computer executing the DynoWare software completed the acquisition chain. The sampling frequency was set at 40 kHz, and no filter was applied to the data. The clamping of the parts left visible marks on their cylinders without inducing cracks. Indeed, the binder weight percentage was high enough to allow the part to be deformed without breaking.

The surface topography was evaluated using a DH-6 roughness measurement instrument from Diavite (Bülach, Switzerland). The resulting arithmetic and total roughness (Ra and Rt, respectively) were computed. A qualitative evaluation of the surface topography was also performed using a digital microscope AM7013MZT from Dino-lite (Torrance, CA, USA). The images were acquired using the DinoCapture software version 2.0 and allowed the identification of the generation of material pull-out.

Each selected cutting condition was tested with three different repetitions. During each repetition, three passes were realised. The surface topography was evaluated after the third pass of each repetition while the cutting forces were recorded continuously for each individual pass. In total, 287 tests of three passes were conducted in this study.

The ceramic oxide composing the feedstock was very abrasive and can lead to accelerated wear of the cutting tool. As a consequence, the wear of each tool was monitored after each pass using the digital microscope AM7013MZT from Dino-lite (Torrance, CA, USA). The presence of uniform or localised wear was assessed using the ISO 8688-2 criteria. No significant flank wear was observed during the tests. This ensured there was no influence of tool wear on the results.

2.3. Tool–Material Couple Standard Application

The tool–material couple standard [23] was used to organise the experiments. The qualification test, determination of $v_{c,min}$ and the range h_{min}–h_{max} were performed. However, the other tests (determination of limiting data ($A_{D,max}$, and Q_{max}), wear tests, and the determination of auxiliary parameters) were not conducted since this paper aims to find the operating condition for the finishing of parts. Indeed, the goal of this kind of operation is to obtain the required surface topography and tolerances. In this framework, $A_{D,max}$ and Q_{max} are of little interest. The wear tests were not followed as in the standard, but the wear of the tool was qualitatively monitored.

2.3.1. Starting Cutting Parameters

Since the selected tools were not directly dedicated to the material milled in this article, no data was available in the supplier catalogues. For each tool, several tests were conducted before beginning the experiments to find a suitable starting cutting speed. Variations of the spindle speed from 3500 rpm to 22,000 rpm by steps of 3500 rpm were conducted (resulting in cutting speed from 66 m/min to 415 m/min). The other cutting parameters a_e, a_p and f_z were set at 0.5 mm, 3 mm and 0.03 mm/tooth, respectively. The criteria to select the cutting speed were: the repeatability of the cutting forces and a surface topography below 1.6 µm Ra without pull-out. Table 2 gives the starting cutting parameters used for each tool.

Table 2. Starting parameters for each tool.

Tool	v_c (m/min)	f_z (mm/tooth)	a_p (mm)	a_e (mm)
Thermoplastics	383			
Composites	324	0.030	3	0.5
Universal	270			

2.3.2. Qualification Test

The qualification test consisted of using the starting conditions and modifying them slightly (between 15% and 20% for this paper) to check if the result of the cutting operation was still acceptable. Table 3 gives the experimental plan that was followed for this test. As for the other tests, each set of parameters was tested with three repetitions, each made of three passes.

Table 3. Experimental plan of the qualification test.

Tool	v_c (m/min)	f_z (mm/tooth)	a_p (mm)	a_e (mm)
Thermoplastics	383	0.030	3	0.5
	443	0.030	3	0.5
	306	0.030	3	0.5
	383	0.036	3	0.5
	383	0.024	3	0.5
	383	0.030	3.6	0.5
	383	0.030	2.4	0.5
	383	0.030	3	0.6
	383	0.030	3	0.4

Table 3. *Cont.*

Tool	v_c (m/min)	f_z (mm/tooth)	a_p (mm)	a_e (mm)
Composites	324	0.030	3	0.5
	389	0.030	3	0.5
	259	0.030	3	0.5
	324	0.036	3	0.5
	324	0.024	3	0.5
	324	0.030	3.6	0.5
	324	0.030	2.4	0.5
	324	0.030	3	0.6
	324	0.030	3	0.4
	324	0.030	3	0.5
Universal	270	0.030	3	0.5
	309	0.030	3	0.5
	228	0.030	3	0.5
	270	0.036	3	0.5
	270	0.024	3	0.5
	270	0.030	3.6	0.5
	270	0.030	2.4	0.5
	270	0.030	3	0.6
	270	0.030	3	0.4

2.3.3. Minimal Cutting Speed

The determination of the minimal cutting speed $v_{c,min}$ required varying the cutting speed while keeping the other parameters (f_z, a_p and a_e) constant. In this paper, the choice was made to scan most of the possible range of the spindle speed (from 3500 rpm to 22,000 rpm). Table 4 gives the different spindle and cutting speeds considered for the determination of $v_{c,min}$. The same range was considered for all the tools.

Table 4. Range of N (in rpm) and v_c (in m/min) considered for the determination of $v_{c,min}$.

N (rpm)	v_c (m/min)
3500	66
7000	132
11,000	207
14,300	270
16,400	309
17,200	324
18,600	351
20,300	383
22,000	415

2.3.4. Mean Chip Thickness Range

Finally, the determination of the mean chip thickness (h_m) range was explored by modifying the feed per tooth (f_z) while keeping the other parameters (v_c, a_p and a_e) constant. The range of variation of f_z, as well as the resulting chip thickness h_m, is given in Table 5.

Table 5. Range of f_z (in mm/tooth) and h_m (in mm) considered for the determination of the mean chip thickness range.

f_z (mm/tooth)	h_m (mm)
0.003	0.001
0.005	0.003
0.010	0.006
0.015	0.008
0.021	0.012
0.030	0.017
0.039	0.022
0.051	0.028
0.066	0.036

2.4. Computation of the Specific Cutting Energy

All the conducted tests required the computation of the specific cutting energy. This specific energy is the ratio between the cutting power and the material removal rate. As the cutting forces were low during the tests (mainly lower than 20 N), it was not possible to record the cutting power during the experiments with an acceptable signal-to-noise ratio. Moreover, the reference frame of the dynamometer was not the same as the tool. Consequently, the total cutting forces along the X, Y, and Z axis of the dynamometer were used to determine the specific cutting energy instead of using the results of the feed and tangential cutting forces. Indeed, there is a proportional link between the tangential component of the force and its radial and axial components, as demonstrated in the recent paper of Demarbaix et al. [22]. The RMS value of the total resulting force for each pass was then used as a representative value.

The total resulting force (in N) was computed as follows:

$$F_{tot} = \sqrt{F_x^2 + F_y^2 + F_z^2} \qquad (1)$$

Finally, as finishing operations were conducted by milling straight shoulders, the radial depth of cut was always less than half of the tool diameter. The mean chip thickness h_m (in mm) can then be computed as:

$$h_m = 2 \cdot f_z \cdot \sqrt{(\frac{a_e}{D}) \cdot (1 - \frac{a_e}{D})} \qquad (2)$$

The resulting specific cutting energy (in $\frac{W \cdot min}{cm^3}$) can be expressed as:

$$W_c = \beta \cdot \frac{F_{tot}}{60 \cdot a_p \cdot h_m} \qquad (3)$$

where β is a real number. For the sake of simplicity, the right-hand side of the equation divided by β will be called specific cutting energy in the present paper since it is directly proportional to it.

3. Results and Discussion

3.1. Qualification Test

For the three tools, the results of the surface topography and specific cutting energy for the qualification test are given in Figures 6–15. In Figures 6, 10, and 13, two red lines give the domain of the 1.6 μm Ra class (from 0.8 μm to 1.6 μm). Each point of data was given $\pm \sigma$ error bars. Indeed, each point on the graph shows the mean value of three measurements carried out with the same cutting parameters. The considered cutting parameters are given with v_c (m/min), f_z (mm/tooth), a_p (mm), and a_e (mm). Figures 8, 9, 12, and 15 show qualitatively the surface topography generated with the milling operations for the qualification point of each tool. In these Figures, only the vertical surface (according to the X and Z axes) generated by the tool is shown. Indeed, in a finishing context, the horizontal surface (according to the X and Y axes) will not exist.

For the thermoplastic tool (Figure 6), all the results are below the 0.8 μm threshold bar. Indeed, all the results belong to the 0.8 μm instead of the 1.6 μm Ra class. The results range from 0.48 μm to 0.71 μm. The repeatability of results is good, with a dispersion of about 15% around the mean for all experiments. In terms of specific cutting energy (Figure 7), the results vary between 2.67 W·min/cm³ and 6.98 W·min/cm³. There are variations up to 100% between the different cutting conditions. The high variation of results can be explained by the relatively high variation in cutting conditions imposed on the tool. Indeed, v_c (m/min), f_z (mm/tooth), a_p (mm), and a_e (mm) were modified by 20%. Even though there were significant variations between the different cutting conditions, the repeatability of results for a given set of cutting parameters was very good (about 4% of the mean).

All the different tests produced a surface topography without material pull-out, as shown in Figure 8 for the parameters v_c = 383 m/min, f_z = 0.030 mm/tooth, a_p = 3 mm, and

a_e = 0.5 mm. The milled surface is shown with the red arrow. As depicted in the picture, the surface topography was shiny and without pull-out as required.

However, there is a zone of approximatively 0.85 mm where the material was not completely cut (see the blue arrow in Figure 8). Figure 9 depicts a side view of the milled straight shoulder. The uncut material is shown again with a blue arrow. All passes performed with the tool dedicated to thermoplastics exhibited the same uncut zone. This, therefore, has to be taken into account when performing the milling of parts. Indeed, by shifting the tool further down, this uncut material can be removed. Consequently, as the goals to achieve were ensuring repeatable results while respecting a Ra class of 1.6 µm without material pull-out, the selected cutting conditions were validated.

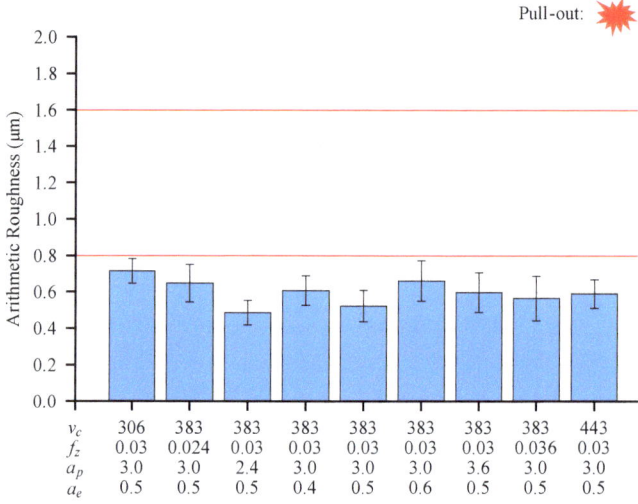

Figure 6. Qualification test of the thermoplastic tool, the evolution of Ra (µm) for different v_c (m/min), f_z (mm/tooth), a_p (mm), and a_e (mm). No material pull-out was observed.

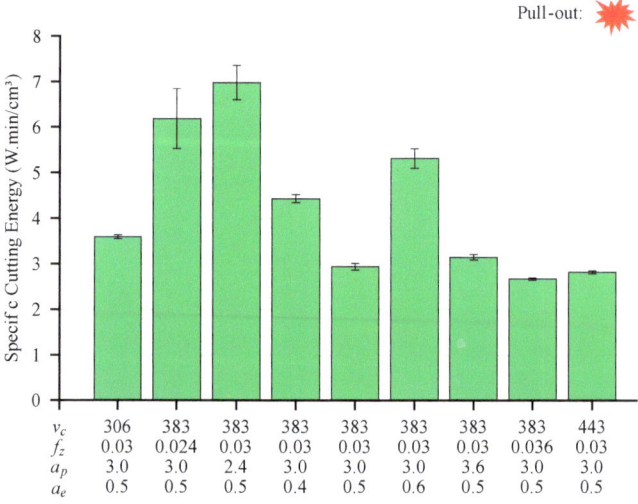

Figure 7. Qualification test of the thermoplastic tool, evolution of W_c (W·min/cm³) for different v_c (m/min), f_z (mm/tooth), a_p (mm), and a_e (mm). No material pull-out was observed.

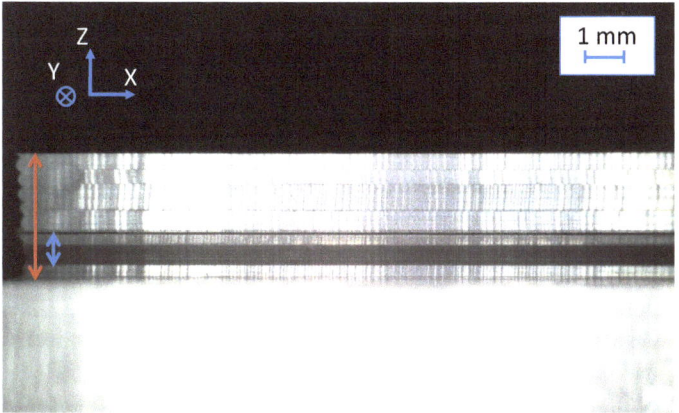

Figure 8. Surface topography of the qualification point for the tool dedicated to thermoplastics, v_c = 383 m/min, f_z = 0.030 mm/tooth, a_p = 3 mm, and a_e = 0.5 mm. No material pull-out was observed.

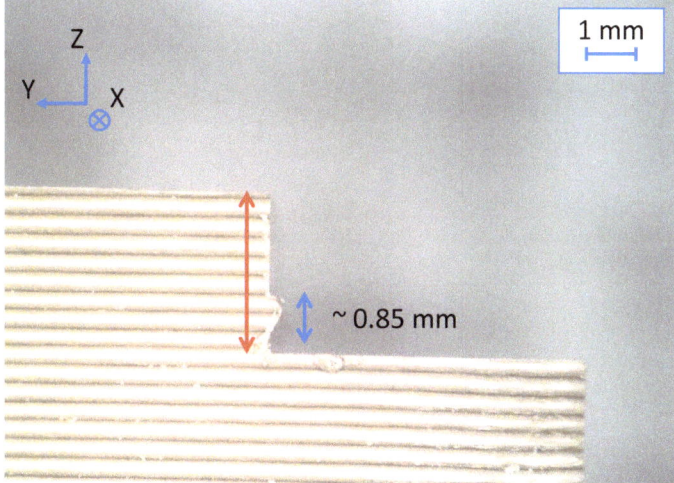

Figure 9. Side view of the milled straight shoulder for the thermoplastic tool.

The qualification test of the composite tool is depicted in Figure 10 for the surface topography results. As can be seen in the graph, all the results of Ra are strictly inside the 1.6 µm Ra class delimited by the red lines. The results go from 0.88 µm to 1.22 µm with a dispersion for each condition representing, on average, 10% of the mean value. As shown in Figure 11, the specific cutting energy ranges from 1.75 W·min/cm^3 and 2.96 W·min/cm^3. The variations between different cutting conditions reached a maximum of 70% and were lower than for the thermoplastic tool. On average, the dispersion of measurements for a given set of cutting parameters reached 5%. This means again that repeatable results were obtained.

All the different tests produced a smooth surface topography without material pull-out, as depicted in Figure 12, for v_c = 324 m/min, f_z = 0.030 mm/tooth, a_p = 3 mm, and a_e = 0.5 mm. Again, the milled surface is shown with the red arrow. As depicted in the picture, there was an absence of pull-out, but there was a shiny finish. In contrast with the tool dedicated to the thermoplastics, all the desired material to remove was cut. Again, the working point selected was validated.

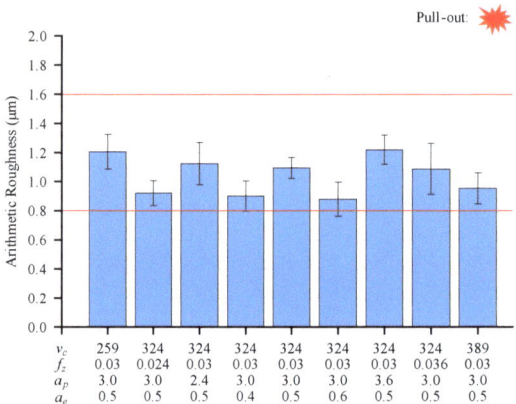

Figure 10. Qualification test of the composite tool, the evolution of Ra (μm) for different v_c (m/min), f_z (mm/tooth), a_p (mm), and a_e (mm). No material pull-out was observed.

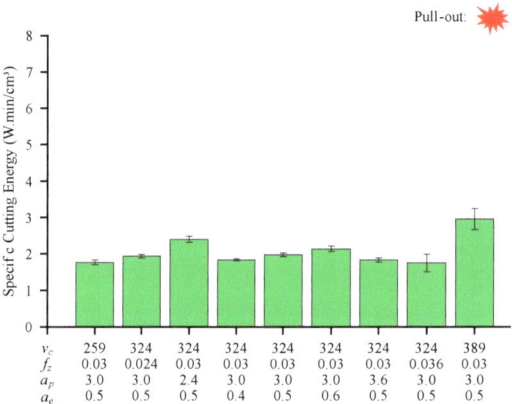

Figure 11. Qualification test of the composite tool, the evolution of W_c (W·min/cm^3) for different v_c (m/min), f_z (mm/tooth), a_p (mm), and a_e (mm). No material pull-out was observed.

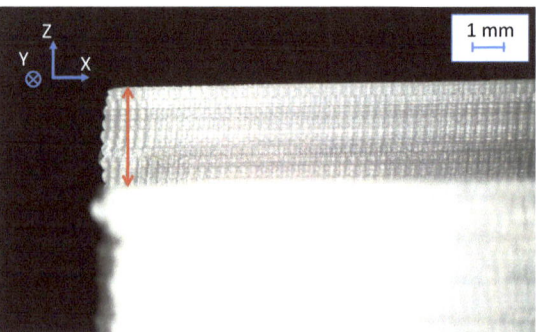

Figure 12. Surface topography of the qualification point for the tool dedicated to composites, v_c = 324 m/min, f_z = 0.030 mm/tooth, a_p = 3 mm, and a_e = 0.5 mm. No material pull-out was observed.

The working point of the universal tool was evaluated as exhibited in Figure 13. For this tool, the results of surface topography were between 0.68 μm and 0.86 μm. The dispersion of each cutting condition represents, as for the other tools, about 13% of the measured Ra. Some of the results were included in the 1.6 μm Ra class, while others were inside the 0.8 µm Ra class. The specific cutting energy of the universal tool, as depicted in Figure 14, went from 3.49 W·min/cm^3 to 5.35 W·min/cm^3. As for the thermoplastic tool, high variations were measured for different cutting conditions. Again, their origin probably comes from the high variations imposed on the cutting parameters. However, the repeatability of the measurements was the best among the three tools, with variations of only 1% around the mean value for a given set of cutting parameters.

The surface topography of the qualification point for the universal tool is depicted in Figure 15 (v_c = 270 m/min, f_z = 0.030 mm/tooth, a_p = 3 mm, and a_e = 0.5 mm). As can be seen in the Figure, material pull-out occurred during the pass (circled in yellow). It should be noted that only a few tests were free from these defects. The absence of material pull-out was a requirement for the tool to be selected. This means that the universal tool cannot be used for the finishing of green ceramics obtained using the PAM process. Even with this information, the other tests of the method were performed.

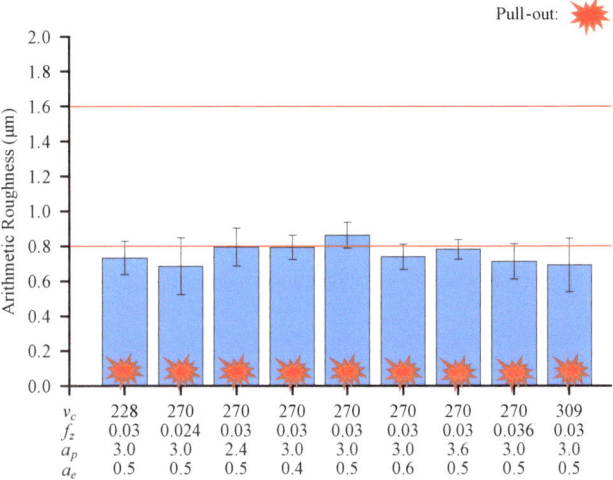

Figure 13. Qualification test of the universal tool, the evolution of Ra (µm) for different v_c (m/min), f_z (mm/tooth), a_p (mm), and a_e (mm).

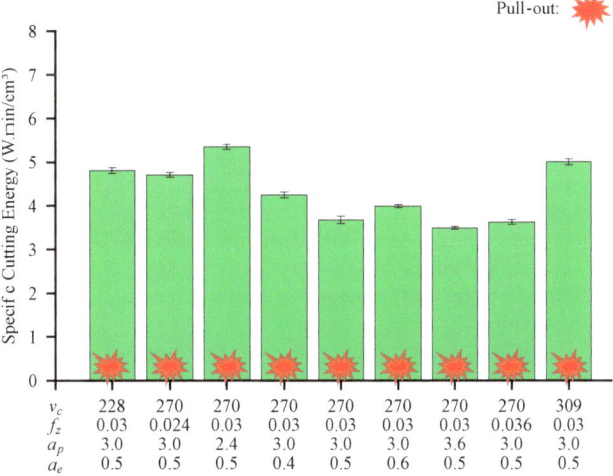

Figure 14. Qualification test of the universal tool, the evolution of W_c (W·min/cm^3) for different v_c (m/min), f_z (mm/tooth), a_p (mm), and a_e (mm).

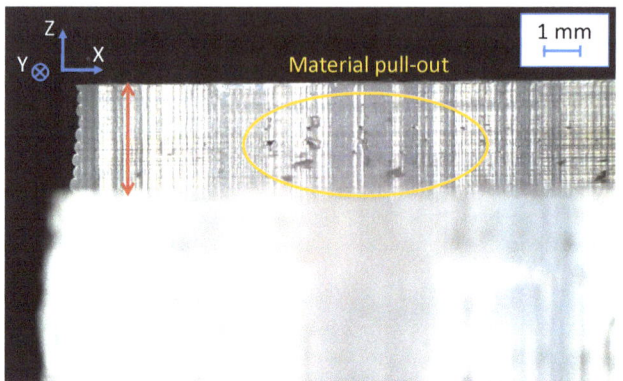

Figure 15. Surface topography of the qualification point for the universal tool, v_c = 270 m/min, f_z = 0.030 mm/tooth, a_p = 3 mm, and a_e = 0.5 mm.

The reference points of the thermoplastic and composite tools were validated since they allowed a smooth surface finish to be obtained while respecting a Ra class of 1.6 μm and producing repeatable results. Conversely, even though the universal tool exhibited the best repeatability results of specific cutting energy, it produced material pull-out for almost all tested cutting conditions. Consequently, the universal tool cannot be qualified for the green zirconia used in this study.

3.2. Determination of the Minimal Cutting Speed

Figure 16 give the arithmetic roughness for different values of v_c, for the thermoplastic tool. All the other cutting parameters were constant (f_z = 0.030 mm/tooth, a_p = 3 mm, and a_e = 0.5 mm). Two red lines delimit the domain of the 1.6 μm Ra class. The specific cutting energy is also depicted in Figure 17 for the same tool. As before, each point of measurement is the average of three measurements and is depicted on the graph with a ±σ error bar. Figures 18 and 19 give the arithmetic roughness and specific cutting energy for the composite tool, respectively. So do Figures 20 and 21 for the universal tool.

The tool–material couple standard [23] gives the expected trend of the specific cutting energy for the minimal cutting speed determination. By increasing the cutting speed and keeping all other parameters constant, the specific cutting energy should exhibit a sudden drop and should then continue to decrease at a lower rate. In some cases, the specific energy does not follow this trend and requires taking into account other parameters, such as the surface topography.

Figure 16 gives the achieved Ra for the thermoplastic tool. As depicted on the graph, the general trend decreased as required by the tool–material couple standard [23]. All the results were below the 1.6 μm threshold and even below the 0.8 μm threshold when the cutting speed was higher than 309 m/min. From 309 m/min to 415 m/min, the results of Ra tended to reach a plateau. However, at 309 m/min, the repeatability was lower than for higher cutting speeds. Indeed, at this speed, the dispersion of measurements achieved 31% around the mean value while it was only 16% on average from 324 m/min to 383 m/min. All the tested cutting speeds produced a smooth surface topography without pull-out as required.

The specific cutting energy did not follow the same trend. Indeed, as depicted in Figure 17, the cutting energy varied between 1.88 W·min/cm³ and 4.04 W·min/cm³ but without a globally decreasing trend. Consequently, the specific cutting energy could not be used as the only indicator to determine the minimal cutting speed. As a consequence, the minimal cutting speed was selected thanks to the arithmetic roughness evolution. The value selected was 324 m/min since, after this speed, the dispersion was lower than for the 309 m/min cutting speed.

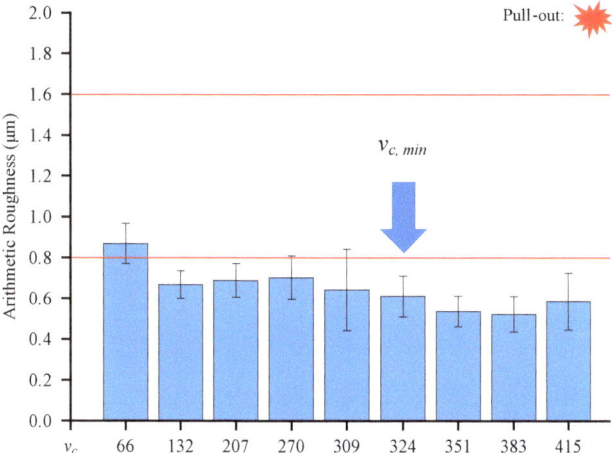

Figure 16. Determination of the minimal cutting speed of the thermoplastic tool, evolution of the arithmetic roughness (μm) for different v_c (m/min) with f_z = 0.030 mm/tooth, a_p = 3 mm, and a_e = 0.5 mm. No material pull-out was observed.

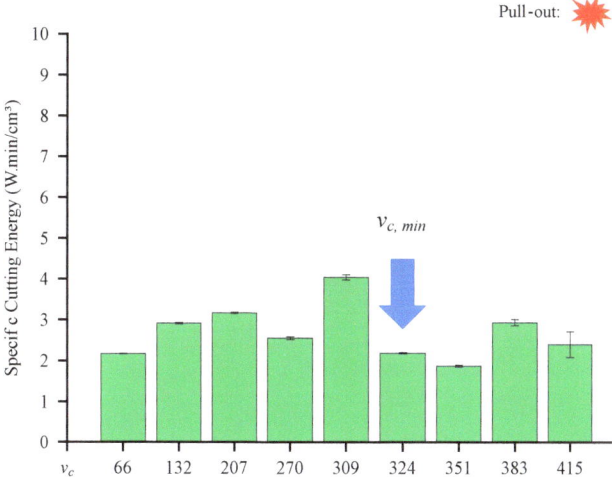

Figure 17. Determination of the minimal cutting speed of the thermoplastic tool, the evolution of the specific cutting energy (W·min/cm³) for different v_c (m/min) with f_z = 0.030 mm/tooth, a_p = 3 mm, and a_e = 0.5 mm. No material pull-out was observed.

The arithmetic roughness of the composite tool for different cutting speeds is given in Figure 18. As depicted in the graph, there is a decreasing trend, as expected. All the values were within the 1.6 μm Ra class. After 351 m/min, the results stabilize and achieve a plateau. The dispersion of measurements does not decrease dramatically when the cutting speed increases. Indeed, it represents, on average, 9.5% of the mean value. The machined surface exhibited pull-out only for the 66 m/min cutting speed. All the other points produced a smooth surface topography.

As for the thermoplastic tool, the specific cutting energy did not show a decreasing trend when the cutting speed increased, as presented in Figure 19. Again, the specific cutting energy could not be used alone to determine the minimal cutting speed. Consequently, 351 m/min was selected as the minimal cutting speed according to the arithmetic roughness evolution.

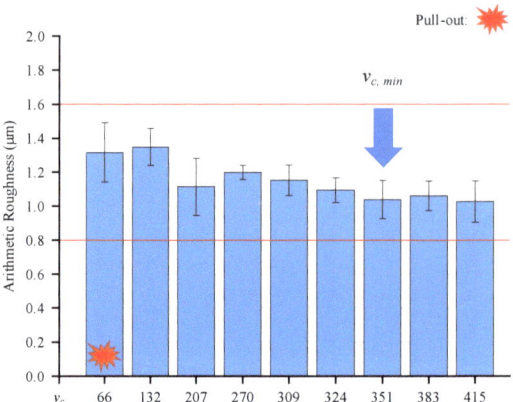

Figure 18. Determination of the minimal cutting speed of the composite tool, evolution of the arithmetic roughness (μm) for different v_c (m/min) with f_z = 0.030 mm/tooth, a_p = 3 mm, and a_e = 0.5 mm.

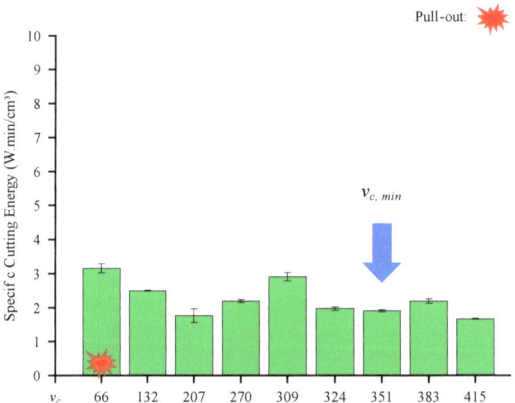

Figure 19. Determination of the minimal cutting speed of the composite tool, evolution of the specific cutting energy (W·min/cm³) for different v_c (m/min) with f_z = 0.030 mm/tooth, a_p = 3 mm, and a_e = 0.5 mm.

The evolution of the arithmetic roughness of the universal tool for different cutting speeds is depicted in Figure 20. As can be seen in the graph, high variations of results were recorded compared to the two other tools. Surprisingly, the results of arithmetic roughness were better (<0.8 μm) for the lowest cutting speeds (66 m/min and 132 m/min). However, the repeatability of results was lower than for the other tested tools. Indeed, for each considered cutting speed, one or two tests produced pull-out. As a consequence, the universal tool could not be used with this material. Indeed, the repeatability of results was one of the required conditions for selecting one of the tools.

The specific cutting energy showed the same trend as the arithmetic roughness. High variations were recorded across the domain of the tested cutting speeds. As for the surface topography, the lowest tested cutting speeds exhibited the best results of specific energy. Difficulty in evacuating the chips may explain the higher specific cutting energy at higher cutting speeds. However, with the evolution of arithmetic roughness, a minimal cutting speed of 351 m/min can be selected. Indeed, after 324 m/min, the arithmetic roughness dramatically decreased, as well as the measurement dispersion. For the same cutting speed as the reference point (v_c = 270 m/min), the Ra results of Figure 20 are higher. Nevertheless, the results of the qualification test are included within the error bars of

Figure 20. The selected minimal cutting speed is also higher than the cutting speed of the reference point. This is the result of the standard method, which first tests a reference point before selecting the minimal cutting speed. However, even with a higher cutting speed for the qualification test, the universal tool produced a surface topography with material pull-out. The conclusions are then the same.

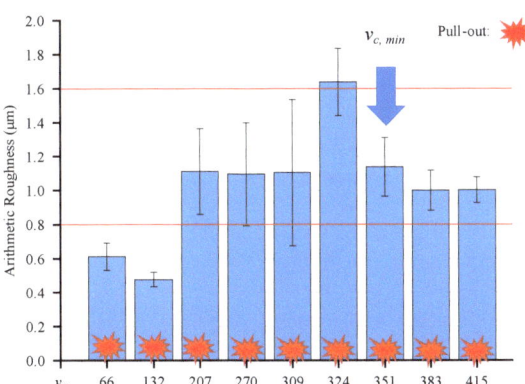

Figure 20. Determination of the minimal cutting speed of the universal tool, the evolution of the arithmetic roughness (μm) for different v_c (m/min) with f_z = 0.030 mm/tooth, a_p = 3 mm, and a_e = 0.5 mm.

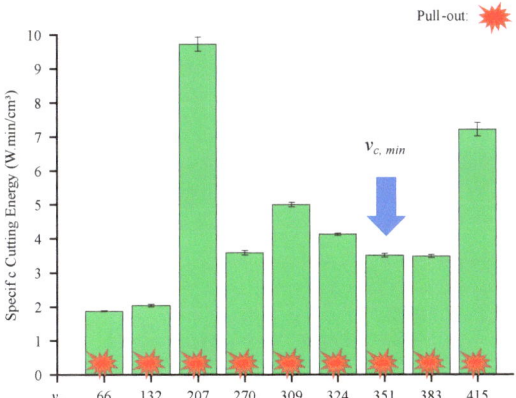

Figure 21. Determination of the minimal cutting speed of the universal tool, the evolution of the specific cutting energy (W·min/cm^3) for different v_c (m/min) with f_z = 0.030 mm/tooth, a_p = 3 mm, and a_e = 0.5 mm.

As depicted in Figures 16–21, and except for the universal tool, the arithmetic roughness showed the expected decreasing trend when the cutting speed increased. However, the specific cutting energy did not exhibit the same behaviour. The minimal cutting speed was, therefore, selected solely based on the arithmetic roughness. The composite tool allowed results within the 1.6 μm Ra class, while the thermoplastic tool reached better results within the 0.8 μm Ra class. Repeatable results and a surface without material pull-out were obtained for both tools. Conversely, the universal tool exhibited higher variations of results, as well as material pull-out on each of the considered cutting conditions. For the specific energy, again, the thermoplastic and composite tools produced repeatable results while the universal tool did not. As a result, the universal tool cannot be selected to realize finishing operations on zirconia green ceramics, while the thermoplastic and composite tools can be good candidates.

3.3. Determination of the Minimal and Maximal Chip Thickness

Figures 22–24 show, for the three selected tools, the evolution of W_c (W·min/cm³, in green) and Ra (µm, in blue) for different average chip thicknesses h_m (mm). Each point of measurement was given a $\pm\sigma$ error bar showing the measurement dispersion.

As described in the tool–material couple standard [23], by varying the chip thickness while keeping all other parameters constant, the specific energy should exhibit a decreasing trend with a sudden drop. The chip thickness at this drop is the lower limit of chip thickness (h_{min}). After this value of h_m, the specific energy should continue to decrease at a slower pace. The determination of the high limit (h_{max}) of the chip thickness requires taking into account the surface topography and the apparition of interfering phenomena such as material pull-out or exceeding the maximum allowed arithmetic roughness. Indeed, when the chip thickness increases, the surface topography degrades progressively and can result in the apparition of material pull-out.

The evolution of the specific cutting energy and arithmetic roughness of the thermoplastic tool is given for different values of h_m (mm) in Figure 22. For these tests, the v_c chosen was 324 m/min (equal to $v_{c,min}$), while a_p and a_e were set at 3 mm and 0.5 mm, respectively. As depicted in the graph, the specific cutting energy decreased when the average chip thickness increased while the arithmetic roughness increased. At 0.048 mm of h_m, the Ra was still below the 1.6 µm threshold, but pull-out appeared in all tests, and the dispersion of measurements increased dramatically. The previously tested value of h_m (0.036 mm) was then selected as the high limit of chip thickness. The specific cutting energy showed a sudden decrease for an average chip thickness of 0.003 mm. This value was then selected as the low limit of chip thickness. For the thermoplastic tool, the chip thickness range ranged between 0.003 mm and 0.036 mm, as depicted with a grey background in Figure 22. These two values of h_m corresponded to feed rates (v_f) of 263 mm/min and 3401 mm/min, respectively.

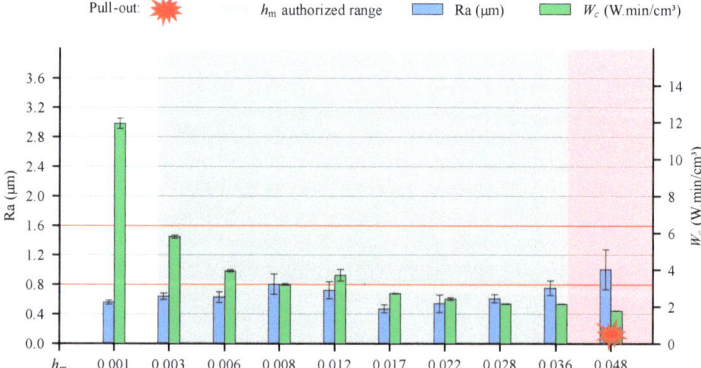

Figure 22. Evolution of the specific cutting energy (W·min/cm³) and Ra (µm) of the thermoplastic tool for different h_m (mm) with v_c = 324 m/min, a_p = 3 mm, and a_e = 0.5 mm.

Figure 23 allows the determination of the low and high limits of chip thickness for the composite tool. As for the thermoplastic tool, v_c was chosen equal to $v_{c,min}$ (351 m/min), while a_p and a_e were set at 3 mm and 0.5 mm, respectively. As expected, W_c and its measurement dispersion decreased when h_m increased. In contrast with the thermoplastic tool, the arithmetic roughness achieved values higher than the 1.6 µm threshold for low values of h_m. The first chip thickness leading to a Ra below the threshold was 0.003 mm. As for the thermoplastic tool, this value was selected as the low limit for h_m. The arithmetic roughness also allowed the selection of the high value of h_m. Indeed, the value of Ra and its dispersion were higher than the 1.6 µm threshold for 0.028 mm of h_m. As a consequence, 0.022 mm of h_m was selected as the high value. It should be noted that no pull-out was generated for all the tested values of h_m. The range of h_m ranges then from 0.003 mm to

0.022 mm, as shown in grey in Figure 23. The corresponding feed rates were 957 mm/min and 9430 mm/min.

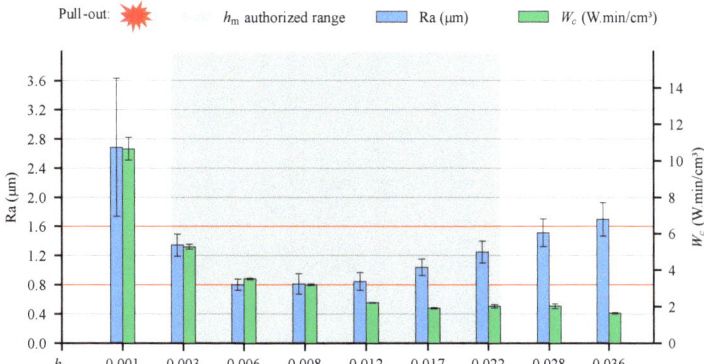

Figure 23. Evolution of the specific cutting energy (W·min/cm^3) and Ra (μm) of the composite tool for different h_m (mm) with v_c = 351 m/min, a_p = 3 mm and a_e = 0.5 mm No material pull-out was observed.

The high and low limits of h_m for the universal tool can be obtained from Figure 24. The cutting speed v_c selected was the same as the qualification point (v_c = 270 m/min), while a_p and a_e were set at 3 mm and 0.5 mm, respectively. As for the two other tools, the specific cutting energy showed a decreasing trend, while the surface topography deteriorated when h_m increased. At an average chip thickness of 0.017 mm, there was a sudden drop in specific cutting energy. All the results of Ra were below the 1.6 μm threshold. Nevertheless, pull-out was produced for all the tested h_m, except for the values of 0.048 mm and 0.062 mm. These two values can be taken as low and high values of chip thickness as shown in grey in Figure 24. These correspond to feed rates of 3689 mm/min and 4779 mm/min. Again, the universal tool shows a lower potential for being used to mill green zirconia parts since it exhibits the lowest range of eligible h_m compared to the thermoplastic and composite tools.

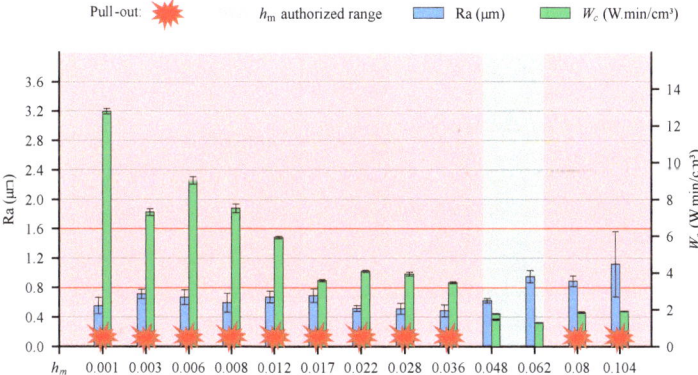

Figure 24. Evolution of the specific cutting energy (W·min/cm^3) and Ra (μm) of the universal tool for different h_m (mm) with v_c = 270 m/min, a_p = 3 mm and a_e = 0.5 mm

3.4. Tool Selection

Only a few of the tests carried out with the universal tool were without material pull-out and with repeatable results. This complicated the use of the tool since only a few cutting conditions could be used to obtain the desired surface topography. As a consequence, this tool could not be used to finish the green ceramics obtained using the PAM process.

The tools dedicated to the machining of composites and thermoplastics allowed the desired surface topography in terms of arithmetic roughness and the absence of material pull-out to be obtained. Their minimal and maximal average chip thickness range allowed their use with different feeds per tooth to be foreseen. However, the maximal average chip thickness of the thermoplastic tool was higher than for the composite tool (0.036 mm vs. 0.022 mm) while maintaining the arithmetic roughness at a lower level. Indeed, at an average chip thickness of 0.036 mm, the tool dedicated to the thermoplastics achieved a Ra of 0.75 µm on average, while the tool for composites reached a Ra of 1.25 µm for an average chip thickness of 0.022 mm.

Nevertheless, the material removal rate of the thermoplastic tool is two times lower than for the composite tool at the maximal average thickness (5.06 cm^3/min vs. 10.96 cm^3/min). Indeed, the composite tool has ten teeth, while the thermoplastic tool has only three. In terms of cost, the thermoplastic tool is about four times more affordable than the composite tool.

The lower cost, achievable Ra, absence of pull-out and larger chip thickness range of the thermoplastic tool make it the best compromise to ensure the finishing of additively manufactured green zirconia. Nevertheless, the uncut zone generated must be taken into account when planning the milling operations.

Table 6 gives the main cutting parameter limits of the thermoplastic and composite tools. Both can be used for the low-cost finishing of additively manufactured zirconia green parts. Though, the thermoplastic tool appears to be the best compromise between price and performance.

Table 6. Main cutting parameter limits of the thermoplastic and composite tools.

	Thermoplastic Tool	Composite Tool
$v_{c,min}$ (m/min)	324	351
$h_{m,min}$ (mm)/$f_{z,min}$ (mm)	0.003/0.005	0.003/0.005
$h_{m,max}$ (mm)/$f_{z,max}$ (mm)	0.036/0.066	0.022/0.039

4. Conclusions

Even though the PAM process is bringing new possibilities in terms of feedstock, it still suffers from rough surface topography and large geometrical and dimensional tolerances. Finishing operations are then required to obtain a part with a smooth surface and tight tolerances. Three standard tools (dedicated to thermoplastics, composites, and a universal tool) were tested according to the tool–material couple standard. The allowed cutting conditions were determined with the goal of ensuring the finishing operation (repeatable results of Ra < 1.6 µm and absence of material pull-out) of green zirconia parts obtained by the PAM process at a lower cost.

The main findings are the following:

- Even though the tool–material couple standard is dedicated to ductile materials, its systematic approach and objective methodology allow us to determine experimentally the cutting parameters in additively manufactured zirconia. This methodology can be used to determine the finishing cutting parameters of other additively manufactured materials using a thermoplastic binder system. Indeed, a wide variety of materials (Ti6Al4V, AISI 316L, Al_2O_3, etc.) are already available as injection moulding feedstocks using a thermoplastic binder and can be shaped with the PAM process, as the green zirconia used in this study.
- The thermoplastic tool is the most affordable while exhibiting the largest chip thickness range and the best achievable surface topography (Ra < 0.8 µm).
- The tool dedicated to composites is four times more expensive than the thermoplastic tool. However, it cannot achieve a surface topography as low as the thermoplastic tool (minimal Ra standing at 0.8 µm while the thermoplastic tool can achieve a minimal Ra of 0.47 µm).
- Repeatable results of surface topography and specific cutting energy were obtained for the thermoplastic and composite tools.

- Only a few tests were without material pull-out for the universal tool, while a large dispersion of measurements was observed. This tool cannot be selected since the absence of material pull-out was a requirement, as well as repeatable results. The other tools only exhibited material pull-out at the extreme limits of their usage domain (maximal average chip thickness, for example).

The recommended cutting conditions for the finishing of AM green zirconia parts using the thermoplastic tool are $v_{c,min} > 324$ m/min, f_z between 0.005 mm/tooth and 0.066 mm/tooth, while the composite tool should be used with $v_{c,min} > 351$ m/min, f_z between 0.005 mm/tooth and 0.039 mm/tooth.

Author Contributions: Conceptualization, F.D., E.R.-L. and L.S.; methodology, L.S.; software, L.S.; validation, F.D., E.R.-L. and L.S.; formal analysis, L.S.; investigation, L.S. and H.D.; resources, F.D., E.R.-L. and J.B.; data curation, L.S. and H.D.; writing—original draft preparation, L.S.; writing—review and editing, F.D., E.R.-L. and L.S.; visualization, L.S.; supervision, F.D. and E.R.-L.; project administration, L.S., F.D. and E.R.-L.; funding acquisition, F.D. and E.R.-L. All authors have read and agreed to the published version of the manuscript.

Funding: This research was funded by the Wallonian regional government thanks to a Win²Wal funding instrument (HyProPAM research project, grant number: 2110084).

Data Availability Statement: The data presented in this study are available on request from the corresponding author.

Acknowledgments: The authors would like to thank the Belgium Ceramic Research Centre (BCRC) for the use of their Pollen AM Series MC printer.

Conflicts of Interest: The authors declare no conflict of interest. The funder had no role in the design of the study; in the collection, analyses, or interpretation of data; in the writing of the manuscript; or in the decision to publish the results.

References

1. Galusek, D.; Ghillányová, K. Ceramic Oxides. In *Ceramics Science and Technology*; Riedel, R., Chen, I.W., Eds.; Wiley-VCH Verlag GmbH & Co. KGaA: Weinheim, Germany, 2014; pp. 1–58. [CrossRef]
2. Altıparmak, S.C.; Yardley, V.A.; Shi, Z.; Lin, J. Extrusion-based additive manufacturing technologies: State of the art and future perspectives. *J. Manuf. Process.* **2022**, *83*, 607–636. [CrossRef]
3. Ferraris, E.; Vleugels, J.; Guo, Y.; Bourell, D.; Kruth, J.P.; Lauwers, B. Shaping of engineering ceramics by electro, chemical and physical processes. *CIRP Ann.-Manuf. Technol.* **2016**, *65*, 761–784. [CrossRef]
4. Demarbaix, A.; Mulliez, M.; Rivière-Lorphèvre, E.; Spitaels, L.; Duterte, C.; Preux, N.; Petit, F.; Ducobu, F. Green Ceramic Machining: Determination of the Recommended Feed Rate for Y-TZP Milling. *J. Compos. Sci.* **2021**, *5*, 231. [CrossRef]
5. Parenti, P.; Cataldo, S.; Grigis, A.; Covelli, M.; Annoni, M. Implementation of hybrid additive manufacturing based on extrusion of feedstock and milling. *Procedia Manuf.* **2019**, *34*, 738–746. [CrossRef]
6. Ferrage, L.; Bertrand, G.; Lenormand, P.; Grossin, D.; Ben-Nissan, B. A review of the additive manufacturing (3DP) of bioceramics: Alumina, zirconia (PSZ) and hydroxyapatite. *J. Aust. Ceram. Soc.* **2017**, *53*, 11–20. [CrossRef]
7. Gonzalez-Gutierrez, J.; Cano, S.; Schuschnigg, S.; Kukla, C.; Sapkota, J.; Holzer, C. Additive manufacturing of metallic and ceramic components by the material extrusion of highly-filled polymers: A review and future perspectives. *Materials* **2018**, *11*, 840. [CrossRef]
8. Krolikowski, M.A.; Krawczyk, M.B. Does Metal Additive Manufacturing in Industry 4.0 Reinforce the Role of Substractive Machining. In *Advances in Manufacturing II*; Trojanowska, J., Ciszak, O., Machado, J.M., Pavlenko, I., Eds.; Springer International Publishing: Cham, Switzerland, 2019; pp. 150–164. [CrossRef]
9. Galante, R.; Figueiredo-Pina, C.G.; Serro, A.P. Additive manufacturing of ceramics for dental applications: A review. *Dent. Mater.* **2019**, *35*, 825–846. [CrossRef]
10. Bourell, D.; Kruth, J.P.; Leu, M.; Levy, G.; Rosen, D.; Beese, A.M.; Clare, A. Materials for additive manufacturing. *CIRP Ann.* **2017**, *66*, 659–681. [CrossRef]
11. Thompson, M.K.; Moroni, G.; Vaneker, T.; Fadel, G.; Campbell, R.I.; Gibson, I.; Bernard, A.; Schulz, J.; Graf, P.; Ahuja, B.; et al. Design for Additive Manufacturing: Trends, opportunities, considerations, and constraints. *CIRP Ann.-Manuf. Technol.* **2016**, *65*, 737–760. [CrossRef]
12. Analysis, S. Ceramics Additive Manufacturing Markets 2017–2028, an Opportunity Analysis and Ten-Year Market Forecast, Technical Report. 2018. Available online: https://www.smartechanalysis.com/reports/ceramics-additive-manufacturing-markets-2017-2028/ (accessed on 6 March 2023).

13. Moinard, D.; Rigollet, C. Procédés de Frittage PIM. Techniques de l'Ingénieur. M3320 v1. 2011. Available online: https://www.techniques-ingenieur.fr/base-documentaire/materiaux-th11/ceramiques-proprietes-et-elaboration-42578210/procedes-de-frittage-pim-m3320/ (accessed on 6 March 2023).
14. Rane, K.; Strano, M. A comprehensive review of extrusion-based additive manufacturing processes for rapid production of metallic and ceramic parts. In *Advances in Manufacturing*; Shanghai University: Shanghai, China, 2019; Volume 7, pp. 155–173. [CrossRef]
15. Turner, B.N.; Gold, S.A. A review of melt extrusion additive manufacturing processes: II. Materials, dimensional accuracy, and surface roughness. *Rapid Prototyp. J.* **2015**, *21*, 250–261. [CrossRef]
16. Kumbhar, N.N.; Mulay, A.V. Post Processing Methods used to Improve Surface Finish of Products which are Manufactured by Additive Manufacturing Technologies: A Review. *J. Inst. Eng. India Ser. C* **2018**, *99*, 481–487. [CrossRef]
17. Hung, W. Post-Processing of Additively Manufactured Metal Parts. In *Additive Manufacturing Processes*; Bourell, D.L., Frazier, W., Kuhn, H., Seifi, M., Eds.; ASM International: Geauga, OH, USA, 2020; pp. 298–315.
18. Spitaels, L.; Rivière-Lorphèvre, E.; Díaz, M.C.; Duquesnoy, J.; Ducobu, F. Surface finishing of EBM parts by (electro-)chemical etching. *Procedia CIRP* **2022**, *108*, 112–117. [CrossRef]
19. Uçak, N.; Çiçek, A.; Aslantas, K. Machinability of 3D printed metallic materials fabricated by selective laser melting and electron beam melting: A review. *J. Manuf. Process.* **2022**, *80*, 414–457. [CrossRef]
20. Flynn, J.M.; Shokrani, A.; Newman, S.T.; Dhokia, V. Hybrid additive and subtractive machine tools – Research and industrial developments. *Int. J. Mach. Tools Manuf.* **2016**, *101*, 79–101. [CrossRef]
21. Mordor Intelligence. Hybrid Additive Manufacturing Machine Market—Growth, Trends, and Forecast (2020–2025), Technical Report. Available online: https://www.mordorintelligence.com/ (accessed on 6 March 2023).
22. Demarbaix, A.; Rivière-Lorphèvre, E.; Ducobu, F.; Filippi, E.; Petit, F.; Preux, N. Behaviour of pre-sintered Y-TZP during machining operations: Determination of recommended cutting parameters. *J. Manuf. Process.* **2018**, *32*, 85–92. [CrossRef]
23. NF E66-520-6; Working Zones of Cutting Tools. Couple Tool–Material. Association Française de Normalisation AFNOR: Paris, France, 1999.

Disclaimer/Publisher's Note: The statements, opinions and data contained in all publications are solely those of the individual author(s) and contributor(s) and not of MDPI and/or the editor(s). MDPI and/or the editor(s) disclaim responsibility for any injury to people or property resulting from any ideas, methods, instructions or products referred to in the content.

Journal of Composites Science

Article

Temperature-Dependent Deformation Behavior of "γ-austenite/δ-ferrite" Composite Obtained through Electron Beam Additive Manufacturing with Austenitic Stainless-Steel Wire

Elena Astafurova *, Galina Maier, Evgenii Melnikov, Sergey Astafurov, Marina Panchenko, Kseniya Reunova, Andrey Luchin and Evgenii Kolubaev

Institute of Strength Physics and Materials Science, Siberian Branch of Russian Academy of Sciences, 634055 Tomsk, Russia
* Correspondence: elena.g.astafurova@ispms.ru

Abstract: Temperature dependence of tensile deformation behavior and mechanical properties (yield strength, ultimate tensile strength, and an elongation-to-failure) of the dual-phase (γ-austenite/δ-ferrite) specimens, obtained through electron-beam additive manufacturing, has been explored for the first time in a wide temperature range $T = (77–300)$ K. The dual-phase structures with a dendritic morphology of δ-ferrite (γ + 14%δ) and with a coarse globular δ-phase (γ + 6%δ) are typical of the as-built specimens and those subjected to a post-production solid–solution treatment, respectively. In material with lower δ-ferrite content, the lower values of the yield strength in the whole temperature range and the higher elongation of the specimens at $T > 250$ K have been revealed. Tensile strength and stages of plastic flow of the materials do not depend on the δ-ferrite fraction and its morphology, but the characteristics of strain-induced $\gamma \rightarrow \alpha'$ and $\gamma \rightarrow \varepsilon \rightarrow \alpha'$ martensitic transformations and strain-hardening values are different for two types of the specimens. A new approach has been applied for the analysis of deformation behavior of additively fabricated Cr-Ni steels. Mechanical properties and plastic deformation of the dual-phase (γ + δ) steels produced through electron beam additive manufacturing have been described from the point of view of composite materials. Both types of the δ-ferrite inclusions, dendritic lamellae and globular coarse particles, change the stress distribution in the bulk of the materials during tensile testing, assist the defect accumulation and partially suppress strain-induced martensitic transformation.

Keywords: stainless steel; additive manufacturing; composite material; martensitic transformation; temperature; strain hardening

1. Introduction

The current industrial and technological development requires new manufacturing methods with high productivity, rapid prototyping and the ability to produce complex-shaped parts at a relatively low cost. Additive manufacturing (AM) is among the novel industrial technologies for fast prototyping of the complex parts made from different materials [1–3]. All modern AM methods are based on a layer-by-layer melting and solidification of the feedstock material (wire or powder) using different sources of energy (laser, plasma, electric arc, or electron beam). Selective laser melting/sintering, electron beam and arc additive manufacturing are the mostly well-known AM methods. The wire–feed electron beam additive manufacturing (EBAM) has some advantages over the powder-bed laser-based AM methods: high deposition rates and the ability to produce large components. However, the as-built EBAM-fabricated parts often require post-built machining and heat treatments to achieve a desired quality [1–3].

Conventionally produced austenitic stainless steels of the 300-series have good weldability and high corrosion resistance, being widely used in industrial, infrastructural and medical applications [4–6]. Austenitic stainless steel is a material appropriate for the EBAM process. Unfortunately, the elemental and phase compositions of the as-built parts do not coincide with those in steel wire used in the EBAM process. Typical elemental composition of austenitic stainless steels includes about 9% of nickel for stabilization of the austenitic structure [4]. At the expense of the high heat input, depletion of a melting pool with nickel occurs during the EBAM process, and, independently from the processing parameters, EBAM-fabricated parts always contain the residual δ-ferrite. Therefore, the main reason for δ-ferrite formation is the EBAM-assisted variation of a Cr/Ni equivalent of steel and, consequently, the change in solidification mode [7–10]. Additionally, columnar coarse austenitic grains usually grow during the additive manufacturing and provide the high anisotropy of the mechanical properties in the as-built material [8–10]. An anisotropic two-phase (γ-austenite + δ-ferrite) microstructure is formed due to the nonequilibrium solidification and crystallization conditions during layer-by-layer deposition of the material and complex thermal history of the resulting bulk product [8–12]. Post-production heat treatment of the additively fabricated stainless steel partially eliminates these effects, i.e, reduces the fraction of δ-ferrite but does not completely dissolve it and retains grain size unaffected [8,9,11]. In fact, EBAM-fabricated bulk products made from stainless steels type AISI304 or AISI321 possess dual-phase composite structures because the content of δ-ferrite in them could be as high as 20% [8–10].

High fraction of dendritic or globular "hard" δ-phase, which is randomly distributed in "soft" austenitic matrix, can influence all microstructure-dependent mechanical properties, stages of plastic flow, deformation mechanisms, and fracture of the additively manufactured steels. Metastable austenitic steels typically undergo a strain-induced $\gamma \to \alpha'$ or $\gamma \to \varepsilon \to \alpha'$ martensitic transformation (MT) [13,14]. Kinetics and the resulting volume fraction of the strain-induced martensite (SIM) are determined by the chemical composition and structure of steel (stacking fault energy (SFE), grain size, phase composition, etc.) and deformation regime (strain rate, temperature, deformation mode, etc.) [13–20]. In austenitic stainless steels (Fe–18%Cr-(8–14)%Ni, mass.%) with low SFE (below 20 mJ/m^2), the strain-induced $\gamma \to \varepsilon \to \alpha'$ MT is realized at $T < 300$ K [14,15,17–20]. The nucleation and growth of the α'-martensitic phase can be realized both with or without intermediate twinning or ε-martensite [15,21,22]. The amount of the α'-SIM in the samples is a temperature-dependent characteristic: if the deformation temperature decreases, the kinetics of the $\gamma \to \alpha'$, $\gamma \to \varepsilon \to \alpha'$ MTs speed up, which assists with a higher strain-hardening rate and higher tensile strength [14,18–20,23]. The mechanisms of the MT development in austenitic stainless steels obtained using the conventional methods have been studied in detail, including the influence of the deformation temperature on tensile properties and the deformation behavior of such steels [13–23].

Since austenitic steels obtained through additive manufacturing are promising materials for operation at low temperatures, the characteristics of the MTs in such materials are of special interest but they have not been studied yet. Y. Hong and coauthors [24] have shown that high density of low-angle boundaries and cellular microstructure suppress dislocation slip and deformation twinning in austenitic stainless steel (fabricated by selective laser melting, SLM). Both factors reduce the number of nucleation sites for SIM, therefore enhancing austenite stability against MTs. Unlike the EBAM method, the δ-phase does not arise in the SLM-produced austenitic stainless steels. To date, the effect of the δ-ferrite on the characteristics of strain-induced martensitic transformation in EBAM-fabricated steels has not been revealed.

In this paper, we study the temperature dependence of the strength properties and SIM-assisted deformation behavior in stainless steel samples obtained using the EBAM method.

2. Materials and Methods

Steel billets with the linear dimensions of $110 \times 33 \times 6$ mm^3 were produced using the EBAM method (ISPMS SB RAS, Tomsk, Russia). Stainless-steel wire type AISI 321 with a diameter of 1 mm was used as a feedstock material. A carbon steel plate with the dimensions of 140 × 75 × 10 mm³ was used as a substrate, which was not cooled during the EBAM process. The additive manufacturing process was carried out in a vacuum chamber, and the following parameters were used: accelerating voltage—30 kV; beam current—40–50 mA; wire–feed speed—180 mm/min; beam sweep—4.5 × 4.5 mm²; scanning frequency—1 kHz. The samples were studied (1) after the additive manufacturing (AM) and (2) after the post-built solid–solution treatment that consisted of annealing for 1 h at a temperature of 1100 °C and water quenching (AM + SST). For comparative analysis, the conventionally produced steel samples type AISI 321 were used (solution-treated for the similar regime, hereinafter called "as-cast").

Mechanical tests for the uniaxial static tension were carried out using flat proportional dumbbell-shaped samples with the gauge sections of 12 × 2.6 × 1.3 mm³. Tensile axis of the samples coincided with a building direction of the EBAM-fabricated billet. After mechanical grinding of the samples, an electrolytic polishing was carried out in a supersaturated solution of chromium anhydride CrO_3 in orthophosphoric acid. Due to the different compositions of austenitic and ferritic phases, the ferritic one was etched during the polishing of the specimens. After that, it can be clearly identified with either light or scanning electron microscopy.

Tension was carried out at temperatures of 77, 183, 223, 273 and 300 K and an initial strain rate of 5×10^{-4} s^{-1} using an electromechanical testing system (model 1185, Instron, High Wycombe, GB) with a low-temperature chamber. At least five samples of the AM and AM + SST steels were tested at each temperature. As no extensometer could be used in the low-temperature chamber, the force (P) and the displacement of the traverse ($\Delta l = l - l_0$, where l is the length of the deformed sample, and l_0 is the initial length of the sample) were collected during the tensile tests. These parameters were converted to the engineering stress ($\sigma_e = P/S_0$, S_0 is the initial cross-sectional area of the sample) and engineering strain ($\varepsilon_e = \Delta l / l_0$). Then, true stress ($\sigma_t = \sigma_e (1 + \varepsilon_e)$)–true strain ($\varepsilon_e = \ln(l/l_0)$) plots were reconstructed assuming a uniform deformation of the sample at strains lower than what corresponded to an ultimate tensile stress (UTS). For higher strain, true diagrams could not be rearranged because the plastic deformation was localized in the neck. To ensure that samples deform uniformly before UTS is reached, some tensile tests were interrupted at different strains and visual control of the sample shape was carried out. Necks formed at the latter stages of plastic flow, after the UTS was reached. Using true diagrams, a strain-hardening rate, SHR = $d\sigma_t/d\varepsilon_t$, was calculated.

The microstructure of the samples was studied using scanning electron microscopy (SEM, Zeiss Leo Evo 50 (Zeiss, Oberkochen, Germany) with a supply for energy dispersive X-ray spectroscopy, EDS, accelerating voltage—30 kV, SE (secondary electrons) regime). Using SEM images, the sizes of austenitic grains and the widths of δ-ferrite lamellae were measured using a linear interception method. X-ray structural and phase analysis of the samples were carried out on a DRON-7 diffractometer (Bragg–Brentano geometry, Co-K_α radiation, accelerating voltage,15 kV, current, 15 mA, 2θ range, 40–120°, Bourevestnik, Saint Petersburg, Russia). The volume fraction of the magnetic δ-ferrite and α'-phase was determined using a multi-functional eddy current device MVP-2M (Kropus, Moscow, Russia) with a limit of the ferritic phase detection of 0.2%.

3. Results

Figure 1 shows the characteristic SEM images of the microstructure in the AM and AM + SST samples. After AM, the microstructure of the samples consists of the coarse austenitic grains with colonies of dendritic δ-ferrite, which are oriented mainly along elongated austenitic grains (this direction coincides with the building direction of the EBAM billet and tensile axis of the samples). While the transverse grain size of austenite is 100–500 microns, the longitudinal one could reach several millimeters. The volume fraction of δ-phase is 14% as it was measured in the magnetic phase analysis. The thickness of the δ-ferrite lamellae varies between 0.5 and 1.5 μm according to the SEM analysis (Figure 1a).

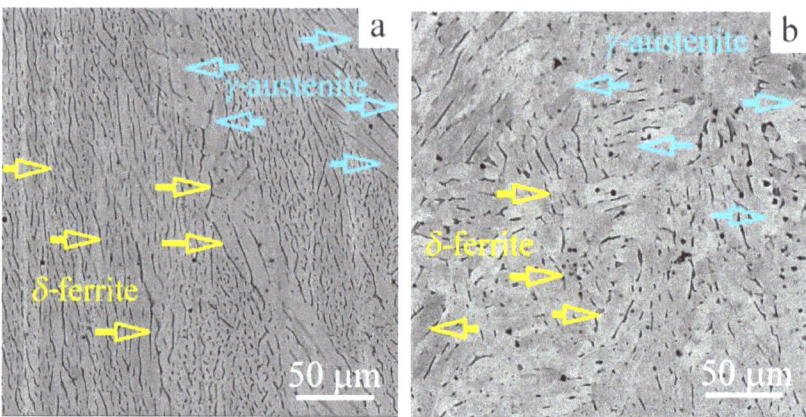

Figure 1. Typical SEM images of the microstructure in AM (**a**) and AM + SST (**b**) samples. Dark gray areas are δ-ferrite lamellae (some of them are marked by yellow arrows) in the austenitic matrix (light grey areas are marked by turquoise arrows). The applied magnification is ×200.

Solid–solution treatment of the additively manufactured steel stimulates a δ→γ phase transformation in the samples. Unfortunately, the complete dissolution of the δ-ferrite does not occur, while the morphology of the δ phase changes drastically. On SEM images, the globular, non-equiaxed δ-ferrite lamellae are observed (Figure 1b). After the solid–solution treatment, the volume content of the ferritic phase is 6%. At the same time, no visible changes were found in the grain structure of the austenitic phase, which is in accordance with our previous data [8,9].

The tensile diagrams obtained for the AM and AM + SST samples depending on the test temperature are shown in Figure 2. Engineering diagrams show the complete plastic flow of the material, including the stage of macroscopic strain localization (in the assumption of the invariable cross-section of the sample during straining). Maximum applied stresses on the engineering diagrams correspond to the UTS, and samples undergo uniform deformation at $\varepsilon_e < \varepsilon_{UTS}$. The common view of the non-deformed samples and those tensile-tested at room temperature and 77 K up to the 10% strain are shown in Figure 2g. At strains higher than ε_{UTS}, plastic deformation is localized in a rather narrow region (necks are seen in Figure 2g for the samples tensile tested to failure) both at room temperature and in a low-temperature deformation regime. The stage of the macroscopical localization of plastic flow in a pre-neck deformation regime is seen in every engineering diagram (Figure 2a,c,e) but it has not been reconstructed in true stress–true strain diagrams (Figure 2b,d,f) and has not been studied in this research.

At the assumption of the uniform decrease of the cross-sectional area of the samples, true stress–true strain diagrams show deformation behavior of the steels' excluding stages at $\varepsilon_t > \varepsilon_{UTS}$ (Figure 2b,d). At room temperature deformation, plastic flow of the AM and AM + SST samples develops in three main stages (Figure 2b,d), which could be clearly identified by the strain-hardening rate (SHR = $d\sigma/d\varepsilon$) variations with strain and stress (Figure 3). After a rather long intermediate stage between elastic and the macroscopical plastic deformation of the samples, the stage I starts at stresses of about 400 MPa (0.05 true strain). The value of SHR gradually decreases with strain and stresses from 2 GPa at the beginning down to 1–1.5 GPa at the end of stage I (about 0.3 true strain). At higher strain, stage II starts, which is almost linear for the AM samples and shows insignificant growth of the SHR for the AM + SST samples. At strains higher than 0.4 (at true stress higher than 800 MPa), stage III starts. It is characterized by high SHR that gradually decreases with strain (stress) (Figure 3a–c).

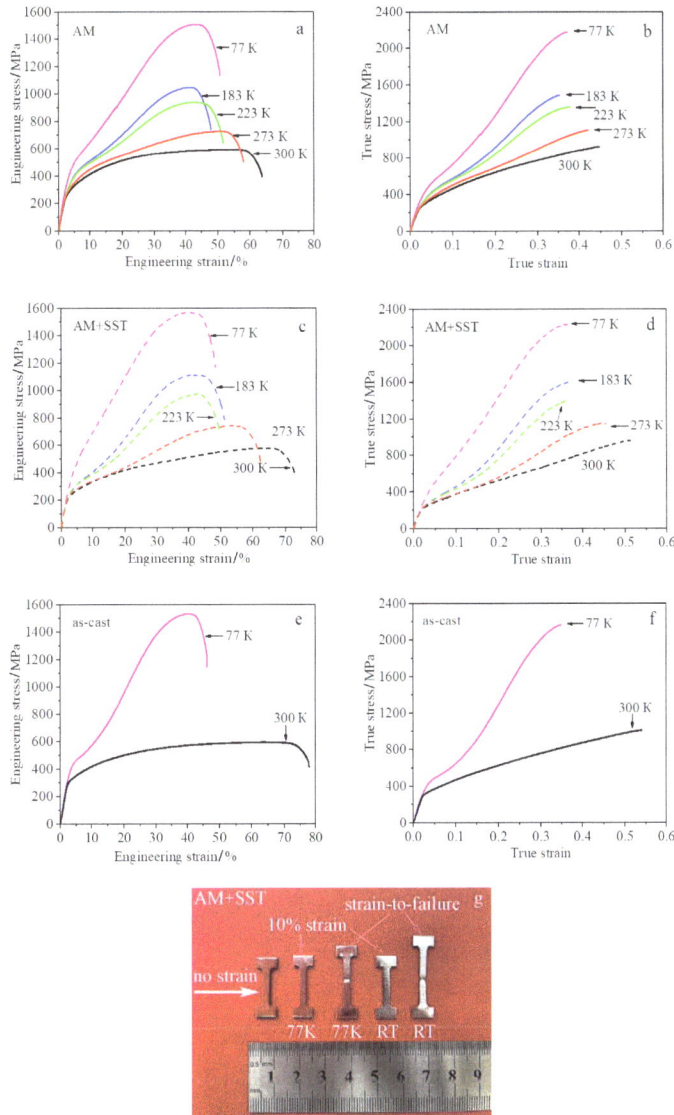

Figure 2. Tensile diagrams in the engineering (**a**,**c**,**e**) and true (**b**,**d**,**f**) coordinates for the AM (**a**,**b**), AM + SST (**c**,**d**) and the as-cast (**e**,**f**) samples. Common view of the AM + SST samples before strain, after 10% strain and after the failure at room temperature (RT) and at 77 K (**g**). Test temperatures are given in the figures.

At room temperature, the strain-hardening behavior of AM + SST samples is very similar to that of the as-cast austenitic steel except for the differences in stage II: almost linear hardening is observed in the as-cast samples, and the increase in SHR with strain is typical of the AM + SST ones (Figure 3c–f). Contrarily, the SHR values for the AM composite specimens are much higher than those in the as-cast steel in stage I, but the deformation behavior in stages II and III are very similar except for the length of the stages: the AM samples show lower elongation-to-failure and shorter lengths of the stages (Figure 3a,b,e,f).

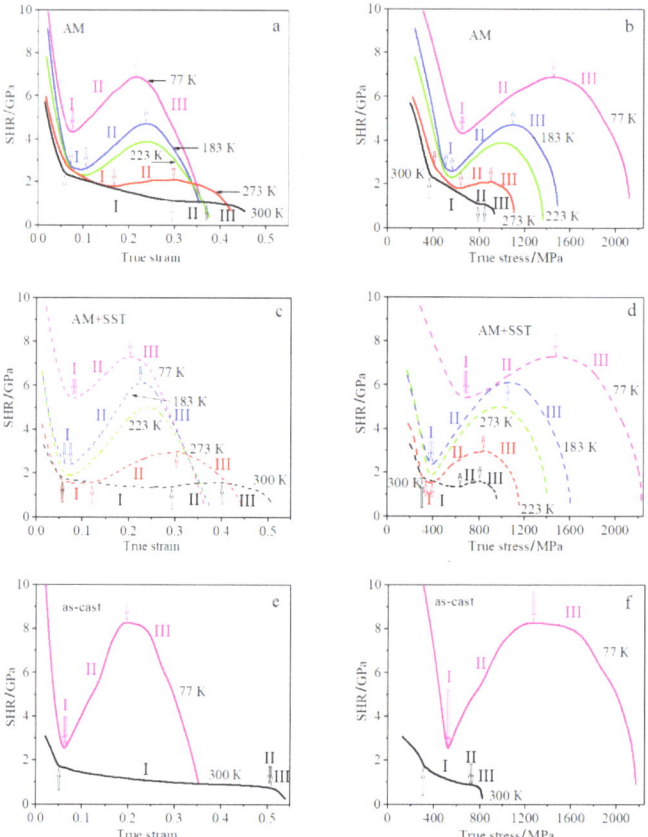

Figure 3. The strain-hardening rate (SHR) vs. true strain (**a,c,e**) and the SHR vs. true stress (**b,d,f**) for the AM (**a,b**), the AM + SST (**c,d**) and the as-cast (**e,f**) samples. Test temperatures are given in the figures.

A decrease in test temperature is accompanied by a change in the shape of the flow curve for all samples. During tensile deformation of the AM and AM + SST samples at 273 K, stage I is much shorter (strain hardening is similar) and stages II and III start much earlier than those at room temperature deformation (Figure 3a–d). Regardless of the phase composition of the samples, a pronounced increase in flow stresses with strain (stress) is observed in stage II, and the SHR value at this stage increases in comparison with that at room temperature. Again, the increase in SHR with strain and stresses in stage II is more pronounced for AM + SST-samples (Figure 3a–d).

When the test temperature is lowered to 223 K, 183 K, and 77 K, stage I becomes very short, and stage II starts very close to the beginning of the plastic flow. Flow stresses and SHR values at stage II increase drastically with strain. Tensile diagrams take a pronounced S-shape (Figure 2a–d). The slopes of the dependences SHR(ε) and SHR(σ) in stage II are greater for the AM + SST samples in comparison with the AM samples, in which the fraction of δ-ferrite is the highest (Figure 3a–d). If one compares the stages of plastic flow at 77 K for the as-cast and additively fabricated materials, two main distinctions can be highlighted. First, despite the close strains at which stage II starts, the corresponded stresses and strain-hardening rate values are much lower in the as-cast material. Second, strain hardening increments $\Delta SHR/\Delta\varepsilon$ and $\Delta SHR/\Delta\sigma$ in stage II are much higher for pure austenitic steel (Figures 2 and 3).

Figure 4 shows the temperature dependence of the yield strength (YS), the UTS and the elongation-to-failure for all steel samples. A decrease in test temperature is accompanied by an increase in the values of the YS and the UTS and by a decrease in total elongation of the steels in the temperature range (77–300) K. When comparing the values of the yield strength for the AM and the AM + SST samples, one can notice that with the decrease in the fraction of δ-ferrite in the microstructure, the YS decreases in the whole temperature range (Figure 4a). Simultaneously, the UTSs for these specimens are equal and the total elongations are different only at room temperature deformation (Figure 4b,c). The non-obvious result is that the YSs for as-cast materials (pure austenite) coincide with those for the composite AM samples containing 14% of δ-ferrite (Figure 4a).

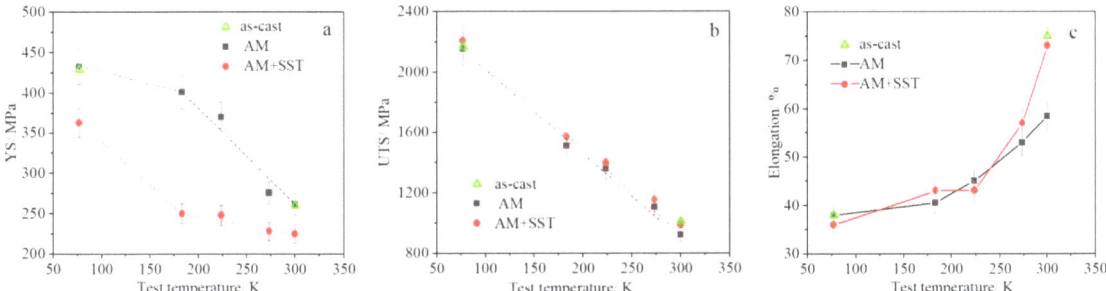

Figure 4. The temperature dependencies of the yield strength (YS) (**a**), the ultimate tensile strength (UTS, estimated using true diagrams) (**b**) and elongation-to-failure (total elongation estimated using engineering diagrams) (**c**) for the AM, the AM + SST and the as-cast samples.

4. Discussion

The experimental data described above indicate that the content of δ-ferrite affects the yield strength of the studied samples. It is surprising that the YSs of the pure austenitic as-cast samples and dual-phase AM samples are equal at temperatures of 77 K and 300 K. The grain sizes (d) of these two materials are different (15μm for as-cast samples and hundreds of micrometers for AM samples). According to the well-known Hall–Petch relationship (YS~$K_{H-P} \times d^{-1/2}$, K_{H-P} is a constant ≈ 400 MPa × μm$^{1/2}$ [25]), the YS of the coarse-grained AM samples must be much lower than that of the as-cast material in the whole temperature range (the decrease in grain size from 100–500 μm down to 15 μm is accompanied by a growth of the YS in 65–85 MPa). If one assumes the grain size of austenitic phase in the AM material, this idea is not supported by the data in Figure 4a.

In fact, due to the high volume fraction of δ-phase and its dendritic morphology, the plastic deformation of the AM samples should be considered as that for the composite material: "hard" inclusions of δ-phase in "soft" austenitic matrix. When we estimate the YS of the AM samples, it is correct to use a "rule-of-mixtures" [26]: $YS_{\gamma+\delta} = 0.86 \times YS_{\gamma\text{-phase}} + 0.14 \times YS_{\delta\text{-phase}}$. This rule is generally applied for the aligned fiber reinforced metal matrix composite under a load in the direction of the fibers (δ-ferrite primary arms in our case). In our previous paper [9], we have considered that, in the very beginning of plastic flow of the AM samples, a free pass of dislocation glide is restricted by the γ/δ interphase boundaries. Therefore, the mean distance between ferritic lamellae could be assumed as d in the Hall–Petch relationship for the resultant value $YS_{\gamma\text{-phase}}$. This assumption allowed us to describe the experimentally observed decrease in the YS value in the AM + SST samples (Figure 4a) due to the partial dissolution of δ-ferrite arms during the SST and due to the increase in a mean-free pass for the dislocation glide. In this approach, the $YS_{\gamma+\delta}$ of the AM samples would exceed the YS of the as-cast material because: (1) the mean distance between ferritic lamellae is one order value lower than the grain size of austenite in the as-cast samples (Figure 1a), and the term $0.86 \times YS_{\gamma\text{-phase}}$ for the AM sample could be higher than the $YS_{\gamma\text{-phase}}$ for the as-cast material; (2) the additional term $0.14 \times YS_{\delta\text{-phase}}$ would increase the YS$_{\gamma+\delta}$ value. This difference is compensated by the δ-ferrite-assisted change in the local stress state in a bulk of the materials. Due to the different elastic properties of the austenite and ferrite, the "hard" δ-ferrite lamellae play the role of stress concentrators under external loading [26–29] and can initiate a macroplastic flow at stresses, which are lower than the YS$_{\gamma+\delta}$ calculated using the "rule of mixtures". This question is still open and needs a precise calculation of the YS in the framework of a separate paper. Nevertheless, the above discussion is supported by the temperature dependences of the YS for the AM and AM + SST samples. The dependence YS(T) for the AM + SST material in Figure 4a has a view typical of fcc materials: the substantial growth of the YS at $T < 200$ K in a temperature range of a thermally activated dislocation glide and a plateau at $T > 200$ K [30]. For the AM samples with high fraction of the δ-ferrite, the slope $\Delta YS/\Delta T$ decreases in a low-temperature deformation regime. This is because the temperature dependence of the YS for bcc ferrite is much higher than that for the fcc austenite [31]. Therefore, in the lower test temperature, the stronger δ-phase reduces the YS$_{\gamma+\delta}$ of the composite material.

The γ/δ interphase boundaries act as barriers for dislocations, and "hard" inclusions of δ-ferrite change stress distribution in the bulk of material, making it very inhomogeneous during tensile testing. As a result, δ-ferrite can assist the activation of dislocation sources in a primary slip system and those with non-maximum Schmid factors (secondary slip systems), enhance the multiple slip and promote the accumulation of dislocations in the interdendritic austenite. We confirmed the similar effects using transmission electron microscopical studies of a high-nitrogen steel and the multicomponent alloys with the coarse "hard" particles [29,32]. The stress-assisted multiple slip promotes high strain hardening in stage I for the AM and the AM + SST samples as compared to single-phase as-cast material (Figures 2 and 3).

The transition to the stage II is associated with the activation of strain-induced ($\gamma \rightarrow \alpha'$) and ($\gamma \rightarrow \varepsilon \rightarrow \alpha'$) martensitic transformations in austenitic CrNi stainless steels with low SFE [13,15,18–20]. Figure 5 shows the X-ray diffraction patterns obtained for the AM and AM + SST samples tensile-tested at room temperature and at 77 K to the strains, corresponding with the beginning of stage II. After room temperature deformation up to 30% strain, austenite is the main phase in the AM and AM + SST specimens, but weak ($\delta + \alpha'$) reflections are seen in the diffraction patterns (both δ and α' phases possess a bcc crystal lattice and close lattice parameters; therefore they cannot be separated in X-ray diffraction patterns; Figure 5a). The only α' phase could arise during the deformation, while fraction of δ-ferrite does not vary. Therefore, the insignificant fraction of SIM was formed at room temperature deformation in stage I. The weak reflections of the ε-phase and relatively high lines of α'-martensite are clearly seen in Figure 5b. This proves the ($\gamma \rightarrow \varepsilon \rightarrow \alpha'$) sequence of the transformation at 77 K. The growth of the α'-martensitic phase could be realized either with or without intermediate ε-martensite, and the temperature dependence of the strain-induced transformation sequence is in accordance with [15,18,21,22].

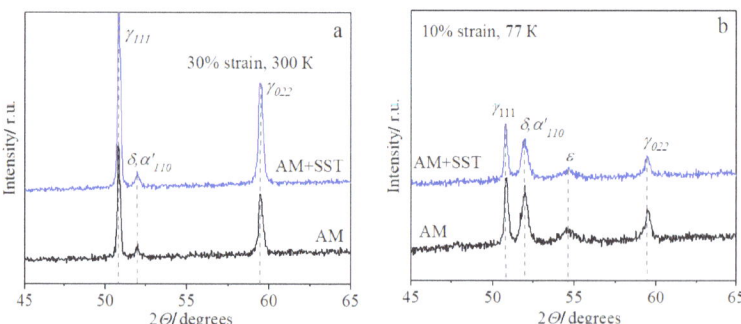

Figure 5. X-ray diffraction patterns for the AM, AM + SST samples tensile-tested with 30% strain (**a**) and 10% strain (**b**) at room temperature (**a**) and at 77 K (**b**).

The macroscopical ($\gamma \rightarrow \alpha'$) transformation is realized in stage II. The strain-hardening rate at stage II is typically directly proportional to the fraction of SIM [19,20]. Deformation behavior of the samples does not allow one to compare the difference in the ($\gamma \rightarrow \alpha'$) transformation at room temperature due to the slow transformation kinetics. However, for low-temperature deformation regimes, the analysis of the deformation stage II in Figure 3 and Table 1 shows that:

(i) The composite dual-phase steels (AM and AM + SST samples) possess slower kinetics of the strain-induced ($\gamma \rightarrow \varepsilon \rightarrow \alpha'$) martensitic transformation relative to the single-phase austenitic as-cast material. Additionally, the higher the fraction of δ-phase in samples, the slower the SIM kinetics;

(ii) δ-ferrite fraction weakly influences strain for the start and length of stage II, which corresponds to the macroscopic growth of the α' SIM. In the dual-phase steels, stage II starts mostly at higher stresses and SHRs compared to pure austenitic as-cast steel. Therefore, the start of the transformation needs higher stresses and strain hardening for the composite structure than for pure austenite.

Table 1. The characteristics of the stage II of the plastic flow for the AM, AM + SST and the as-cast samples.

Characteristic of Tensile Diagram	Test Temperature, K	AM	AM + SST	As-Cast
Strain at start of stage II (in values of true strain)	77	0.07	0.08	0.06
	183	0.10	0.07	–
	223	0.11	0.08	–
	273	0.17	0.12	–
	300	0.31	0.29	0.35
Stress at start/finish of stage II, MPa	77	650/1450	690/1470	520/1230
	183	570/1100	390/1050	–
	223	570/1000	390/970	–
	273	640/910	420/850	–
	300	810/840	650/800	710/760
Length of stage II (in values of true strain)	77	0.14	0.12	0.13
	183	0.14	0.17	–
	223	0.13	0.16	–
	273	0.13	0.18	–
	300	0.07	0.10	0.06
SHR at start/finish of stage II, MPa	77	4330/6900	5440/7320	2580/8250
	183	2600/4720	2380/6140	–
	223	2320/3900	1900/5000	–
	273	1820/2160	1580/3000	–
	300	1080/1070	1360/1590	920/870

These results directly show that dendritic lamellae and globular coarse particles of δ-ferrite partially suppress the strain-induced martensitic transformation in stainless steel type 321. These data are in accordance with the results reported by Y. Hong [24], who obtained similar SIM behavior for the SLM-fabricated austenitic stainless steel. The authors concluded that low-angle grain boundaries and cellular microstructure inhibit dislocation slip, the formation of dislocation bands and twins. All these factors restrict the nucleation of α' SIM. In our case, the reason for the reduced kinetics of the SIM transformation in the

EBAM-fabricated samples could be similar: a δ-ferrite-assisted stress distribution stimulates dislocation gliding in the multiple slip systems in the stage I, providing a specific highly defective microstructure in interdendritic regions and inhibiting α' nucleation and growth.

5. Conclusions

For the first time, strength properties, deformation behavior, and development of the strain-induced $\gamma \rightarrow \alpha'$ and $\gamma \rightarrow \varepsilon \rightarrow \alpha'$ MT in samples of stainless steel, fabricated using electron beam additive manufacturing, have been explored in a wide temperature range. The uniaxial tensile testing at the temperatures of 77, 183, 223, 273 and 300 K has been carried out for two types of the samples: after additive manufacturing process (as-built) and after post-processing solid solution treatment. The samples have the dual-phase ($\gamma + \delta$) structures with different contents of δ-ferrite: 14% and 6%, respectively.

In the studied temperature range, the samples with higher content of δ-ferrite have higher values of yield strength. The composite dual-phase steels (AM and AM + SST samples) possess slower kinetics of strain-induced ($\gamma \rightarrow \varepsilon \rightarrow \alpha'$) martensitic transformation in deformation stage II relative to the single-phase austenitic as-cast material. A novel approach has been proposed for the interpretation of the results. The plastic deformation of the additively manufactured samples should be considered for the composite materials. The YS of the samples could be described by the "rule of mixtures" assuming that "hard" δ-ferrite lamellae play the role of stress concentrators in a "soft" austenitic matrix, and they can reduce the yield strength and partially suppress strain-induced martensitic transformation in a composite structure under loading.

Author Contributions: Conceptualization, E.A. and E.K.; methodology, S.A. and E.A.; investigation, S.A., E.M., M.P., K.R. and A.L.; data curation, E.A. and E.M.; writing—original draft preparation, G.M., E.A. and A.L.; writing—review and editing, E.A. and E.K.; visualization, G.M.; supervision, E.A. and E.K.; project administration, E.A. and E.K.; funding acquisition, E.A. and E.K. All authors have read and agreed to the published version of the manuscript.

Funding: This research was funded by the Government research assignment for ISPMS SB RAS, project FWRW-2022-0005.

Data Availability Statement: Data available on request.

Acknowledgments: The study was conducted using the equipment of ISPMS SB RAS ("Nanotech" center) and Tomsk State University (Tomsk Materials Science Center of Collective Use). The authors thank V. Rubtsov and S. Nikonov for their assistance with the additive manufacturing of the steels.

Conflicts of Interest: The authors declare no conflict of interest.

References

1. Li, N.; Huang, S.; Zhang, G.; Qin, R.; Liu, W.; Xiong, H.; Shi, G.; Blackburn, J. Progress in Additive Manufacturing on New Materials: A Review. *J. Mater. Sci. Technol.* **2019**, *35*, 242–269. [CrossRef]
2. Ngo, T.D.; Kashani, A.; Imbalzano, G.; Nguyen, K.T.Q.; Hui, D. Additive manufacturing (3D printing): A review of materials, metals, applications and challenges. *Compos. Part B Eng.* **2018**, *143*, 172–196. [CrossRef]
3. Frazier, W.E. Metal additive manufacturing: A review. *J. Mater. Eng. Perform.* **2014**, *23*, 1917–1928. [CrossRef]
4. Lo, K.H.; Shek, C.H.; Lai, J.K.L. Recent Developments in Stainless Steels. *Mater. Sci. Eng. R* **2009**, *65*, 39–104. [CrossRef]
5. Tian, K.V.; Passaretti, F.; Nespoli, A.; Placidi, E.; Condò, R.; Andreani, C.; Licoccia, S.; Chass, G.A.; Senesi, R.; Cozza, P. Composition—Nanostructure Steered Performance Predictions in Steel Wires. *Nanomaterials* **2019**, *9*, 1119. [CrossRef] [PubMed]
6. Tian, K.V.; Festa, G.; Basoli, F.; Laganá, G.; Scherillo, A.; Andreani, C.; Bollero, P.; Licoccia, S.; Senesi, R.; Cozza, P. Orthodontic archwire composition and phase analysis by neutron spectroscopy. *Dent. Mater. J.* **2017**, *36*, 282–288. [CrossRef]
7. Bajaj, P.; Hariharan, A.; Kini, A.; Kurnsteiner, P.; Raabe, D.; Jagle, E.A. Steels in additive manufacturing: A review of their microstructure and properties. *Mater. Sci. Eng. A* **2020**, *772*, 138633. [CrossRef]
8. Melnikov, E.V.; Astafurova, E.G.; Astafurov, S.V.; Maier, G.G.; Moskvina, V.A.; Panchenko, M.Y.; Fortuna, S.V.; Rubtsov, V.E.; Kolubaev, E.A. Anisotropy of the tensile properties in austenitic stainless steel obtained by wire-feed electron beam additive growth. *Lett. Mater.* **2019**, *9*, 460–464. [CrossRef]
9. Astafurova, E.G.; Panchenko, M.Y.; Moskvina, V.A.; Maier, G.G.; Astafurov, S.V.; Melnikov, E.V.; Fortuna, A.S.; Reunova, K.A.; Rubtsov, V.E.; Kolubaev, E.A. Microstructure and grain growth inhomogeneity in austenitic steel produced by wire-feed electron beam melting: The effect of post-building solid-solution treatment. *J. Mater. Sci.* **2020**, *55*, 9211–9224. [CrossRef]

10. Astafurov, S.V.; Astafurova, E.G. Phase composition of austenitic stainless steels in additive manufacturing: A review. *Metals* **2021**, *11*, 1052. [CrossRef]
11. Chen, X.; Li, J.; Cheng, X.; Wang, H.; Huang, Z. Effect of heat treatment on microstructure, mechanical properties and corrosion properties of austenitic stainless steel 316L using arc additive manufacturing. *Mater. Sci. Eng. A* **2018**, *715*, 307–314. [CrossRef]
12. Tarasov, S.Y.; Filippov, A.V.; Shamarin, N.N.; Fortuna, S.V.; Maier, G.G.; Kolubaev, E.A. Microstructural evolution and chemical corrosion of electron beam wire-feed additive manufactured AISI 304 stainless steel. *J. Alloy. Compd.* **2019**, *803*, 364–370. [CrossRef]
13. Tamamura, I.; Wayman, C.M. Martensitic Transformations and Mechanical Effects. In *Martensite*; Olson, G.B., Owen, W.S., Eds.; ASM International: Russel Township, OH, USA, 1992; pp. 227–242.
14. Sohrabi, M.J.; Naghizadeh, M.; Mirzadeh, H. Deformation induced martensite in austenitic stainless steels: A review. *Arch. Civil Mech. Eng.* **2020**, *20*, 124. [CrossRef]
15. Kireeva, I.; Chumlyakov, Y.I. The orientation dependence of γ-α' martensitic transformation in austenitic stainless steel single crystals with low stacking fault energy. *Mater. Sci. Eng. A* **2008**, *481*, 737–741. [CrossRef]
16. Litovchenko, I.Y.; Tyumentsev, A.N.; Akkuzin, S.A.; Naiden, E.P.; Korznikov, A.V. Martensitic transformations and the evolution of the defect microstructure of metastable austenitic steels during severe plastic transformations by high pressure torsion. *Phys. Met. Metall.* **2016**, *117*, 847–856. [CrossRef]
17. Choi, J.-Y.; Jin, W. Strain induced martensite formation and its effect on strain hardening behavior in the cold drawn 304 austenitic stainless steels. *Scr. Mater.* **1997**, *36*, 99–104. [CrossRef]
18. Spencer, K.; Veron, M.; Yu-Zhang, K.; Embury, J.D. The strain induced martensite transformation in austenitic stainless steels. Part 1–Influence of temperature and strain history. *Mater. Sci. Technol.* **2009**, *25*, 7–17. [CrossRef]
19. Seetharaman, V.; Krishnan, R. Influence of the martensitic transformation on the deformation behaviour of an AISI 316 stainless steel at low temperature. *J. Mater. Sci.* **1981**, *16*, 523–530. [CrossRef]
20. De, A.K.; Speer, J.G.; Matlock, D.K.; Matlock, D.K.; Murdock, D.C.; Mataya, M.C.; Comstock, R.J., Jr. Deformation-induced phase transformation and strain hardening in type 304 austenitic stainless steel. *Metall. Mater. Trans. A* **2006**, *37*, 1875–1886. [CrossRef]
21. Li, X.-F.; Ding, W.; Cao, J.; Ye, L.-Y.; Chen, J. In situ TEM observation on martensitic transformation during tensile deformation of SUS304 metastable austenitic stainless steel. *Acta Metall. Sin. (Engl. Lett.)* **2015**, *28*, 302–306. [CrossRef]
22. Kaoumi, D.; Liu, J. Deformation induced martensitic transformation in 304 austenitic stainless steel: In-situ vs. ex-situ transmission electron microscopy characterization. *Mater. Sci. Eng. A* **2018**, *715*, 73–82. [CrossRef]
23. Nagy, E.; Mertinger, V.; Tranta, F.; Solyom, J. Deformation induced martensitic transformation in stainless steels. *Mater. Sci. Eng. A* **2004**, *378*, 308–313. [CrossRef]
24. Hong, Y.; Zhou, C.; Zheng, Y.; Zhang, L.; Zheng, J.; Chen, X.; An, B. Formation of strain-induced martensite in selective laser melting austenitic stainless steel. *Mater. Sci. Eng. A* **2019**, *740–741*, 420–426. [CrossRef]
25. Astafurov, S.V.; Maier, G.G.; Melnikov, E.V.; Moskvina, V.A.; Panchenko, M.Y.; Astafurova, E.G. The strain-rate dependence of the Hall-Petch effect in two austenitic stainless steels with different stacking fault energies. *Mater. Sci. Eng. A* **2019**, *756*, 365–372. [CrossRef]
26. Chawla, N.; Chawla, K.K. *Metal Matrix Composites*; Springer: New York, NY, USA, 2006.
27. Astafurov, S.V.; Shilko, E.V.; Ovcharenko, V.E. Influence of Phase Interface Properties on Mechanical Characteristics of Metal Ceramic Composites. *Phys. Mesomech.* **2014**, *17*, 282–291. [CrossRef]
28. Psakhie, S.G.; Shilko, E.V.; Grigoriev, A.S.; Astafurov, S.V.; Dimaki, A.V.; Smolin, A.Y. A mathematical model of particle–particle interaction for discrete element based modeling of deformation and fracture of heterogeneous elastic–plastic materials. *Eng. Fract. Mech.* **2014**, *130*, 96–115. [CrossRef]
29. Astafurova, E.G.; Melnikov, E.V.; Reunova, K.A.; Moskvina, V.A.; Astafurov, S.V.; Panchenko, M.Y.; Mikhno, A.S.; Tumbusova, I. Temperature Dependence of Mechanical Properties and Plastic Flow Behavior of Cast Multicomponent Alloys $Fe_{20}Cr_{20}Mn_{20}Ni_{20}Co_{20-x}C_x$ (x = 0, 1, 3, 5). *Phys. Mesomech.* **2021**, *24*, 674–683. [CrossRef]
30. Chumlyakov, Y.I.; Kireeva, I.V.; Sehitoglu, H.; Litvinova, E.I.; Zaharova, E.G.; Luzginova, N.V. High-Strength Single Crystals of Austenitic Stainless Steel with Nitrogen Content: Mechanisms of Deformation and Fracture. *Mater. Sci. Forum* **1999**, *318–320*, 395–400.
31. Honeycombe, R.W.K. *The Plastic Deformation of Metals*; Edward Arnold (Publ.): London, UK, 1968; 447p.
32. Astafurova, E.G.; Moskvina, V.A.; Maier, G.G.; Gordienko, A.I.; Burlachenko, A.G.; Smirnov, A.I.; Bataev, V.A.; Galchenko, N.K.; Astafurov, S.V. Low-temperature tensile ductility by V-alloying of high-nitrogen CrMn and CrNiMn steels: Characterization of deformation microstructure and fracture micromechanisms. *Mater. Sci. Eng. A* **2019**, *745*, 265–278. [CrossRef]

Disclaimer/Publisher's Note: The statements, opinions and data contained in all publications are solely those of the individual author(s) and contributor(s) and not of MDPI and/or the editor(s). MDPI and/or the editor(s) disclaim responsibility for any injury to people or property resulting from any ideas, methods, instructions or products referred to in the content.

Article

Mechanical Properties of PLA Specimens Obtained by Additive Manufacturing Process Reinforced with Flax Fibers

Ana Paulo [1], Jorge Santos [1], João da Rocha [1], Rui Lima [2,3] and João Ribeiro [1,4,5,*]

1. Instituto Politécnico de Bragança, Campus de Santa Apolónia, 5300-253 Bragança, Portugal
2. MEtRICs, Mechanical Engineering Department, Campus de Azurém, University of Minho, 4800-058 Guimarães, Portugal
3. CEFT, Faculty of Engineering of the University of Porto (FEUP), Rua Dr. Roberto Frias, 4200-465 Porto, Portugal
4. Centro de Investigação de Montanha (CIMO), Instituto Politécnico de Bragança, Campus de Santa Apolónia, 5300-253 Bragança, Portugal
5. Laboratório Associado para a Sustentabilidade e Tecnologia em Regiões de Montanha (SusTEC), Instituto Politécnico de Bragança, Campus de Santa Apolónia, 5300-253 Bragança, Portugal
* Correspondence: jribeiro@ipb.pt

Abstract: Although polylactic acid (PLA) is one of the most used materials in additive manufacturing, its mechanical properties are quite limiting for its practical application, therefore, to improve these properties it is frequent to add fibers and, in this way, create a more resistant composite material. In this paper, the authors developed PLA composites reinforced with flax fibers to evaluate the improvement of tensile and flexural strength. The experimental design of experiments was based on the L18 Taguchi array where the control factors were the extruder temperature (three levels), number of strands (three levels), infill percentage of the specimens (three levels), and whether the flax fiber had surface chemical treatment. The tensile and flexural specimens were made on a 3D printing machine and was a mold was developed to fix and align the fiber strands during the printing process. The tensile and flexural experimental tests were performed in agreement with ASTM D638.14 and ISO 14125 standards, respectively. Analyzing the results, it was verified that the surface chemical treatment (NaOH) of the fiber did not show any influence in the mechanical properties of the composites; in contrast, the infill density demonstrated a huge influence for the improvement of mechanical strength. The maximum values of tensile and bending stress were 50 MPa and 73 MPa, respectively. The natural fiber reinforcement can improve the mechanical properties of the PLA composites.

Keywords: composite with natural fibers; flax fibers; PLA; additive manufacturing; mechanical properties

Citation: Paulo, A.; Santos, J.; da Rocha, J.; Lima, R.; Ribeiro, J. Mechanical Properties of PLA Specimens Obtained by Additive Manufacturing Process Reinforced with Flax Fibers. *J. Compos. Sci.* **2023**, *7*, 27. https://doi.org/10.3390/jcs7010027

Academic Editor: Yuan Chen

Received: 15 November 2022
Revised: 31 December 2022
Accepted: 6 January 2023
Published: 10 January 2023

Copyright: © 2023 by the authors. Licensee MDPI, Basel, Switzerland. This article is an open access article distributed under the terms and conditions of the Creative Commons Attribution (CC BY) license (https://creativecommons.org/licenses/by/4.0/).

1. Introduction

In the last decade, additive manufacturing processes have been increasingly used in different applications [1] which can range from prototype manufacturing (first applications) [2] to industry [3,4], through leisure [5], to scientific research [6,7], among other applications [8,9]. There are many additive manufacturing processes [10]; however, one of the most common is the fused deposition modelling (FDM) [11]. FDM, also known as fused filament fabrication, is a process within the field of material extrusion. FDM employs filaments made of thermoplastic polymers and creates parts layer by layer by selectively de-positing melted material along a predetermined path [10]. The most popular thermo-plastic polymers used in FDM are the PLA (polylactic acid) and ABS (acrylonitrile butadiene styrene) [4]. ABS is a polymeric material that derives from petroleum and its composition has volatile organic compounds that can cause damage to the environment and human health [12]. On the other hand, PLA is a biodegradable polymer, highly interesting in technological terms due to its applications in the environmental field [13]. It is a type of impact modified filament for the 3D printer, which is sustainable; it does not use volatile organic compounds, and allows the final product to have an accelerated degradation time through the action of humidity, temperature, light, and soil microorganisms, at the end of its useful life. As this is a thermoplastic polymer, which comes from renewable sugar-based raw materials, it may be an alternative to the use of non-biodegradable polymers or polymers with long-term degradation [14,15]. Despite the advantages of these materials, such as good adaptability to the FDM process, low cost and the obtained parts having a good resolution [16], the mechanical strength is relatively low [17] for more demanding applications. As a result, it would be interesting to improve some mechanical properties, namely, tensile and flexural strength, one possibility being to reinforce the parts with fibers (natural or synthetic).

Natural fibers are fibers that are not synthetic or manufactured. They come mainly from animals or plants [18]. Animal fibers consist of proteins (wool or silk), while plant fibers consist of cellulose [19]. Currently, natural fibers that originate from plants are the most widely used because they are suitable for use in composites with structural requirements. Moreover, plant fiber can be grown in many countries and can be harvested after short periods of time. In addition to cellulose, natural fibers are composed of lignin, hemicellulose, pectin, and waxes, and can be considered as natural composites containing mainly cellulose fibrils embedded in a lignin matrix. The nature of the cellulose and its crystallinity play an important role in reinforcing the efficiency of the natural fiber. The cellulose fibrils are aligned along the length direction of the fiber, ensuring maximum tensile and flexural strength, and providing stiffness [20]. Natural fibers have unique characteristics such as abundance, non-toxicity, high performance, versatility, and easy processing at low cost.

The natural fiber reinforced polymer composites (NFPCs) have several applications: besides the automotive industry, they arealso used in the construction industry due to their strength, low density, biodegradability, and high lifetime [21]. The fibers mostly used in industrial applications are flax, knaf and hemp because of the fibers' strength properties [21]. The properties of natural fibers vary, as they depend on the type of fiber, its source, and the moisture conditions. They depend on the fiber composition, the microfibril angle (i.e., angle of orientation of the micro-fibril in relation to the main fiber axis [22]), structure, defects, cell dimensions, physical properties, chemical properties, and also the fiber matrix [18]. Among the different plant natural fibers used in NFPCs, flax is one of the plant species that has been cultivated for longer in the world. It is a member of the genus Linum in the family Linaceae; Linum usitatissimum L. is the most common species among the 298 different species that are known. Flax fibers are found near the stem and are the mechanical support of the plant that is very thin [23–26]. They are considered one of the strongest fibers, because of their very complex structure [27]. These fibers are made up of a series of polyhedra that form overlapping elementary fibers over a considerable length, held together by an interface consisting mainly of hemicellulose and pectin. The typical diameter of an elemental flax fiber is between 10 and 15μm, although technical flax fibers range between 35 and 150μm [13].

As the need for FDM in industries grows over time, many researchers are drawn to it to improve the filler quality [28,29]. Synthetic or carbon fibers have often been used to reinforce the filler, although these fibers are harmful to the environment. Consequently, many researchers suggested using natural fiber instead of synthetic ones as the reinforcement, which can also be blended with a bio-polymer matrix, namely thermoplastics, as the polymer matrix in FDM industries. Multiple experimental tests have been performed to demonstrate the potential of natural fibers as the leading material in composite industries [30–34]. From these many pieces of research, it was possible to verify that there are some requirements which a natural fiber reinforced polymer composite (NFRPC) needs to fulfill to be manufactured by AM, namely (1) fiber homogeneity; (2) fiber alignment; (3) types of reinforcements and matrices; (4) adequate fiber-matrix bonding; (5) good interlayer bonding; and (6) minimal porosity [35]. Both the matrix materials, which hold the fibers in place, and reinforcement need to be compatible with the selected 3D printing technique. Fiber distribution homogeneity is crucial to guarantee reliable properties throughout the printed part. The possibility to control fiber alignment and distribution in a predetermined location and direction enables the strengthening of sections of an object. Fiber reinforcement of proper length, size and shape must be selected to suit the intended purpose of the part. Fiber-loading is also essential for getting AM composites with good mechanical properties. A good interlayer fusion is necessary to prevent delamination. Ultimately the unwanted voids should be minimized because they would affect the mechanical properties of the NFRPC [35].

Based on these requirements, the authors of this paper developed a few experimental tests to determine an optimal combination of parameters to improve the mechanical properties of NFRPC made by FDM. To achieve the adequate fiber-matrix and interlayer bonding, many authors suggest the use of a surface chemical and physical treatment of fibers which enhances the adhesion properties of the interface between the fibers and the matrix, and decreases the absorption of water by the fibers [15,36,37]. These processes can be considered as modifiers of the properties of these fibers [38]. One of the most used chemical treatments is the alkaline one, in which the fibers are submerged in an alkaline solution, namely NaOH (sodium hydroxide), for a short period [39]. This increases the fiber's surface roughness and improves its mechanical properties. At the level of chemical bonding with the matrix material, it is possible to expose more cellulose on the fiber surface [39,40]. The most important change in this treatment is the breaking of the hydrogen bonds in the lattice structure, thus causing the surface roughness to increase. It also removes a certain amount of lignin, wax and oils that cover the outer surface of the fiber cell wall, exposing the short length crystals and depolymerizing the cellulose [41]. A consequence of a lower fiber-matrix bonding and interlayer bonding is the delamination. It was for this reason that the the influence of nozzle temperature was analyzed, or, in other words, the melting temperature of the PLA [35]. To evaluate the fiber homogeneity and alignment, long flax fibers were integrated into the specimen with the same direction of the applied load, varying the number of stands (fibers). According to Motru et al. [42], for PLA specimens reinforced with flax fibers, the ultimate tensile strength of the composite laminates tends to increase linearly with respect to the increase in weight percentage of the fiber. Therefore, specimens with three levels of fibers' percentage were tested and to guarantee the same position and distance of these fibers a mold was settled in which the fibers were fixed with glue tape. This type of fibers fixation has also another advantage that is the imposition of a pre-load in the fibers, despite the low value, that could improve the mechanical properties [43]. To evaluate the requirement of types of reinforcements and matrices, despite the materials referred to previously, it also depends on the raster orientation and infill level. The lowest ultimate tensile strength was reduced, proximally, by 55% in the specimen with the raster direction of 90° [44] and a higher infill density increases the tensile and yield strength [44].

The study presented in this paper aims to optimize the combination of flax fiber quantity, nozzle temperature, infill density and fiber surface chemical treatment to achieve the higher value of tensile and flexural strength of flax fiber reinforced composites.

2. Materials and Methods

2.1. Design of Experiments

Based on the literature referred to in the previous section, there are different manufacturing parameters that have a great influence on the properties of obtained products. Hence, the quality of the parts made by additive manufacturing are strongly influenced by the nozzle temperature and the infill density. In its turn, the mechanical properties of NFPCs vary considerably due to the fiber volume fraction (or number of strands) and the fiber surface treatment. For these reasons, the authors decided to use these parameters as control factors with different levels. In Table 1 the control factors with their respective levels are presented.

Table 1. Control factors of the composite.

Symbol	Control Factor	Level 1	Level 2	Level 3
A	Fiber surface treatment	NaOH treatment	No treatment	
B	Nozzle temperature	190 °C	200 °C	220 °C
C	Number of strands	10	15	20
D	Infill density	25%	50%	100%

With the control factors and their respective levels previously defined, it was possible to create a Taguchi L18 orthogonal array (Table 2). This array was used for the implementation of both groups of experimental tests.

Table 2. Taguchi L18 orthogonal array.

Test Number	A Fiber Surface Treatment	B Temperature	C Number of Strands	D Infill Density
1	1	1	1	1
2	1	1	2	2
3	1	1	3	3
4	1	2	1	1
5	1	2	2	2
6	1	2	3	3
7	1	3	1	2
8	1	3	2	3
9	1	3	3	1
10	2	1	1	3
11	2	1	2	1
12	2	1	3	2
13	2	2	1	2
14	2	2	2	3
15	2	2	3	1
16	2	3	1	3
17	2	3	2	1
18	2	3	3	2

2.2. Materials

2.2.1. Specimens Manufacturing

To determine the tensile and flexural strength, it was necessary to undergo tensile tests and flexural tests, according to the standards ASTM D638.14 [45] and ISO 14125 [16], respectively. Consequently, the geometries and dimensions of these specimens had to follow these standards. In Figure 1 the drawing of the two types of specimens is presented. Table 3 shows the values of specimen dimensions.

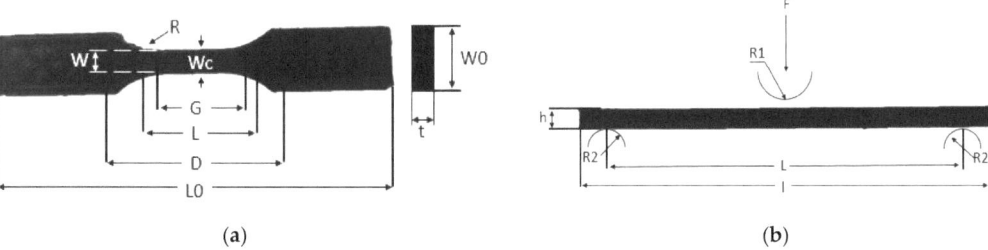

(a) (b)

Figure 1. (**a**) Tensile specimen geometry (based on ASTM D638.14) and (**b**) Flexural specimen geometry (based on ISO 14125).

Table 3. Tensile and flexural specimen dimensions.

Tensile Test Piece Dimensions [mm]		Flexural Test Piece Dimensions [mm]	
D	30	F	-
G	10	h	4
L	19	l	80
L0	63	L	64
R	13	R1	10
t	4	R2	10
W	4		
W0	10		
WC	4		

The flax fibers are used to reinforce the 3D printed specimens and to manufacture the specimens the PLA (EasyFil PLA from FormFutura) which is a biodegradable thermoplastic was chosen. This material has other advantages, such as ease of printing, low deformation and good adhesion between layers. To spatially position the fibers in the defined locations, it was necessary to develop special molds. These molds were manufactured from wood (MDF) and boards were cut using a cutting laser machine (GCC X252). Figure 2 shows the molds used in this work (mold (a) used in tensile specimens and mold (b) used in flexural specimens). The fibers were attached to the grooves of the molds and the mold-fiber system was placed on the table of the 3D printer (Anycubic 3D printer) where the PLA deposition takes place.

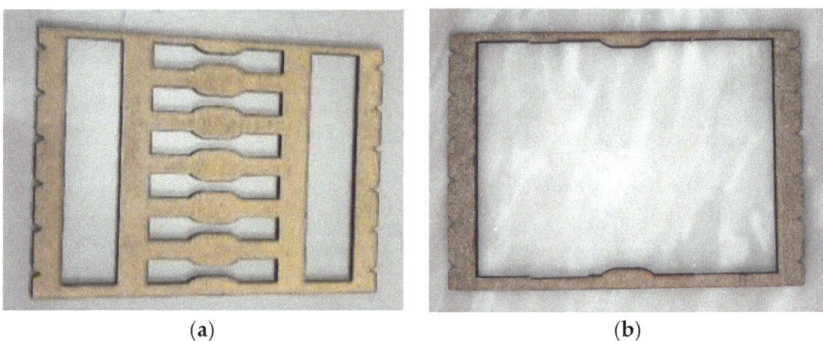

(a) (b)

Figure 2. Molds used for tensile specimens (**a**) and flexural specimens (**b**).

The process began by joining each of the fibers with small knots, wrapping and attaching them to the grooves of the mold, according to the number of fiber strands that would be wanted, i.e., with 10, 15, or 20 strands. To improve the fixation of the fibers to the mold and guarantee their right position, pieces of tape (3 M) were used. This procedure also allowed us to apply a pre-load in the fibers.

The specimens manufacturing process begins with printing, approximately, half of specimen thickness directly on the printer table. When the printing reaches 50% of the thickness specimen, the printer is paused, and the mold with the fibers is placed in the middle of the printed specimen. Printing continues for the remaining 50% of the specimens. After the whole process, which takes about 2 h for the tensile tests and 4 h for the bending tests, it is necessary to let the samples and the machine cool down to remove the specimens (Figure 3a,b). After finishing the process, the 6 specimens from the tensile tests and the 12 specimens from the flexural tests are wrapped with adhesive tape.

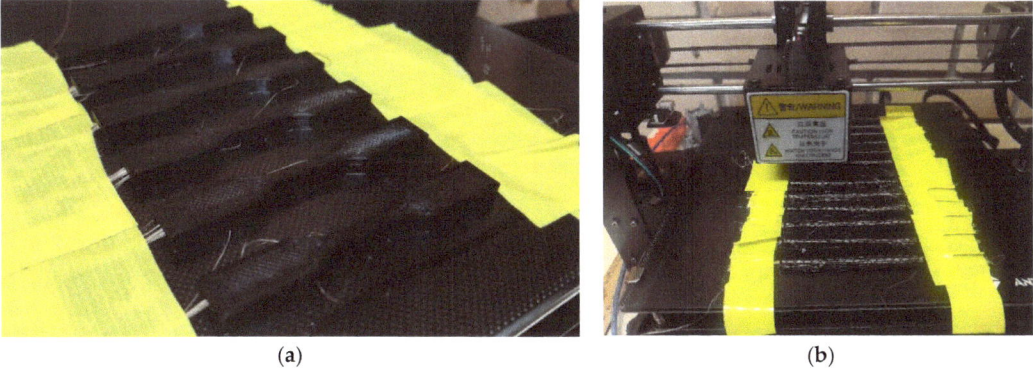

Figure 3. PLA printed specimens for tensile (**a**) and bending tests (**b**).

For the tensile specimens, 6 were made for each combination; thus, for the 18 combinations defined (L18), the total number of specimens for the tensile tests was 108. Likewise, for the bending tests, 6 specimens were produced for each combination; as such, 108 specimens were obtained for the 18 combinations. To evaluate the improvement or decrease of mechanical properties, PLA specimens without any reinforcement were manufactured, and, consequently, three groups of specimens with 25%, 50% and 100% of infill density were printed. For each group, 6 specimens were made which makes a total of 18.

2.2.2. Chemical Surface Treatment

To improve the surface properties of the flax fibers, they were subjected to a sodium hydroxide treatment. The flax fibers were immersed in a 5% sodium hydroxide (NaOH) solution for 3 h at room temperature. After this treatment, the fibers were washed with 5% acetic acid to neutralize the NaOH. To remove all residues of acetic acid, the flax fibers were washed with distilled water and then they were dried in a 120 °C oven (Scientific Series 9000 Oven) for 2 h. The result from drying at room temperature (approximately 20 °C) for 24 h can be seen in Figure 4b.

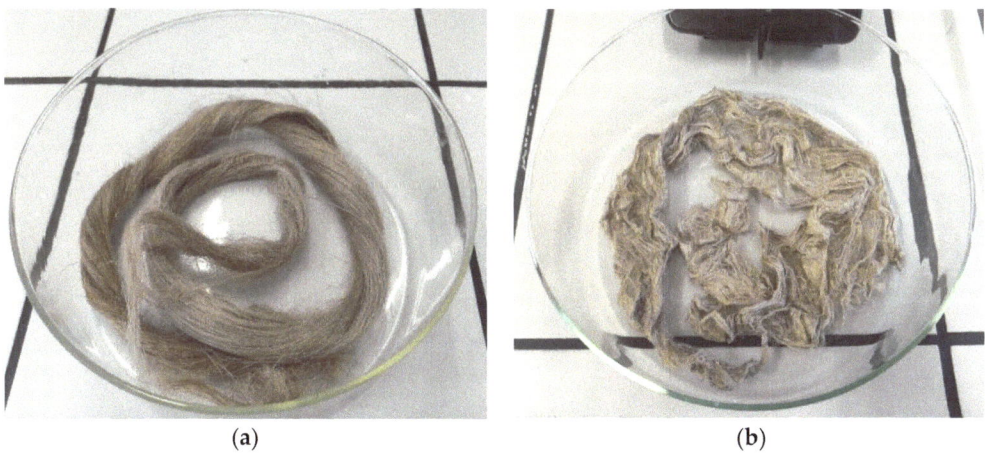

Figure 4. Untreated flax (**a**) and treated flax (**b**).

2.3. *Tensile Tests*

The tensile test took place in the laboratory of structures and strength of materials. To perform the experimental test the ASTM D638.14 [45] standard was used, in which the test speed was 10 mm/min. Thus, a specimen (Figure 5a) was inserted in the machine, placed at 30 mm between the clamps (Figure 5b).

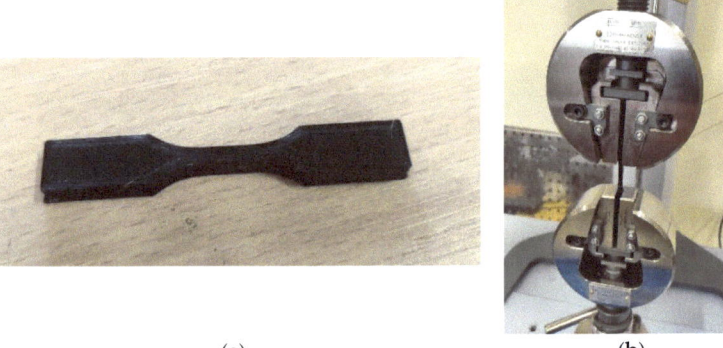

(a) (b)

Figure 5. Tensile specimen (**a**) and tensile tests (**b**).

The tensile strength can be calculated from the maximum load and the transverse area. To represent the stress-strain curve, it is necessary to determine the respective values, for each point. The stress is calculated by dividing the applied load by the average cross-sectional area.

2.4. Flexural Tests

For this experimental work the three-point flexural test was used (Figure 6).

Figure 6. Bending test.

The standard used was ISO 14125 [46], as well as a test speed at 1.7 mm/min.
The test speed was calculated using the Equation (1) [46]:

$$V = \frac{\varepsilon' L^2}{6h} \quad (1)$$

where:

V: Test speed (mm/min)
ε': Deformation rate of 1%/min
L: Outer span (mm)
h: Thickness (mm)

Thus, the specimen was inserted in the machine according to the standard ISO 14125. After finishing all the tests, it was possible to check the graphs and the obtained data.

For the three-point flexural test (Figure 6), it was necessary to use a test specimen, that is, a specimen in the shape of a flat beam of constant rectangular cross section. In three-point flexural, the maximum bending stress occurs at the outer surface of the specimen and is given by Equation (2) [46]:

$$\sigma = \frac{3PL}{2bh^2} \quad (2)$$

where:

σ: Bending stress at the outer surface (MPa)
P: Load applied at the point (N)
L: Specimen length (mm)
b: Specimen width (mm)
h: Specimen thickness (mm))

The strain is calculated from the Equation (3) [46]:

$$\varepsilon = \frac{6\delta h}{L^2} \quad (3)$$

where:

ε: Deformation
δ: Deflection—Distance of the lower or upper surface of the specimen in the middle of the span that has deviated from the initial position (mm)
h: Specimen thickness (mm)
L: Specimen length (mm)

3. Results and Discussion

After analyzing the results obtained through the universal testing machine, the stress and strain were computed, and a graph was drawn for each of the 108 tests for tensile and flexural tests. In addition to these tests, the same was done for the specimens without fibers, with a temperature of 200 °C and a fill percentage of 50%, corresponding to test 19* (for the tensile and flexural tests). An analysis of each graph was done, and the following graphs of the averages for each test were obtained.

Figure 7 presents the stress-strain curve for tensile tests. Each curve (ex. Test 1, Test 2, etc.) represents the average of the six specimens for each L18 combination.

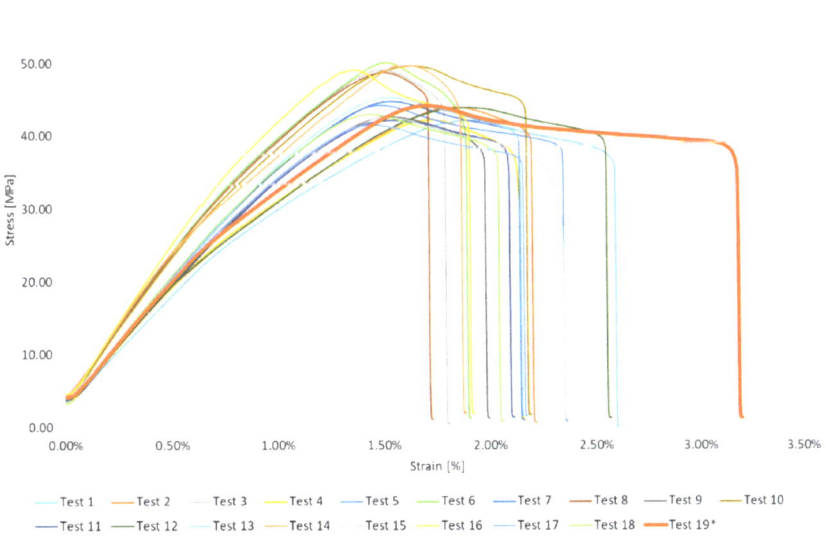

Figure 7. Stress-strain for tensile strength.

Figure 7 shows that test 6 achieved the best results with a maximum tensile stress, followed by tests 3 and 8. First, test 6 represents the chemically treated specimens, at a temperature of 200 °C, with 20 fibers and an infill percentage of 100%. Secondly, test 10 depicts the specimens without chemical treatment, at a temperature of 190 °C, with 10 fibers and an infill percentage of 100%. Finally, test 14 represents the specimens without chemical treatment, at a temperature of 200 °C, with 15 fibers and an infill percentage of 100%. In other words, both tests share the fact that the infill percentage is total (100%).

Figure 8 presents the stress-strain curve for flexural tests. Each curve represents the average of the six specimens for each combination defined by the L18 Taguchi array.

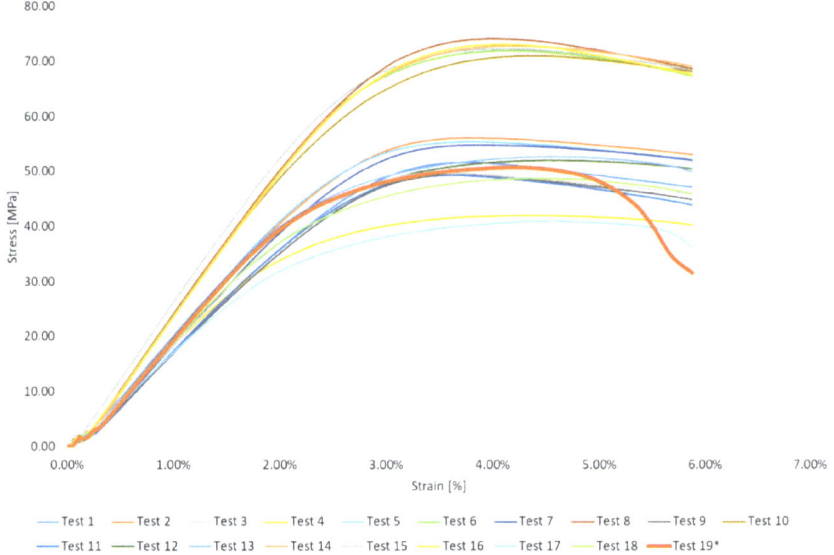

Figure 8. Stress-strain for flexural strength.

Figure 8 shows that test 8 revealed the best results, followed by tests 16 and 14. To begin with, test 8 represents the chemically treated specimens, at a temperature of 220 °C, with 15 fibers and an infill percentage of 100%. Then, test 16 represents the specimens without chemical treatment, at a temperature of 220 °C, with 10 fibers and an infill percentage of 100%. Finally, test 14 represents the specimens without chemical treatment, at a temperature of 200 °C, with 15 fibers and an infill percentage of 100%. This means that both tests only share the percentage of infill, which was the maximum.

Table 4 shows the mean values of the maximum stresses for the 6 specimens in each test, as well as their standard deviation, the last line of this table (experiment 19*) represents the experimental test for specimens with pure PLA. As represented graphically, it can be seen that essay 6 obtained the highest ultimate strength for the tensile test which has the value of 49.96 MPa and a standard deviation of 0.89 MPa. For the bending tests, the one with the highest stress was test 8 with a value of 72.94 MPa and a standard deviation of 2.09 MPa. In both cases, the values of standard deviation are very low which is an indication that all specimen groups have a very similar behavior to the average. Related to the test performed without fibers, the tensile test obtained a value of 43.43 MPa with a standard deviation of 1.33 MPa. For the bending, it achieved a value of 53.55 MPa and a standard deviation of 4.59 MPa. These results demonstrate that, with the reinforcement of natural fibers, the composite becomes more resistant.

Table 4. Values of the mean maximum stresses and their standard deviation.

Experiment Number	Average Tensile Strength [MPa]	Standard Deviation [MPa]	Average Flexural Strength [MPa]	Standard Deviation [MPa]
1	41.99	0.72	51.04	0.78
2	44.02	0.52	54.13	3.38
3	48.25	1.59	72.59	0.79
4	41.99	0.49	44.32	3.96
5	44.01	0.70	54.22	2.79
6	49.96	0.89	71.86	1.13
7	44.55	1.02	52.80	3.52
8	49.16	0.88	72.94	2.09
9	41.86	1.09	46.71	3.68
10	49.55	0.68	70.97	0.50
11	42.11	0.50	46.30	3.81
12	44.41	0.59	53.57	3.05
13	45.03	0.83	54.08	2.85
14	49.25	1.02	72.13	1.71
15	42.69	0.54	48.89	2.75
16	48.85	0.65	72.41	1.69
17	41.09	0.48	41.78	2.75
18	43.26	1.37	50.97	3.82
19 *	43.43	1.33	53.55	4.59

Analyzing the results shown in Table 4, is possible to observe that the fibers can improve or worsen mechanical strength (tensile and flexural). Thus, the average tensile strength for the specimens without fibers is 43.43 MPa while the best result for the NFPCs is 49.96 MPa (Test 6), however, for some tests, like test 17, the tensile strength decreases (41.09 MPa). The same observation happened for the flexural tests, the average flexural strength of the pure PLA specimen is 53.55 MPa and for test 3 the value is 72.59 MPa; still, for test 17 the flexural strength is 41.78 MPa.

The experimental results can be converted into a signal-to-noise ratio (S/N). Taguchi suggests using the S/N ratio to determine the quality characteristics that deviate from the desired values.

The Signal-to-Noise (S/N) ratio developed by Taguchi is a performance measure for choosing the control levels that best handle noise. In this method the term "signal" symbolizes the desired value for the output characteristic, and the term "noise" symbolizes the undesired value. The Signal-to-Noise ratio takes into account the mean and variance, that is, it is the ratio between the mean (signal) and the standard deviation (noise). The S/N equation depends on the criteria for the quality characteristic to be optimized. There are three categories of the quality characteristics in the S/N ratio analysis, which are the lowest better, the highest better and the nominal best. Regardless of the quality category of the features, a higher S/N ratio will correspond to better quality features. Thus, the level with the highest S/N ratio is the optimal level of the control factors. With the analysis of variance, it is possible to see which control factors are statistically feasible, using the results obtained in the S/N and ANOVA analyses, the optimal combination of control factors and the prediction of their levels. The goal of this study is to maximize tensile and flexural strength, thus, the quality feature category for the S/N ratio is the highest-best:

$$S/N = -10 \log \left(\frac{1}{n} \sum_{i=1}^{n} \frac{1}{y_i^2} \right) \quad (4)$$

where n is the number of observations and y_i are the observed data [20,47].

The S/N ratios for tensile and flexural strength are shown in Table 5.

Table 5. S/N ratios for tensile and flexural strength.

Test Number	A	B	C	D	S/N$_{ts}$ Ratio [db]	S/N$_{fs}$ Ratio [db]
1	1	1	1	1	32.46	34.15
2	1	1	2	2	32.87	34.61
3	1	1	3	3	33.66	37.22
4	1	2	1	1	32.46	32.84
5	1	2	2	2	32.87	34.65
6	1	2	3	3	33.97	37.13
7	1	3	1	2	32.97	34.39
8	1	3	2	3	33.83	37.25
9	1	3	3	1	32.43	33.30
10	2	1	1	3	33.90	37.02
11	2	1	2	1	32.49	33.22
12	2	1	3	2	32.95	34.54
13	2	2	1	2	33.07	34.63
14	2	2	2	3	33.84	37.15
15	2	2	3	1	32.61	33.74
16	2	3	1	3	33.78	37.19
17	2	3	2	1	32.27	32.37
18	2	3	3	2	32.71	34.07

In Table 5, S/Nts is the S/N ratio for tensile strength and S/Nfs is the S/N ratio for flexural strength, where the S/N results for the 18 combinations are represented.

For a higher S/N ratio, the best category is applied in order to maximize the response (tensile and flexural strength). The average S/N ratio for the control factors of levels 1, 2 and 3 can be calculated by averaging the S/N ratios of the corresponding tests. In Tables 6 and 7, the average S/N ratio for each level of control factor, i.e., the response, is shown.

Table 6. Mean response table of S/N ratio for tensile strength and significant interaction.

Symbol	Control Factor	Mean S/N Ratio [db]		
		Level 1	Level 2	Level 3
A	Fiber surface treatment	33.06	33.07	
B	Nozzle temperature	33.05	33.14	33.00
C	Number of strands	33.11	33.03	33.05
D	Infill density	32.45	32.91	33.83

Table 7. Mean response table of the S/N ratio for flexural strength and significant interaction.

Symbol	Control Factor	Mean S/N Ratio [db]		
		Level 1	Level 2	Level 3
A	Fiber surface treatment	35.06	34.88	
B	Nozzle temperature	35.13	35.02	34.76
C	Number of strands	35.04	34.88	35.00
D	Infill density	33.27	34.48	37.16

Figures 9 and 10 show the response graph of the S/N ratio for tensile and flexural strength, respectively.

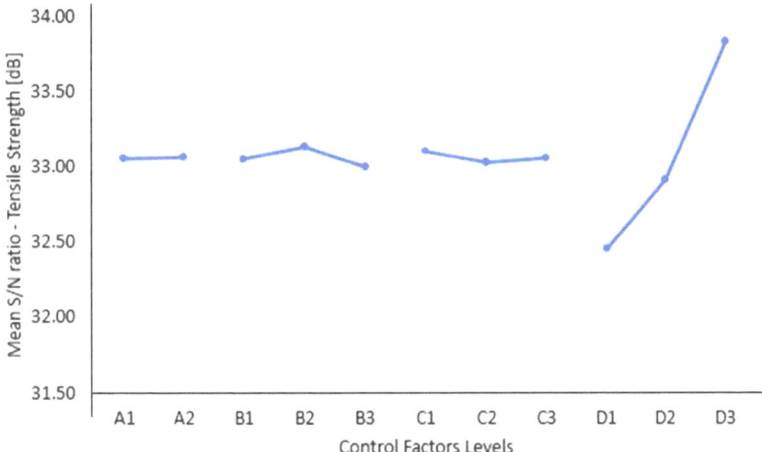

Figure 9. S/N ratio response graph for tensile strength.

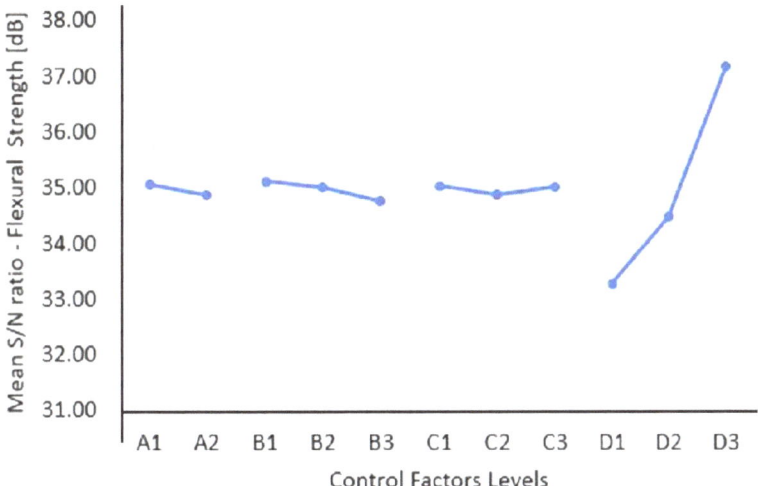

Figure 10. S/N ratio response graph for flexural strength.

Through the analysis of Figure 9, it is possible to determine the optimal combination to increase the tensile strength of the flax fiber reinforced composite. Regarding the tensile tests, the ideal combination is A1/A2, B2, C1 and D3, i.e., with/without chemical treatment, with a temperature of 200 °C, 10 fibers and an infill percentage of 100%.

By analyzing Figure 10, it is possible to determine the optimal combination to increase the flexural strength of the flax fiber reinforced composite. Concerning the bending tests, the optimal combination is A1, B1, C1 and D3, i.e., with chemical treatment, a temperature of 190 °C, 10 fibers and an infill percentage of 100%.

The relative importance of the control factors for tensile and flexural strength needs to be identified more precisely using ANOVA analysis to determine the levels of importance of the control factors. In Tables 8 and 9, it is clear which control factor influences more the tensile and flexural strength values.

Table 8. ANOVA for tensile strength.

Source	DF	Adj SS	Adj MS	F-Value	p-Value	Contribution [%]
A	1	0.00	0.00	0.00	0.96	0.00%
B	2	0.07	0.03	2.85	0.11	1.12%
C	2	0.01	0.01	0.57	0.58	0.22%
D	2	5.80	2.90	245.99	0.00	96.69%
Error	10	0.12	0.01			1.97%
Total	17	6.00				100.00%

Table 9. ANOVA for flexural strength.

Source	DF	Adj SS	Adj MS	F-Value	p-Value	Contribution [%]
A	1	0.14	0.14	0.88	0.37	0.29%
B	2	0.43	0.21	1.31	0.31	0.85%
C	2	0.08	0.04	0.26	0.78	0.17%
D	2	47.52	23.76	146.00	0.00	95.42%
Error	10	1.63	0.16			3.27%
Total	17	49.80				100.00%

In Tables 8 and 9, the variance results for each control factor can be checked, where DF is the degree of Freedom, Adj SS is the sum of squares and Adj MS are the mean squares. The F-test is a statistical tool to check which parameters significantly affect the quality of the characteristics, that is, it is defined as the ratio of the mean square deviations to the mean square error.

After analyzing the results of the F-test value, it is possible to verify that, for tensile and flexural strength, the most significant control factors are the nozzle temperature and the infill percentage of the specimen, with the remaining factors varying. For tensile strength, the infill percentage was very significant with 96.69%, followed by the nozzle temperature, number of strands, and the use or not of fiber treatment, with 1.12%, 0.22%, and 0%, respectively. For flexural strength, the infill percentage was also the most significant with 95.42%, followed by the nozzle temperature, the use or not of fiber treatment, and the number of strands with 0.85%, 0.29% and 0.17%, respectively.

Analyzing these results, is possible to observe that the most important parameter that contributes to the improvement of studied mechanical properties is the infill density (or percentage), because the increase of this parameter is directly related to the growth of the specimen strength and resistance. On other hand, the influence of the fibers and their chemical surface treatment is very low. The reason for this happening might be related to the adhesion between the PLA and the fibers. The temperature to ensure that the PLA has sufficient fluidity to cover the entire outer area of the fiber is not high enough, whereas molten PLA does not allow good wettability to fill the entire fiber.

As the optimal combinations of parameters and levels are different from the L18 analysis, confirmation tests are required.

Confirmation Tests

Given the results obtained, the tests were performed again for the optimal combinations, i.e., for the tensile tests, 3 specimens were made without chemical treatment, with a nozzle temperature of 200 °C, 10 fibers, and an infill percentage of 100%. For the bending tests, 3 specimens were also made, but with chemical treatment, a nozzle temperature of 190 °C, 10 fibers, and an infill percentage of 100%. Only 3 specimens were produced, because the standard deviation of the previous tests was quite small. Figure 11 graphically represents the optimal combinations of tensile strength.

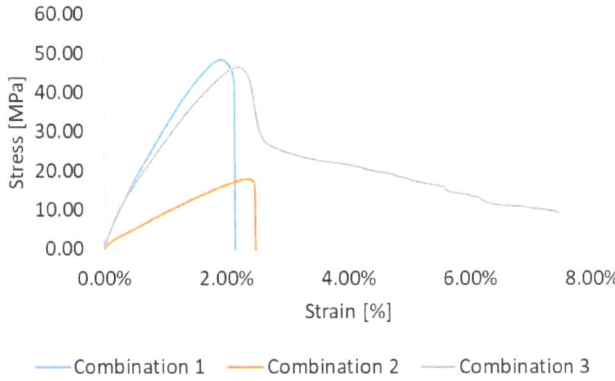

Figure 11. Optimum combinations of tensile strength.

A maximum stress of 48.43 MPa can be observed, i.e., it is very close to the stresses of the best tests performed previously. Figure 12 graphically represents the optimal combinations of flexural strength.

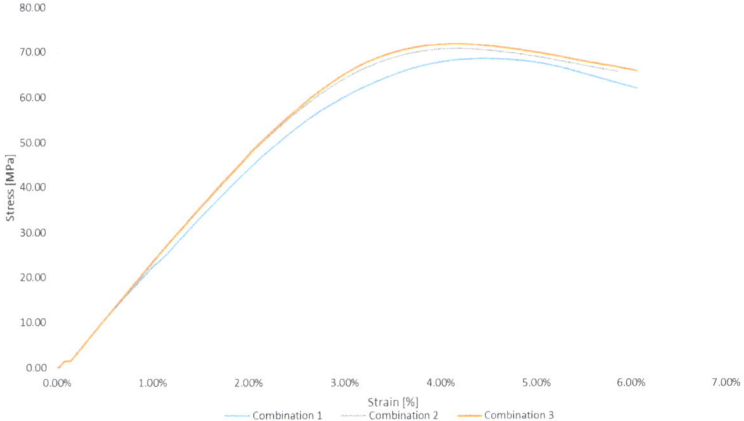

Figure 12. Optimum combinations of flexural strength.

A maximum stress of 71.90 MPa can be observed, which is quite similar to the best tests performed previously.

4. Conclusions

The main conclusions of this work are as follows:
- The maximum stress for the tensile test is 49.96 MPa and for the bending is 72.94 MPa with a standard deviation of 0.89 MPa and 2.09 MPa, respectively. The values of the standard deviation are very low which is a sign that all specimens within groups have a very similar behavior. This observation could be an interesting feature for industrial applications, because it is possible to guarantee very similar properties for manufactured products by this process.
- The natural fiber reinforcement, for many combinations of control factors, improve the mechanical strength of the composite. Comparing the tensile strength of pure PLA with best result of the composite the values are 43.43 MPa and 49.96 MPa, respectively. The flexural test obtained a value of 53.55 MPa (PLA without fibers) compared to the maximum value of the flax fiber reinforced composite of 72.94 MPa. However, the

- pure PLA can be stronger than some combinations of NFPCs, thus, for example, the test 17 showed low values of mechanical strength (tensile and flexural) (41.09 MPa and 41.78 MPa, respectively).
- With the Taguchi method, it was possible to determine the optimal combinations for maximum tensile and flexural strength. For tensile, the optimum combination is A1/A2, B2, C1 and D3, i.e., with/without chemical treatment, with a nozzle temperature of 200 °C, 10 fibers and an infill percentage of 100%. For bending, the optimal combination is A1, B1, C1 and D3, i.e., with chemical treatment, a nozzle temperature of 190 °C, 10 fibers and infill percentage of 100%.
- With the analysis of variance, it was found that the infill percentage was the parameter with the greatest contribution to the increase of tensile and flexural strength with percentages of 96.69% and 95.42%, respectively.
- After performing the confirmation tests with the optimal combinations obtained previously, the maximum values for the tensile and flexural strength were determined at 48.43 MPa and 71.90 MPa, respectively.

Author Contributions: Conceptualization, J.R. and J.d.R.; Data curation, A.P. and J.S.; Investigation, A.P., J.S. and R.L.; Methodology, A.P., J.S., J.d.R. and J.R.; Resources, J.R., J.d.R. and R.L.; Software, A.P.; Supervision, J.d.R., R.L. and J.R.; Validation, A.P. and J.S.; Writing—original draft, A.P., J.S., J.R. and J.d.R.; Writing—review and editing, A.P. and J.R. All authors have read and agreed to the published version of the manuscript.

Funding: Financial support was provided by Portugal's national funding FCT/MCTES (PIDDAC) to Centro de Investigação de Montanha (CIMO) (UIDB/00690/2020 and UIDP/00690/2020) and SusTEC (LA/P/0007/2020).

Data Availability Statement: Not applicable.

Conflicts of Interest: The authors declare no conflict of interest.

References

1. Gibson, I.; Rosen, D.; Brent Stucker, M.K. *Additive Manufacturing Technologies*, 3rd ed.; Springer: Cham, Switzerland, 2021.
2. Horn, T.J.; Harrysson, O.L.A. Overview of Current Additive Manufacturing Technologies and Selected Applications. *Sci. Prog.* **2012**, *95*, 255–282. [CrossRef] [PubMed]
3. Krimi, I.; Lafhaj, Z.; Ducoulombier, L. Prospective Study on the Integration of Additive Manufacturing to Building Industry—Case of a French Construction Company. *Addit. Manuf.* **2017**, *16*, 107–114. [CrossRef]
4. Hernandez Korner, M.E.; Lambán, M.P.; Albajez, J.A.; Santolaria, J.; Ng Corrales, L.d.C.; Royo, J. Systematic Literature Review: Integration of Additive Manufacturing and Industry 4.0. *Metals* **2020**, *10*, 1061. [CrossRef]
5. Matos, F.; Godina, R.; Jacinto, C.; Carvalho, H.; Ribeiro, I.; Peças, P. Additive Manufacturing: Exploring the Social Changes and Impacts. *Sustainability* **2019**, *11*, 3757. [CrossRef]
6. Souza, A.; Souza, M.S.; Pinho, D.; Agujetas, R.; Ferrera, C.; Lima, R.; Puga, H.; Ribeiro, J. 3D Manufacturing of Intracranial Aneurysm Biomodels for Flow Visualizations: Low Cost Fabrication Processes. *Mech. Res. Commun.* **2020**, *107*, 103535. [CrossRef]
7. Ivey, M.; Melenka, G.W.; Carey, J.P.; Ayranci, C. Characterizing Short-Fiber-Reinforced Composites Produced Using Additive Manufacturing. *Adv. Manuf. Polym. Compos. Sci.* **2017**, *3*, 81–91. [CrossRef]
8. Razvi, S.S.; Feng, S.; Narayanan, A.; Lee, Y.-T.T.; Witherell, P. A Review of Machine Learning Applications in Additive Manufacturing. In Proceedings of the 39th Computers and Information in Engineering Conference, Anaheim, CA, USA, 18–21 August 2019; American Society of Mechanical Engineers: New York, NY, USA, 2019; Volume 1.
9. Singh, S.; Ramakrishna, S. Biomedical Applications of Additive Manufacturing: Present and Future. *Curr. Opin. Biomed. Eng.* **2017**, *2*, 105–115. [CrossRef]
10. Frandsen, C.S.; Nielsen, M.M.; Chaudhuri, A.; Jayaram, J.; Govindan, K. In Search for Classification and Selection of Spare Parts Suitable for Additive Manufacturing: A Literature Review. *Int. J. Prod. Res.* **2020**, *58*, 970–996. [CrossRef]
11. Lee, C.H.; Padzil, F.N.B.M.; Lee, S.H.; Ainun, Z.M.A.; Abdullah, L.C. Potential for Natural Fiber Reinforcement in PLA Polymer Filaments for Fused Deposition Modeling (FDM) Additive Manufacturing: A Review. *Polymers* **2021**, *13*, 1407. [CrossRef]
12. Tseng, T.K.; Chu, H. The Catalytic Incineration of Styrene over an Mn_2O_3/Fe_2O_3 Catalyst. *J. Air Waste Manag. Assoc.* **2002**, *52*, 1153–1160. [CrossRef]
13. Aliotta, L.; Gigante, V.; Coltelli, M.B.; Cinelli, P.; Lazzeri, A.; Seggiani, M. Thermo-Mechanical Properties of PLA/Short Flax Fiber Biocomposites. *Appl. Sci.* **2019**, *9*, 3797. [CrossRef]
14. Soleimani, M.; Tabil, L.; Panigrahi, S.; Opoku, A. The Effect of Fiber Pretreatment and Compatibilizer on Mechanical and Physical Properties of Flax Fiber-Polypropylene Composites. *J. Polym. Environ.* **2008**, *16*, 74–82. [CrossRef]

15. Bénard, Q.; Fois, M.; Grisel, M. Roughness and Fibre Reinforcement Effect onto Wettability of Composite Surfaces. *Appl. Surf. Sci.* **2007**, *253*, 4753–4758. [CrossRef]
16. García-Martínez, H.; Ávila-Navarro, E.; Torregrosa-Penalva, G.; Rodríguez-Martínez, A.; Blanco-Angulo, C.; de la Casa-Lillo, M.A. de la Low-Cost Additive Manufacturing Techniques Applied to the Design of Planar Microwave Circuits by Fused Deposition Modeling. *Polymers* **2020**, *12*, 1946. [CrossRef] [PubMed]
17. Oliveira, C.; Rocha, J.; Ribeiro, J.E. *Mechanical and Physical Characterization of Parts Manufactured by 3D Printing*; Springer: Cham, Switzerland, 2023; pp. 77–88.
18. Mohammed, L.; Ansari, M.N.M.; Pua, G.; Jawaid, M.; Islam, M.S. A Review on Natural Fiber Reinforced Polymer Composite and Its Applications. *Int. J. Polym. Sci.* **2015**, *2015*, 243947. [CrossRef]
19. John, M.J.; Thomas, S. Biofibres and Biocomposites. *Carbohydr. Polym.* **2008**, *71*, 343–364. [CrossRef]
20. Ribeiro, J.E.; Rocha, J.; Queijo, L.; Polyester, N.C.H.S. The Influence of Manufacturing Factors in the Short-Fiber Non-Woven Chestnut Hedgehog Spine- Reinforced Polyester Composite Performance The Influence of Manufacturing Factors in the Short-Fiber. *J. Nat. Fibers* **2021**, *18*, 1307–1319. [CrossRef]
21. Kumar, R.; Ul Haq, M.I.; Raina, A.; Anand, A. Industrial Applications of Natural Fibre-Reinforced Polymer Composites–Challenges and Opportunities. *Int. J. Sustain. Eng.* **2019**, *12*, 212–220. [CrossRef]
22. Müssig, J.; Haag, K. *The Use of Flax Fibres as Reinforcements in Composites*; Woodhead Publishing: Cambridge, UK, 2015; ISBN 9781782421276.
23. Baley, C.; Gomina, M.; Breard, J.; Bourmaud, A.; Davies, P. Variability of Mechanical Properties of Flax Fibres for Composite Reinforcement. A Review. *Ind. Crops Prod.* **2020**, *145*, 111984. [CrossRef]
24. Morvan, C.; Andème-Onzighi, C.; Girault, R.; Himmelsbach, D.S.; Driouich, A.; Akin, D.E. Building Flax Fibres: More than One Brick in the Walls. *Plant Physiol. Biochem.* **2003**, *41*, 935–944. [CrossRef]
25. Baley, C.; Gomina, M.; Breard, J.; Bourmaud, A.; Drapier, S.; Ferreira, M.; Le Duigou, A.; Liotier, P.J.; Ouagne, P.; Soulat, D.; et al. Specific Features of Flax Fibres Used to Manufacture Composite Materials. *Int. J. Mater. Form.* **2019**, *12*, 1023–1052. [CrossRef]
26. Goudenhooft, C.; Alméras, T.; Bourmaud, A.; Baley, C. The Remarkable Slenderness of Flax Plant and Pertinent Factors Affecting Its Mechanical Stability. *Biosyst. Eng.* **2019**, *178*, 1–8. [CrossRef]
27. Bos, H.L.; Müssig, J.; van den Oever, M.J.A. Mechanical Properties of Short-Flax-Fibre Reinforced Compounds. *Compos. Part A Appl. Sci. Manuf.* **2006**, *37*, 1591–1604. [CrossRef]
28. Aida, H.J.; Nadlene, R.; Mastura, M.T.; Yusriah, L.; Sivakumar, D.; Ilyas, R.A. Natural Fibre Filament for Fused Deposition Modelling (FDM): A Review. *Int. J. Sustain. Eng.* **2021**, *14*, 1988–2008. [CrossRef]
29. De Bortoli, L.S.; de Farias, R.; Mezalira, D.Z.; Schabbach, L.M.; Fredel, M.C. Functionalized Carbon Nanotubes for 3D-Printed PLA-Nanocomposites: Effects on Thermal and Mechanical Properties. *Mater. Today Commun.* **2022**, *31*, 103402. [CrossRef]
30. Stoof, D.; Pickering, K.; Zhang, Y. Fused Deposition Modelling of Natural Fibre/Polylactic Acid Composites. *J. Compos. Sci.* **2017**, *1*, 8. [CrossRef]
31. Jamadi, A.H.; Razali, N.; Petrů, M.; Taha, M.M.; Muhammad, N.; Ilyas, R.A. Effect of Chemically Treated Kenaf Fibre on Mechanical and Thermal Properties of PLA Composites Prepared through Fused Deposition Modeling (FDM). *Polymers* **2021**, *13*, 3299. [CrossRef]
32. Obada, D.O.; Kuburi, L.S.; Dauda, M.; Umaru, S.; Dodoo-Arhin, D.; Balogun, M.B.; Iliyasu, I.; Iorpenda, M.J. Effect of Variation in Frequencies on the Viscoelastic Properties of Coir and Coconut Husk Powder Reinforced Polymer Composites. *J. King Saud Univ.-Eng. Sci.* **2020**, *32*, 148–157. [CrossRef]
33. Zhong, W.; Li, F.; Zhang, Z.; Song, L.; Li, Z. Short Fiber Reinforced Composites for Fused Deposition Modeling. *Mater. Sci. Eng. A* **2001**, *301*, 125–130. [CrossRef]
34. Mangat, A.S.; Singh, S.; Gupta, M.; Sharma, R. Experimental Investigations on Natural Fiber Embedded Additive Manufacturing-Based Biodegradable Structures for Biomedical Applications. *Rapid Prototyp. J.* **2018**, *24*, 1221–1234. [CrossRef]
35. Goh, G.D.; Yap, Y.L.; Agarwala, S.; Yeong, W.Y. Recent Progress in Additive Manufacturing of Fiber Reinforced Polymer Composite. *Adv. Mater. Technol.* **2019**, *4*, 1800271. [CrossRef]
36. Liu, Z.T.; Sun, C.; Liu, Z.W.; Lu, J. Adjustable Wettability of Methyl Methacrylate Modified Ramie Fiber. *J. Appl. Polym. Sci.* **2008**, *109*, 2888–2894. [CrossRef]
37. Sinha, E.; Panigrahi, S. Effect of Plasma Treatment on Structure, Wettability of Jute Fiber and Flexural Strength of Its Composite. *J. Compos. Mater.* **2009**, *43*, 1791–1802. [CrossRef]
38. Ahmad, F.; Choi, H.S.; Park, M.K. A Review: Natural Fiber Composites Selection in View of Mechanical, Light Weight, and Economic Properties. *Macromol. Mater. Eng.* **2015**, *300*, 10–24. [CrossRef]
39. Asumani, O.M.L.; Reid, R.G.; Paskaramoorthy, R. The Effects of Alkali-Silane Treatment on the Tensile and Flexural Properties of Short Fibre Non-Woven Kenaf Reinforced Polypropylene Composites. *Compos. Part A Appl. Sci. Manuf.* **2012**, *43*, 1431–1440. [CrossRef]
40. Li, X.; Tabil, L.G.; Panigrahi, S. Chemical Treatments of Natural Fiber for Use in Natural Fiber-Reinforced Composites: A Review. *J. Polym. Environ.* **2007**, *15*, 25–33. [CrossRef]
41. Agrawal, R.; Saxena, N.; Sharma, K.; Thomas, S.; Sreekala, M. Activation Energy and Crystallization Kinetics of Untreated and Treated Oil Palm Fibre Reinforced Phenol Formaldehyde Composites. *Mater. Sci. Eng. A* **2000**, *277*, 77–82. [CrossRef]

42. Motru, S.; Adithyakrishna, V.H.; Bharath, J.; Guruprasad, R. Development and Evaluation of Mechanical Properties of Biodegradable PLA/Flax Fiber Green Composite Laminates. *Mater. Today Proc.* **2020**, *24*, 641–649. [CrossRef]
43. Rout, J.; Tripathy, S.S.; Misra, M.; Mohanty, A.K.; Nayak, S.K. The Influence of Fiber Surface Modification on the Mechanical Properties of Coir-Polyester Composites. *Polym. Compos.* **2001**, *22*, 468–476. [CrossRef]
44. Khosravani, M.R.; Berto, F.; Ayatollahi, M.R.; Reinicke, T. Characterization of 3D-Printed PLA Parts with Different Raster Orientations and Printing Speeds. *Sci. Rep.* **2022**, *12*, 1016. [CrossRef]
45. Materials, P.; Materials, E.I.; Matrix, P.; Materials, C.; Specimens, P. Standard Test Method for Tensile Properties of Plastics 1. *Open J. Compos. Mater.* **2006**, *3*, 1–15. [CrossRef]
46. *ISO 14125:1998*; Fibre-Reinforced Plastic Composites. British Standards Institution: London, UK, 1998.
47. Senesathit, S.; Deng, S.; Zhang, S.; Mohammed, Y.; Al Rubaei, A.; Yuan, X.; Wang, T.; Duan, J. Study and Investigate Effects of Cutting Surface in CNC Milling Process for Aluminum Based on Taguchi Design Method. *Int. J. Eng. Res.* **2019**, *8*, 288–295. [CrossRef]

Disclaimer/Publisher's Note: The statements, opinions and data contained in all publications are solely those of the individual author(s) and contributor(s) and not of MDPI and/or the editor(s). MDPI and/or the editor(s) disclaim responsibility for any injury to people or property resulting from any ideas, methods, instructions or products referred to in the content.

Article

Fabrication Temperature-Related Porosity Effects on the Mechanical Properties of Additively Manufactured CFRP Composites

Olusanmi Adeniran [1,*], Norman Osa-uwagboe [2,3], Weilong Cong [1] and Monsuru Ramoni [4]

[1] Department of Industrial, Manufacturing, and Systems Engineering, Texas Tech University, Lubbock, TX 79409, USA
[2] Wolfson School of Mechanical, Electrical, and Manufacturing Engineering, Loughborough University, Loughborough, Leicestershire LE11 3TU, UK
[3] Air Force Research and Development Centre, Nigerian Air Force Base, Kaduna PMB 2104, Nigeria
[4] Department of Industrial Engineering, Navajo Technical University, Crownpoint, NM 87313, USA
* Correspondence: olusanmi.adeniran@ttu.edu; Tel.: +1-(682)-561-4015

Abstract: The use of additive manufacturing in fabricating composite components has been gaining traction in the past decade. However, some issues with mechanical performance still need to be resolved. The issue of material porosity remains a pertinent one which needs more understanding to be able to come up with more viable solutions. Different researchers have examined the subject; however, more research to quantitatively determine fabrication temperatures effects at the micro-scale are still needed. This study employed micro-CT scan analysis to quantitatively compare fabrication temperatures effect at 230 °C, 250 °C, 270 °C, and 290 °C on the mechanical properties of AM fabricated carbon-fiber-reinforces plastic (CFRP) composites, testing carbon fiber-reinforced polyamide (CF-PA) and carbon fiber-reinforced acrylonitrile butadiene styrene (CF-ABS) samples. This micro-CT examination followed an SEM evaluation, which was used to determine temperature effects on interlayer and intralayer porosity generation. The porosity volume was related to the mechanical properties, in which it was determined how temperatures influence porosity volumes. It was also determined that fabrication temperature generally affects semicrystalline composites more than amorphous composites. The overall porosity volumes from the interlayer and intralayer voids were determined, with the interlayer voids being more influential in influencing the mechanical properties.

Keywords: additive manufacturing; fabrication temperature; porosity effects; carbon fiber-reinforced polymer composites; mechanical properties; micro-CT scan; scanning electron microscopy

1. Introduction

1.1. Additively Manufactured CFRP Composites

The rapid increase in demand for products with diverse applicability, accuracy, and ease of manufacturing has led to the advancement of flexible manufacturing techniques. In the past two decades, the concept of AM has grown into a reliable method of producing complex structures for several industries, such as aerospace, biomedical, automobile, telecommunications, defense, and renewable energies [1]. These expanding applications can be traced to the advantage of manufacturing prototypes quickly at lower costs and without specialized tooling [2–4].

A distinguishing feature of CFRP composites fabricated by AM is interlayers. These contribute immensely to the degree of porosity in the material, which determines its mechanical performance. The interlayers are characterized by relatively large triangular voids of similar sizes, which are formed by gaps between the print beads during deposition. This often compounds the issue of porosity, resulting in poorer mechanical properties than for parts fabricated through other manufacturing methods. Hence, the importance of a better understanding of fabrication temperature influenced interlayer and intralayer porosity effects. To advance the manufacturing technique, a greater understanding of the process effect is still needed to improve the viability of the manufacturing method.

1.2. Current Understanding of Deposition Temperature-Related Porosity Effects in AM Fabricated CFRP Composites

Different studies have conducted investigations on porosity effects on the mechanical performance of AM-fabricated CFRP composites [5–12]. However, there have been limited investigations on the quantitative comparison of fabrication temperature-related porosity effects. A better understanding provided by more detailed micro-CT analysis is still needed to cultivate knowledge of how deposition temperatures influence porosity features and affect the mechanical performance of AM-fabricated CFRP composites. The investigations that examined porosity effects did not evaluate deposition temperature effects, and those that examined temperature effects did not qualitatively compare porosity effects.

Zhang et al. [5] examined interlayer porosities as influenced by printing raster angles in AM-fabricated CFRP composites but did not consider the effects of deposition temperatures. Petro et al. [9] explored an X-ray CT scan to determine porosity volumes in a 41% fiber content continuous CFRP composite, where they compared the results to the ASTM D3171 standard. However, they did not consider fabrication temperature-related porosity effects. Tekinalp et al. [6] investigated the effect of fiber volumes on tensile properties and the degree of porosity. However, they did not examine the effect of printing temperatures or provide a clear measurement procedure to arrive at their reported porosity values. Ning et al. provided porosity measurements in two of their reported investigations [7,13] that examined the effects of fiber content and the effects of reinforcement materials on porosities. However, they applied a rudimentary measurement procedure, where they performed manual weighing of samples and subtraction from a calculated solid mass. In another study [14] where the authors tested the effects of process factors including fabrication temperatures, they did not measure or compare porosity volumes.

Other reported investigations on the effects of process temperatures on the mechanical performance of AM-fabricated CFRP composites have been devoted to layer strength from matrix material viscoelasticity fluidity rather than addressing the topic of temperature-related porosity effects [15–19]. Ajinjeru et al. [15] investigated the viscoelasticity of polyetherimide (PEI) matrix at different process temperatures to gain more understanding of the effects of matrix material fluidity on the ease of processing but did not examine porosity effects. These researchers also investigated acrylonitrile butadiene styrene (ABS) and polyphenyl sulfone (PPSU) to determine suitable processing conditions [16], but also did not investigate porosity. Kishore et al. [17] investigated interlayer strength improvement by using infrared heating to increase the surface temperature of the printed layer just before the deposition of the succeeding layer, thus improving surface properties; however, they did not examine temperature-influenced porosity. Yang et al. [19] investigated a novel process consisting of continuous fiber hot-dipping, matrix material melting, and impregnated composite extrusion to improve composite properties, but did not discuss the topic of temperature-related porosity effects.

1.3. Research Motivation

Of all the known challenges to the successful application of AM-fabricated CFRP composites, the issue of interlayer and intralayer porosity is one of the most pertinent, requiring better understanding. However, a detailed quantitative understanding of the effect of printing temperature-influenced porosity on the mechanical performance of the composites is not well developed. This motivated our study to investigate the effects of printing temperature-related porosity on the mechanical performance of the composite for samples fabricated using the fused deposition modeling (FDM) technique. The investigations included evaluating the relationships between printing temperatures, mesostructure formation, fabrication porosities, and the mechanical properties of the composite. The understanding gained contributes to knowledge of fabrication temperature process control for the improved mechanical performance of CFRP composites fabricated by AM.

2. Materials and Methods

2.1. AM Workpiece Fabrication Procedure

Test workpieces used to conduct the investigations were fabricated from two different CFRP composite materials with manufactured carbon fibers with an average diameter of 7 µm and length of less than 400 µm: 15% CF-PA6 filament compounded from high-modulus short carbon fiber in a PA6 copolymer (3D XTEC, Grand Rapids, MI, USA) and 15% CF-ABS filament compounded from high-modulus short carbon fiber in Sabic MG-94 ABS (3D XTEC, Grand Rapids, MI, USA).

A 3D printer (Prusa Mk3 i3, Prague, Czech Republic) was used for the FDM fabrication in a modified control printing enclosure fixture (Creality 3D, Shenzhen, China) to maintain a 45 ± 5 °C printing environment temperature and relative humidity of lower than 20% RH. The 230 °C–290 °C printing temperature range was chosen to determine how fabrication temperatures influence porosity volume because of the material viscosities and flowability within that temperature range. Table 1 shows the printing parameters used to fabricate the test workpieces and Figure 1 illustrates the 3D printer setup employed.

Table 1. Material processing parameters.

Parameter	Unit	Value
Infill Density	%	100
Printer Enclosed Temperature	°C	50 ± 5
Bed Temperature	°C	100
Raster Angle	Deg	0, 90
Layer Thickness	mm	0.25
Printing Speed	mm/sec	30
Nozzle Temperature	°C	230, 250, 270, 290

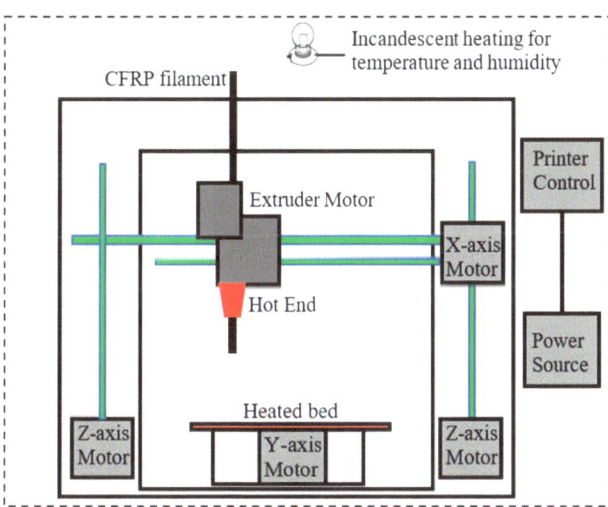

Figure 1. Setup of a 3D printer to control immediate printing environment temperature and humidity. Based on Adeniran et al. [20].

2.2. Measurement Procedure

2.2.1. Mechanical Measurements

Tensile samples that consisted of four workpieces per data point for the AM-fabricated CF-PA and CF-ABS composites were tested to quantitatively compare fabricating temperature-related porosities and the resulting effects on the mechanical properties of the composites. A 10 kN load cell in a universal mechanical tester (MTS Criterion Model 45, Eden Prairie, MN, USA) was used to conduct tensile tests, while a 10 kg minor load and a 100 kg major load were used in the Rockwell tester (Clark C12A, Novi, MI, USA) to conduct the hardness test. Table 2 shows the test procedures applied in the study.

Table 2. Test standards and equipment.

Test	ASTM Standard	Equipment	Test Speed	Unit
Tensile	D638 (Type I)	MTS Criterion Model 45	5.0	mm/min
Rockwell Hardness	D785	Clark Tester C12A	-	-
Scanning Electron Microscopy	-	Thermo Scientific Phenom XL	100×	Magnification
Micro-CT	-	Nikon X-Tex XTH	2	Proj/sec

2.2.2. Scanning Electron Microscopy

Cross-sections of fractured tensile workpieces of the CF-PA and CF-ABS composite samples printed at 230 °C, 250 °C, 270 °C, and 290 °C were examined at 100× magnification in a scanning electron microscope (Thermo Fisher Phenon XL, Waltham, MA, USA) to evaluate the mesostructure formation of the composites as determined by the deposition temperature and matrix materials.

2.2.3. Micro-CT Microscopy

To improve the accuracy of the temperature influenced-porosity determination, 3 samples, each with cross-section volumes of 25 mm × 9 mm × 2.5 mm, were scanned for each data point of the 230 °C, 250 °C, 270 °C, and 290 °C composite samples. Each cross-sectional volume was divided into 6 segments (1A–3B), as illustrated in Figure 2, with the average mean porosity values determined over the segmented volumes while Figure 3 shows the micro CT microscope (Nikon XTH X-Tex 160Xi, France) configuration used for the measurement.

The scan was performed on the CT scan microscope with the scanning point-to-point resolution set to 2.5 µm at an exposure time of 500 ms with a total of 3016 tiff images created per sample scan. The acquired microscopy data were transferred and post-processed using commercial software (Dragonfly ORS, Adelaide, South Australia, Australia). The automatic features of the software were used to determine the optimum center of rotation, with all generated slices combined to develop the volumetric image in the reconstruction phase. Characterization noise generated with the volumetric imaging was reduced using Gaussian filters, which, in the case of this investigation, resulted in minimal data loss from the sample materials' homogeneity. The generated tiff files were incorporated into the software with voxel analysis to determine volumetric porosity.

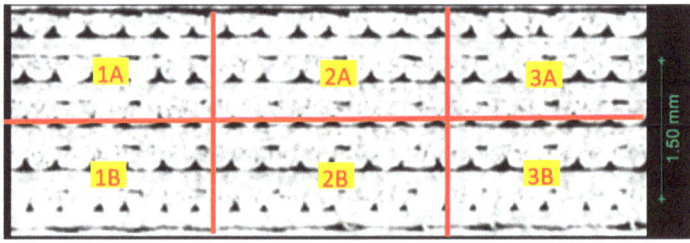

Figure 2. Illustration of segmented volume areas used in micro-CT porosity determination of AM-fabricated CFRP composites.

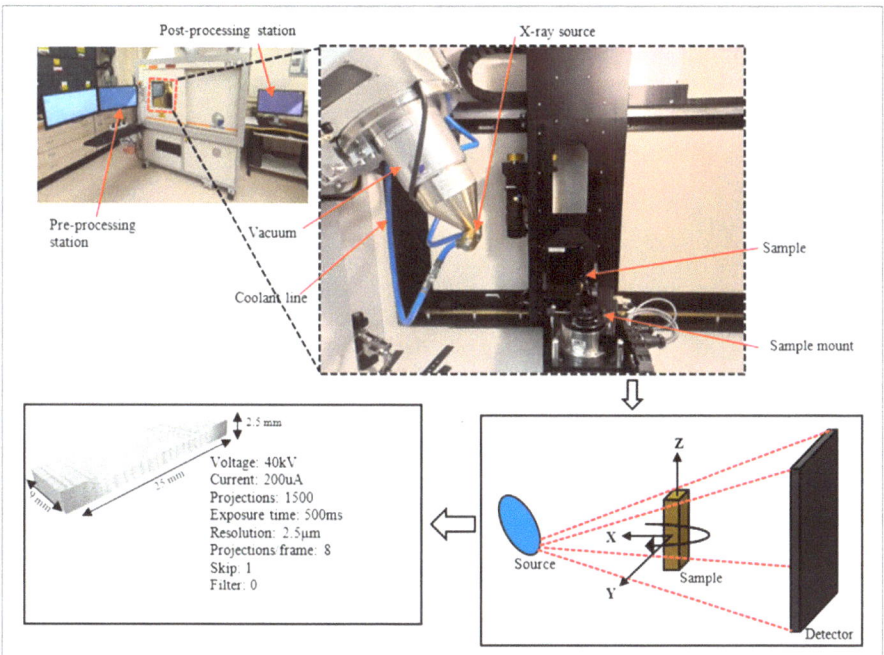

Figure 3. X-ray micro-CT setup and detailed scanning parameters.

3. Results

3.1. Scanning Electron Microscopy Results

SEM determination was used to interpret the effects of printing temperatures of 230 °C to 290 °C on interlayer and intralayer porosity generation for the composites. The thermal history of adjoining melt layers as determined by deposition temperatures influences the extrusion melt solidification, bonding, and the interlayer feature formed including the degree of porosity.

3.1.1. SEM Result for CF-PA Composite

Figure 4 shows the SEM observations of the mesostructure features for the AM-fabricated CF-PA composites.

The SEM examination of the AM fabricated CF-PA composite shows increasing fiber matrix coalescence with nozzle head temperature for the 230 °C–290 °C temperature range examined. A spongy-like mesostructure with less regular layer formation was observed, which can be attributed to the excessive shrinkage of the semicrystalline matrix recrystallization on cooling from fabrication. The low melt viscosity of the semicrystalline morphology can also be ascribed to contribute to the irregular interlayer mesostructure formation. According to Vaes and Puyvelde [21], porosities could be an issue in semicrystalline matrix AM fabrication due to the heterogenous self-nucleation of their structural crystals from insufficient heat transfer, melting, and high shear deformations during fabrication to influence the material's porosities.

The fusion bonding theory applies to AM composite fabrication, where the increasing heat from the 230 °C–90 °C nozzle head temperatures in this case improved the carbon fiber matrix coalescence to reduce interlayer porosities. This is because the higher the fabrication temperature is above the glass transition temperature, the more even the mechanisms of chain rearrangement in the print are during the heating and solidification processes [22].

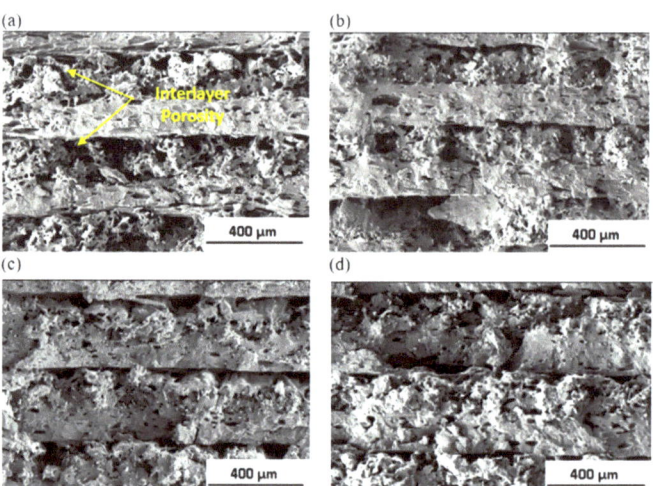

Figure 4. Mesostructure of CF-PA composites printed at (**a**) 230 °C, (**b**) 250 °C, (**c**) 270 °C, (**d**) 290 °C [100×].

3.1.2. SEM Result for CF-ABS Composite

Figure 5 shows the SEM observation of the mesostructure features for the AM-fabricated CF-ABS composites.

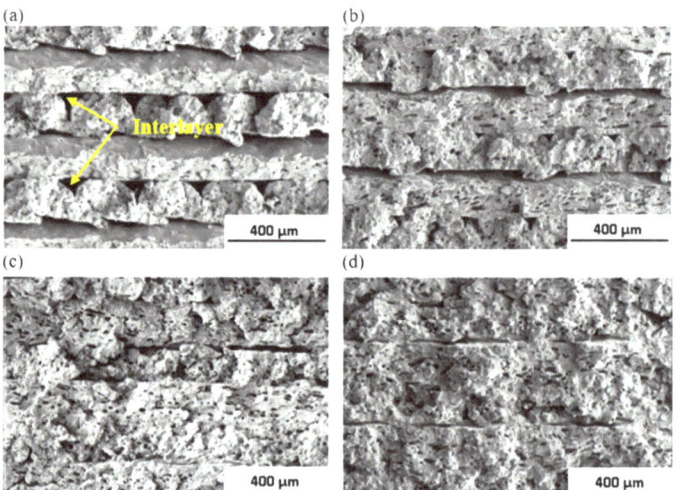

Figure 5. Mesostructure of CF-ABS composites printed at (**a**) 230 °C, (**b**) 250 °C, (**c**) 270 °C, (**d**) 290 °C [100×].

The SEM evaluation in Figure 5 shows that the CF-ABS composite exhibits a regular arrangement of the interlayers in the mesostructure. This is because of the high fluidity of the ABS matrix during solidification [23]. Increasing the fabrication temperatures between 270 °C–290 °C showed improved mesostructure formation and reduced interlayer porosities that were attributable to the improved melt flow at the increased temperatures. However, the steeper temperature gradient on cooling due to the higher temperatures resulted in greater intralayer porosity.

Similar to Figure 4, plastic material fusion bonding theory can also be applied to explain how increased heating from nozzle head temperatures can improve carbon fiber matrix coalescence and reduce interlayer porosities in the CF-ABS composites. However, at more elevated temperatures, temperature increases become detrimental and may result in degraded molecular chain strength, resulting in reduced mechanical performance, as was seen for the composite at the 290 °C fabrication temperature.

3.2. Micro-CT Scan Result

The micro-CT scan provides a more detailed quantitative evaluation of porosity features. Through the scan, porosity volume comparisons for the different fabrication temperatures were determined. Figure 6 presents reconstructed CT scan void images for the CF-PA and CF-ABS composites, which show a general void reduction with the increasing fabrication temperature under study.

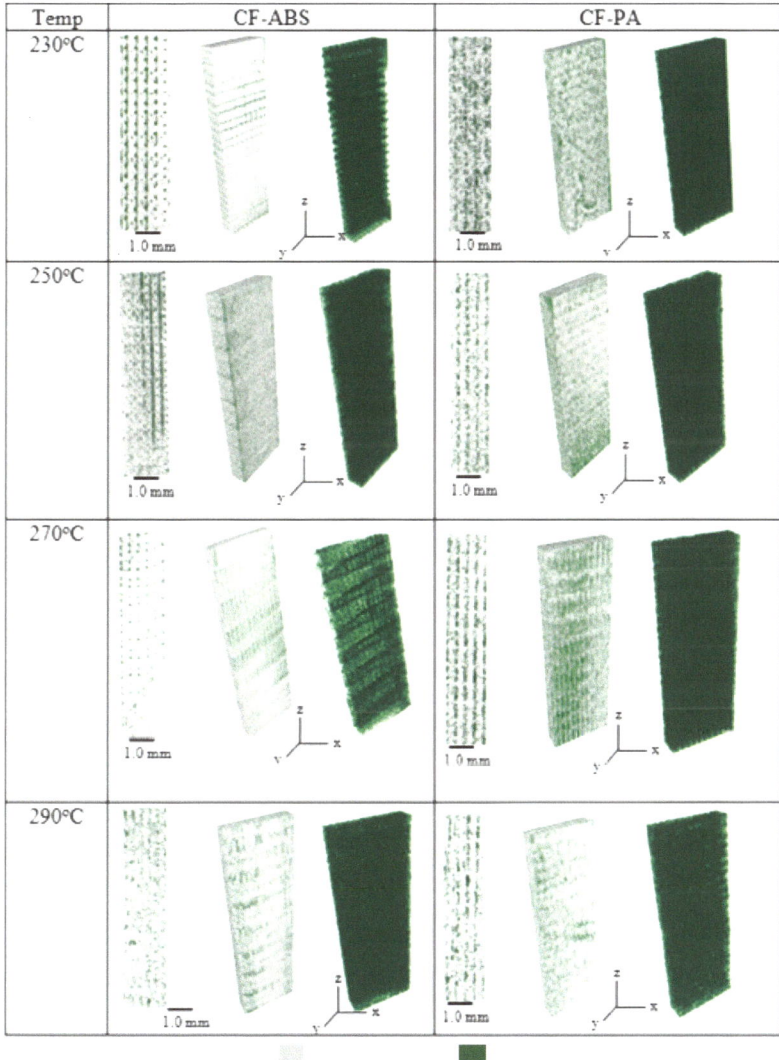

Figure 6. Reconstructed X-ray CT void for CF-ABS and CF-PA composite specimens for the different fabrication temperatures under study.

The variation is visualized with the grey pixels representing the solid volume of the composite and the green pixels representing the porosities. Understanding the effects of fabrication temperature on the degree of porosity is vital to characterizing the composites' mechanical performance to gain insights into how they affect material properties such as strength, modulus, ductility, toughness, hardness, etc. This is needed to determine the range of fabrication temperatures required to achieve the required material properties for different applications. Figures 7 and 8 show a plot of the test results for the CF-PA and CF-ABS composites, respectively, fabricated between 230 °C and 290 °C.

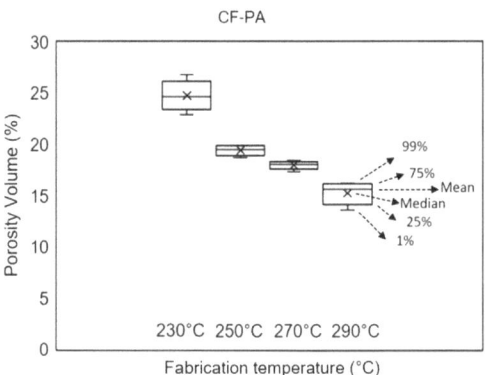

Figure 7. Fabrication temperature-related porosity volume in AM-fabricated CF-PA composite.

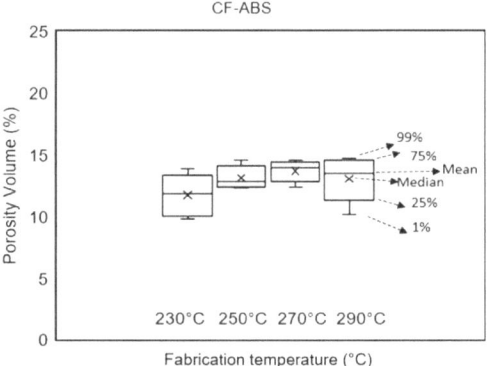

Figure 8. Fabrication temperature-influenced porosity volume in AM-fabricated CF-ABS composite.

3.2.1. Micro-CT Scan Porosity Results for CF-PA Composite

The results show an overall trend of reduced porosity from the combined inter and intralayer porosities effect when increasing the fabrication temperature from 230 °C to 290 °C.

Figure 7 shows a gradual decrease in the porosity of the CF-PA composite as the fabrication temperature increased from 230 °C to 290 °C, with a plot of test data distribution in percentiles within the overall data set. The composites show a consistent reduction in porosity volumes with mean values of 24.7%, 19.4%, 18.0%, and 15.3% for 230 °C, 250 °C, 270 °C, and 290 °C, respectively. A reduced porosity was observed with increasing temperature due to the fiber matrix's increasing coalescence, resulting in more solid mesostructure formation with reduced interlayer features.

3.2.2. Micro-CT Scan Porosity Results for CF-ABS Composite

Figure 8 shows the results for the overall porosity trends for CF-ABS within the 230 °C to 290 °C fabrication temperature range under study.

Even though the SEM evaluations of the composite showed pronounced interlayer porosity differences with fabrication temperature, the effect on the overall porosity volume for the amorphous CF-ABS was not quite pronounced when quantitatively measured in the CT scan. Figure 8 shows the overall porosity volume differences were insignificant with mean values of 11.7% for the 230 °C, 13.1% for the 250 °C, 13.9% for the 270 °C, and 13.1% for the 290 °C samples. Increasing the deposition temperature from 230 °C to 290 °C resulted in more interlayer coalescence with reduced interlayer porosities. However, intralayer porosities increased as a result of the steeper temperature gradient on cooling. These overall effects from the inter and intra layers were seen to account for the cause of the insignificant differences in the porosity volumes observed across the 230 °C to 290 °C deposition temperature range. The study by Adeniran et al. [24] provided more insight into the porosity types generated in CF-ABS composites fabricated by AM, where they discussed two fabrication-generated porosity types. The interlayer porosities are explained as triangular gaps in the non-contact areas of the print beads during layer-upon-layer material build-up while intralayer porosities are the void formation inside individual print beads due to gas evolution in the course of the semi-liquid to extrusion of the print material.

3.3. Fabrication Temperature-Related Porosity Effects on Mechanical Properties

3.3.1. Mechanical Properties of CF-PA Composite

A generally significant effect of fabrication temperature-influenced porosity effects within a 230 °C to 290 °C range on the tensile performance of AM-fabricated CF-PA composite was observed. This is in line with the CT scan microscopy results in Section 3.2.1, which showed the overall porosity reduced by as much as 38% from 230 °C to 290 °C for the samples investigated. Figure 9 shows the different mechanical properties examined for the AM-fabricated CF-PA composite.

Tensile Strength: The material tensile strength shown in Figure 9a is an indication of the maximum stress that can be applied while stretched in tension before material failure by elastic or plastic deformation. Printing temperatures played some role in influencing tensile strength, which may be related to the interlayer porosity feature as shown by the mesostructure formation and micro-CT scan evaluation in Sections 3.1.1 and 3.2.1, respectively. A significant effect was observed for the fabricated CF-PA composite, which displayed up to 37% increase in tensile strength at an optimized 290 °C printing temperature compared to the 230 °C baseline.

Fabrication temperature effects can be explained with better fiber matrix and layer-to-layer molecular bonding at higher temperatures, which promotes better inter-laminar and cross-laminar chemical bonding between print layers. This follows the explanation by Brenken et al. [25] that fabrication temperatures have strong effects on the melt viscosity of plastic matrices for bond formation between adjacent layers. Higher temperatures further from the glass transition temperature promote the diffusion-based fusion of adjacent layers after interfaces are established, while lower temperatures usually result in decreased molecular mobility, which hinders the molecular diffusion process.

Tensile Modulus: The modulus values define the stiffness properties of the material. As seen in Figure 9b, fabrication temperatures have insignificant effects on the CF-PA composite. Less than 10% variation was observed in the modulus properties for the 230 °C to 290 °C print temperature range. The insignificant effects of fabrication temperature can be explained by the theory propounded by Vaes and Puyvelde [21], which explained the effect of temperature in determining material modulus. They discussed modulus as influenced by the heterogenous self-nucleation of the structural crystals of the composite, which are influenced by heat transfer, melting, and shear deformations during the fabrication process. However, no great effect was observed for the 230 °C to 290 °C temperature range examined.

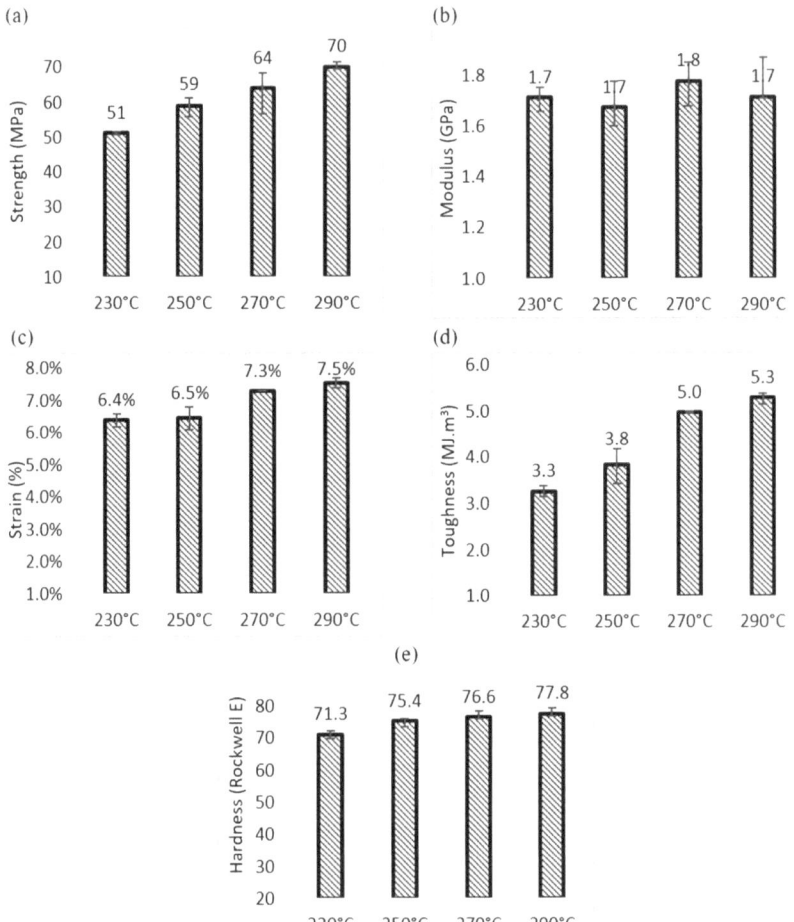

Figure 9. Tensile properties of semi-crystalline CF-PA composites at different printing temperatures: (**a**) ultimate tensile strength, (**b**) modulus, (**c**) ductility, (**d**) toughness, (**e**) hardness.

Tensile Ductility: The property of ductility measures the material's characteristics for plastic deformation before fracture. The results in Figure 9c show fabricating temperature trends similar to those observed for tensile strength in Figure 9a. Up to a 17% increase in ductility properties was observed at the 290 °C fabrication temperature, which offered the optimal temperature compared to the 230 °C temperature. The reduced degree of porosity volumes with increasing fabrication temperatures can be used to explain this increase in ductility with temperature when the materials are subjected to strain.

Tensile Toughness: The composite material's ability to absorb energy and plastically deform without fracturing is provided by its toughness values. As seen in Figure 9d, similar to the strength and ductility properties, fabrication temperature has some significant effects on the toughness properties of the CF-PA composite, with an approximate 60% increase in value observed at the highest 290 °C temperature over the 230 °C baseline. The toughness property, generally determined by the material strength and ductility characteristics, has similar determining features to the material strength and ductility, which also influence the toughness property.

Hardness: The Rockwell hardness number directly relates to the indentation hardness of the material and defines the resistance of a material to localized plastic deformation.

Figure 9e shows the Rockwell E values for the CF-PA composite exhibiting similar deposition temperatures related to hardness trends as the other mechanical properties. Although not as significant as the other properties, there was an upward increase in value with increasing deposition temperatures from 230 °C to 290 °C, which may also be related to the decreased porosities with increasing fabrication temperatures up to 290 °C.

3.3.2. Mechanical Properties of CF-ABS Composite

A generally insignificant effect of fabrication temperature-influenced porosity within a 230 °C to 290 °C range was observed on the tensile performance of AM-fabricated CF-ABS composite. This is in line with the CT scan microscopy results in Section 3.2.2, which showed the overall porosity for the most part to be within the range of 15%, except for the 270 °C fabricating temperature that showed a deviated range of up to 43% from the three other temperature data points under examination. Figure 10 shows the different mechanical properties examined for the CF-ABS composite.

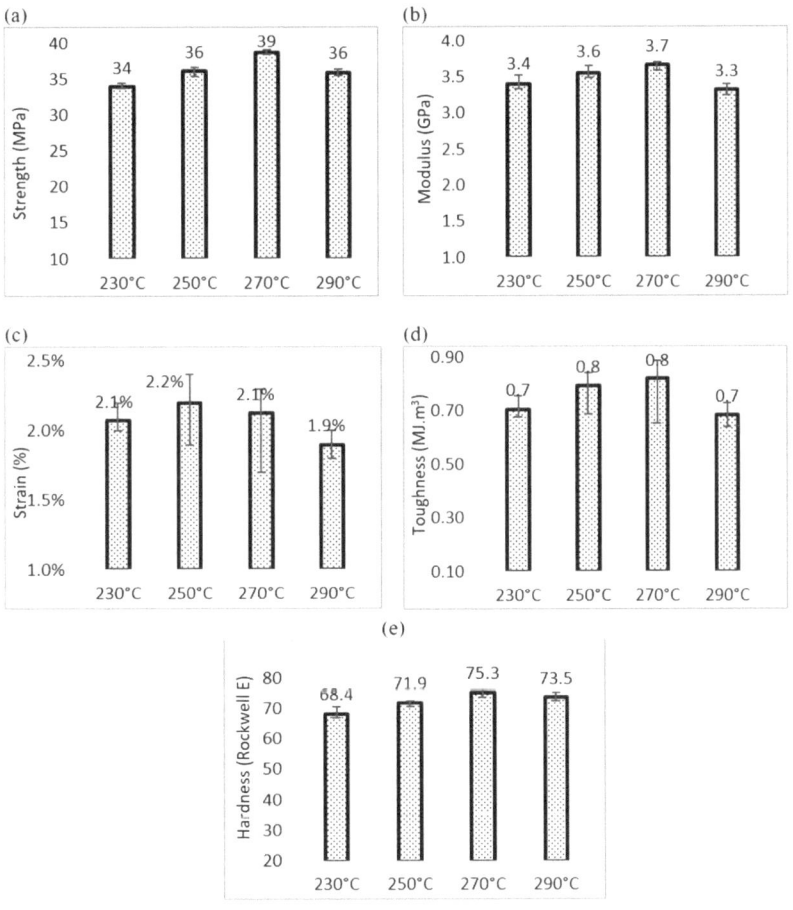

Figure 10. Tensile properties of amorphous CF-ABS composite at different printing temperatures: (**a**) ultimate tensile strength, (**b**) modulus, (**c**) ductility, (**d**) toughness, (**e**) hardness.

Tensile Strength: As seen in Figure 10a, fabrication temperature played some role in influencing the tensile strength of the CF-ABS composite. However, not so significant. An up to 15% increase in strength was observed at 270 °C, which was found to be optimal for the fabrication temperature range under examination.

Similar to the CF-PA composite, the temperature effects can be explained by the theory of better fiber matrix and layer-to-layer molecular bonding with higher printing temperatures, due to their promotion of good inter-laminar and cross-laminar chemical bonding between print layers. This similarly follows the explanation by Brenken et al. [25] that fabrication temperatures strongly influence the melt viscosity of plastic matrices for bond formation between interlayers. High temperatures are needed for the fusion bonding of adjacent beads following interface formation, while lower temperatures usually decrease molecular mobility and limit the molecular diffusion process. However, excessive temperatures can also be detrimental, leading to alteration of the material composition and degradation of the material strength as seen with the CF-ABS composite beyond 270 °C fabrication temperature.

Tensile Modulus: As seen with the tensile strength results for the AM-fabricated CF-ABS composite, fabricating temperature has insignificant effects. Figure 10b shows less than 10% variation for the 230 °C to 290 °C fabrication temperatures examined. The insignificant effect on the tensile modulus can be ascribed to the similarity of overall porosity volumes observed from the CT scan results, which are determinant to the stiffness features that determine the modulus properties of the composite.

Tensile Ductility: As observed for the other tensile properties of the AM-fabricated CF-ABS composite, the material ductility shown in Figure 10c has similar trends for the effect of fabrication temperature. The effects were found to be insignificant within a 10% range for the 230 °C to 290 °C fabrication temperatures examined, which can also be attributed to the insubstantial differences in the overall porosities generated at the different fabrication temperatures.

Tensile Toughness: Figure 10d shows the material toughness, which exhibits similar trends in strength and ductility. Fabrication temperature has an insignificant effect on AM-fabricated CF-ABS composite (less than 10%), with the material strength and ductility properties also known to have an overall effect on this property and similar intrinsic material features that determine strength and ductility also determining the material toughness characteristics.

Hardness Properties: Similar to the tensile strength, ductility, and toughness, the hardness properties for the AM-fabricated CF-ABS composite as seen in Figure 10a show the Rockwell E values, which exhibit similar deposition temperature-related hardness trends. The effect was insignificant but still shows the property value peaking at 270 °C as seen for the other tensile properties. The insignificant differences can also be attributed to the insubstantial effects of the deposition temperature on the overall interlayer and intralayer porosity volume differences for the 230 °C to 290 °C deposition temperatures examined.

4. Conclusions

In this study, deposition temperature-related porosity effects on mechanical properties—strength, modulus, ductility, toughness, and hardness—of AM-fabricated CFRP composites using semicrystalline CF-PA and amorphous CF-ABS samples were examined. The investigations were conducted by relating deposition temperature trends to mesostructure formation, porosities, and mechanical properties, and the following conclusions were drawn:

(1) Deposition temperatures have some effects on porosity volumes in AM-fabricated CFRP composites, with semicrystalline CF-PA much more significantly affected than the amorphous CF-ABS composite.
(2) The degree of porosity is largely determined by the characteristics of the matrix material.
(3) A direct relationship exists between CFRP composites' porosity and mechanical properties.
(4) The overall porosity volumes are determinant of the interlayer and intralayer voids, but the interlayer voids play a greater role in determining the mechanical properties.
(5) Semicrystalline composites exhibit higher porosity volumes than amorphous composites due to rapid recrystallization as the chains rearrange during the cooling of the print beads.

Author Contributions: Conceptualization, O.A.; methodology, O.A. and N.O.-u.; validation, O.A.; formal analysis, O.A. and N.O.-u.; investigation, O.A. and N.O.-u.; resources, O.A., M.R. and W.C.; data curation, O.A. and N.O.-u.; writing—original draft preparation, O.A.; writing—review and editing, O.A., M.R. and W.C.; supervision, W.C. All authors have read and agreed to the published version of the manuscript.

Funding: This research received no external funding.

Data Availability Statement: Data used in this research are available upon request in Microsoft txt, Docx, and jpg file formats.

Conflicts of Interest: The authors declare that there is no conflict of interest.

References

1. Frketic, J.; Dickens, T.; Ramakrishnan, S. Automated Manufacturing and Processing of Fiber-Reinforced Polymer (FRP) Composites: An Additive Review of Contemporary and Modern Techniques for Advanced Materials Manufacturing. *Addit. Manuf.* **2017**, *14*, 69–86. [CrossRef]
2. Melchels, F.P.W.; Feijen, J.; Grijpma, D.W. A Review on Stereolithography and Its Applications in Biomedical Engineering. *Biomaterials* **2010**, *31*, 6121–6130. [CrossRef]
3. Dimas, L.S.; Buehler, M.J. Modeling and Additive Manufacturing of Bio-Inspired Composites with Tunable Fracture Mechanical Properties. *Soft Matter* **2014**, *10*, 4436–4442. [CrossRef]
4. Mazzoli, A. Selective Laser Sintering in Biomedical Engineering. *Med. Biol. Eng. Comput.* **2013**, *51*, 245–256. [CrossRef]
5. Zhang, W.; Cotton, C.; Sun, J.; Heider, D.; Gu, B.; Sun, B.; Chou, T.W. Interfacial Bonding Strength of Short Carbon Fiber/Acrylonitrile-Butadiene-Styrene Composites Fabricated by Fused Deposition Modeling. *Compos. Part B Eng.* **2018**, *137*, 51–59. [CrossRef]
6. Tekinalp, H.L.; Kunc, V.; Velez-Garcia, G.M.; Duty, C.E.; Love, L.J.; Naskar, A.K.; Blue, C.A.; Ozcan, S. Highly Oriented Carbon Fiber-Polymer Composites via Additive Manufacturing. *Compos. Sci. Technol.* **2014**, *105*, 144–150. [CrossRef]
7. Ning, F.; Cong, W.; Qiu, J.; Wei, J.; Wang, S. Additive Manufacturing of Carbon Fiber Reinforced Thermoplastic Composites Using Fused Deposition Modeling. *Compos. Part B Eng.* **2015**, *80*, 369–378. [CrossRef]
8. Adeniran, O.; Cong, W.; Bediako, E.; Aladesanmi, V. Additive Manufacturing of Carbon Fiber Reinforced Plastic Composites: The Effect of Fiber Content on Compressive Properties. *J. Compos. Sci.* **2021**, *5*, 325. [CrossRef]
9. Petrò, S.; Reina, C.; Moroni, G. X-Ray CT-Based Defect Evaluation of Continuous CFRP Additive Manufacturing. *J. Nondestr. Eval.* **2021**, *40*, 7. [CrossRef]
10. Tagscherer, N.; Schromm, T.; Drechsler, K. Foundational Investigation on the Characterization of Porosity and Fiber Orientation Using XCT in Large-Scale Extrusion Additive Manufacturing. *Materials* **2022**, *15*, 2290. [CrossRef] [PubMed]
11. Adeniran, O.; Cong, W.; Aremu, A.; Oluwole, O. Forces in Mechanics Finite Element Model of Fiber Volume Effect on the Mechanical Performance of Additively Manufactured Carbon Fiber Reinforced Plastic Composites. *Forces Mech.* **2023**, *10*, 100160. [CrossRef]
12. Adeniran, O. Mechanical Performance of Carbon Fiber Reinforced Plastic Composites Fabricated by Additive Manufacturing. Ph.D. Thesis, Texas Tech University, Lubbock, TX, USA, 2022.
13. Ning, F.; Cong, W.; Hu, Z.; Huang, K. Additive Manufacturing of Thermoplastic Matrix Composites Using Fused Deposition Modeling: A Comparison of Two Reinforcements. *J. Compos. Mater.* **2017**, *51*, 3733–3742. [CrossRef]
14. Ning, F.; Cong, W.; Hu, Y.; Wang, H. Additive Manufacturing of Carbon Fiber-Reinforced Plastic Composites Using Fused Deposition Modeling: Effects of Process Parameters on Tensile Properties. *J. Compos. Mater.* **2017**, *51*, 451–462. [CrossRef]
15. Ajinjeru, C.; Kishore, V.; Lindahl, J.; Sudbury, Z.; Hassen, A.A.; Post, B.; Love, L.; Kunc, V.; Duty, C. The Influence of Dynamic Rheological Properties on Carbon Fiber-Reinforced Polyetherimide for Large-Scale Extrusion-Based Additive Manufacturing. *Int. J. Adv. Manuf. Technol.* **2018**, *99*, 411–418. [CrossRef]
16. Ajinjeru, C.; Kishore, V.; Liu, P.; Lindahl, J.; Hassen, A.A.; Kunc, V.; Post, B.; Love, L.; Duty, C. Determination of Melt Processing Conditions for High-Performance Amorphous Thermoplastics for Large Format Additive Manufacturing. *Addit. Manuf.* **2018**, *21*, 125–132. [CrossRef]
17. Kishore, V.; Ajinjeru, C.; Nycz, A.; Post, B.; Lindahl, J.; Kunc, V.; Duty, C. Infrared Preheating to Improve Interlayer Strength of Big Area Additive Manufacturing (BAAM) Components. *Addit. Manuf.* **2017**, *14*, 7–12. [CrossRef]
18. Cinquin, J.; Chabert, B.; Chauchard, J.; Morel, E.; Trotignon, J.P. Characterization of a Thermoplastic (Polyamide 66) Reinforced with Unidirectional Glass Fibres. Matrix Additives and Fibres Surface Treatment Influence on the Mechanical and Viscoelastic Properties. *Composites* **1990**, *21*, 141–147. [CrossRef]
19. Yang, C.; Tian, X.; Liu, T.; Cao, Y.; Li, D. 3D Printing for Continuous Fiber Reinforced Thermoplastic Composites: Mechanism and Performance. *Rapid Prototyp. J.* **2017**, *23*, 209–215. [CrossRef]
20. Adeniran, O.; Cong, W.; Aremu, A. Material Design Factors in the Additive Manufacturing of Short Carbon Fiber Reinforced Plastic Composites: A State-of-the-Art-Review. *Adv. Ind. Manuf. Eng.* **2022**, *5*, 100100. [CrossRef]

21. Vaes, D.; van Puyvelde, P. Semi-Crystalline Feedstock for Filament-Based 3D Printing of Polymers. *Prog. Polym. Sci.* **2021**, *118*, 101411. [CrossRef]
22. Jiang, Z.; Diggle, B.; Tan, M.L.; Viktorova, J.; Bennett, C.W.; Connal, L.A. Extrusion 3D Printing of Polymeric Materials with Advanced Properties. *Adv. Sci.* **2020**, *7*, 1–32. [CrossRef] [PubMed]
23. Gofman, I.v.; Yudin, V.E.; Orell, O.; Vuorinen, J.; Grigoriev, A.Y.; Svetlichnyi, V.M. Influence of the Degree of Crystallinity on the Mechanical and Tribological Properties of High-Performance Thermoplastics over a Wide Range of Temperatures: From Room Temperature up to 250 °C. *J. Macromol. Sci. Part B Phys.* **2013**, *52*, 1848–1860. [CrossRef]
24. Adeniran, O.; Cong, W.; Bediako, E.; Adu, S.P. Environmental Affected Mechanical Performance of Additively Manufactured Carbon Fiber–Reinforced Plastic Composites. *J. Compos. Mater.* **2022**, *56*, 1139–1150. [CrossRef]
25. Brenken, B.; Dissertation, A. Extrusion Deposition Additive Manufacturing of Fiber Reinforced Semi-Crystalline Polymers. Ph.D. Thesis, Purdue University, West Lafayette, IN, USA, 2017.

Disclaimer/Publisher's Note: The statements, opinions and data contained in all publications are solely those of the individual author(s) and contributor(s) and not of MDPI and/or the editor(s). MDPI and/or the editor(s) disclaim responsibility for any injury to people or property resulting from any ideas, methods, instructions or products referred to in the content.

Article

Additive Manufacturing of C/C-SiC Ceramic Matrix Composites by Automated Fiber Placement of Continuous Fiber Tow in Polymer with Pyrolysis and Reactive Silicon Melt Infiltration

Corson L. Cramer [1,*], Bola Yoon [2], Michael J. Lance [2], Ercan Cakmak [2], Quinn A. Campbell [1] and David J. Mitchell [2]

[1] Manufacturing Science Division, Oak Ridge National Laboratory, Oak Ridge, TN 37831, USA
[2] Materials Science and Technology Division, Oak Ridge National Laboratory, Oak Ridge, TN 37831, USA
* Correspondence: cramercl@ornl.gov

Abstract: An additive manufacturing process for fabricating ceramic matrix composites has been developed based on the C/C-SiC system. Automated fiber placement of the continuous carbon fibers in a polyether ether ketone matrix was performed to consolidate the carbon fibers into a printed preform. Pyrolysis was performed to convert the polymer matrix to porous carbon, and then Si was introduced by reactive melt infiltration to convert a portion of the carbon matrix to silicon carbide. The densities and microstructures were characterized after each step during the processing, and the mechanical properties were measured. The C/C-SiC composites exhibited a porosity of 10–20%, characteristic flexural strength of 234.91 MPa, and Weibull modulus of 3.21. The composites displayed toughness via a significant displacement to failure.

Keywords: 3D printing; ceramic matrix composites; reactive melt infiltration

1. Introduction

High temperature materials such as carbon (C) and silicon carbide (SiC) have unique combinations of properties for high temperature applications. Carbon monoliths and carbon fiber–carbon matrix (C/C) composites have superior mechanical properties at elevated temperature [1,2], but they will oxidize via conversion to gaseous species in environments containing oxygen. SiC is robust up to 1700 °C [3], forms an oxygen-protective silica scale [4–6], has low density, has low thermal expansion [7], has high thermal conductivity [8], has a high elastic modulus [9], has high strength [10], and retains its strength at elevated temperature [9,10]. While monolithic SiC exhibits brittle failure behavior indicative of ceramics, it is a prime candidate matrix material for high temperature composites.

Inherently brittle materials such as SiC can exhibit damage tolerance when formed into ceramic matrix composites (CMCs) because the cracking in the SiC matrix deflects at the fiber–matrix interfaces (thin boron nitride (BN) or C layers), the load is transferred to the fibers, and the energy is absorbed in individual fiber cracking and pullout [11–13]. Several decades of research have been performed on the processing and testing of C/SiC composites, employing a SiC matrix reinforced with carbon fibers (Cf). While early versions of C/SiC CMCs did not have an interface coating, current state-of-the-art C/SiC CMCs are fabricated with Cf cloths that have been coated with interface layers and, subsequently, densified with SiC using chemical vapor infiltration [14–16], forming a C/SiC composite with strain to failure greater than monolithic SiC. Another closely related composite is C/C-SiC, in which there is some of the carbon matrix protecting the Cf, while the remainder of the matrix is densified with SiC. In C/C-SiC composites, the fiber interface coating is less critical because a less strong carbon matrix provides damage tolerance. C/C-SiC composites have been fabricated with fiber cloths, chemical vapor deposition (CVD) or resin build-up of carbon, and silicon (Si) reactive melt infiltration (RMI) [17–21] and have

Citation: Cramer, C.L.; Yoon, B.; Lance, M.J.; Cakmak, E.; Campbell, Q.A.; Mitchell, D.J. Additive Manufacturing of C/C-SiC Ceramic Matrix Composites by Automated Fiber Placement of Continuous Fiber Tow in Polymer with Pyrolysis and Reactive Silicon Melt Infiltration. *J. Compos. Sci.* **2022**, *6*, 359. https://doi.org/10.3390/jcs6120359

Academic Editor: Yuan Chen

Received: 20 October 2022
Accepted: 17 November 2022
Published: 23 November 2022

Publisher's Note: MDPI stays neutral with regard to jurisdictional claims in published maps and institutional affiliations.

Copyright: © 2022 by the authors. Licensee MDPI, Basel, Switzerland. This article is an open access article distributed under the terms and conditions of the Creative Commons Attribution (CC BY) license (https://creativecommons.org/licenses/by/4.0/).

also shown significant improvements in strain to failure compared with monolithic SiC and C materials.

One method to obtain the partial carbon matrix is to use polyether ether ketone (PEEK) as the fiber binder and pyrolyze it to form carbon, and PEEK's high char yield ensures a high amount of carbon in the matrix [22]. Compression molding (or warm pressing) of C/C-SiC composites has been achieved starting with discontinuous carbon fibers with 6–24 mm lengths [23] or 0/90° 2D plain weave carbon fiber [24] with PEEK (because of its high char yield) and Si RMI. Both studies reported toughness and fiber pullout. However, while providing composites with excellent properties, these fabrication methods require expensive tooling and intricate manual labor and are time- and cost-intensive.

In other developments, the advent of automated fiber placement (AFP) fabrication of polymer matrix composites (PMCs) has been demonstrated, incorporating single tows or tapes uniformly by melting and resolidifying thermoplastic-impregnated continuous carbon fibers [25–29]. The methods, termed fused deposition modeling (FDM) and fused filament fabrication (FFF), respectively, are used interchangeably, and AFP is a new concept derivative of these techniques, specifically using continuous fibers. For designs that do not require weaves and can utilize the 0/90° 2D application of unidirectional tapes, this drastically reduces the time and cost to make PMCs. Following the AFP process, composites can be removed from build plates and used directly or transitioned to further machining processes. Continuous-fiber AFP for CMCs was first reported in the literature with Cf and modified preceramic polymer (liquid or thermosetting resin) as the printing matrix [30]. The printer was a modified FDM-type machine that melts the printing matrix. In addition, an in situ screw-driven extrusion delivery system has been used to pre-impregnate fiber tows prior to printing, further densified with CVI [31]. Additionally, there are efforts to develop FDM processes with short fiber for fabricating CMCs [32,33], but these composites tend to have low fiber volume fraction and fibers with short lengths.

In this work, information on previous studies of composites made via PMC AFP and Cf-PEEK pyrolysis with RMI were leveraged and combined to form CMCs on a small scale. The manufacturing process consisted of 3D printing automatically placed fiber preforms in a polymer matrix. Once the PMCs were printed, the polymer matrix was pyrolyzed to form carbon, resulting in porous Cf-C (sometimes called C/C) composites, and then an additional step of RMI of Si was performed to fabricate a C/C-SiC composite. This novel manufacturing technique for CMCs allows for faster and more controlled processing, since fiber preforms were made by robot, and no tooling was necessary in the processing. This paper details the processing steps, characterization of the material after each processing step, and mechanical properties of the C/C-SiC composites processed without pressure (free-formed).

2. Materials and Methods

2.1. Materials

FBR-MP0007 from Desktop Metal, Inc. (Burlington, MA, USA) was used as the printing filament. This filament consists of 60 vol.% continuous Cf tow impregnated with PEEK to high density (1.57 g/cc), i.e., less than 3% porosity. Silicon metal powder (US Nanomaterial, Inc., Houston, TX, USA) with purity of 99.9% and 20 micrometer particle size was used for Si RMI.

2.2. Printing

The printing of the Cf-PEEK composites was done using an FFF printer called the Fiber from Desktop Metal (Burlington, MA). This printing style is known as micro automated fiber placement (μAFP), where tapes of previously impregnated Cf tows are forced through the heated nozzle and deposited onto the build platform. Tapes were laid down in 10 layers with alternating 0/90° direction. Each tape layer was 0.25 mm thick and 3 mm wide. The temperature of the build plate during printing was 150 °C. The temperature of the printing nozzle during printing was 350 °C.

2.3. Pyrolysis

Samples were pyrolyzed to convert PEEK to amorphous carbon, preparing the preform for reactive melt infiltration. Pyrolysis was conducted at 850 °C for 30 min at a heating rate of 2 °C/min, in 500 standard cubic centimeters per minute (sccm) of nitrogen gas (99.9% purity from Airgas) in a graphite furnace.

2.4. Silicon Reactive Melt Infiltration

Pressureless RMI of silicon was performed on the composites after pyrolysis [34]. The mass of silicon required to fill the pores (Equation (1)) was calculated for each sample, and only that amount of Si was added for RMI, in order to achieve high density, limited residual Si, and high SiC content from the reaction. The mass of silicon to use for the infiltration is denoted as $mass_{Si}$. $V_{Cf/C\ Part}$ is the volume of the pyrolyzed carbon fiber-carbon matrix part. The mass of carbon fiber-carbon matrix part is denoted as $mass_{Cf/C\ Part}$. ρ_{Si} is the mass density of silicon, and 1.8 is the density value used for carbon. Half of the predetermined amount of powdered Si was placed onto BN plates in a graphite crucible, the panels were placed on top of the Si powder, and the other half of the Si powder was placed on top of the porous C/C samples.

$$mass_{Si} = V_{Cf/C\ Part} \cdot \left(1 - \frac{\frac{mass_{Cf/C\ Part}}{V_{Cf/C\ Part}}}{1.8}\right) \cdot \rho_{Si} \quad (1)$$

The heating cycle for reactive melt infiltration of silicon was done in two steps in one vacuum furnace. The first step heated samples to 1450 °C at 10 °C/min holding for 30 min under vacuum of ~10 Pa, followed by the second step, in which the chamber is filled with Ar to a total pressure of ~100,000 Pa to minimize silicon volatilization, and temperature increased to 1670 °C at 10 °C/min followed by holding for 1 h. The purpose of the second step at 1670 °C was to allow enough reaction time between Si and C to convert as much Si as possible to reaction-bonded SiC.

2.5. Characterization and Mechanical Properties

Archimedes, geometrical, and areal densities were measured where applicable. For pore volume, optical cross-sections were analyzed for areal density, and that value was assumed to hold in 3D, so that the correct amount of silicon could be added. Optical microscopy of specimen cross sections was performed with a Keyence VHX-1000 system. Raman spectroscopy was used to identify the two carbon phases (fibers and matrix from converted PEEK), residual Si metal, and SiC. Raman spectra were obtained with a Renishaw Invia Raman microprobe using a Nd-YAG laser operating at 532 nm and a spot size of ~1 μm. Phase identification was conducted using the principal component analysis within the vendor's Wire 5.2 software.

X-ray computed tomography (XCT) was done using a Zeiss Metrotom (Carl Zeiss Industrial Metrology, Maple Grove, MN, USA) to help identify delamination, pores, defects, and distortion. The scans were done with a 0.5 mm Cu filter at 180 kV, 97 μA, and 1000 ms with 16 μm resolution in the X, Y, and Z directions. All of the XCT data were reconstructed using VGstudio Max version 3.2.

Thermogravimetric analysis (TGA) using a TA Q50 (TA Instruments, New Castle, DE, USA) was used to determine the decomposition, mass loss, and char yield of the Cf-PEEK printing tow or filament. Thermal treatment was performed at a 10 °C/min heating rate from room temperature to 900 °C in air and flowing nitrogen (N_2) at a rate of 50 mL/min.

Four-point bending tests were performed (Equation (2)) on samples with dimensions of 50 mm × 4 mm × 3 mm to measure the displacement and flexural strength, following the procedure outlined in ASTM C1161 using an electromechanical test frame (Instru-Met, Union, NJ, USA) and a silicon carbide fixture with an inner span of 20 mm and outer support span of 40 mm. The strain rate was 0.01 mm/s at 50 Hz collection rate. σ_f is the flexural strength, F is the load, L is the support span, b is the width of the test specimen, and d is the thickness of the sample.

$$\sigma_f = \frac{3FL}{4bd^2} \qquad (2)$$

3. Results and Discussion

Figure 1 shows the TGA curve for the Cf-PEEK filament in air and N_2. In both cases, the PEEK started to decompose at 600 °C. The comparison is made to show that PEEK and carbon fibers will completely decompose (i.e., reduce in mass to near zero) if heated to 900 °C in air. Heating in the N_2 environment resulted in 80 wt.% retention in the material, which is not a direct char yield because carbon fibers were already embedded in the PEEK as a tow, but it shows that most of the material is retained after the pyrolysis. More imaging analysis is needed to elucidate the microstructure of the resulting Cf-C material.

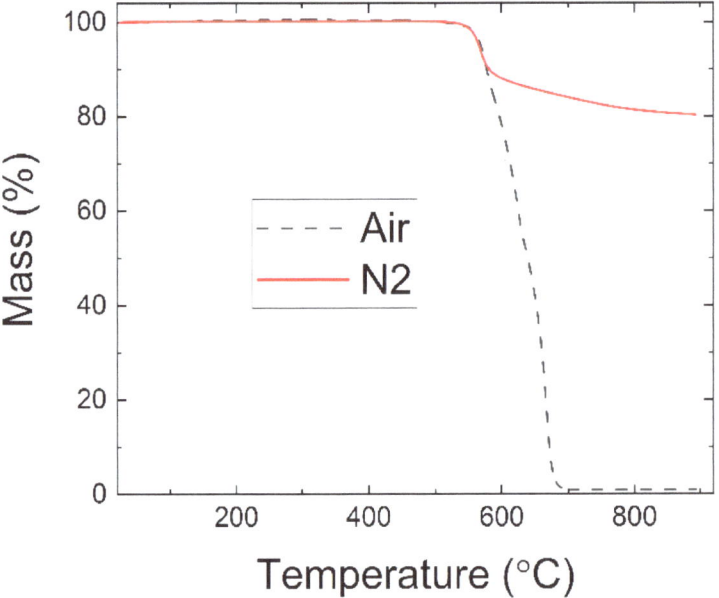

Figure 1. TGA data of printing filament run in air and under nitrogen flow.

Figure 2 shows an optical image of a printed Cf-PEEK part. The sample was highly dense, and there were no detectable pores. That is because the PEEK worked its way into the part well and adheres the fiber layers. Moreover, the CT scan resolution is in the micron range, so no pores in the submicron range can be detected. Since these two materials, PEEK and carbon, have very low density, it is difficult to obtain scans where the two are distinctly identified versus the background ambient air.

Figure 3 shows an optical image of a pyrolyzed Cf-PEEK part. The pyrolyzed part consists of Cf, shown by the circles in the 2D image and the carbon matrix around the Cf from the conversion process during pyrolysis. The newly formed carbon matrix holds the fiber tows together and mostly surrounds the Cf. There are several areas of cracking and debonding of the carbon matrix around the fibers, which provides a pathway for the subsequent Si infiltrant. The mass and density of the samples were measured before and after pyrolysis and are shown in Table 1. The dense Cf-PEEK lost mass during pyrolysis and decreased in density. During pyrolysis in an inert atmosphere, predominantly non-carbon elements are removed as volatiles, leaving predominantly carbon in the matrix. Therefore, the density decrease is from two events: (1) the loss of matter from the decomposition of the

PEEK thermoplastic structure and (2) the potential bloating of the structure due to pressure built up during the pyrolysis phase.

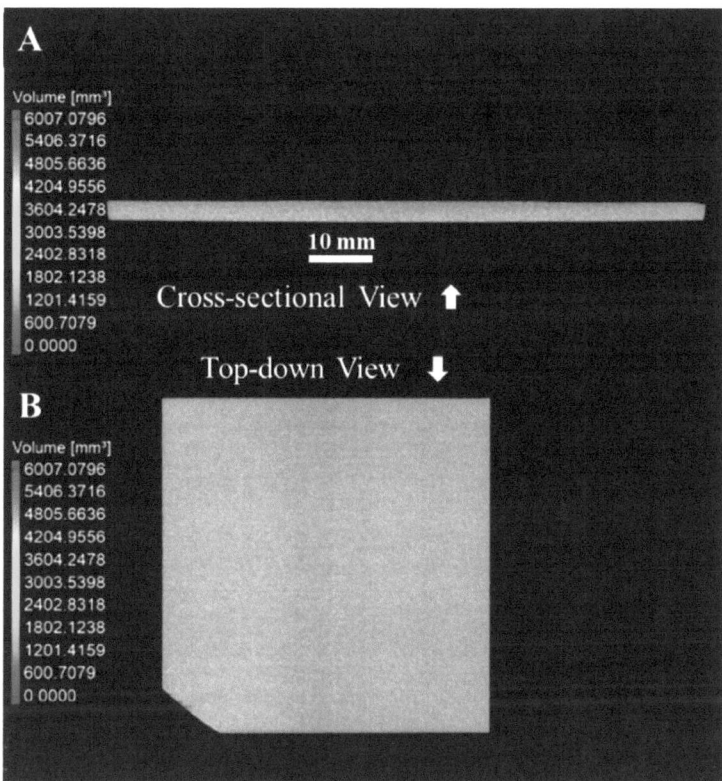

Figure 2. XCT scans of (**A**) cross-section and (**B**) top-down view of printed Cf-PEEK sample.

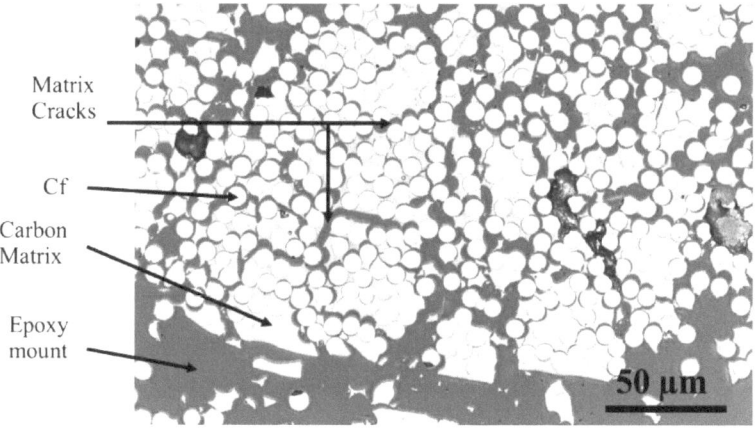

Figure 3. Optical image of a cross-section of pyrolyzed Cf-PEEK sample showing the microstructure and porosity of the Cf-C composite.

Table 1. Information on preforms, pyrolyzed samples, and MI samples.

Property	As-Printed Cf-PEEK	After Pyrolysis	After RMI
Mass (g)	24.82	20.81	34.27
Density (g/cc)	1.52	1.31	1.93
Porosity (%)	3	40	20

An XCT scan of Cf-C in Figure 4 was performed to show the volumetric porosity and defects of the bulk samples when converting the Cf-PEEK samples into porous Cf-C samples. There is some delamination between tows, and there is porosity between and within the tows. The porosity of the samples was measured to be about 40 vol.%. This verified that the carbon matrix had enough cracks and open porosity to enable infiltration by the liquid silicon material.

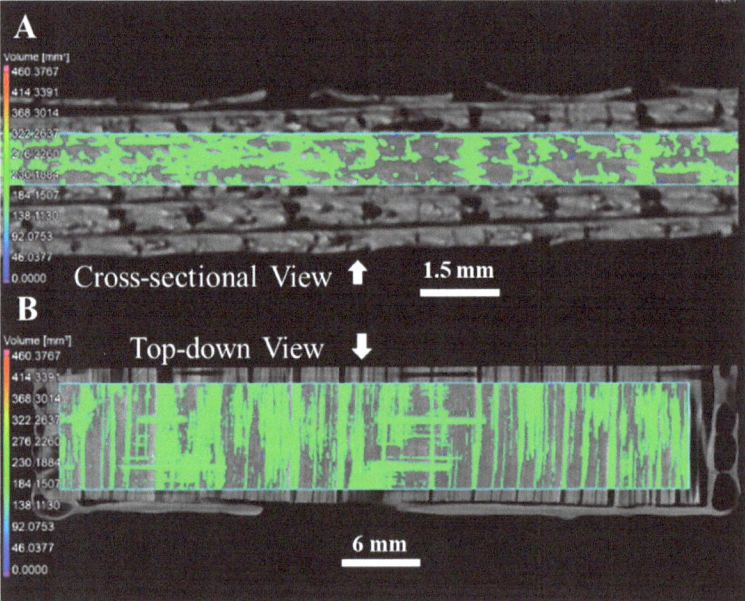

Figure 4. XCT scans of (**A**) cross-section and (**B**) top-down view of pyrolyzed Cf-PEEK sample showing the porosity of the pyrolyzed Cf-C composite.

Figure 5A shows an image of a C/C-SiC part after processing the Cf-PEEK samples through pyrolysis to convert PEEK to carbon and the Si melt infiltration step to convert some of the carbon matrix to SiC. The panel was primarily flat with some deviation (~5%), likely due to a minor difference between the volume increase experienced during the formation of SiC and the initial level of porosity in the Cf-C preform. There is residual Si on the surface, so XCT scans were performed to quantify and spatially resolve the pores and defects in the microstructure. A sample was subjected to impact damage by driving a nail through it (Figure 5B). The composite demonstrated toughness, as the sample did not catastrophically fail. The structural integrity was maintained in all but the affected area with the local impact of the nail. Panels were machined into bend bars to further quantify the strength and toughness of the composites.

Figure 5. Images of (**A**) fully processed C/C-SiC panel and (**B**) C/C-SiC panel after the nail impact toughness test, demonstrating qualitative toughness.

Figure 6 shows optical micrograph images of the cross-sections of a C/C-SiC composite plate. Regions of delamination can be observed in Figure 6A. Causes of delamination must be identified, and the extent of delamination must be minimized in order for these composites to be useful as structural materials. The images in Figure 6B and C show that the thickness of the layers in the C/C-SiC composite is approximately 200 μm. It can be observed in Figure 6B and C that following Si RMI, the microstructure is a mixture of carbon fibers, SiC, Si, and porosity, indicating that the melt infiltration was successful, but the process could be further optimized. The microstructures of the C/C-SiC composites processed for this study resemble those of similar materials [23,24]. The results from previous studies indicate that applying pressure during fabrication may reduce delamination. Delamination areas or gaps can occur in parts when the fiber tows are not placed close enough together during the printing deposition process.

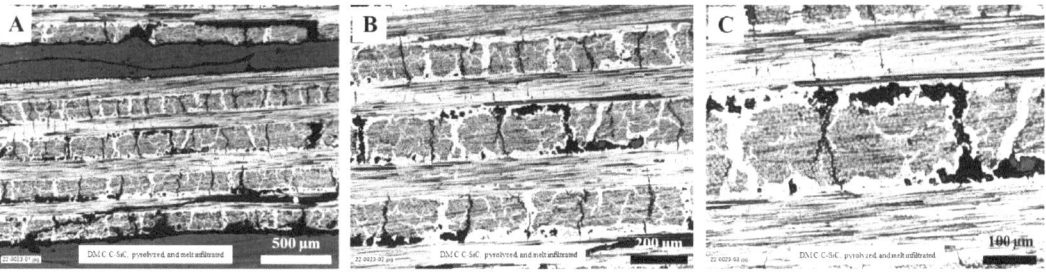

Figure 6. Optical images of a cross-section of pyrolyzed and Si-melt-infiltrated C/C-SiC composites, with increasing magnification (**A**–**C**).

An XCT scan of the C/C-SiC preform was obtained to observe the volumetric porosity and defects in the final composite. The results of the XCT analysis are shown in Figure 7. The areas of delamination and pores between tows can be identified in the scans. The total amount of porosity in the composites was estimated to be approximately 10–20 vol.%, although delamination could have introduced variations in the estimated value. In addition, the total warpage was minimal and was under 5% deviation from the printed part. An animation of the CT scan reconstruction showing the part volume and defects is shown in Supplementary Video S1.

Figure 8 shows SEM images of cross-sections of the final C/C-SiC. Figure 8A shows several printing layers or tows where the in-plane and out-of-plane tows can be identified. The materials were identified by EDS, where the areas of SiC and unreacted residual silicon can be seen. Figure 8B shows a more magnified cross-sectional area of the in-plane carbon fibers and also shows the reaction-formed SiC in between the carbon fibers. Figure 8C shows a more magnified cross-sectional area of the out-of-plane fibers. The tightly packed carbon fibers are surrounded by the carbon matrix, and there are some areas of the cracks that were filled with reaction-bonded SiC and Si.

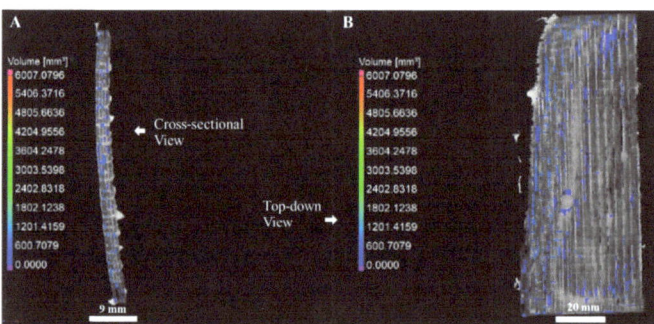

Figure 7. XCT scans of (**A**) cross-section and (**B**) top-down views showing the total porosity in the C/C-SiC composite.

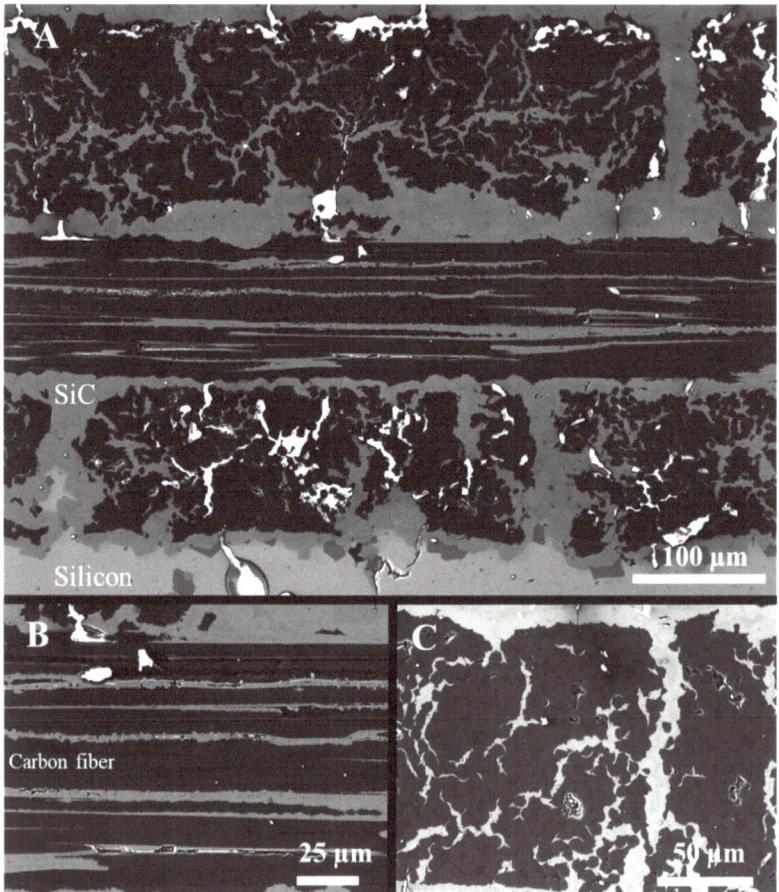

Figure 8. (**A**) SEM of a cross-section showing several layers, (**B**) higher magnification of the in-plane fibers, and (**C**) higher magnification of the out-of-plane fibers of a pyrolyzed and Si-melt-infiltrated Cf-PEEK sample showing the total porosity of the final C/C-SiC composite.

Figure 9 shows the results of Raman spectroscopy from a cross-section of the final C/C-SiC composite. The white region is the matrix carbon, gray is carbon fibers, red is residual silicon, and green is SiC. The Raman spectra collected from the SiC phase had broad peaks suggestive of a highly disordered structure. The polytype could not be identified [35].

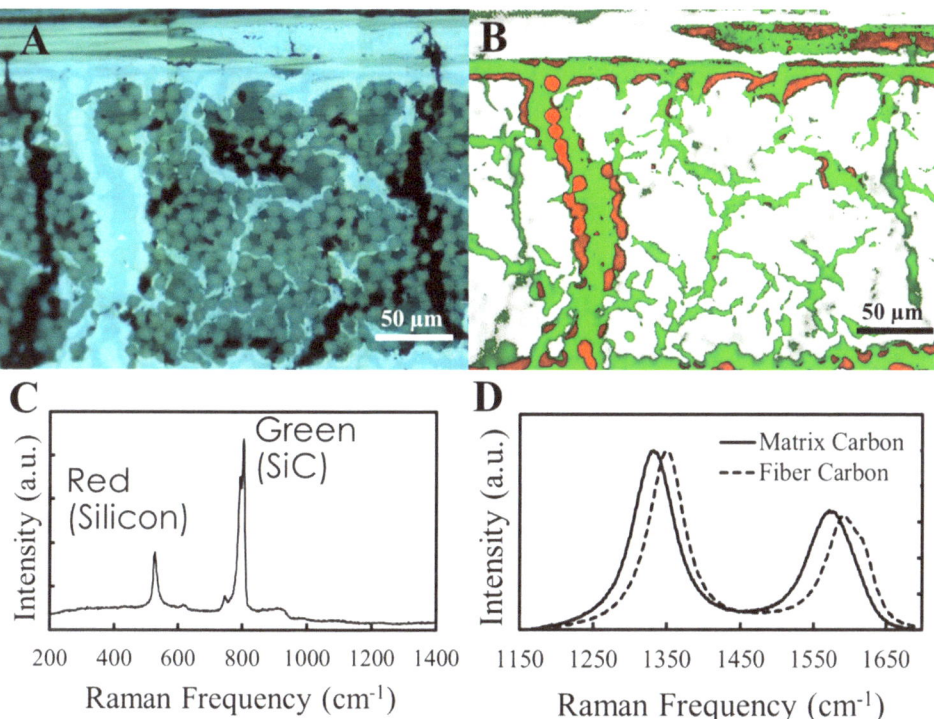

Figure 9. (**A**) Optical image of cross-section and (**B**) overlay of Raman maps, (**C**) average Raman spectra of the final C/C-SiC composite showing Si and SiC, and (**D**) average Raman spectra of the final C/C-SiC composite showing carbon matrix and carbon fibers.

The carbon fibers and the adjacent matrix have nearly identical spectra, showing a phase that is nanocrystalline graphite. The difference in contrast is likely due to the fibers being more nanocrystalline or dense than the carbon matrix pyrolyzed from PEEK. The Raman peaks of the matrix carbon are at a lower frequency compared to those of the fiber carbon, indicating a more disordered structure [36]. The spatially resolved map shows that the silicon penetrated the tows and reacted to make SiC in the porous regions of the Cf-C preform. There is residual silicon on some of the surfaces, and a minor reaction occurred between the Si and the carbon fibers. This was due to a templating effect, where the reacted SiC was observed to be the size and shape of the carbon fibers. The areas where SiC was formed showed residual silicon, and they may even be mixed at a lower length scale. The carbon fibers were, to a large part, protected by the partial carbon matrix, with a sheath of SiC forming around the Cf-C tow.

Figure 10 shows the mechanical data in the form of strength–displacement curves for the C/C-SiC composites tested in four-point bending and the Weibull statistics. The samples exhibited a large range of strengths but a relatively high strain to failure. A characteristic strength of 234.91 MPa was determined via Weibull analysis, with a Weibull modulus of 3.21 found using linear regression. The variability most likely arises from the porosity and the delamination. The displacement to failure was high, similar to well-formed CMCs. This is a result of the composites yielding via matrix cracking, the fibers beginning to bear more of the load, the matrix cracks deflecting along the fiber surfaces, the fibers pulling out of the matrix, and the fibers cracking in succession, rather than all occurring at one time. Images of the fracture surfaces were acquired to verify the fracture modes relative to the observed mechanical behavior.

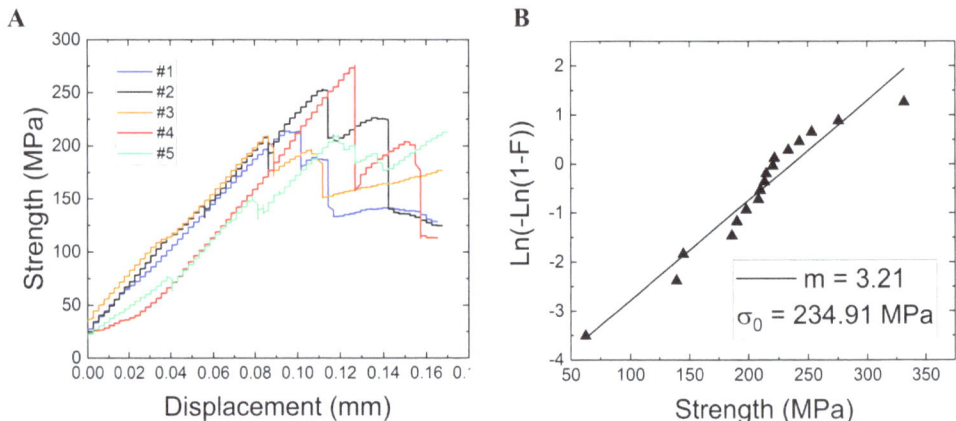

Figure 10. Mechanical data from C/C-SiC composite samples subjected to (**A**) four-point bend tests (stress versus. displacement) and (**B**) Weibull statistics.

Figure 11 shows the optical and SEM images of the fracture surfaces of a sample following four-point bend testing. The samples tested for this study typically exhibited failure behavior similar to that shown in Figure 11. The optical image in Figure 11A shows the side view of the fractured specimen, where no delamination is observed, but fiber pullout occurred. A side-by-side SEM cross-section view of the fracture surface is shown in Figure 11B. Regions where there are gaps between the fiber tows can be identified (circle in Figure 11B). Such defects can be a source of failure at a lower-than-expected stress.

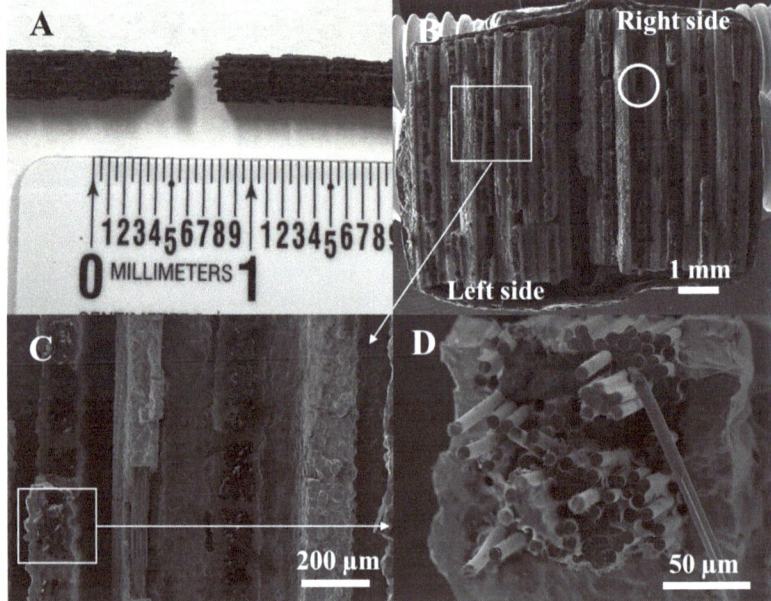

Figure 11. (**A**) Macro and (**B–D**) SEM images of fracture surfaces after four-point bend failure.

The samples showed similar behavior to that observed in Figure 11. It is evident that no delamination occurred, and most of the failure involved fiber pullout. The macro image in Figure 11A shows the side view of the fracture surface, where no delamination can be observed, and fiber pull out is visible. The side-by-side cross-section view, Figure 11B, shows gaps, as pointed out by a circle, where the tows were either completely pulled out, were not fully consolidated with matrix, or were printed with a small gap leaving the large open area in the fracture surface. Figure 11C shows the area more clearly, where the transverse fiber tows are encased in a material, and the out-of-plane tows show many areas of Cf pullout. Figure 11D shows this pullout, and this is where the high toughness comes from.

Figure 12 shows the SEM/EDS data from a fracture surface. Some of the fracture surfaces demonstrated this morphology, where the SiC reaction bond has encased the entire tow of Cf-C. Those areas show up in the white or lighter hue, and they are a region from which cracks may initiate and propagate. The areas of carbon, including the carbon fibers, can also be identified in Figure 12. There was some oxygen detected by EDS, primarily located on the surface of the SiC. There are cracks going through the SiC that have been arrested in the areas where there are a significant amount of carbon fibers or in regions of porosity. The crack deflection and arrest provide the toughness in the composites.

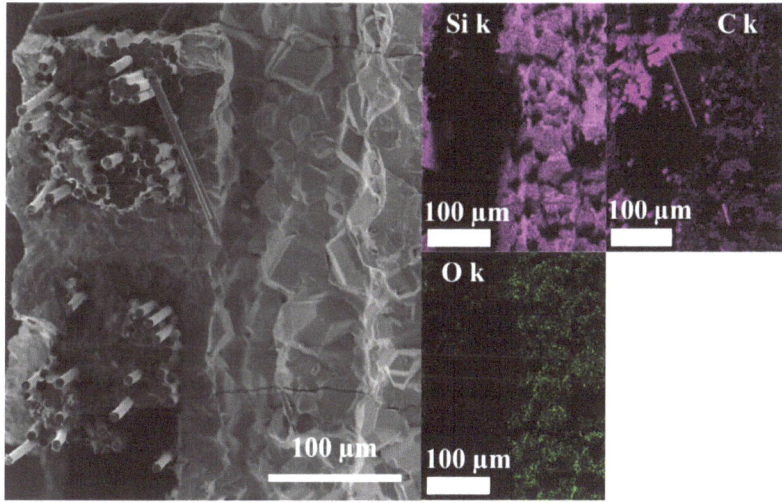

Figure 12. SEM/EDS of a fracture surface showing the presence of SiC, Cf, and carbon.

4. Summary and Discussion

Miniature AFP or FDM 3D printing is viable to fabricate fiber-reinforced ceramic matrix composites, but more development is required to optimize the mechanical performance. With continued development, larger-scale AFP systems could be applied to fabricating SiC and other CMCs. One aspect likely causing delamination is the heating and curing of each layer during the sequential printing process. The immediate layer of melted thermoplastic gets deposited on material that has already been melted and resolidified, which may have reacted with the ambient atmosphere (oxidized and absorbed H_2O/CO_2). However, the densification process following the additive manufacturing of the fiber preform is the same as that used in traditional CMC fabrication methods, and, therefore, the microstructures resemble those from the literature. Large defects and delaminations are known to result in low mechanical strengths, so process development is required to minimize such features in the final microstructure. CT scans with higher resolution or even nano-CT are necessary to identify smaller pores and defects. Variant energy sources should also be explored to differentiate low-density phases and materials.

Applying pressure during curing and/or pyrolysis may improve the layer-to-layer bonding and inhibit delamination. The pyrolyzed PEEK in the porous carbon matrix protected the carbon fibers from any significant reaction and provided enough difference in bonding, so that a fiber interface coating was not needed. While a fiber-interface coating might be beneficial, it has been found that C/C composites with no fiber coating may exhibit damage tolerance if weak interfaces are achieved via processing, by making the first matrix infiltrations porous or less crystalline [37–39]. The mechanical performance of C/C-SiC may also benefit from such a differential in material properties at the fiber/matrix interface [40–42]. The composites in the current study with pyrolyzed PEEK provide a weaker interface with the carbon fibers, by providing a differential in nanocrystallinity and porosity from that of the neighboring fibers and matrix, which enables the damage-tolerant mechanical behavior. Since melt infiltration is typically performed at ambient pressure, a reduction in delamination would be best accomplished via improvements in pyrolysis and new materials. The damage tolerance of these composites comes from the significant fiber pullout and crack arrest, which could be improved if the tows were printed closer together, providing for a more continuous composite microstructure. In the composites fabricated for this initial study, there are small gaps from tows that are not placed directly next to each other, leading to delamination and porosity artifacts that will decrease strength. However, the composite demonstrated damage tolerance regardless of the microstructural defects. This composite fabrication method is amenable to any continuous fiber tow and polymer filament that can be impregnated and melted, so many other thermoplastic materials may be employed. In addition, other means of densification can be applied, such as chemical vapor infiltration (CVI) and/or polymer infiltration and pyrolysis (PIP). This demonstrated manufacturing technique has the potential to enable low-cost, time-efficient, and reduced-hand-touch manufacturing of CMCs.

5. Conclusions

C/C-SiC CMCs were fabricated via additive manufacturing, which opens up the design space and provides the potential for faster and more cost-effective fabrication of composites. The AFP of polymer matrix composites for PEEK thermoplastic and continuous fibers was accomplished. The Cf-PEEK fibrous preforms were pyrolyzed to form a porous Cf-C sample, followed by performing Si melt infiltration to form SiC. CT scans showed that a dense CMC was fabricated, but there were some porosity and delamination defects in the microstructure. SiC was formed and surrounded the fiber tows as a protective partial matrix, which provides the potential for oxidation resistance. The characteristic strength was 234.91 MPa with a high displacement to failure, demonstrating toughness in the composites.

Supplementary Materials: The following supporting information can be downloaded at: https://www.mdpi.com/article/10.3390/jcs6120359/s1, Video S1: CT scan reconstruction.

Author Contributions: Conceptualization, C.L.C. and D.J.M.; methodology, C.L.C. and B.Y.; validation, D.J.M.; formal analysis, C.L.C., M.J.L., E.C., Q.A.C. and D.J.M.; data curation, C.L.C., B.Y., M.J.L., E.C. and Q.A.C.; writing—original draft preparation, C.L.C.; writing—review and editing, C.L.C., B.Y., M.J.L., E.C., Q.A.C. and D.J.M. All authors have read and agreed to the published version of the manuscript.

Funding: This manuscript has been authored by UT-Battelle LLC under Contract No. DE-AC05-00OR22725 with the U.S. Department of Energy. The United States Government retains and the publisher, by accepting the article for publication, acknowledges that the United States Government retains a non-exclusive, paid-up, irrevocable, world-wide license to publish or reproduce the published form of this manuscript, or allow others to do so, for United States Government purposes. The Department of Energy will provide public access to these results of federally sponsored research in accordance with the DOE Public Access Plan (http://energy.gov/downloads/doe-public-access-plan(accessed on 19 October 2022)).

Institutional Review Board Statement: This manuscript was reviewed and approved internally with the tracking number PUB ID 179108.

Acknowledgments: This manuscript has been authored by UT-Battelle LLC under Contract No. DE-AC05-00OR22725 with the U.S. Department of Energy, and this material is based upon work supported by the Oak Ridge National Laboratory SPP system. Access to the Raman spectrometer was provided by the Nuclear Nonproliferation Division, Oak Ridge National Laboratory. The authors would also like to give a special thanks to Richard A. Lowden for their commitment and outreach to education about ceramic matrix composites and Dana McClurg for the mechanical testing of the composites.

Conflicts of Interest: The authors declare no conflict of interest.

References

1. Fielda, J.; Swain, M. The Indentation Characterisation of the Mechanical Properties of Various Carbon Materials: Glassy Carbon, Coke and Pyrolytic Graphite. *Carbon* **1996**, *34*, 1357–1366. [CrossRef]
2. Windhorst, T.; Blount, G. Carbon-Carbon Composites: A Summary of Recent Developments and Applications. *Mater. Des.* **1997**, *18*, 11–15. [CrossRef]
3. Harris, G. *Properties of Silicon Carbide*; INSPEC, Institution of Electrical Engineers: London, UK, 1995.
4. Opila, E.J.; Smialek, J.L.; Robinson, R.C.; Fox, D.S.; Jacobson, N.S. SiC Recession Caused by SiO_2 Scale Volatility under Combustion Conditions: II. Thermodynamics and Gaseous-Diffusion Model. *J. Am. Ceram. Soc.* **1999**, *82*, 1826–1834. [CrossRef]
5. Terrani, K.A.; Pint, B.A.; Parish, C.M.; Silva, C.M.; Snead, L.L.; Katoh, Y. Silicon Carbide Oxidation in Steam up to 2 MPa. *J. Am. Ceram. Soc.* **2014**, *97*, 2331–2352. [CrossRef]
6. Ogbuji, L.U.J.T.; Opila, E.J. A Comparison of the Oxidation Kinetics of SiC and Si_3N_4. *J. Electrochem. Soc.* **1995**, *142*, 925–930. [CrossRef]
7. Li, Z.; Bradt, R.C. Thermal Expansion and Thermal Expansion Anisotropy of SiC Polytypes. *J. Am. Ceram. Soc.* **1987**, *70*, 445–448. [CrossRef]
8. Watari, K.; Shinde, S. High Thermal Conductivity Materials. *MRS Bull.* **2001**, *26*, 440–444. [CrossRef]
9. Snead, L.L.; Nozawa, T.; Katoh, Y.; Byun, T.-S.; Kondo, S.; Petti, D.A. Handbook of SiC Properties for Fuel Performance Modeling. *J. Nucl. Mater.* **2007**, *371*, 329–377. [CrossRef]
10. Yamada, K.; Mohri, M. Properties and Applications of Silicon Carbide Ceramics. In *Silicon Carbide Ceramics—1*; Springer: Dordrecht, The Netherlands, 1991; pp. 13–44. [CrossRef]
11. Bansal, N.; Lamon, J. *Ceramic Matrix Composites: Materials, Modeling and Technology*; Wiley: New York, NY, USA, 2014.
12. Lowden, R. *Characterization and Control of the Fiber-Matrix Interface in Ceramic Matrix Composites*; Oak Ridge National Laboratory: Oak Ridge, NT, USA, 1989. [CrossRef]
13. Moeller, H.H.; Long, W.G.; Caputo, A.J.; Lowden, R.A. Fiber-Reinforced Ceramic Composites. *Int. SAMPE Symp. Exhib.* **1987**, *32*, 977–984.
14. Noda, T.; Araki, H.; Abe, F.; Okada, M. Microstructure and Mechanical Properties of CVI Carbon Fiber/ SiC Composites. *J. Nucl. Mater.* **1992**, *191*, 539–543. [CrossRef]
15. Xu, Y.; Cheng, L.; Zhang, L. Carbon/Silicon Carbide Composites Prepared by Chemical Vapor Infiltration Combined with Silicon Melt Infiltration. *Carbon* **1999**, *37*, 1179–1187. [CrossRef]
16. Staehler, J.M.; Mall, S.; Zawada, L.P. Frequency Dependence of High-Cycle Fatigue Behavior of CVI C/SiC at Room Temperature. *Compos. Sci. Technol.* **2003**, *63*, 2121–2131. [CrossRef]
17. Fan, S.; Zhang, L.; Xu, Y.; Cheng, L.; Lou, J.; Zhang, J.; Yu, L. Microstructure and Properties of 3D Needle-Punched carbon/Silicon Carbide Brake Materials. *Compos. Sci. Technol.* **2007**, *67*, 2390–2398. [CrossRef]
18. Si-Zhou, J.; Xiang, X.; Zhao-Ke, C.; Peng, X.; Bai-Yun, H. Influence Factors of C/C–SiC Dual Matrix Composites Prepared by Reactive Melt Infiltration. *Mater. Des.* **2009**, *30*, 3738–3742. [CrossRef]
19. Patel, M.; Saurabh, K.; Prasad, V.V.B.; Subrahmanyam, J. High Temperature C/C-SiC Composite by Liquid Silicon Infiltration: A Literature Review. *Bull. Mater. Sci.* **2012**, *35*, 67–77. [CrossRef]
20. Jiang, G.; Yang, J.; Xu, Y.; Gao, J.; Zhang, J.; Zhang, L.; Cheng, L.; Lou, J. Effect of Graphitization on Microstructure and Tribological Properties of C/SiC Composites Prepared by Reactive Melt Infiltration. *Compos. Sci. Technol.* **2008**, *68*, 2468–2473. [CrossRef]
21. Krenkel, W.; Berndt, F.J.M.E. C/C–SiC Composites for Space Applications and Advanced Friction Systems. *Mater. Sci. Eng. A* **2005**, *412*, 177–181. [CrossRef]
22. Day, M.; Cooney, J.D.; Wiles, D.M. The Thermal Degradation of Poly(Aryl-Ether-Ether-Ketone) (PEEK) as Monitored by Pyrolysis—GC/MS and TG/MS. *J. Anal. Appl. Pyrolysis* **1990**, *18*, 163–173.
23. Liensdorf, T.; Langhof, N.; Krenkel, W. Mechanical Properties of PEEK-Derived C/SiC Composites with Different Fiber Lengths. *Adv. Eng. Mater.* **2018**, *21*, 1800835. [CrossRef]
24. Reichert, F.; Langhof, N.; Krenkel, W. Influence of Thermal Fiber Pretreatment on Microstructure and Mechanical Properties of C/C-SiC with Thermoplastic Polymer-Derived Matrices. *Adv. Eng. Mater.* **2015**, *17*, 1119–1126. [CrossRef]

25. Kabir, S.M.F.; Mathur, K.; Seyam, A.F.M. A Critical Review on 3D Printed Continuous Fiber-Reinforced Composites: History, Mechanism, Materials and Properties. *Compos. Struct.* **2020**, *232*, 111476. [CrossRef]
26. Oromiehie, E.; Prusty, B.G.; Compston, P.; Rajan, G. Automated Fibre Placement Based Composite Structures: Review on the Defects, Impacts and Inspections Techniques. *Compos. Struct.* **2019**, *224*, 110987. [CrossRef]
27. Forcellese, A.; Mancia, T.; Russo, A.; Simoncini, M.; Vita, A. Robotic Automated Fiber Placement of Carbon Fiber Towpregs. *Mater. Manuf. Process.* **2021**, *37*, 539–547. [CrossRef]
28. Wu, Y.; Wang, K.; Neto, V.; Peng, Y.; Valente, R.; Ahzi, S. Interfacial Behaviors of Continuous Carbon Fiber Reinforced Polymers Manufactured by Fused Filament Fabrication: A Review and Prospect. *Int. J. Mater. Form.* **2022**, *15*, 18. [CrossRef]
29. Mashayekhi, F.; Bardon, J.; Berthé, V.; Perrin, H.; Westermann, S.; Addiego, F. Fused Filament Fabrication of Polymers and Continuous Fiber-Reinforced Polymer Composites: Advances in Structure Optimization and Health Monitoring. *Polymers* **2021**, *13*, 789. [CrossRef] [PubMed]
30. Mei, H.; Yan, Y.; Feng, L.; Dassios, K.G.; Zhang, H.; Cheng, L. First Printing of Continuous Fibers into Ceramics. *J. Am. Ceram. Soc.* **2018**, *102*, 3244–3255. [CrossRef]
31. Yan, Y.; Mei, H.; Zhang, M.; Jin, Z.; Fan, Y.; Cheng, L.; Zhang, L. Key Role of Interphase in Continuous Fiber 3D Printed Ceramic Matrix Composites. *Compos. Part A Appl. Sci. Manuf.* **2022**, *162*, 107127. [CrossRef]
32. Freudenberg, W.; Wich, F.; Langhof, N.; Schafföner, S. Additive Manufacturing of Carbon Fiber Reinforced Ceramic Matrix Composites Based on Fused Filament Fabrication. *J. Eur. Ceram. Soc.* **2021**, *42*, 1822–1828. [CrossRef]
33. Abel, J.; Kunz, W.; Michaelis, A.; Singh, M.; Klemm, H. Non-Oxide CMC Fabricated by Fused Filament Fabrication (FFF). *Int. J. Appl. Ceram. Technol.* **2021**, *19*, 1148–1155. [CrossRef]
34. Cramer, C.L.; Elliott, A.M.; Lara-Curzio, E.; Flores-Betancourt, A.; Lance, M.J.; Han, L.; Blacker, J.; Trofimov, A.A.; Wang, H.; Cakmak, E.; et al. Properties of SiC-Si Made via Binder Jet 3D Printing of SiC Powder, Carbon Addition, and Silicon Melt Infiltration. *J. Am. Ceram. Soc.* **2021**, *104*, 5467–5478. [CrossRef]
35. Nakashima, S.I.; Harima, H. Raman Investigation of SiC Polytypes. *Phys. Status Solidi A* **1997**, *162*, 39–64. [CrossRef]
36. Ferrari, A.C.; Robertson, J. Interpretation of Raman Spectra of Disordered and Amorphous Carbon. *Phys. Rev. B* **2000**, *61*, 14095–14107. [CrossRef]
37. Sun, J.; Li, H.; Han, L.; Song, Q. Enhancing Both Strength and Toughness of Carbon/Carbon Composites by Heat-Treated Interface Modification. *J. Mater. Sci. Technol.* **2018**, *35*, 383–393. [CrossRef]
38. Huang, R.; Zeng, F.; Peng, Y.; Liu, H.; Gao, Y.; Yang, Q. Influence of Fiber/Matrix Interface on the Texture Evolution and Fracture Toughness of C/C. *Diam. Relat. Mater.* **2022**, *127*, 109112. [CrossRef]
39. Chowdhury, P.; Sehitoglu, H.; Rateick, R. Damage Tolerance of Carbon-Carbon Composites in Aerospace Application. *Carbon* **2018**, *126*, 382–393. [CrossRef]
40. Després, J.-F.; Monthioux, M. Mechanical Properties of C/SiC Composites as Explained from Their Interfacial Features. *J. Eur. Ceram. Soc.* **1995**, *15*, 209–224. [CrossRef]
41. Xu, J.; Guo, L.; Wang, H.; Li, W.; Liu, N.; Wang, T. Influence of Graphitization Temperature on Microstructure and Mechanical Property of C/C-SiC Composites with Highly Textured Pyrolytic Carbon. *J. Eur. Ceram. Soc.* **2022**, *42*, 1893–1903. [CrossRef]
42. Chen, Z.; Fang, G.; Xie, J.; Liang, J. Experimental Study of High-Temperature Tensile Mechanical Properties of 3D Needled C/C-SiC Composites. *Mater. Sci. Eng. A* **2016**, *654*, 271–277. [CrossRef]

Article

Design and Construction of a Low-Cost-High-Accessibility 3D Printing Machine for Producing Plastic Components

Kajogbola R. Ajao [1], Segun E. Ibitoye [1,2,*], Adedire D. Adesiji [1] and Esther T. Akinlabi [3]

[1] Department of Mechanical Engineering, Faculty of Engineering and Technology, University of Ilorin, PMB 1515, Ilorin 240222, Nigeria
[2] Department of Mechanical Engineering Science, Faculty of Engineering and the Built Environment, University of Johannesburg, P.O. Box 524, Auckland Park 2006, South Africa
[3] Department of Mechanical and Construction Engineering, Faculty of Engineering and Environment, Northumbria University, Newcastle NE1 8ST, UK
* Correspondence: ibitoye.s@unilorin.edu.ng

Citation: Ajao, K.R.; Ibitoye, S.E.; Adesiji, A.D.; Akinlabi, E.T. Design and Construction of a Low-Cost-High-Accessibility 3D Printing Machine for Producing Plastic Components. *J. Compos. Sci.* **2022**, *6*, 265. https://doi.org/10.3390/jcs6090265

Academic Editors: Yuan Chen and Francesco Tornabene

Received: 29 June 2022
Accepted: 1 August 2022
Published: 9 September 2022

Publisher's Note: MDPI stays neutral with regard to jurisdictional claims in published maps and institutional affiliations.

Copyright: © 2022 by the authors. Licensee MDPI, Basel, Switzerland. This article is an open access article distributed under the terms and conditions of the Creative Commons Attribution (CC BY) license (https://creativecommons.org/licenses/by/4.0/).

Abstract: The additive manufacturing process creates objects directly by stacking layers of material on each other until the required product is obtained. The application of additive manufacturing technology for teaching and research purposes is still limited and unpopular in developing countries, due to costs and lack of accessibility. In this study, an extruding-based 3D printing additive manufacturing technology was employed to design and construct a low-cost-high-accessibility 3D printing machine to manufacture plastic objects. The machine was designed using SolidWorks 2020 version with a $10 \times 10 \times 10$ cm^3 build volume. The fabrication was carried out using locally available materials, such as PVC pipes for the frame, plywood for the bed, and Zinc Oxide plaster for the bed surface. Repetier firmware was the operating environment for devices running on the computer operating system. Cura was used as the slicing software. The fabricated machine was tested, and the printer produced 3D components with desired structural dimensions. The fabricated 3D printer was used to manufacture some plastic objects using PLA filament. The recommended distance between the nozzle tip and the bed is 0.1 mm. The constructed 3D printer is affordable and accessible, especially in developing nations where 3D printing applications are limited and unpopular.

Keywords: 3D printer; additive manufacturing; slicing software; firmware

1. Introduction

For decades, machines and parts have been constructed using traditional (subtractive) methods, such as welding, folding, soldering, machining, etc. The conventional machining methods include milling, drilling, grilling, and turning [1,2]. The methods fabricate products and parts by removing material from a block of material to achieve the required geometry. The machining methods have helped humankind to create different things [3]. However, several constraints have limited its performance and efficiency—it is challenging to produce intricate parts, fixtures, time and energy-consuming, low precision and accuracy, expensive, and waste materials. The introduction of non-conventional machinings, such as electric discharge and chemical machining and the utilization of computer technology, helps to minimize these drawbacks [1,4].

In contrast, additive manufacturing is constructing an object by making a successive layer of materials to form the object [5–8]. An example of additive manufacturing is a three-dimensional (3D) printing process - a process of fabricating 3D objects from a numerical file [1,9]. Three-dimensional printing consists of computer-assisted design, added material, and the machine [10]. The printing process starts by making a digital object design using computer-aided design (CAD) software or capturing the object with a 3D scanner. Popular software choices include SolidWorks, AutoCAD, Inventor, etc. [11–13]. The CAD design defines the geometry of the object [1,14], and the CAD design quality significantly affects the printer's production and precision [3]. CAD designs are prepared for printing by slicing the design into a series of horizontal layers using slicing software while uploading into the 3D printer. The slicing software also generates a G-code for each sliced layer and creates a specific tool-path instruction for the extruder head that deposits the material layer by layer [8,15,16]. The American Society for additive manufacturing established standards that categorize the additive manufacturing processes into seven bases: material extrusion, directed energy deposition, vat photopolymerization, binder jetting, sheet lamination, material jetting, and powder bed fusion [17].

Three-dimensional printers use different technology to fabricate the final product layer by layer. Some technology melts the material, while some employ a high-powered ultraviolet laser to cure photo-reactive resin, creating the layer to manufacture the final products [1,17]. Examples of this technology include fused deposition modeling, stereolithography, and selective laser sintering [1,3,10]. The 3D printing technology allows the fabrication of complex and intricate objects without assembly. It saves time, labor, cost, and energy. Three-dimensional printing was initially used for prototyping, but recently it has found more applications in manufacturing processes [14]. Applications of 3D printing can be found in medicine, entertainment, maquettes, industrial and biomedical engineering, architectural model, aerospace, defense, automotive, etc. [1,18].

Several research efforts on manufacturing 3D printers and printing 3D objects have been reported in the literature, including studies on the mechanics and materials characteristics [19–23]. Raja et al. [24] fabricated a calcium-deficient hydroxyapatite bone scaffold at low temperature. The 3D printer conditions were optimized, and outstanding mechanical characteristics were achieved at 70% porosity without sintering after the 3D printing. The essential factor required to complete low-temperature fabrication of calcium phosphate via 3D printing were enumerated in the report. A similar study was conducted by Xiao et al. [25]. They fabricated porous polycaprolactone scaffolds through 3D printing at low temperatures. Polymer printing inks were prepared, and their rheological characteristics printing inks were assessed. The performance of the printing was evaluated. Different porous polycaprolactone scaffolds were constructed using printing inks at different feed ratios. Analyses of the results revealed that the printing ink formulation is significantly affected by the performance of the 3D printer.

A surface made of a flexible antiadhesive polymer was constructed using a 3D printer [15]. A fused deposition model type was used to make the molds with microgrid arrangements. The shape of the arrangement was controlled by moving the mold (where the filament is printed) circularly. Twelve different antiadhesive surfaces were fabricated by changing the direction of the mold. The adhesive characterization showed that mold with microgrid Y with an internal angle and resolution of 60° and 800 μm, respectively, displayed the best adhesive characteristics. A reduction in adhesive forces of 92.6% and 99.5% was recorded when the Kapton tape was immediately attached and detached, and when the Kapton tape was separated after eighteen days, respectively.

Sevvel et al. [26] designed and constructed a desktop 3D printer by adopting the Fused Deposition Modeling (FDM) principle. The major component of the machine includes an x-y component, timer, extruder, Arduino board, extruder nozzle, extruder head, and finfitted extruder nozzle. The display system could show the timer, extruder temperature, and axis position. It functions by utilizing three software: Cura, SolidWorks, and Pronterface. It was reported that the machine performed satisfactorily.

A comprehensive review composed of the basics, types, techniques, advantages and disadvantages of 3D printing can be found in Bhusnure et al. [3]. Also, a detailed report on the comparison of different kinds of 3D printing technologies can be found in Jasveer and Jianbin [2], while a review on the prospects in 3D printing for nanocomposite materials is presented in Patel et al. [27].

Low-cost 3D printers range from $100 to $1000, whereas budget printers cost less than $300, and advanced hobbyist printers' prices are between $300 and $1000 [28]. On the other side of the spectrum, professional printers cost between $1000 and $5000, and industrial printers cost over $5000. The high-end printers offer more speed, print quality, and reliability. The use of low-cost desktop 3D printers is increasing due to their applications in research, education, and sustainable manufacturing [28]. The most common FDM 3D printing materials are polylactic acid (PLA), acrylonitrile butadiene styrene (ABS), and their different fusions. High-end printers utilize more advanced materials with better properties, such as higher heat resistance, flexural strength, tensile strength, impact resistance, and rigidity.

Ozsoy et al. [29] investigated the mechanical properties of PLA and ABS. A tensile, compression, and 3-point bending test was conducted. It was discovered that both materials' mechanical properties depend on the infill settings. The mechanical properties were greatly improved by increasing the filling density. PLA had higher compressive strength than ABS, however, ABS showed more ductility due to their difference in molecular structure. Raj et al. [30] studied PLA components to determine if they can be a suitable replacement for ABS components. The PLA components were subjected to various mechanical and degradation tests. It was concluded that PLA was eco-friendly and possessed the strength needed to produce plastic components, especially biomedical ones.

Surface finish is a concern when it comes to using Fused Deposition Modeling (FDM) 3D printers. Objects printed usually need post-processing surface finishing, such as sanding. Optimization of print parameters from the slicing software settings can substantially improve the object surface finish. Gordeev et al. [31] evaluated the impact of significant printing parameters on the surface quality of the resulting print. The parameters considered were wall thickness, type of the internal filling of the walls, print speed, extrusion temperature, and filament flow rate. The experiment showed that thin-walled printed objects needed infills to have a better surface finish. Conclusively, different print parameters need to be tweaked based on the object's geometry to have a great surface finish that will not require much post-processing.

There are still just a few applications of additive manufacturing, such as 3D printing, in developing nations, despite the growing global switch from subtractive to additive manufacturing. This is because teaching and research in additive manufacturing are unpopular and challenging in higher education institutions in underdeveloped countries, due to the high cost of purchasing and importing 3D printing machines. It is also frequently challenging to maintain imported 3D printing equipment and replace damaged parts. When the need occurs, people often struggle to repair and replace accessories, particularly plastic ones that deteriorate quickly. Therefore, it is necessary to design and build low-cost, highly accessible 3D printing machines utilizing readily available materials to create plastic components and accessories for teaching and research purposes in higher education institutions in developing countries.

2. Methodology

2.1. Materials

Table 1 presents the specifications of the component and materials used to fabricate the 3D printing machine. The components were selected based on design specifications, durability, and availability.

Table 1. Component and material used for the fabrication of the 3D printing machine.

Item	Quantity	Specification	Function
Frame (PVC pipe and fittings)	17	As determined by the calculations and print volume	Serves as chassis on which components are placed
Stepper motor	5	NEMA 17, 420N mm, 600 rpm	Generates torque for the movement of parts
Filament	1 Spool	Poly-lactic Acid (PLA)	Material from which part is printed
Extruder	1	MK8, 0.4 mm	Deposits the plastic filament on the bed

Table 1. Cont.

Item	Quantity	Specification	Function
Bed	1	150 × 120 × 10 mm of wood	A platform where the molten plastic is deposited
Timing belt	2	V-Belt, 5 mm wide	Transfers motor drive to move bed and extruder along y and x-axis, respectively
Pulley	2	2 mm pitch, 12.5 diameter	Serves as a point of attachment to transfer motion to the belt
Leadscrews	2	2 mm lead, 240 mm	To move the extruder head along z-axis
Steel rods	6	8 mm and 6 mm	Serves as a rail on which the bed and extruder move
Ball bearings	2	5 × 16 × 5 mm	Supports rotating belt
Linear bearings	7	8 mm and 6 mm	Carriage is mounted upon them and slides along a steel rod
Coupler	2	5 mm to 8 mm	Transfers the motor drive to the leadscrew
End stop	3	Mechanical type	Prevents the bed and extruder from moving past their range
Fan	1	DC 12V	Provides active cooling of the top printed layer
User interface and connectivity	1	LCD	Controls the 3D printer without a computer connection
Controller board	1	ATMEGA1284P	Directs the motion components based on commands sent from a computer and interprets input from the sensors
Printed Parts	1 set	As determined by the structure of the attached components	Holds components in place

2.2. Methods

2.2.1. Design Consideration

The essential factors considered in the design of the 3D printing machine are the cost of construction, size, sturdiness, assembling, availability of materials, and application. The machine's overall cost must be low to be affordable for people in developing countries. The design of the 3D printer was carried out considering the print build volume of $10 \times 10 \times 10$ cm^3, which is the maximum geometry size of an obtained object. The machine was designed to print objects with polylactic acid (PLA), a bio-plastic and thermoplastic monomer made from agricultural produce, such as corn starch or sugar cane. Since it does not emit toxic fumes when heated, PLA is considered safer to use for experiments on 3D printing in schools and colleges.

Semi-permanent assembly method was adopted in the design. This ensures the machine's structural stability, easy maintenance, and durability. Materials that are readily and locally available were considered for the fabrication of the machine. This is to ensure the easy replacement of parts when the need arises. Also, materials that can withstand the vibratory forces due to the motion of the electric motion were considered. The design stage also considered the machine's durability and structural stability.

2.2.2. Design Calculation

Design of the Stepper Motor

The linear velocity v of the stepper motor is calculated using Equation (1) [32].

$$v = r\omega; \quad \omega = \frac{2\pi N}{60} \qquad (1)$$

where r is the radius of curvature of circular path = 6.25 mm; ω is the angular speed of the motor; N is the constant speed of the motor = 600 rpm. By computation, v = 392.69 mm/s.

The force due to the electric motor is calculated using Equation (2).

$$\text{Force} = \frac{\text{Torque}}{\text{The radius of curvature of the circular path}} \qquad (2)$$

By computation, force = 67.2 N.

Therefore 6.72 kg can be pulled over a distance of 392.69 mm in one second using NEMA 17.

2.2.3. Design of Timing Belt

The width of the belt is calculated using Equation (3) [32].

$$w = \frac{Ff}{\sigma t} \quad (3)$$

where w is the width of the belt; σ is the ultimate strength of polyurethane = 20.77 MPa; F is the force of the electric motor; f is the factor of safety = 2; t is the belt thickness = 1.3 mm. By computation, w = 4.98 mm. The standard width = 5 mm.

The allowable tensile load on this belt is 134.4 N, and since the force exerted by the motor is 67.2 N, which is far less than the allowable tensile load, the selected belt would not fail under the design conditions.

The length of each belt is calculated using Equation (4) [32].

$$L = \frac{\pi}{2}(D+d) + \sqrt{4c^2 + D^2 + d^2} \quad (4)$$

where D is the diameter of pulley; d = diameter of bearing; c = centre distance between two pulleys; L = length of timing belt; T = thickness of timing belt.

For belt 1 (x-axis), D = 16 mm, d = 12.5 mm, c = 310 mm, and T = 1.3 mm. Using Equation (4), L_1 (x-axis belt) = 665.10 mm.

Belt 2 (y-axis) has slightly different dimensions (D = 16 mm, d = 12.5 mm, c = 290 mm, and T = 1.3 mm). Using Equation (4), L_2 (y-axis belt) = 625.12 mm.

2.2.4. Deflection of Frame

The 3D printer frame was analyzed, and reaction forces in the frame section was determined. The maximum deflection occurred at the x-axis movement assembly section, which carries the extruder head. The natural frequency of a body is directly proportional to the square root of stiffness, and stiffness is inversely proportional to deflection. In other words, deflection is inversely proportional to the natural frequency. The maximum deflection was determined to be 14.3 microns, which is considered negligible [13]. Hence, vibratory forces were ignored.

2.2.5. Slicing Software and Firmware

Firmware is a computer program that provides an operating environment that allows devices to run on the operating system. It helps the device perform all monitoring functions associated with the 3D printing machine. Repetier, open-source firmware for most RepRap family 3D printers, was used as the firmware of the 3D printer. The firmware supports printing from both USB and SD cards with folders, and utilizing look-ahead trajectory planning. It runs via the printer's mainboard. It controls all the activities on the machine and coordinates the buttons, heaters, steppers, lights, sensors, LCD, etc., using G-code as the control language.

Cura, produced by Ultimaker, was used as the slicing software. The software was selected because it supports STL, 3MF, and OBJ 3D file formats. Also, Cura can import and convert 2D images (.JPG, .PNG, .BMP, and .GIF) to 3D extruded models. It allows the user to work and print multiple models. Figures 1 and 2 show the CAD model of the 3D printing machine.

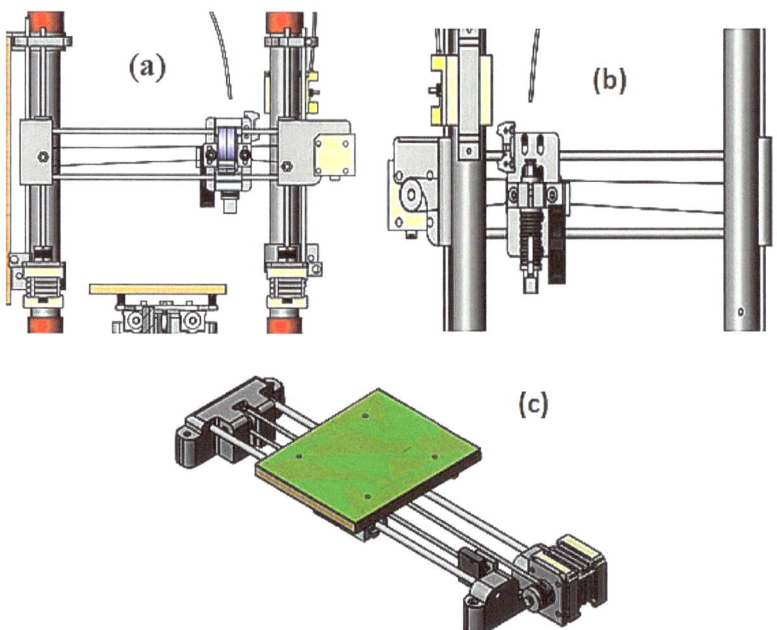

Figure 1. Assembly CAD Model- (**a**) z-axis Movement, (**b**) x-axis Movement, and (**c**) y-axis Movement.

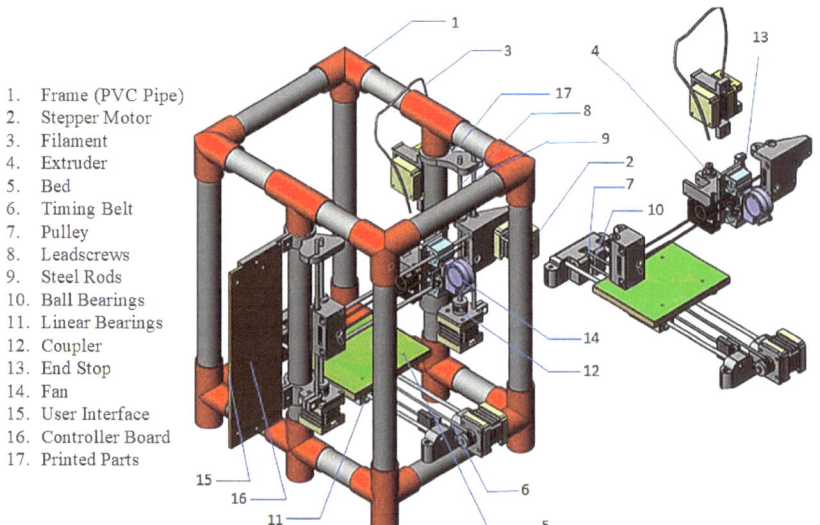

1. Frame (PVC Pipe)
2. Stepper Motor
3. Filament
4. Extruder
5. Bed
6. Timing Belt
7. Pulley
8. Leadscrews
9. Steel Rods
10. Ball Bearings
11. Linear Bearings
12. Coupler
13. End Stop
14. Fan
15. User Interface
16. Controller Board
17. Printed Parts

Figure 2. The isometric view of the 3D printer CAD model.

3. Results and Discussion

Figure 3 shows pictorial views of the constructed 3D printer. The machine was first tested by verifying the motion system, followed by printing an object. While testing the printer's motion system, it was discovered that the primary method of checking the correctness of the x, y, and z axes motion system is to manually move the print head and bed fully through the x, y, and z axes, respectively. A smooth movement with no obstruction shows that the axes' positioning is correct, provided the frame holding the elements is in position.

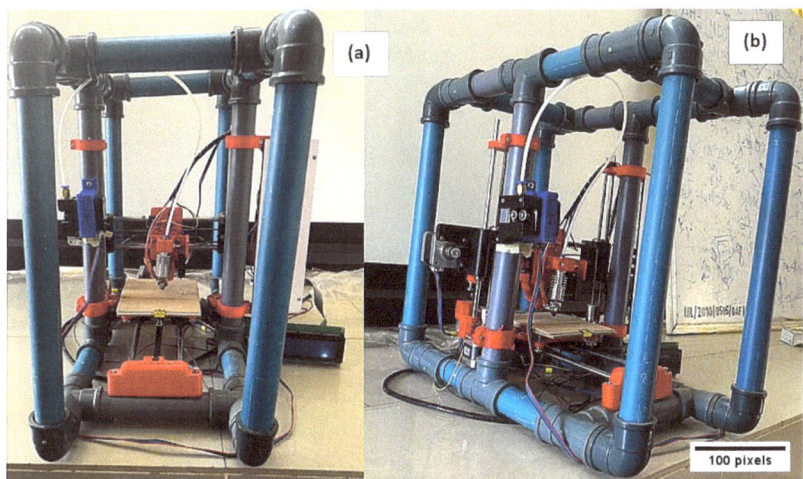

Figure 3. The constructed 3D Printer—(**a**) front view and (**b**) side view.

Operation of the Fabricated 3D Printer

These steps can be followed to print an object using this 3D printer:

i. design the model using any Computer-Aided Design (CAD) software and ensure the model is exported in a 3D printing compatible format such as .STL or .OBJ;
ii. load the model file into a slicing software with the necessary settings, such as the speed, temperature, layer height, shell, thickness, fill, support, and filament settings;
iii. save the model G-Code, and automatically the G-Code of the model following earlier inputted settings will be generated and saved as a file;
iv. transfer the saved file to an SD card and then insert it into the slot on the printer's motherboard;
v. connect the printer to the power source and calibrate the printer by ensuring the bed is leveled and all axes movement mechanisms are functioning correctly;
vi. insert the filament and print the model from the file on the SD card and after the model has finished printing, retrieve it from the bed and, if necessary, post-process the print.

Different test models were printed to determine the accuracy and efficiency of the 3D printer, and it's suitability for replacing the imported 3D printers. Objects and parts from different fields of studies (engineering - gears; medical - prosthetics and nose masks) were printed to evaluate the applicability of the printer, and components of the 3D printer were printed to tests the printer's self-replication. All the printer's parts, except for the electrical components, could be printed.

There are certain geometries that are difficult to print. This is because an FDM 3D printer functions by depositing semi-molten plastic layer by layer to create an object, and the layer supports each new layer underneath it. Therefore, if the preceding layer does not support an overhang structure, the new layer falls, resulting in a catastrophic print. Before printing starts, the object design and geometry are analyzed to determine the need for a support structure. If support is needed, it can be simply incorporated from the slicing software input settings. After the printing, the support structure is carefully removed from the object. The support structures are designed to be broken off easily. Figure 4 shows the human foot and outer ear with their support structures. Samples of other printed objects are shown in Figure 5. The specification of the constructed 3D printing machine is presented in Table 2, while the bill of engineering materials and evaluation is shown in Table 3.

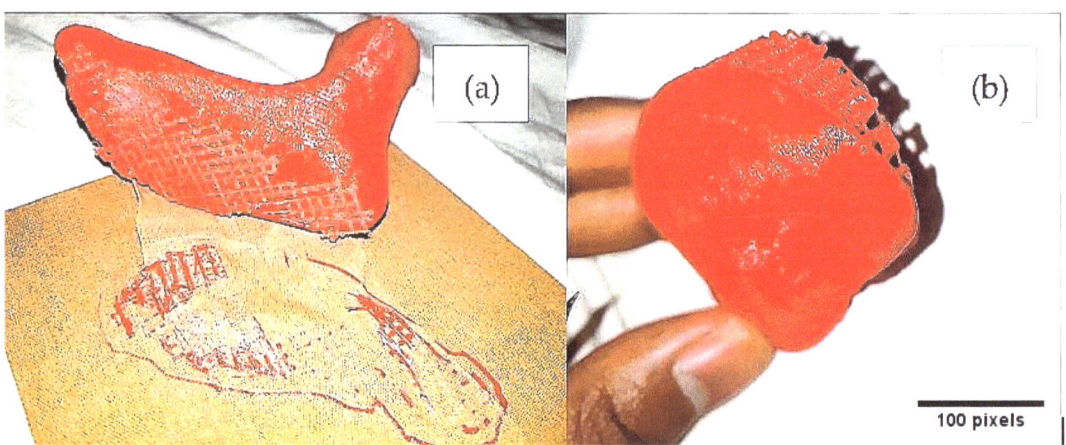

Figure 4. Illustration of overhangs: (**a**) human foot with a support structure; (**b**) human outer ear with a support structure.

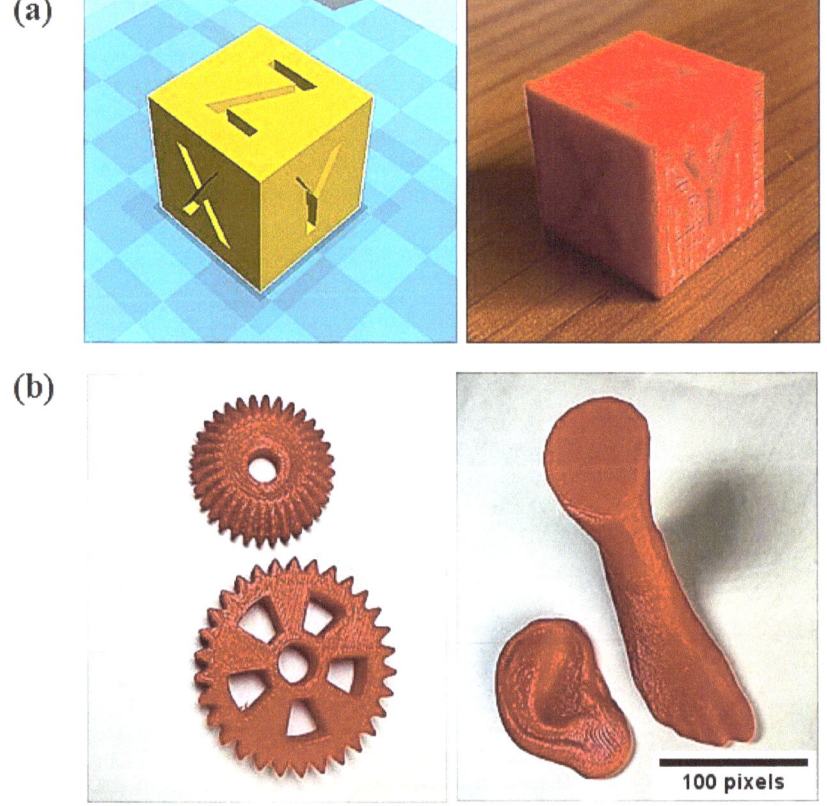

Figure 5. *Cont.*

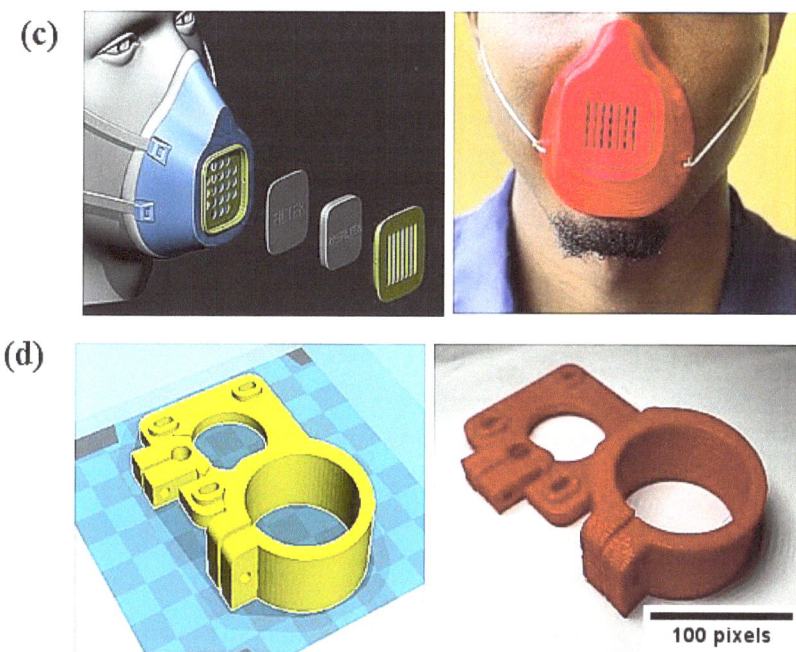

Figure 5. Designed objects printed using the fabricated 3D printing machine: (**a**) calibration cube; (**b**) bevel gear, spur gear, outer human ear, and foot; (**c**) nose mask; (**d**) fixture holding the electric motor controlling the z-axis.

Table 2. Specification of the constructed 3D printer.

Items	Specifications
Build volume	$100 \times 100 \times 100$ mm^3
Method	Fused deposition modelling
Printer size	500 mm (L) $\times$ 380 mm (W) $\times$ 425 mm (H)
Printer weight	3.95 kg
Number of extruders	1
Filament diameter	1.75 mm
Nozzle diameter	0.4 mm
Filament type	PLA
Layer resolution height	100 microns
Power supply	DC12 V, 5.0 A
Power consumption	240 V, 50-60 Hz
Connectivity	USB, SD card

Table 3. The bill of engineering material and evaluation.

Material	Quantity	Cost (Naira, ₦)
Frame (PVC pipe and fittings)	17	4200
NEMA 17 stepper motor	5	15,000
PLA filament	1 Spool	17,000
MK8 extruder	1	3300
Bed	1	1000
Timing belt and pulley	2	1500
Leadscrews	2	3000
Steel rods	6	7200

Table 3. *Cont.*

Material	Quantity	Cost (Naira, ₦)
Ball bearings	2	500
Linear bearings	7	8750
Coupler	2	1000
End stop	3	2200
DC fan	1	500
User interface and connectivity	1	15,500
Printed parts	1 Set	20,000
Screws, bolts, nuts, and springs	1 Set	1000
Miscellaneous		3000
Total		**104,650**

Total price equivalence in dollars is $233 ($1 = ₦450). It should be noted that these are retail prices, and the cost of fabrication could be about 50% cheaper when parts are purchased in bulk for mass production.

4. Conclusions

A low-cost-high-accessibility fused deposition modeling 3D printer was designed and constructed in this study using locally available materials. The following conclusions could be drawn from the study:

i. The design of the frame was made robust using PVC pipes, and commercial off-the-shelf components were used where possible, especially for the electrical parts.
ii. The 3D printer is self-reproducible, which means all parts of the machine may be manufactured using the same 3D printer since the fixtures are made of PLA plastic, except for the PVC frame, wooden bed, and electrical parts.
iii. The recommended distance between the nozzle tip and the bed is 0.1 mm.
iv. The printer's accuracy level was shown by the printed object's dimensions correctness compared to the digital design, which gave a percentage error of 0.74%.
v. The machine performed satisfactorily with a total cost of ₦104,650 ($233) and can be used in place of imported 3D printers in developing nations.

Author Contributions: A.D.A. conceptualized and carried out the construction of the printer. K.R.A. supervised the research. S.E.I. and A.D.A. wrote the article. E.T.A. review and edit the article. All authors have read and agreed to the published version of the manuscript.

Funding: This research received no external funding.

Data Availability Statement: The data supporting this study's findings are available within the article.

Conflicts of Interest: The authors declares no conflict of interest.

References

1. Huang, W.; Zhang, L.; Li, W.; Sun, J.; Liang, W.; Song, X.; Mao, X.; Wang, Z. Various Types and Applications of Additive Manufacturing. *Int. Conf. Appl. Math. Model. Simul. Optim.* **2019**, 377–381. [CrossRef]
2. Jasveer, S.; Jianbin, X. Comparison of Different Types of 3D Printing Technologies. *Int. J. Sci. Res. Publ.* **2018**, *8*, 1–9. [CrossRef]
3. Bhusnure, O.G.; Gholve, S.V.; Sugave, B.K.; Dongre, R.C.; Gore, S.A.; Giram, P.S. 3D Printing & Pharmaceutical Manufacturing: Opportunities and Challenges. *Int. J. Bioassays* **2016**, *5*, 4723–4738.
4. Prabhu, T. Modern Rapid 3D printer—A Design Review. *Int. J. Mech. Eng. Technol.* **2016**, *11*, 29–37.
5. Pekgor, M.; Nikzad, M.; Arablouei, R.; Masood, S. Materials Today: Proceedings Sensor-based filament fabrication with embedded RFID microchips for 3D printing. *Proceedings* **2021**, *46*, 124–130.
6. Carrasco-correa, E.J.; Francisco, E. The emerging role of 3D printing in the fabrication of detection systems. *Trends Anal. Chem.* **2021**, *136*, 116177. [CrossRef]
7. Ruberu, K.; Senadeera, M.; Rana, S.; Gupta, S.; Chung, J.; Yue, Z.; Venkatesh, S.; Wallace, G. Coupling machine learning with 3D bioprinting to fast track optimisation of extrusion printing. *Appl. Mater. Today* **2021**, *22*, 100914. [CrossRef]

8. Farhan, M.; Alam, A.; Ateeb, M.; Saad, M. Materials Today: Proceedings Real-time defect detection in 3D printing using machine learning. *Mater. Today Proc.* **2021**, *42*, 521–528. [CrossRef]
9. Popovski, F.; Mijakovska, S.; Popovska, H.D.; Nalevska, G.P. Creating 3D Models with 3D Printing Process. *Int. J. Comput. Sci. Inf. Technol.* **2021**, *13*, 59–68. [CrossRef]
10. Fettig, A. Purposes, Limitations, and Applications of 3D Printing in Minnesota Public Schools. *Culminating Proj. Inf. Media.* **2017**, *1*, 1–52.
11. Ilhan, E.; Ulag, S.; Sahin, A.; Karademir, B.; Ekren, N.; Kilic, O.; Sengor, M.; Kalaskar, D.M.; Nuzhet, F.; Gunduz, O. Fabrication of tissue-engineered tympanic membrane patches using 3D-Printing technology. *J. Mech. Behav. Biomed. Mater.* **2021**, *114*, 104219. [CrossRef]
12. Xenikakis, I.; Tsongas, K.; Tzimtzimis, E.K.; Zacharis, K.; Theodoroula, N.; Kalogianni, E.P.; Demiri, E.; Vizirianakis, I.S.; Tzetzis, D.; Fatouros, D.G. Fabrication of hollow microneedles using liquid crystal display (LCD) vat polymerization 3D printing technology for transdermal macromolecular delivery. *Int. J. Pharm.* **2021**, *597*, 120303. [CrossRef]
13. Adesiji, A.D. Design and Construction of a 3d Printer. Bachelor's Thesis, Department of Mechanical Engineering, University of Ilorin, Ilorin, Nigeria, 2021.
14. Rayna, T.; Striukova, L. From rapid prototyping to home fabrication: How 3D printing is changing business model innovation. *Technol. Forecast. Soc. Chang.* **2016**, *102*, 214–224. [CrossRef]
15. Sung, J.; So, H. 3D printing-assisted fabrication of microgrid patterns for flexible antiadhesive polymer surfaces. *Surf. Interfaces* **2021**, *23*, 100935. [CrossRef]
16. Zhou, X.; Zhou, G.; Junka, R.; Chang, N.; Anwar, A.; Wang, H.; Yu, X. Fabrication of polylactic acid (PLA) -based porous scaffold through the combination of traditional bio-fabrication and 3D printing technology for bone regeneration. *Colloids Surf. B Biointerfaces* **2021**, *197*, 111420. [CrossRef]
17. Al-maliki, J.Q.; Al-maliki, A.J.Q. The Processes and Technologies of 3D Printing. *Int. J. Adv. Comput. Sci. Technol.* **2015**, *4*, 161–165.
18. Burleson, J.; Dipaola, C. 3D Printing in Spine Surgery. In *3D Printing in Orthopaedics: Subspecialties*; Elsevier: Amsterdam, The Netherland, 2019; pp. 105–122. [CrossRef]
19. Zaghlou, M.Y.M.; Zaghloul, M.M.Y.; Zaghloul, M.M.Y. Developments in polyester composite materials—An in-depth review on natural fibres and nano fillers. *Compos. Struct.* **2021**, *278*, 114698. [CrossRef]
20. Zaghloul, M.M.Y.; Mohamed, Y.S.; El-Gamal, H. Fatigue and tensile behaviors of fiber-reinforced thermosetting composites embedded with nanoparticles. *J. Compos. Mater.* **2019**, *53*, 709–718. [CrossRef]
21. Zaghloul, M.M.Y.; Zaghloul, M.M.Y. Influence of flame retardant magnesium hydroxide on the mechanical properties of high density polyethylene composites. *J. Reinf. Plast. Compos.* **2017**, *36*, 1802–1816. [CrossRef]
22. Zaghloul, M.M.Y.; Zaghloul, M.Y.M.; Zaghloul, M.M.Y. Experimental and modeling analysis of mechanical-electrical behaviors of polypropylene composites filled with graphite and MWCNT fillers. *Polym. Test.* **2017**, *63*, 467–474. [CrossRef]
23. Zaghloul, M.Y.; Zaghloul, M.M.Y.; Zaghloul, M.M.Y. Influence of Stress Level and Fibre Volume Fraction on Fatigue Performance of Glass Fibre-Reinforced Polyester Composites. *Polymers* **2022**, *14*, 2662. [CrossRef]
24. Raja, N.; Sung, A.; Park, H.; Yun, H. Low-temperature fabrication of calcium deficient hydroxyapatite bone scaffold by optimization of 3D printing conditions. *Ceram. Int.* **2021**, *47*, 7005–7016. [CrossRef]
25. Xiao, X.; Jiang, X.; Yang, S.; Lu, Z.; Niu, C.; Xu, Y.; Huang, Z.; Kang, Y.J.; Feng, L. Solvent evaporation induced fabrication of porous polycaprolactone scaffold via low-temperature 3D printing for regeneration medicine researches. *Polymers* **2021**, *217*, 123436. [CrossRef]
26. Sevvel, P.; Srinivasan, D.; Balaji, A.J.; Gowtham, N. Design & Fabrication of Innovative Desktop 3D Printing Machine. *Mater. Today Proc.* **2020**, *22*, 3240–3249.
27. Patel, V.; Joshi, U.; Joshi, A. Opportunities in 3D Printing for Nanocomposite Materials: A Review. *J. Emerg. Technol. Innov. Res.* **2019**, *6*, 367–372.
28. O'Connell, J.; Haines, J. How Much Does a 3D Printer Cost in 2022? 2022. Available online: https://m.all3dp.com/2/how-much-does-a-3d-printer-cost/ (accessed on 20 July 2022).
29. Ozsoy, K.; Ercetin, A.; Cevik, Z.A. Comparison of Mechanical Properties of PLA and ABS Based Structures Produced by Fused Deposition Modelling Additive Manufacturing. *Eur. J. Sci. Technol.* **2021**, *27*, 802–809.
30. Raj, S.A.; Muthukumaran, E.; Jayakrishna, K. A Case Study of 3D Printed PLA and Its Mechanical Properties. *Mater. Today Proc.* **2018**, *5*, 11219–11226. [CrossRef]
31. Gordeev, E.G.; Galushko, A.S.; Ananikov, V.P. Improvement of quality of 3D printed objects by elimination of microscopic structural defects in fused deposition modeling. *PLoS ONE* **2018**, *13*, e0198370.
32. Kuurmi, R.S.; Gupta, J.K. *A Textbook of Machine Design*, 1st ed.; EURASIA Publishing House (PVT.) Ltd: Kolkata, India, 2005.

MDPI AG
Grosspeteranlage 5
4052 Basel
Switzerland
Tel.: +41 61 683 77 34

Journal of Composites Science Editorial Office
E-mail: jcs@mdpi.com
www.mdpi.com/journal/jcs

Disclaimer/Publisher's Note: The statements, opinions and data contained in all publications are solely those of the individual author(s) and contributor(s) and not of MDPI and/or the editor(s). MDPI and/or the editor(s) disclaim responsibility for any injury to people or property resulting from any ideas, methods, instructions or products referred to in the content.

www.ingramcontent.com/pod-product-compliance
Lightning Source LLC
LaVergne TN
LVHW072319090526
838202LV00019B/2311